Single-Variable Student's Solutions Manual

to accompany

Calculus

Concepts and Connections

Robert T. Smith
Millersville University

Roland B. Minton
Roanoke College

Prepared by

Stephen Agard
University of Minnesota
and
David Frank
University of Minnesota

Boston Burr Ridge, IL Dubuque, IA Madison, WI New York San Francisco St. Louis
Bangkok Bogotá Caracas Kuala Lumpur Lisbon London Madrid Mexico City
Milan Montreal New Delhi Santiago Seoul Singapore Sydney Taipei Toronto

The McGraw·Hill Companies

Single-Variable Student's Solutions Manual to accompany
CALCULUS: CONCEPTS AND CONNECTIONS
ROBERT T. SMITH AND ROLAND B. MINTON

Published by McGraw-Hill Higher Education, an imprint of The McGraw-Hill Companies, Inc., 1221 Avenue of the Americas, New York, NY 10020.

This book is printed on acid-free paper.

1 2 3 4 5 6 7 8 9 0 QPD/QPD 0 9 8 7 6 5

ISBN 0-07-283094-8

www.mhhe.com

CONTENTS

Chapter 0

Preliminaries

0.1 Polynomials and Rational Functions

1. (2, 1) and (0, 2):

$$m = \frac{2-1}{0-2} = \frac{1}{-2} = -\frac{1}{2}$$

(0, 2) and (4, 0):

$$m = \frac{0-2}{4-0} = \frac{-2}{4} = -\frac{1}{2}$$

Since the slopes are the same, the points are collinear.

3. (4, 1) and (3, 2):

$$m = \frac{2-1}{3-4} = \frac{1}{-1} = -1$$

(3, 2) and (1, 3):

$$m = \frac{3-2}{1-3} = \frac{1}{-2} = -\frac{1}{2}$$

Since the slopes are not equal, the points are not collinear.

5. $m = \dfrac{6-2}{3-1} = \dfrac{4}{2} = 2$

7. $m = \dfrac{-1-(-6)}{1-3} = \dfrac{5}{-2} = -\dfrac{5}{2}$

9. $m = \dfrac{-0.4-(-1.4)}{-1.1-0.3} = \dfrac{1.0}{-1.4} = -\dfrac{5}{7}$

11. P_2 could be (2, 5)
(among many other choices).

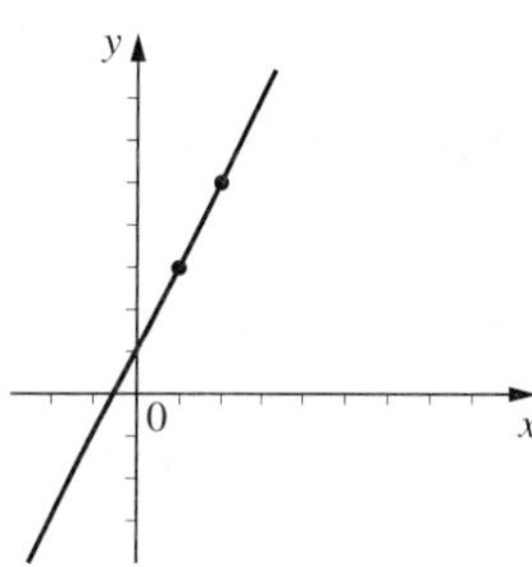

The equation is

$$y = 2(x-1) + 3 = 2x + 1.$$

13. P_2 could be (0, 1).

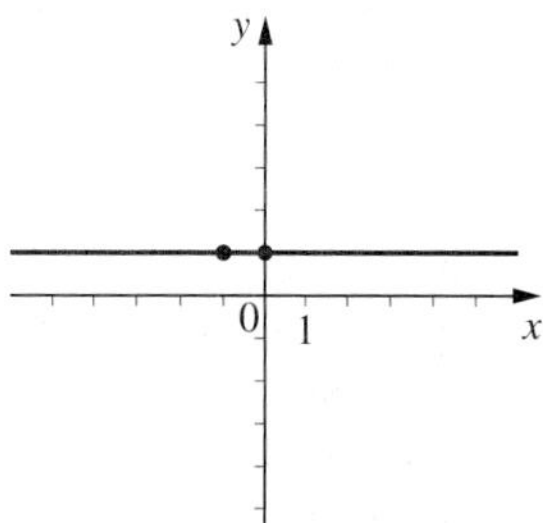

The equation is $y = 1$.

15. P_2 could be (3.3, 2.3).

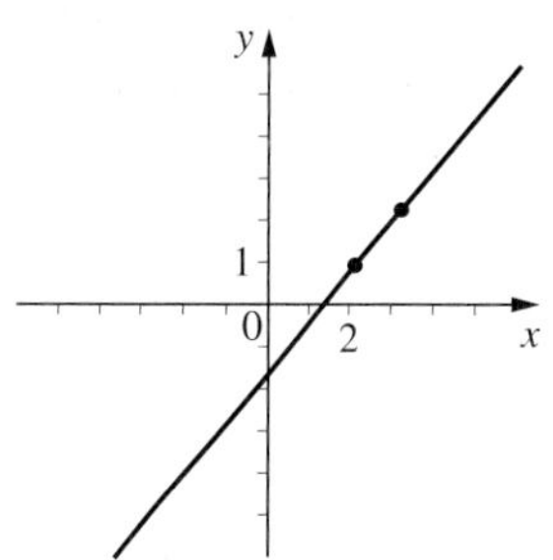

The equation is

$$y = 1.2(x - 2.3) + 1.1 = 1.2x - 1.66.$$

17. These two lines both have slope 3 . They are parallel unless they are coincident, but they are not coincident since (1, 2) is on the first line but not on the second (known by the fact that it does not work in the second equation).

19. perpendicular—slopes are -2 and $1/2$

21. perpendicular—slopes are 3 and $-1/3$

23. (a)

$$y = 2(x - 2) + 1 \text{ or}$$
$$y = 2x - 3$$

(b)

$$y = -\frac{1}{2}(x - 2) + 1 \text{ or}$$
$$y = -\frac{1}{2}x + 2$$

25. (a)

$$y = 2(x - 3) + 1 \text{ or}$$
$$y = 2x - 5$$

(b)

$$y = -\frac{1}{2}(x - 3) + 1 \text{ or}$$
$$y = -\frac{1}{2}x + \frac{5}{2} \text{ (slope-intercept) or}$$
$$2y + x = 5 \text{ (to avoid fractions)}$$

27.

$$P_1 = (1, 1), \ P_2 = (2, 3)$$
$$m = \frac{3 - 1}{2 - 1} = \frac{2}{1} = 2$$
$$y = 2(x - 1) + 1 = 2x - 1$$
$$\text{At } x = 4 : y = 7$$

29.

$$P_1 = (0.5, 4), \ P_2 = (1, 3)$$
$$m = \frac{3 - 4}{1 - 0.5} = \frac{-1}{0.5} = -2$$
$$y = -2(x - 1) + 3 = -2x + 5$$
$$\text{At } x = 4 : y = -3$$

31. This looks like the graph of a function.

33. This is not the graph of a function. The vertical line $x = 0$ meets the curve twice; nearby vertical lines meet it three times.

35. This is clearly a cubic polynomial, and also a rational function because it can be written as

$$f(x) = \frac{x^3 - 4x + 1}{1}.$$

(By this device all polynomials are rational.)

37. This is a rational function (quadratic numerator, linear denominator) but not a polynomial.

39. This is not rational (square roots not permitted).

41. The natural, maximal domain is $x + 2 \geq 0$, i.e., $x \geq -2$ or in "interval language": $[-2, \infty)$.

43. Negatives are permitted inside the cube root. There are no restrictions, hence $(-\infty, \infty)$.

45. The denominator cannot be zero:

$$x^2 - 1 = 0 \text{ implies}$$
$$x^2 = 1$$
$$x = \pm 1$$

Hence the natural, maximal domain requires $x \neq \pm 1$. In "interval language," this is $(-\infty, 1) \cup (-1, 1) \cup (1, \infty)$.

47. $f(x) = x^2 - x - 1$ for *all* x. By straightforward substitution,

$$f(0) = 0^2 - 0 - 1 = -1$$
$$f(2) = 2^2 - 2 - 1 = 1$$
$$f(-3) = (-3)^2 - (-3) - 1 = 11$$
$$f\left(\frac{1}{2}\right) = \left(\frac{1}{2}\right)^2 - \frac{1}{2} - 1 = -\frac{5}{4}.$$

49. Again by straightforward substitution,

$$f(0) = \sqrt{0 + 1} = 1$$
$$f(3) = \sqrt{3 + 1} = 2$$
$$f - (1) = \sqrt{-1 + 1} = 0$$
$$f\left(\frac{1}{2}\right) = \sqrt{\frac{1}{2} + 1} = \sqrt{\frac{3}{2}} = \frac{\sqrt{6}}{2}.$$

In ancient high school tradition, it was considered "good form" to eliminate radicals from the denominator.

51. For exactly one car, a reasonable domain might be $8 \le x \le 12$. The upper endpoint would be expanded for more cars and/or other storage. The open-ended response $x > 0$ is perfectly acceptable.

53. In the absence of any constraints (marketing area, number of outlets, production schedule) one can hardly be more precise than $0 \le x$.

55. We feel that there may well be a positive correlation (more study hours = better grade), but not a functional relation.

57. While not denying a negative correlation (more exercise = less weight), there are too many other factors (metabolic rate, diet) to be able to quantify a person's weight as a function of the amount of exercise.

59. A flat interval corresponds to an interval of constant speed; going up means that the speed is increasing (this *could* be due to going downhill, but that is neither necessary or sufficient); going down means that the speed is decreasing (likely when going uphill).

61. Given that $y = x^2 - 2x - 8$, the y-intercept is when $x = 0$, hence when $y = -8$, or in "point language": $(0, -8)$. The x-intercepts are when $y = 0$, hence

$$x^2 - 2x - 8 = 0$$
$$(x - 4)(x + 2) = 0$$
$$x = 4 \text{ or } x = -2.$$

In "point language": $(4, 0)$ and $(-2, 0)$.

63. Given that $y = x^3 - 8$, the y-intercept is when $x = 0$, hence when $y = -8$, or in "point language": $(0, -8)$. The x-intercepts are when $y = 0$, hence,

$$x^3 - 8 = 0$$
$$(x - 2)(x^2 + 2x + 4) = 0$$

$x = 2$ (the quadratic factor, which can be written in the form $(x + 1)^2 + 3$, is *never* zero); In "point language": $(2, 0)$.

65. Given that $y = \dfrac{x^2 - 4}{x + 1}$, the y-intercept is when $x = 0$, hence when $y = -4$, or in "point language": $(0, -4)$.

The x-intercepts are when $y = 0$, hence,

$$x^2 - 4 = 0$$
$$(x - 2)(x + 2) = 0$$
$$x = 2 \text{ or } x = -2.$$

In "point language": $(2, 0)$ and $(-2, 0)$.

67. $x^2 - 4x + 3$ factors into $(x - 3)(x - 1)$. The "zeros" or *roots*, are $x = 1$ and $x = 3$.

69. For $x^2 - 4x + 2$, the roots by the quadratic formula are

$$\frac{4 \pm \sqrt{16 - 8}}{2} = 2 \pm \sqrt{2},$$

numerically about .5858 and 3.4142.

71. This one turns out to factor completely:

$$x^3 - 3x^2 + 2x =$$
$$x(x^2 - 3x + 2) =$$
$$x(x - 2)(x - 1).$$

The roots are 0, 1, 2.

73. With $t = x^3$, $x^6 + x^3 - 2$ becomes $t^2 + t - 2$ and factors as $(t + 2)(t - 1)$. The expression is zero only if one of the factors is zero, i.e., if $t = 1$ or $t = -2$. With $x = t^{1/3}$, the first occurs only if $x = (1)^{1/3} = 1$. The latter occurs only if $x = -2^{1/3}$, about -1.2599. These are the two roots.

75. If $B(h) = -1.8h + 212$, then solving $B(h) = 98.6$ for h leads to

$$98.6 = -1.8h + 212$$
$$1.8h = 113.4$$
$$h = \frac{113.4}{18} = 63.$$

This (63,000 feet above sea-level, more than double the height of Mt. Everest) would be the elevation at which we humans boil alive in our skins. Of course the cold of space and the near-total lack of external pressure create additional complications that we shall not try to analyze.

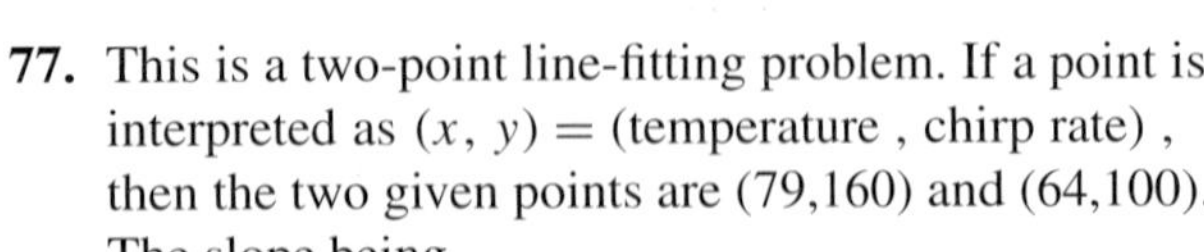

77. This is a two-point line-fitting problem. If a point is interpreted as $(x, y) =$ (temperature , chirp rate) , then the two given points are (79,160) and (64,100). The slope being

$$\frac{160-100}{79-64} = \frac{60}{15} = 4,$$

we could write $y - 100 = 4(x - 64)$, or $y = 4x - 156$, or $x = (y + 156)/4$. If, as seems likely, one wanted to count chirps to estimate temperature, the latter could be expressed in the form

temp = (# chirps in 15 seconds) plus 39.

Here, (1/4)chirp-rate-per-minute = # chirps in 15 seconds, whereas $39 = 156/4$.

0.2 Graphing Calculators and Computer Algebra Systems

1.

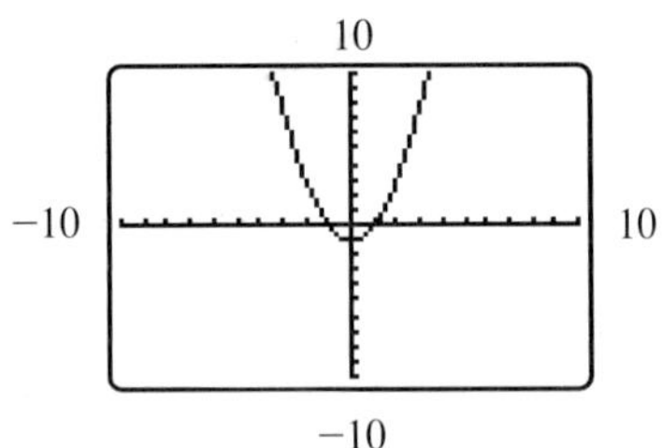

3.

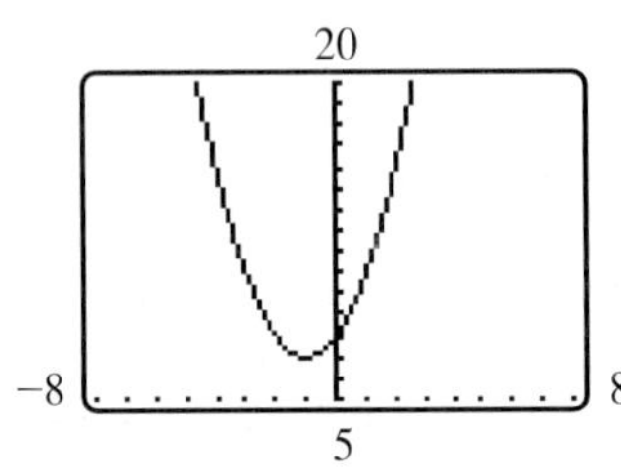

5.

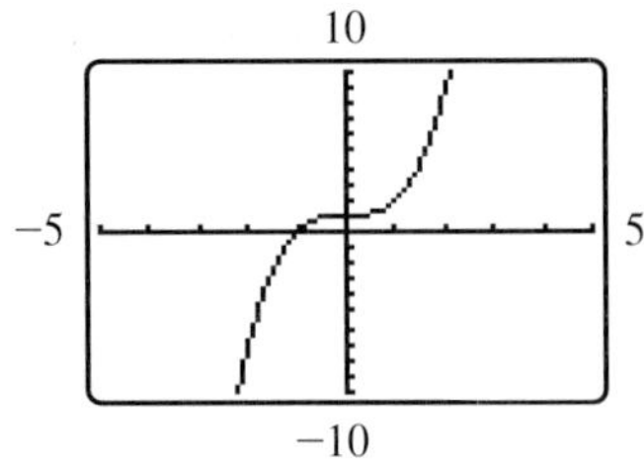

7.

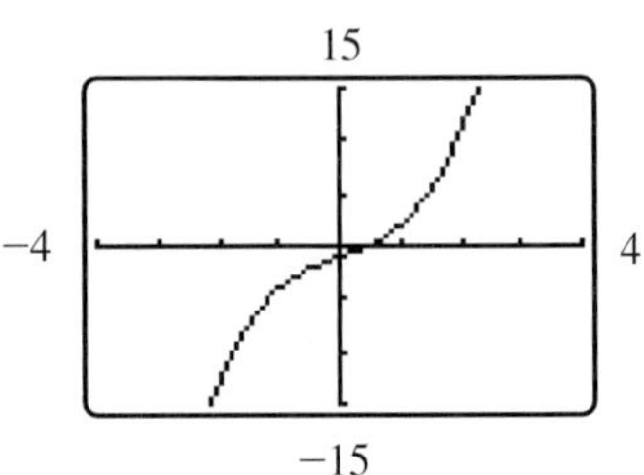

9.

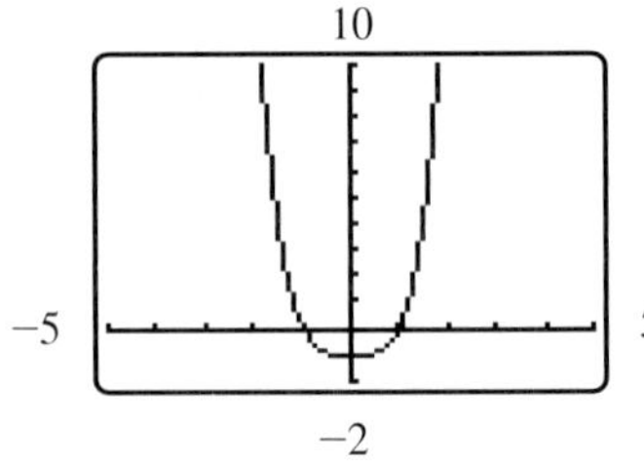

11.

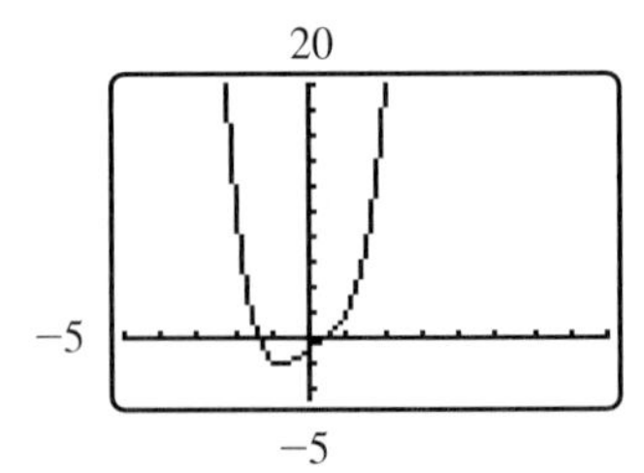

13.

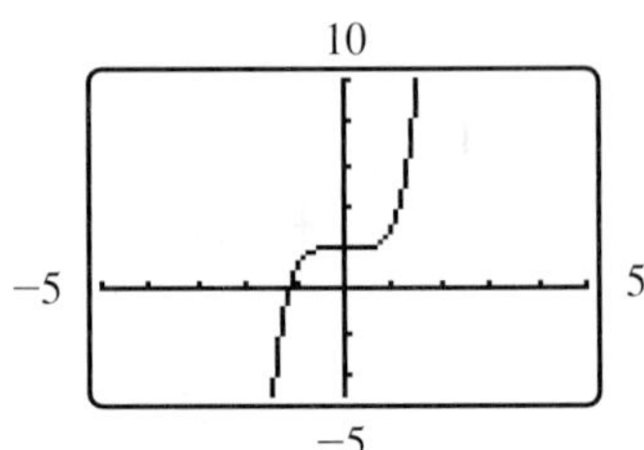

15.

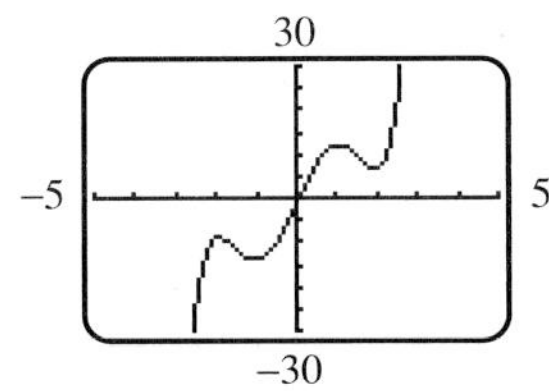

17.

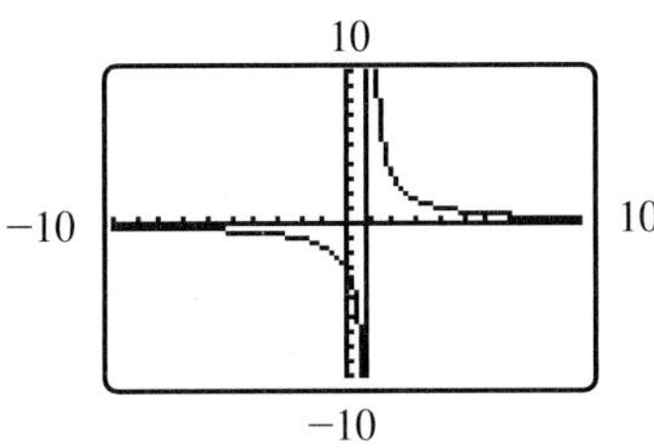

19.

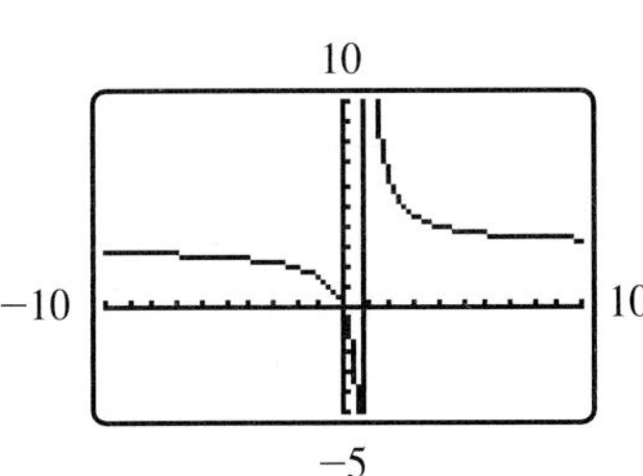

21.

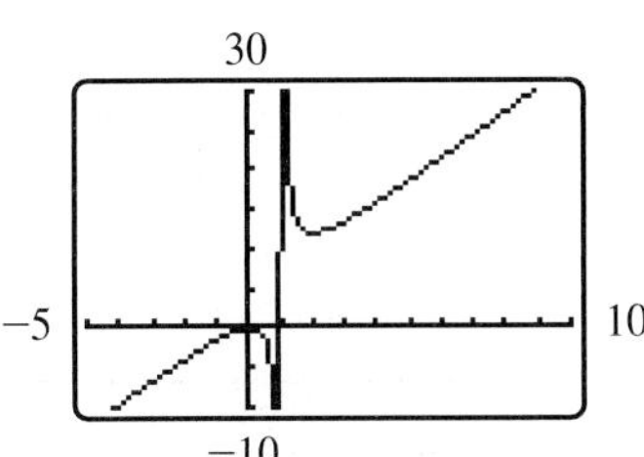

23.

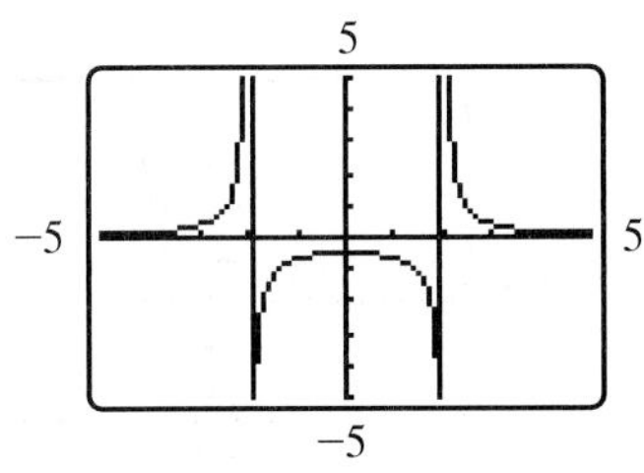

25.

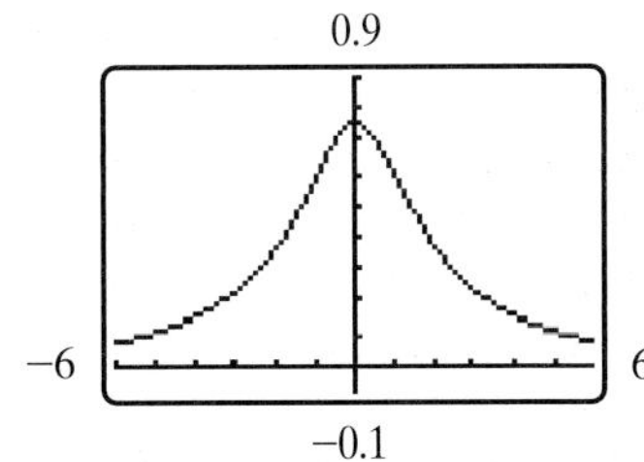

27.

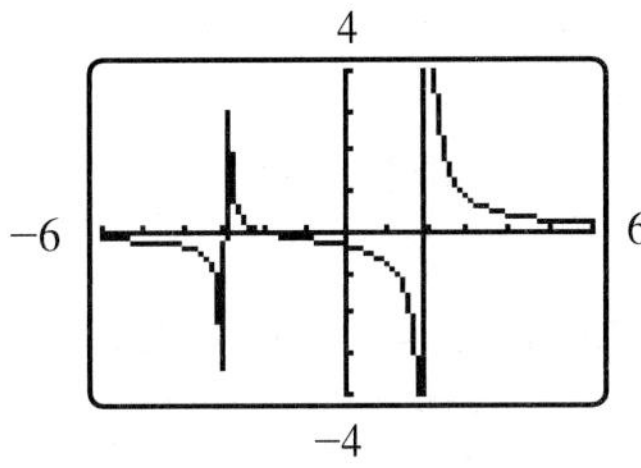

29.

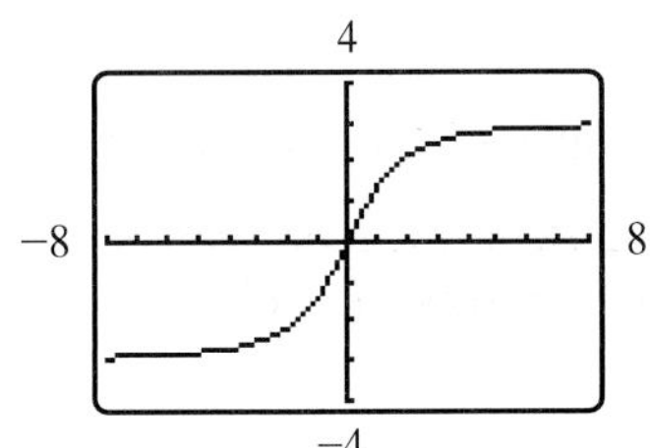

31. A vertical asymptote may occur when the denominator is zero, hence if

$$\begin{aligned} x^2 - 4 &= 0 \\ x^2 &= 4 \\ x &= \pm 2. \end{aligned}$$

See comment below Exercise 37.

33. A vertical asymptote may occur when the denominator is zero, hence if

$$\begin{aligned} x^2 + 3x - 10 &= 0 \\ (x+5)(x-2) &= 0 \\ x &= -5, x = 2. \end{aligned}$$

See comment below Exercise 37.

35. A vertical asymptote may occur when the denominator is zero. This denominator however is never zero, so there are no vertical asymptotes.

37. A vertical asymptote may occur when the denominator is zero, hence if

$$
\begin{aligned}
x^3 + 3x^2 + 2x &= 0\\
x(x^2 + 3x + 2) &= 0\\
x(x+2)(x+1) &= 0\\
x = 0,\ x = -2,\ x &= -1.
\end{aligned}
$$

The *potential* vertical asymptote actually *does* occur if the *numerator* is not zero. This is the case in Exercises 31, 33, 37, 43, 45, but not in the forthcoming Exercise 47.

39.

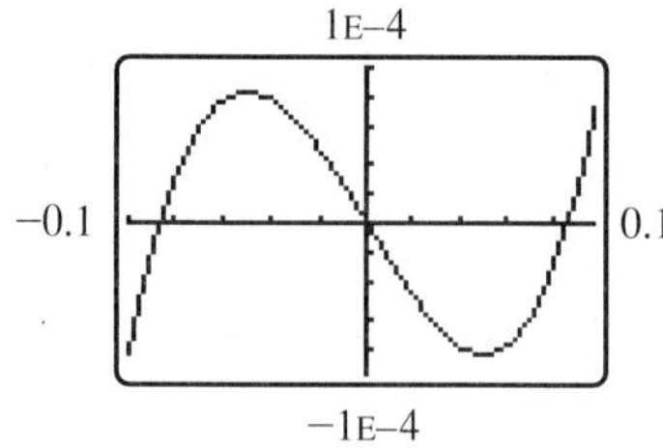

41.

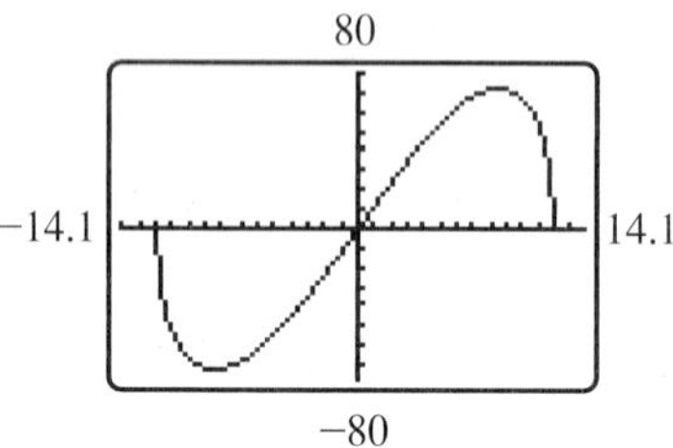

43.

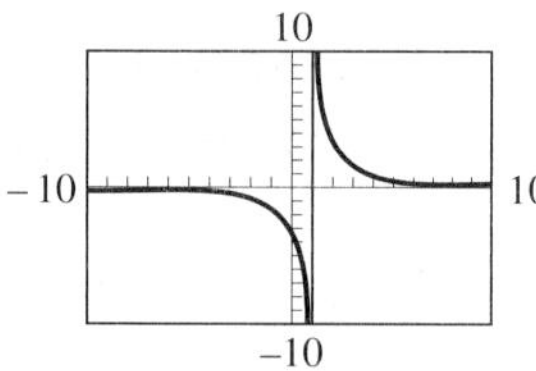

Vertical asymptote: $x = 1$

45.

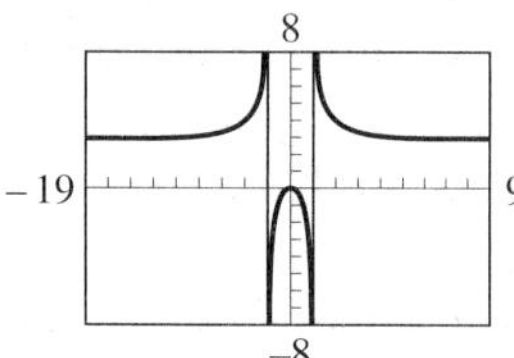

Vertical asymptotes: $x = 1$ and $x = -1$

47. The radicand $x^4 + x = x(x^3 + 1) = x(x+1)(x^2 - x + 1)$ is zero only twice: at $x = 0$ and at $x = -1$. Between these two values, however, it is negative with the effect that the function is not defined there. Thus the graph has a gap over the closed interval $[-1, 0]$. Regarding the endpoints, it would be natural (if premature) to assume that they were vertical asymptotes, and indeed that is true about $x = 0$, since the numerator $x^2 - 1 = (x-1)(x+1)$ is not zero when $x = 0$. However, the numerator *is* zero when $x = -1$, with the effect that the formula can be altered, *when* $x < -1$ *and* $x + 1 = -\sqrt{(x+1)^2}$, to read

$$
\begin{aligned}
&\frac{(x+1)(x-1)}{\sqrt{x(x+1)(x^2 - x + 1)}}\\
&= \frac{-\sqrt{(x+1)^2}(x-1)}{\sqrt{x(x+1)(x^2 - x + 1)}}\\
&= -(x-1)\sqrt{\frac{(x+1)}{x(x^2 - x + 1)}}.
\end{aligned}
$$

Now we can see that when x is near (but still less than) -1, both radicand and radical are near zero. This is not a vertical *asymptote*, but (as we will learn in the fullness of time) a place where the limit from the left is *zero*. The following sketch suggests these features. The curve has the added feature of a *horizontal* asymptote ($y = 1$), but we are not yet ready to fully understand this notion.

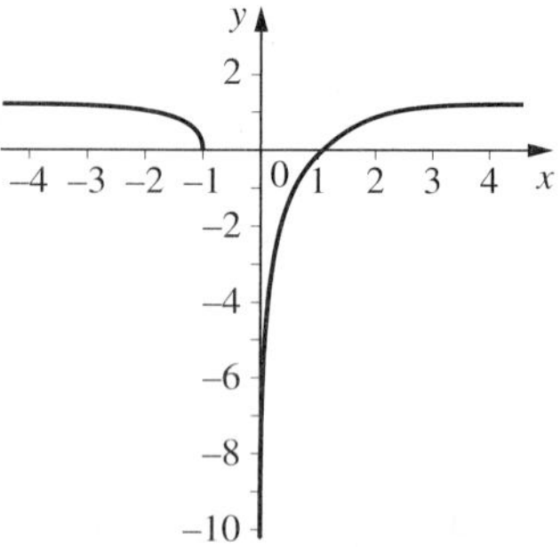

49. The blow-up makes it appear that there are two intersection points. Solving algebraically, the equation is

$$\sqrt{x - 1} = x^2 - 1 (x \geq 1).$$

Squaring we get

$$x - 1 = \{x^2 - 1\}^2 = \{(x - 1)(x + 1)\}^2$$
$$= (x - 1)^2(x + 1)^2.$$

We see that $x = 1$ is one solution (obvious from the start), while for any other, we can cancel one factor of $x - 1$ and find

$$1 = (x - 1)(x + 1)^2 = (x^2 - 1)(x + 1) =$$
$$x^3 + x^2 - x - 1, \text{ hence}$$
$$x^3 + x^2 - x - 2 = 0.$$

By solver or spreadsheet, this equation has only the one solution $x \approx 1.206$.

Note: The last cubic equation *might have had* other solutions (it didn't) but if such an additional solution had been less than one, it would not have counted in this problem, since the original domain required $x \geq 1$.

51. The graph does not clearly show the number of intersection points. Solving algebraically:

$$x^3 - 3x^2 = 1 - 3x$$
$$x^3 - 3x^2 + 3x - 1 = 0$$
$$(x - 1)^3 = 0$$
$$x = 1.$$

There is only one solution: $x = 1$.

53. After zooming out, the graph shows that there are two solutions: one near zero, and one around ten. By algebra, the equation is

$$(x^2 - 1)^{2/3} = 2x + 1$$
$$(x^2 - 1)^2 = (2x + 1)^3$$
$$x^4 - 2x^2 + 1 = 8x^3 + 12x^2 + 6x + 1$$
$$x^4 - 8x^3 - 14x^2 - 6x = 0$$
$$x(x^3 - 8x^2 - 14x - 6) = 0.$$

We confirm the obvious solution $x = 0$, and by solver or spreadsheet, find the second solution $x \approx 9.534$.

55. The graph shows that there are two solutions: $x \approx \pm 1.177$ by calculator or spreadsheet.

57.

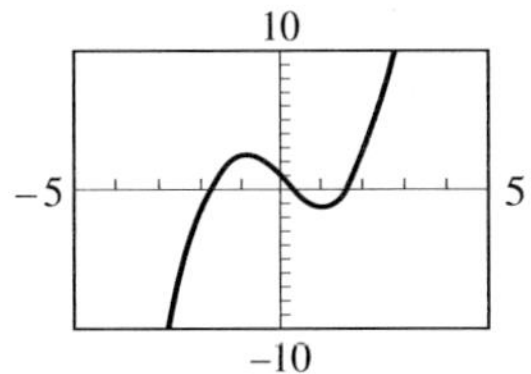

zeros at about -1.879, 0.347, 1.532

59.

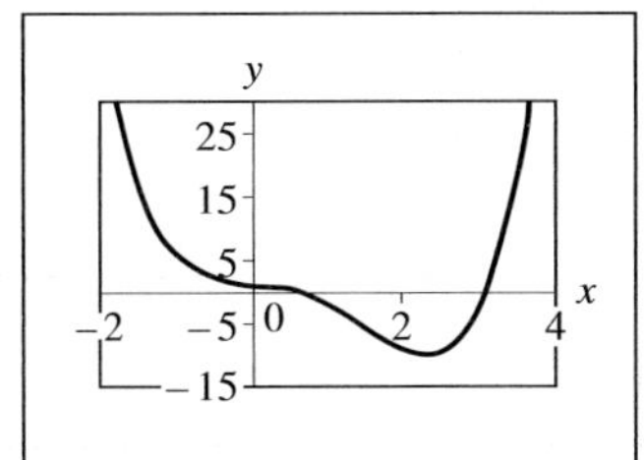

zeros at about .5637, 3.0715

61.

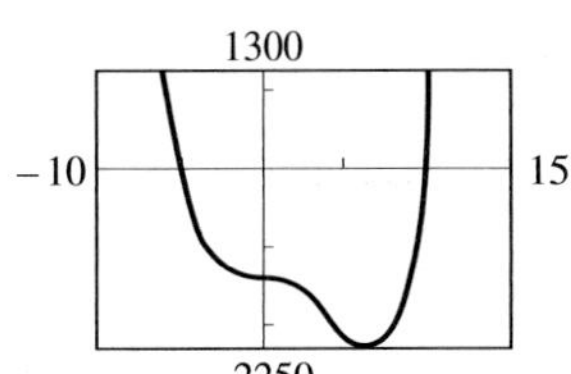

zeros at about -5.248, 10.006

63. The graph of $y = x^2$ on the window $-10 \leq x \leq 10$, $-10 \leq y \leq 10$, appears identical (except for labels) to the graph of $y = 2(x - 1)^2 + 3$ if the latter is drawn on a graphing window centered at the point $(1, 3)$ with

$$1 - 5\sqrt{2} \leq x \leq 1 + 5\sqrt{2}, \ -7 \leq y \leq 13.$$

The reason for this is that the most readily observable features are the vertex (dead center) and the two exit points. These remain in exactly the same proportion in the second graph. The student must take the trouble to actually draw the second graph in order to appreciate this problem.

65. $\sqrt{y^2}$ is the distance from (x, y) to the x-axis. $\sqrt{x^2 + (y - 2)^2}$ is the distance from (x, y) to the point $(0, 2)$. If we require that these be the same, and we square both quantities, we come to

$$\begin{aligned} y^2 &= x^2 + (y-2)^2 \\ y^2 &= x^2 + y^2 - 4y + 4 \\ 4y &= x^2 + 4 \\ y &= \frac{1}{4}x^2 + 1. \end{aligned}$$

In this relation, we see that y is a quadratic function of x. The graph is commonly known as a parabola.

0.3 Inverse Functions

1. $f(g(x)) = f(x^{1/5}) = (x^{1/5})^5 = x^{(5/5)} = x^1 = x$

$g(f(x)) = g(x^5) = (x^5)^{1/5} = x^{(5/5)} = x^1 = x$

3. Noting that $[g(x)]^3 = \dfrac{x-1}{2}$, we can write

$$f(g(x)) = 2[g(x)]^3 + 1 = 2\left[\frac{x-1}{2}\right] + 1 = x.$$

On the other hand,

$$f(x) - 1 = 2x^3, \quad \frac{f(x)-1}{2} = x^3, \text{ and}$$

$$g(f(x)) = \sqrt[3]{\frac{f(x)-1}{2}} = \sqrt[3]{x^3} = x.$$

5. One knows about the graph of the cube-function, that horizontal lines meet it exactly once. The graph of $y = x^3 - 2$ is simply a vertical translation (down two) of the cube-graph. Horizontal lines still meet it exactly once. Therefore f is one-to-one. To find the inverse function, write

$$\begin{aligned} y &= x^3 - 2 \\ y + 2 &= x^3 \\ \sqrt[3]{y+2} &= x. \end{aligned}$$

At this point the left side is a formula for $f^{-1}(y)$. Since the formula is good for all y, it is likewise good for all x, i.e.,

$$f^{-1}(x) \equiv \sqrt[3]{x+2}.$$

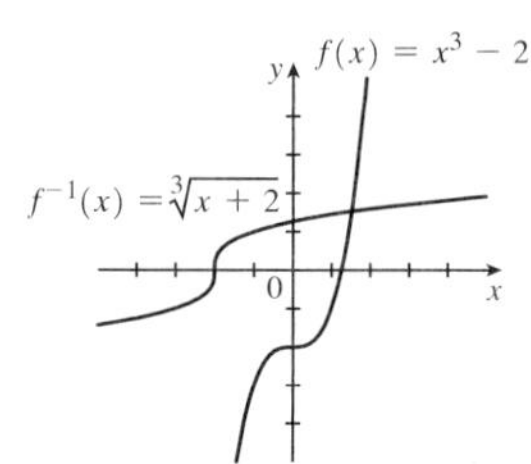

7. The graph of $y = x^5$ is much like the cube-function. All the remarks in Exercise 5 apply here as well. To find a formula for the inverse, write

$$\begin{aligned} y &= x^5 - 1 \\ y + 1 &= x^5 \\ \sqrt[5]{y+1} &= x. \end{aligned}$$

Again, the left side is a formula for $f^{-1}(y)$, good for all y (or for all x) . Therefore,

$$f^{-1}(x) \equiv \sqrt[5]{x+1}.$$

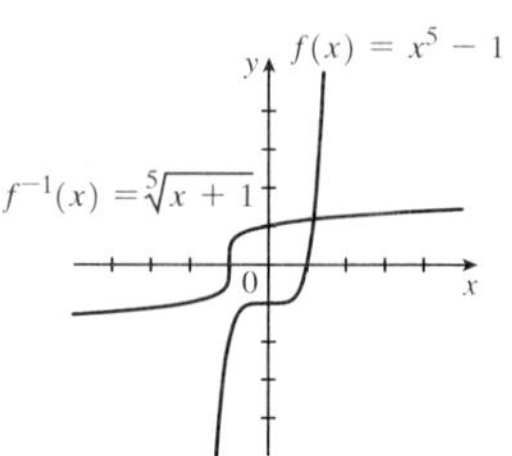

9. The function is not one-to-one since it is an even function ($f(-x) = f(x)$). In particular, $f(2) = 18 = f(-2)$.

11. Here, the natural domain requires that the radicand (the object inside the radical) be nonnegative. Hence $x \geq -1$ is required, while all function-values are nonnegative. Hence, the inverse, if defined at all, will be defined only for nonnegative numbers. Sometimes one can determine the existence of an inverse in the process of trying to find its formula. If all the steps in the process are completely unambiguous, the formula will often speak for itself in retrospect. This is an example: Write

$$y = \sqrt{x^3 + 1}$$
$$y^2 = x^3 + 1$$
$$y^2 - 1 = x^3$$
$$\sqrt[3]{y^2 - 1} = x.$$

There were no question marks in this process, and the left side is a formula for $f^{-1}(y)$, good for $y \geq 0$. Therefore, $f^{-1}(x) = \sqrt[3]{x^2 - 1}$ whenever $x \geq 0$.

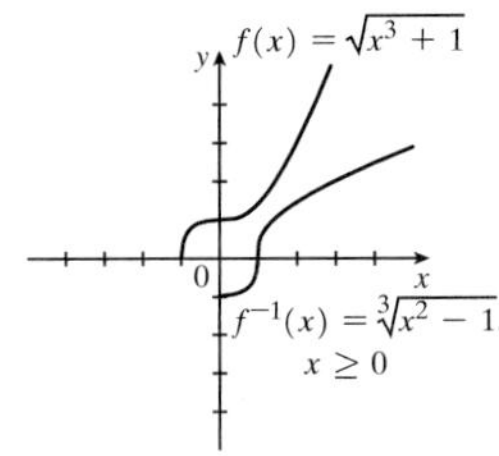

13. (a) Solve $f(x) = -1$, i.e.,

$$x^3 + 4x - 1 = -1.$$
$$\therefore\ x^3 + 4x = 0,$$
$$\text{so } x(x^2 + 4) = 0.$$
$$\therefore\ x = 0 = f^{-1}(-1).$$

(b) Solve $f(x) = 4$, i.e.,

$$x^3 + 4x - 1 = 4.$$
$$\therefore\ x^3 + 4x - 5 = 0.$$

The list of possible rational roots includes 1, which is a solution. By long division, we find

$$x^3 + 4x - 5 = (x - 1)(x^2 + x + 5).$$

The quadratic factor is never zero, so 1 is the only root, and $1 = f^{-1}(4)$.

15. (a) Solve $f(x) = -5$, i.e.,

$$x^5 + 3x^3 + x = -5.$$
$$\therefore\ x^5 + 3x^3 + x + 5 = 0.$$

Of the possible rational roots $\{\pm 1, \pm 5\}$, -1 works, and we find that $x^5 + 3x^3 + x + 5$ factors into

$$(x + 1)(x^4 - x^3 + 4x^2 - 4x + 5).$$
$$\therefore\ x = -1 = f^{-1}(-5).$$

It is not obvious that there might not be another solution, but that is an operative assumption in the problem.

(b) Solve $f(x) = 5$, i.e.,

$$x^5 + 3x^3 + x = 5.$$

It's an odd function, so the solution is the negative of the answer to (a).

$$\therefore\ x = 1 = f^{-1}(5)$$

17. (a) Solve $f(x) = 4$, i.e.,

$$\sqrt{x^3 + 2x + 4} = 4.$$
$$\therefore\ x^3 + 2x + 4 = 16,$$
$$x^3 + 2x - 12 = 0.$$

Out of the list of ten possible rational roots, 2 works. We can write

$$x^3 + 2x - 12 = (x - 2)(x^2 + 2x + 6).$$

The quadratic being expressible as

$$x^2 + 2x + 6 = (x + 1)^2 + 5,$$

we see that it is never zero, and so

$$x = 2 = f^{-1}(4).$$

(b) Solve $f(x) = 2$, i.e.,

$$\sqrt{x^3 + 2x + 4} = 2.$$
$$\therefore\ x^3 + 2x + 4 = 4,$$
$$x^3 + 2x = 0,$$
$$x(x^2 + 2) = 0.$$

Clearly the unique solution is $x = 0$, and so $0 = f^{-1}(2)$.

19.

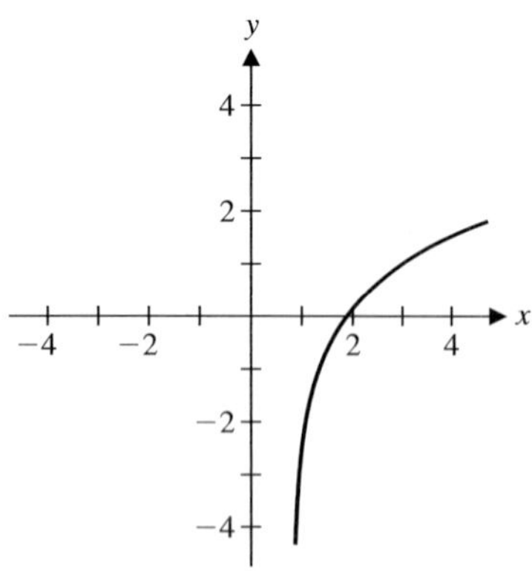

21.

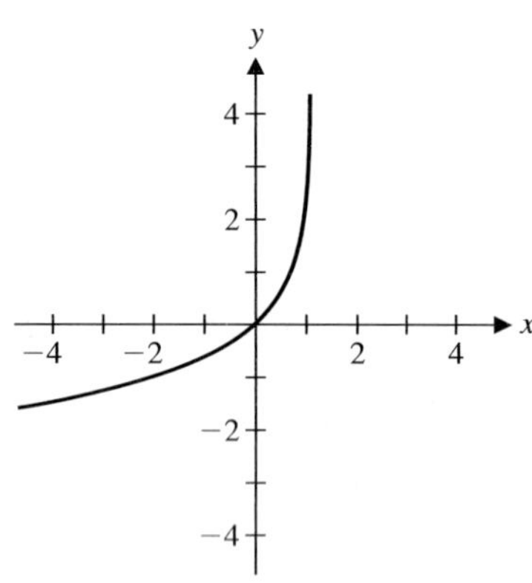

23. Since 23 is halfway between 20 ($= f(2)$) and 26 ($= f(3)$), the x-value for $y = 23$ should be halfway between 2 and 3, i.e., $f^{-1}(23)$ is estimated linearly by 2.5. See the picture below Exercise 25.

Since the lines between the points fall to the right of the apparent true curve of the graph, this estimate is too high.

25. Since 5 is three-quarters of the way from 2 ($= f(3)$) to 6 ($= f(2)$), the x-value should be three-quarters of the way from 3 to 2, i.e., between the points fall to the right of the apparent true curve of the graph, this estimate is too high. $f^{-1}(5)$ is estimated linearly by 2.25.

Since the lines between the points fall to the right of the apparent true curve of the graph, this estimate is too high.

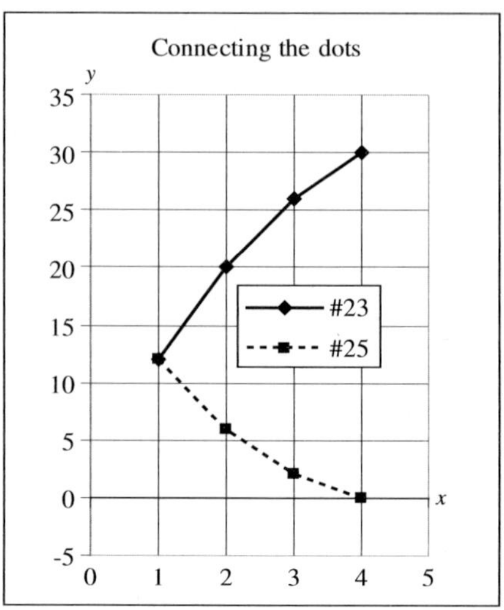

27. If $y = f(x) = x^3 - 5$, then the horizontal line test is passed, and $x = (y + 5)^{1/3}$. The right side of this equation is a formula for $f^{-1}(y)$, which is good for all y. Thus $f^{-1}(x) = (x + 5)^{1/3}$ for all x.

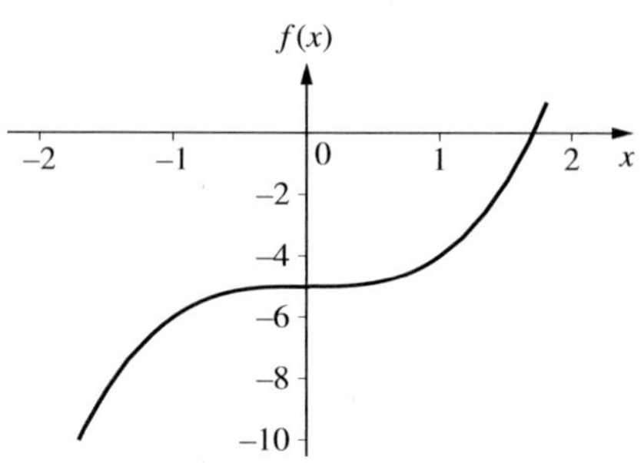

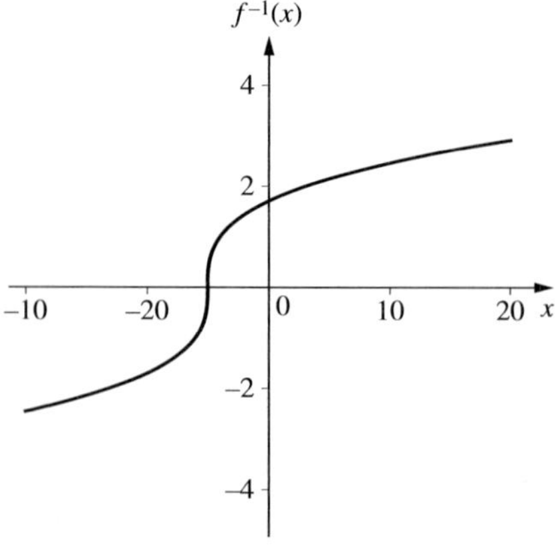

29. This function $f(x) = x^3 + 2x - 1$ easily passes the horizontal line test and is invertible. An old, difficult-to-use, and largely forgotten formula for solving cubic equations by radicals gives the following formula for the inverse function $g : g(x) =$

$$\frac{1}{\sqrt[3]{2}}\frac{1}{\sqrt{3}}\left\{\begin{array}{c}\sqrt[3]{3\sqrt{3}(x+1)+\sqrt{27(x+1)^2+32}}\\+\\\sqrt[3]{3\sqrt{3}(x+1)-\sqrt{27(x+1)^2+32}}\end{array}\right\}.$$

It is by no means obvious that either of the identities $f(g(x)) = x$ or $g(f(x)) = x$ are valid. We do see that $g(-1) = 0$, consistent with $f(0) = -1$, but given that $f(1) = 2$, it is *not* apparent that $g(2) = 1$, i.e., that

$$1 = \frac{1}{\sqrt[3]{2}}\frac{1}{\sqrt{3}}\left\{\begin{array}{c}\sqrt[3]{9\sqrt{3}+5\sqrt{11}}\\+\\\sqrt[3]{9\sqrt{35}\sqrt{11}}\end{array}\right\}.$$

It can be checked by calculator, however.

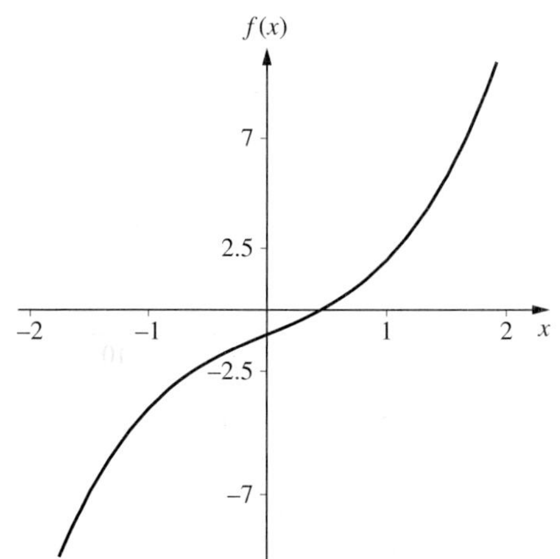

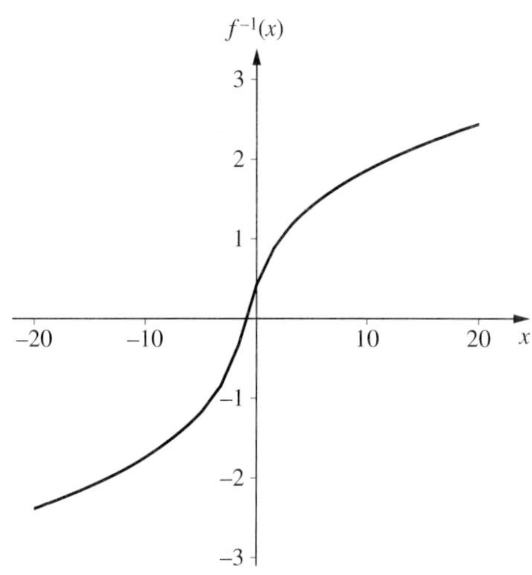

31. Visually, the horizontal line test fails. To prove conclusively that a function as complicated as this is not one-to-one requires considerable experience and the very calculus we are about to begin. However, for the record we note that

$$\begin{aligned} f(\sqrt{3}) &= 3^{5/2} - 3(3^{3/2}) - 1 \\ &= 9\sqrt{3} - 9\sqrt{3} - 1 = -1 = f(0). \end{aligned}$$

This pair of numbers (0 and $\sqrt{3}$), which give the same f-value, suffices to show that f is not one-to-one, but the art of finding these numbers is beyond expectations for the beginning student.

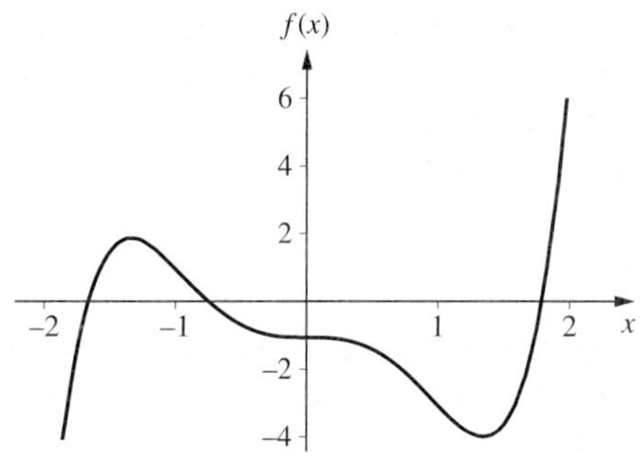

33. If $y = f(x) = 1/(x+1)$, then the horizontal line test is passed, and by algebra, $x = (1-y)/y$, which is a formula good for all y other than $y = 0$. Thus $f^{-1}(x) = (1-x)/x$ for x not zero.

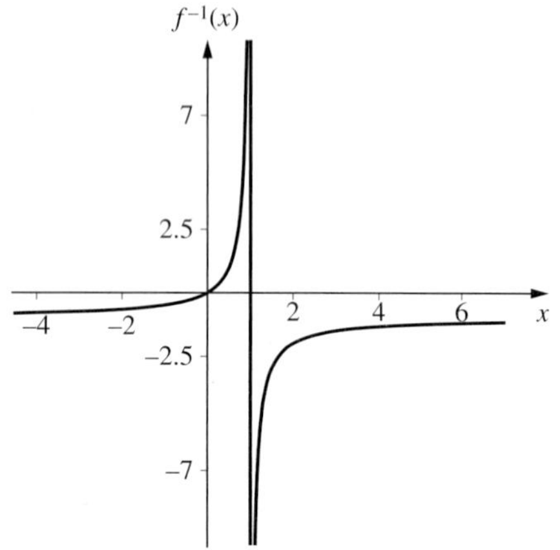

35. Very like Exercise 33; if $y = f(x) = x/(x+4)$, then the horizontal line test is passed, and by algebra, $x = 4y/(1-y)$, which is a formula good for all y other than $y = 1$. Thus $f^{-1}(x) = 4x/(1-x)$ for x not 1.

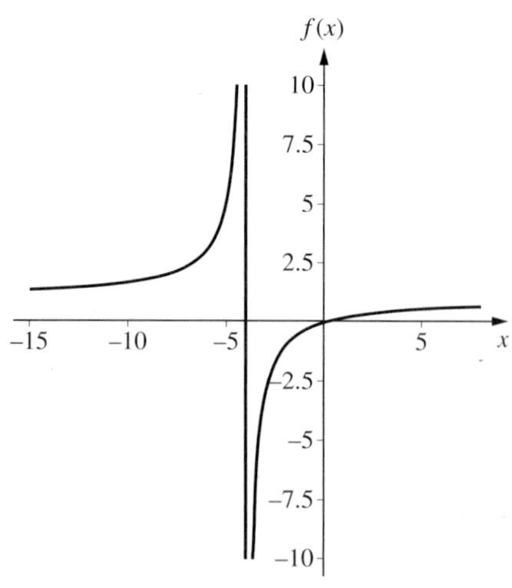

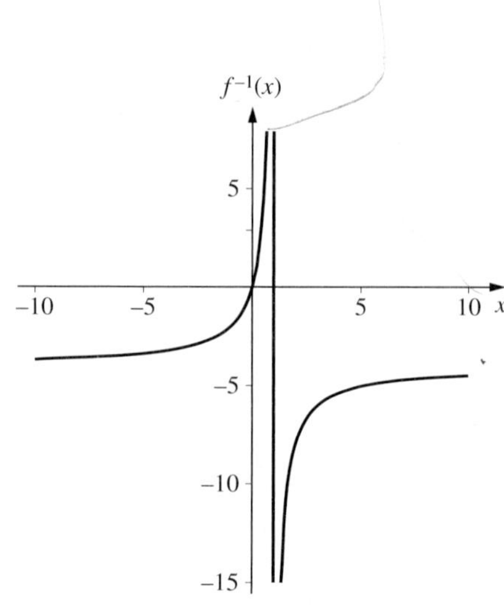

37. With the given formulae,

$$f(g(x)) = [g(x)]^2 = \left[\sqrt{x}\right]^2 = x,$$
$$g(f(x)) = \sqrt{f(x)} = \sqrt{x^2} = |x|.$$

Because $x \geq 0$, the absolute value is the same as x. Thus these functions (both defined only when $x \geq 0$) are an inverse pair.

(In the picture, $y = x^2$ is the one which lies above the other, except for $0 < x < 1$.)

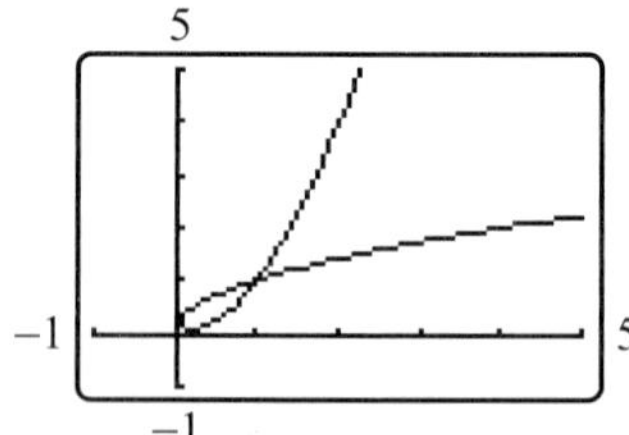

39. With $f(x) = x^2$ defined only for $x \leq 0$ (shown below as dotted), the horizontal line test is easily passed. The formula for the inverse function g is $g(x) = -\sqrt{x}$ shown below as solid and defined only for $x \geq 0$. The verification that these are an inverse pair is routine, much like in Exercise 37.

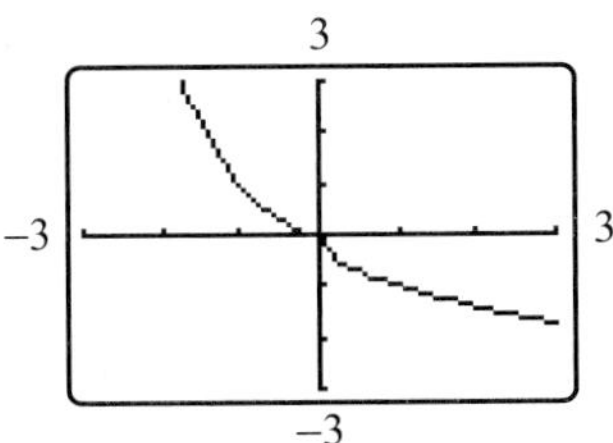

41. The graph of $y = (x - 2)^2$ is a simple parabola with vertex at (2, 0). If we take only the right half $\{x \geq 2\}$ (shown below as the lower right graph), the horizontal line test is easily passed, and the formula for the inverse function g is $g(x) = 2 + \sqrt{x}$ defined only for $x \geq 0$ and shown below as the upper left graph. The verification that these are an inverse pair is routine, much like in Exercise 37.

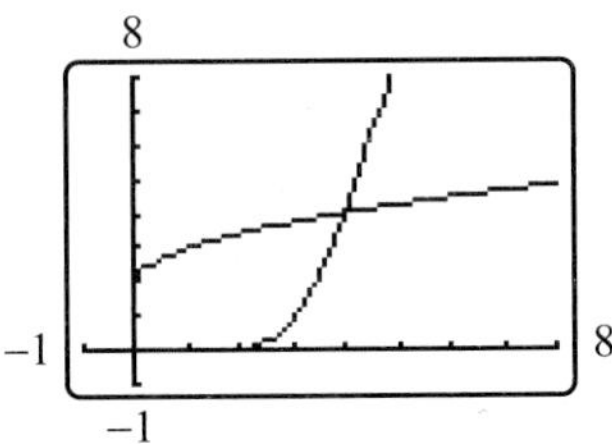

43. In the first place, for $f(x)$ to be defined, the radicand must be nonnegative, i.e., $0 \leq x^2 - 2x = x(x - 2)$, which entails either $x \leq 0$ or $x \geq 2$. One can restrict the domain to either of these intervals and have an invertible function. Taking the latter for convenience, the inverse will be found as follows: start from

$$y = \sqrt{x^2 - 2x}$$
$$y^2 = x^2 - 2x = x^2 - 2x + 1 - 1 = (x - 1)^2 - 1$$
$$y^2 + 1 = (x - 1)^2$$
$$\sqrt{y^2 + 1} = \pm(x - 1).$$

With $x \geq 2$ and the left side nonnegative, we must choose the plus sign. We can then write $x = 1 + \sqrt{y^2 + 1}$.

The right side is now a formula for $f^{-1}(y)$, seemingly good for any y, but we recall from the original formula (as a radical) that y must be nonnegative. We summarize the conclusion:

$$f^{-1}(x) = 1 + \sqrt{x^2 + 1}, \quad (x \geq 0).$$

This is the dotted graph below. The solid graph is the original $y = f(x) = \sqrt{x^2 - 2x}$.

Had we chosen $\{x0\}$, the "other half of the domain," and called the new function h, (same formula as f but a different domain, not shown) we would have come by choosing the minus sign, to the formula

$$h^{-1}(x) = 1 - \sqrt{x^2 + 1}, \ (x \geq 0).$$

The two inverse formulae, if graphed together, fill out the right half of the hyperbola $-x^2 + (y - 1)^2 = 1$.

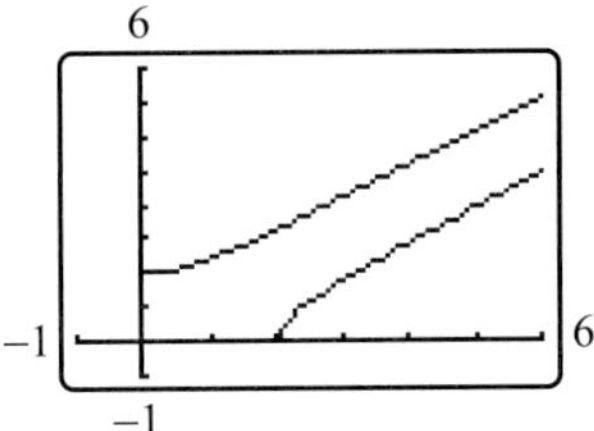

45. The function $\sin(x)$ (solid below) is increasing and one-to-one on the interval

$$-\frac{\pi}{2} \leq x \leq \frac{\pi}{2}.$$

One does not "find" the inverse in the sense of solving the equation $y = \sin(x)$ and obtaining a formula. It is done only in theory or as a graph. The name of the inverse is the "arcsin" function ($y = \arcsin(x)$, shown dotted below), and some of its properties are developed in the next section.

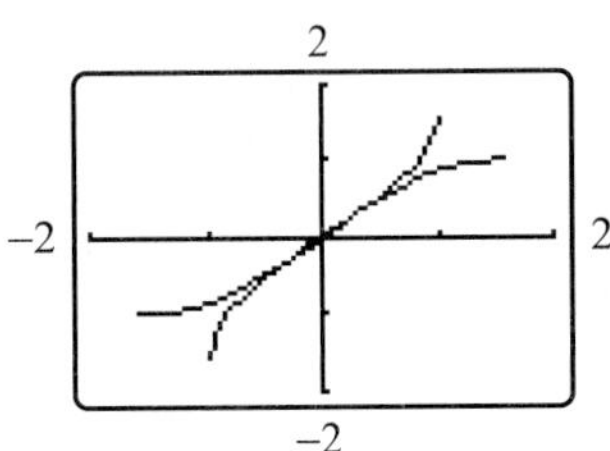

47. A company's income is not in fact a function of time, but a function of a time *interval* (income is defined as the change in *net worth*). When income *is* viewed as a function of time, it is usually after picking a fixed time interval (week, month, quarter, or year) and assigning the income for the period in a consistent manner to either the beginning or the ending date, as in " . . . income for the quarter beginning . . . ." This much said, income more often than not rises and falls over time, so the function is unlikely to be one-to-one. In short, income functions *usually* do not have inverses.

49. During an interval of free fall following a drop, the height is decreasing with time and (barring a powerful up-draft, as with hail) an inverse exists. After impact, if there is a bounce then some of the heights are repeated and the function is no longer one-to-one on the expanded time interval. If there is no bounce and the height remains constantly zero (plugging), then again the function ceases to be one-to-one on any time interval which extends beyond the time of impact. If there is a landing in water, the function may continue to decrease as the object sinks, which could prolong the one-to-one time interval.

51. Two three-dimensional shapes with congruent profiles will cast identical shadows if the congruent profiles face the light source. Such objects need not be fully identical in shape. (For an example, think of a sphere and a hemisphere with the flat side of the latter facing the light.) The shadow as a function of shape is not one-to-one and does not have an inverse.

53. The usual meaning of a "ten percent cut in salary" is that the new salary is 90% of the old. Thus after a ten percent raise the salary is 1.1 times the original, and after a subsequent ten percent cut, the salary is 90% of the raised salary, or .9 times 1.1 times the original salary. The combined effect is 99% of the original, and therefore the ten percent raise and the ten percent cut are not inverse operations.

The 10%-raise function is $y = f(x) = (1.1)x$, and the inverse relation is $x = y/(1.1) = (0.90909\ldots)y$. Thus $f^{-1}(x) = (0.90909)x$ and in the language of cuts, this is a pay cut of fractional value $1 - 0.90909\ldots = 0.090909\ldots$ or $9.0909\ldots$ percent.

0.4 Trigonometric and Inverse Trigonometric Functions

1. **(a)** $$\left(\frac{\pi}{4}\right)\left(\frac{180^\circ}{\pi}\right) = 45^\circ$$

(b) $$\left(\frac{\pi}{3}\right)\left(\frac{180^\circ}{\pi}\right) = 60^\circ$$

(c) $$\left(\frac{\pi}{6}\right)\left(\frac{180^\circ}{\pi}\right) = 30^\circ$$

(d) $$\left(\frac{4\pi}{3}\right)\left(\frac{180^\circ}{\pi}\right) = 240^\circ$$

This particular collection (the first three) of basic angles should be in the student's memory bank, along with $\pi/2$, the right angle (90°) and π, the straight angle (180°). For (d), think $4\pi/3 = \pi + \pi/3 \leftrightarrow 180^\circ + 60^\circ$.

3. **(a)** $$(180)\left(\frac{\pi}{180}\right) = \pi$$

(b) $$(270)\left(\frac{\pi}{180}\right) = \frac{3\pi}{2}$$

(c) $$(120)\left(\frac{\pi}{180}\right) = \frac{2\pi}{3}$$

(d) $$(30)\left(\frac{\pi}{180}\right) = \frac{\pi}{6}$$

5. If $2\cos(x) - 1 = 0$, then

$$2\cos(x) = 1,$$
$$\cos(x) = 1/2,$$
$$x = \frac{\pi}{3} \pm 2n\pi \text{ or } x = -\frac{\pi}{3} \pm 2n\pi$$
$$(n = 0, 1, 2, 3, \ldots).$$

Comment: Considering the quantifications, one could replace $-\pi/3$ by $5\pi/3$ or any other item already in the list ($5\pi/3$ is the case $n = 1$ using the *plus* sign). It is strictly a matter of taste. It will be the same list, as the student will quickly realize if he/she starts to write them down.

7. If $\sqrt{2}\cos(x) - 1 = 0$, then

$$\sqrt{2}\cos(x) = 1,$$
$$\cos(x) = 1/\sqrt{2},$$
$$x = \frac{\pi}{4} \pm 2n\pi \text{ or } x = -\frac{\pi}{4} \pm 2n\pi$$
$$(n = 0, 1, 2, 3, \ldots).$$

9.

$$\sin^2 x - 4\sin x + 3 = 0$$
$$(\sin x - 1)(\sin x - 3) = 0$$
$$\sin x = 1 \text{ or } \sin x = 3$$

There are no solutions for $\sin x = 3$, so the solutions are

$$x = \frac{\pi}{2} \pm 2n\pi$$
$$(n = 0, 1, 2, 3, \ldots).$$

11.

$$\sin^2 x + \cos x - 1 = 0$$
$$(1 - \cos^2 x) + \cos x - 1 = 0$$
$$(\cos x)(\cos x - 1) = 0$$
$$\cos x = 0 \text{ or } \cos x = 1$$
$$x = \pm\frac{\pi}{2} \pm 2n\pi, \quad x = \pm 2n\pi$$
$$(n = 0, 1, 2, 3, \ldots)$$

13.

$$\cos^2 x + \cos x = 0$$
$$(\cos x)(\cos x + 1) = 0$$
$$\cos x = 0 \text{ or } \cos x = 1$$
$$x = \frac{\pi}{2} \pm n\pi, \quad x = \pi \pm 2n\pi$$
$$(n = 0, 1, 2, 3, \ldots)$$

15.

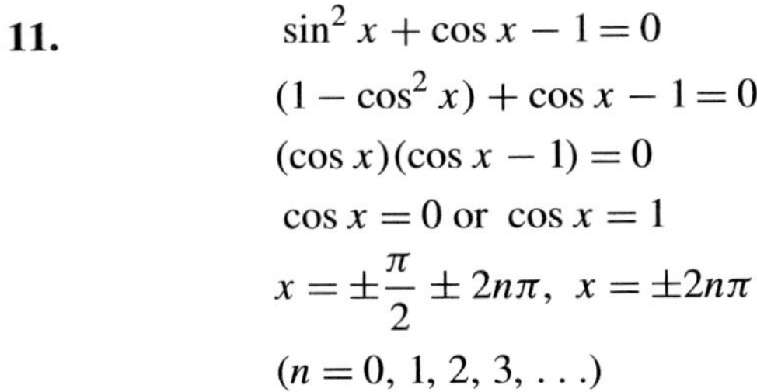

17.

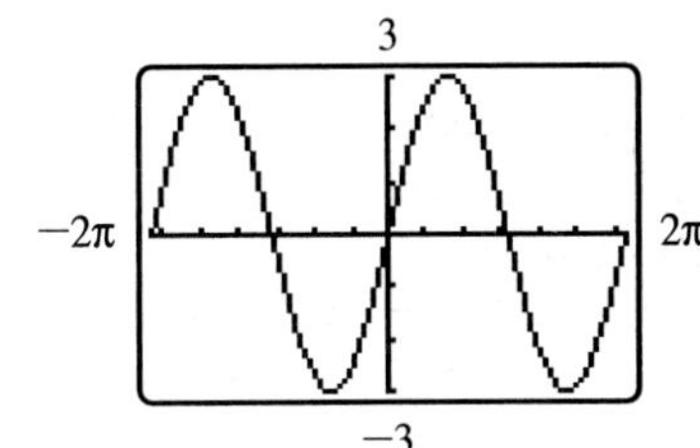

19.

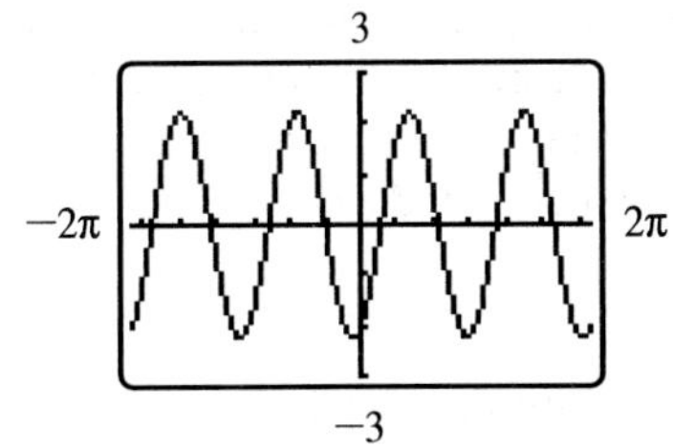

21.

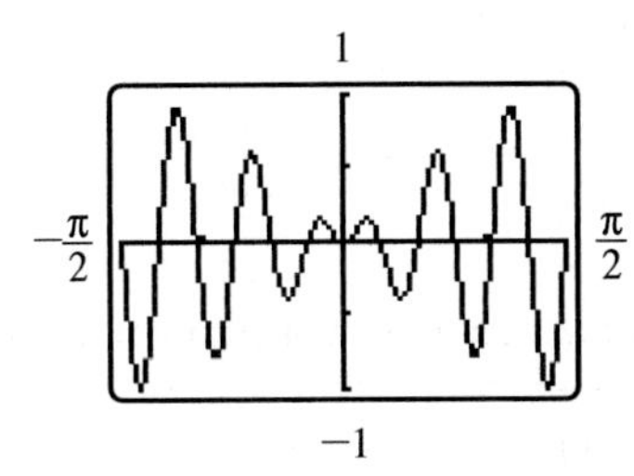

23.

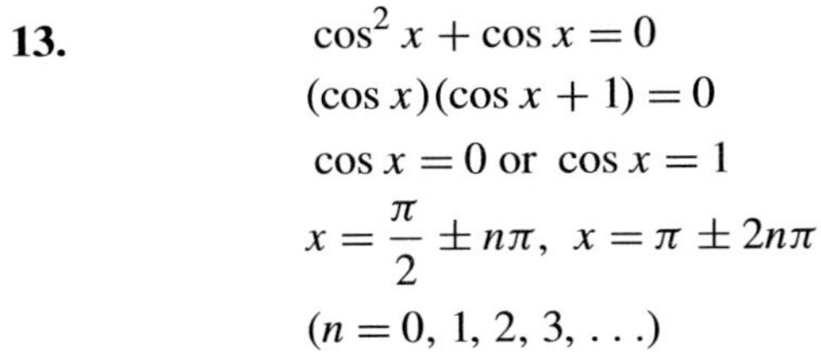

25. amplitude $= 3$

period $= \frac{2\pi}{2} = \pi$ (x-units per cycle)

frequency $= \frac{1}{\text{period}} = \frac{1}{\pi}$ (cycles per x-unit)

Comment: These are a matter of reading the function. There is no work. The quantities "period" and "frequency" are better understood in a context of (e.g.) $\cos(\omega x)$ in which x is *time* (say measured in "seconds"). Since the input to the cosine function *must* be in radians, the frequency parameter ω has the units of "radians per second." The *period* $2\pi/\omega$ is measured in seconds and represents the elapsed time required for a single "cycle." Finally, the *frequency* $\omega/2\pi$ has the units of "cycles per second" and is very natural. When, as in this problem, the x-units in $\cos(\omega x)$ are unspecified, the best we can say is that the period is measured in elapsed units-of-x *per cycle*, and the frequency is in cycles *per unit-of-x*.

27. amplitude $= 5$
period $= \dfrac{2\pi}{3}$ (x-units per cycle)
frequency $= \dfrac{3}{2\pi}$ (cycles per x-unit)

29. amplitude $= 3$
period $= \dfrac{2\pi}{2} = \pi$ (x-units per cycle)
frequency $= \dfrac{1}{\pi}$ (cycles per x-unit)

We are completely ignoring the presence of $-\pi/2$. This has an influence on the so-called "phase shift," which will be studied in Chapter 6.

31. amplitude $= 4$ (ignore the minus sign)
period $= 2\pi$ (x-units per cycle)
frequency $= \frac{1}{2\pi}$ (cycles per x-unit)

33. The student is not expected to produce from scratch a proof of such a fundamental law of trigonometry as this formula for the sine of a *difference*. Indeed, such a proof requires assumptions of great significance which are by no means obvious to the student at this stage of his/her education. All that is expected is for the student to utilize the formula for the sine of a *sum* together with the odd and even properties of some trig functions. It could go like this:

$$\sin(\alpha - \beta) = \sin[\alpha + (-\beta)]$$
$$= \sin\alpha\cos(-\beta) + \sin(-\beta)\cos\alpha$$
$$= \sin\alpha\cos\beta - \sin\beta\cos\alpha.$$

35. **(a)** One can write

$$\cos(2\theta) = \cos(\theta + \theta) =$$
$$\cos(\theta)\cos(\theta) - \sin(\theta)\sin(\theta) =$$
$$\cos^2\theta - \sin^2\theta = \cos^2\theta - [1 - \cos^2\theta]$$
$$2\cos^2\theta - 1.$$

(b) Just continue on, writing

$$\cos(2\theta) = 2\cos^2\theta - 1 =$$
$$2[1 - \sin^2\theta] - 1 = 1 - 2\sin^2\theta.$$

37. On the unit circle, the angle $\theta \in [0, \pi]$ for which $\cos\theta = 0$ is $\theta = \dfrac{\pi}{2}$. Hence, $\cos^{-1}(0) = \dfrac{\pi}{2}$.

39. On the unit circle, the angle $\theta \in \left[-\dfrac{\pi}{2}, \dfrac{\pi}{2}\right]$ for which $\sin\theta = -1$ is $\theta = -\dfrac{\pi}{2}$. Hence, $\sin^{-1}(-1) = -\dfrac{\pi}{2}$.

41. On the unit circle, the angle

$$\theta \in \left[0, \frac{\pi}{2}\right) \cup \left(\frac{\pi}{2}, \pi\right]$$

for which $\sec\theta = 1$ is $\theta = 0$. Hence, $\sec^{-1} 1 = 0$.

43. On the unit circle, the angle

$$\theta \in 0, \left[0, \frac{\pi}{2}\right) \cup \left(\frac{\pi}{2}, \pi\right]$$

for which $\sec\theta = 2$ is $\theta = \frac{\pi}{3}$. Hence, $\sec^{-1} 2 = \frac{\pi}{3}$.

45. On the unit circle, the angle $\theta \in \left[-\dfrac{\pi}{2}, \dfrac{\pi}{2}\right]$ for which $\cot\theta = 1$ is $\theta = \dfrac{\pi}{4}$. Hence, $\cot^{-1}(1) = \dfrac{\pi}{4}$.

47.
$$5\cos(x + \beta) =$$
$$5\cos x\cos\beta - 5\sin x\sin\beta$$

If this is to be $4\cos x - 3\sin x$, we must have

$$5\cos\beta = 4 \quad \text{and} \quad 5\sin\beta = 3$$
$$\cos\beta = \frac{4}{5} \quad \text{and} \quad \sin\beta = \frac{3}{5}$$

This is possible since

$$\left(\frac{4}{5}\right)^2 + \left(\frac{3}{5}\right)^2 = \frac{16}{25} + \frac{9}{25} = 1.$$

β could be

$$\sin^{-1}\left(\frac{3}{5}\right) = \sin^{-1}(0.6) = 0.6435 \text{ (radians)}$$

by calculator. It is not clear what trial-and-error can accomplish. This β is unique up to added multiples of 2π.

49. $\cos(2x)$ has period $\frac{2\pi}{2} = \pi$, so it repeats at intervals of width π, 2π, 3π, etc.
$3\sin(\pi x)$ has period $\frac{2\pi}{\pi} = 2$, so it repeats at intervals of width 2, 4, 6, etc.

Since one list consists of irrational numbers and the other of rational numbers, there can't be any numbers on both lists. $f(x)$ is not periodic.

51. $\sin(2x)$ has period $\frac{2\pi}{2} = \pi$, so it repeats at intervals of width π, 2π, 3π, etc.
$\cos(5x)$ has period $\frac{2\pi}{5}$, so it repeats at intervals of width $\frac{2\pi}{5}$, $\frac{4\pi}{5}$, $\frac{6\pi}{5}$, etc.

Since 2π is the first number to appear on both lists, $f(x)$ is periodic with minimum period 2π.

53.

$$\cos^2\theta = 1 - \sin^2\theta = 1 - \left(\frac{1}{3}\right)^2 = 1 - \frac{1}{9} = \frac{8}{9}.$$

Because θ is in the first quadrant, its cosine is nonnegative. Hence,

$$\cos\theta = \sqrt{\frac{8}{9}} = \frac{2\sqrt{2}}{3} = 0.9428\ldots.$$

55.

$$\cos^2\theta = 1 - \sin^2\theta = 1 - \left(\frac{1}{2}\right)^2 = 1 - \frac{1}{4} = \frac{3}{4}.$$

Because θ is in the second quadrant, its cosine is negative. Hence,

$$\cos\theta = -\sqrt{\frac{3}{4}} = -\frac{\sqrt{3}}{2} = -.8660\ldots.$$

Of course this particular angle should be recognized at once ($5\pi/6$ or $150°$).

57. Assume $0 < x < 1$ and give the temporary name θ to $\sin^{-1}(x)$. In a little right triangle with hypotenuse 1 and one leg of length x, the angle θ will show up opposite the x-side, and the adjacent side will have length $\sqrt{1-x^2}$. Write:

$$\cos(\sin^{-1}(x)) = cos(\theta)$$

$$= \frac{\text{adjacent}}{\text{hypotenuse}} = \frac{\sqrt{1-x^2}}{1} = \sqrt{1-x^2}.$$

The formula is numerically correct in the cases $x = 0$ and $x = 1$, and both sides are *even* functions of x ($f(-x) = f(x)$), so the formula is good for $-1 \le x \le 1$.

59. Assume $1 < x$ and give the temporary name θ to $\sec^{-1}(x)$. In a right triangle with hypotenuse x and one leg of length 1, the angle θ will show up adjacent to the side of length 1, and the opposite side will have length $\sqrt{x^2-1}$. Write:

$$\tan(\sec^{-1}(x)) = \tan(\theta)$$

$$= \frac{\text{opposite}}{\text{adjacent}} = \frac{\sqrt{x^2-1}}{1} = \sqrt{x^2-1}.$$

The formula is numerically correct in the case $x \ge 1$. Dealing with negative x is trickier: assume $x > 1$ for the moment. The key identity is $\sec^{-1}(-x) = \pi - \sec^{-1}(x)$. Taking tangents on both sides and applying the identity

$$\tan(a-b) = \frac{\tan(a)\tan(b)}{1+\tan(a)\tan(b)} \text{ with } a = \pi,$$

$$\tan(a) = 0,\ b = \sec^{-1}(x),\ \text{ we find}$$

$$\tan(\sec^{-1}(-x)) = \frac{0 - \tan(\sec^{-1}(x))}{1+0}$$

$$= -\sqrt{x^2-1} = -\sqrt{(-x)^2-1}.$$

In this identity, $-x$ (on both sides) plays the role of an arbitrary number < -1. Consequently, the final formula is $\tan(\sec^{-1}(x)) = -\sqrt{x^2-1}$ whenever $x \le -1$.

61. Reasoning much as in the two previous, with numbers in place of letters, one easily sees that the angle $\cos^{-1}(1/2)$ is 60°, but even without that, its sine is

$$\sqrt{1-\left(\frac{1}{2}\right)^2} = \frac{\sqrt{3}}{2}.$$

63. Reasoning much as in the three previous, with numbers in place of letters, one easily envisions a right triangle in which the hypotenuse is 5, an acute angle is highlighted, the opposite leg is 4, and the adjacent leg is 3. The highlighted angle has

$$\text{cosine} = \frac{\text{adjacent}}{\text{hypotenuse}} = \frac{3}{5},$$

and its

$$\text{tangent} = \frac{\text{opposite}}{\text{adjacent}} = \frac{4}{3},$$

which is the answer to the problem.

65.

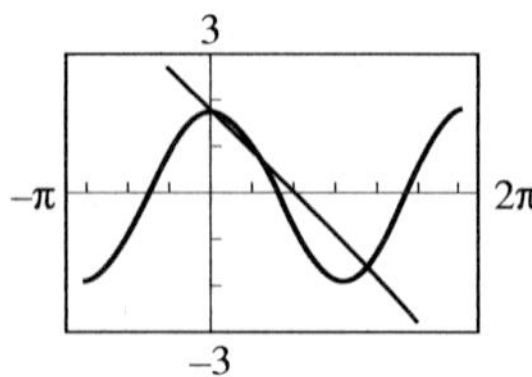

$$3 \text{ solutions} : x = 0,\ x \approx 1.109,\ xx \approx 3.698$$

67.

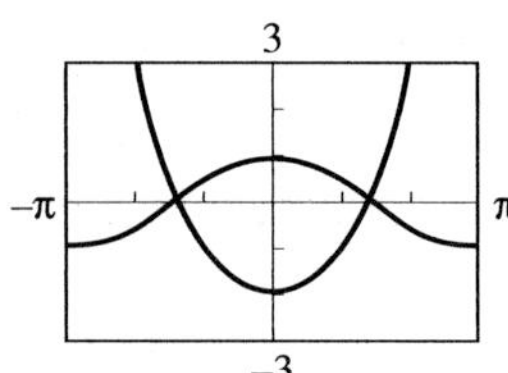

$$2 \text{ solutions } x \approx -1.455,\ x \approx 1.455$$

69.

$$\frac{h}{2} = \tan 20°$$
$$h = 2\tan 20° \approx 0.73 \text{ (miles)}$$

71.

$$\frac{h}{80+20} = \tan 50°$$
$$h = 100\tan 50° \approx 119.2 \text{ (feet).}$$

73. Since we are not given the vertical distance from the man's eye to the top of his head, we assume initially that the line of sight from the eye to the bottom of the picture frame is horizontal. Using feet as the measuring standard, we find

$$\tan A = \frac{\text{opposite}}{\text{adjacent}} = \frac{20/12}{x} = \frac{5}{3x}$$
$$A = \tan^{-1}\left(\frac{5}{3x}\right).$$

Now assuming that the vertical spread from eye to top of head is y (inches), we have the more complicated formula

$$A = \tan^{-1}\left(\frac{20+y}{12x}\right) - \tan^{-1}\left(\frac{y}{12x}\right)$$

The previous is the special case $y = 0$, and is all that was expected of the student.

75. Presumably, the given amplitude (170) is the same as the "peak voltage" (v_p). Recalling an earlier discussion (Exercise 25 this section): the role of ω there is played by $2\pi f$ here, the frequency in cycles per second (Hz) was $\omega/2\pi$, which is now the f-parameter $(2\pi f/2\pi)$. The period was $2\pi/\omega$ (which is now $1/f$), given in this case to be $\pi/30$ (seconds). So, apparently, the frequency is $f = 30/\pi$ (cycles per second) and the meter voltage

$$\left(\frac{v_p}{\sqrt{2}}\right) \quad \text{is} \quad \frac{170}{\sqrt{2}} \approx 120.2.$$

77. There seems to be a certain slowly increasing base for sales $(110 + 2t)$, and given that the sine function has period

$$\frac{2\pi}{\pi/6} = 12 \text{ (months)} = \text{one year},$$

the sine term apparently represents some sort of seasonally cyclic pattern. If we assume that travel peaks at Thanksgiving, the effect is that time zero would correspond to a time one quarter-period (3 months) prior to Thanksgiving, or very late August.

The annual increase for the year beginning at time t is given by $s(t+12)-s(t)$ and automatically ignores both the seasonal factor and the basic 110, and indeed it is the constant $2 \times 12 = 24$ (in thousands of dollar per year and independent of the reference point t).

79. As luck would have it, the trig functions csc and cot, being reciprocals respectively of sine and tangent, have inverses almost exactly where the other two do, both on the interval $\left[-\frac{\pi}{2}, \frac{\pi}{2}\right]$ but excluding the origin where neither is defined, and excluding the lower endpoint in the case of the cotangent. The range for the sine is $[-1, 1]$, hence the range for the csc is $\{|x| \geq 1\}$ and this is the domain for $\csc^{-1}$. The tangent assumes all values, and so does the cot (zero included as a value by convention when $x = \pi/2$ or $-\pi/2$), so the domain for $\cot^{-1}$ is universal. Finally, we simply copy the language of the others:

$y = csc^{-1}(x)$ if $|x| \geq 1$, y lies in $\left[-\frac{\pi}{2}, \frac{\pi}{2}\right]$

and $x = \csc(y)$.

$y = cot^{-1}(x)$ if y lies in $\left(-\frac{\pi}{2}, \frac{\pi}{2}\right]$,

and $x = \cot(y)$.

Comments: The latter procedure assigns the value $y = \pi/2$ to be $\cot^{-1}(0)$. It was necessary to exclude $-\pi/2$ from the domain of the cotangent in order for that function to meet the horizontal line test and be invertible. There are many identities which render these two functions unnecessary, e.g.,

$\cot^{-1} x = \pi/2 - \tan^{-1}(x)$ when $x \geq 0$ and

$\csc^{-1}(x) = \sin^{-1}(1/x)$ when $\{|x| \geq 1\}$.

0.5 Exponential and Logarithmic Functions

1. $2^{-3} = \frac{1}{2^3} = \frac{1}{8}$

3. $3^{1/2} = \sqrt{3}$

5. $5^{2/3} = \sqrt[3]{5^2} = \sqrt[3]{25}$

7. $\frac{1}{x^2} = x^{-2}$

9. $\frac{2}{x^3} = 2x^{-3}$

11. $\frac{1}{2\sqrt{x}} = \frac{1}{2x^{1/2}} = \frac{1}{2}x^{-1/2}$

13. $4^{3/2} = \left(\sqrt{4}\right)^3 = 2^3 = 8$

15. $\frac{\sqrt{8}}{2^{1/2}} = \frac{\sqrt{8}}{\sqrt{2}} = \sqrt{4} = 2$

17. $2e^{-1/2} \approx 1.213$

19. $\frac{12}{e} \approx 4.415$

21.

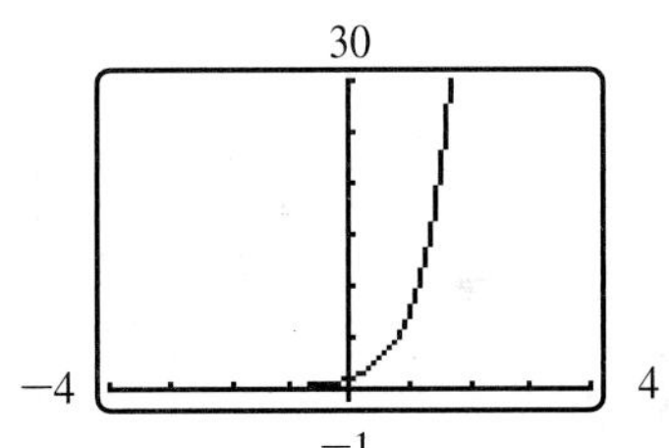

23.

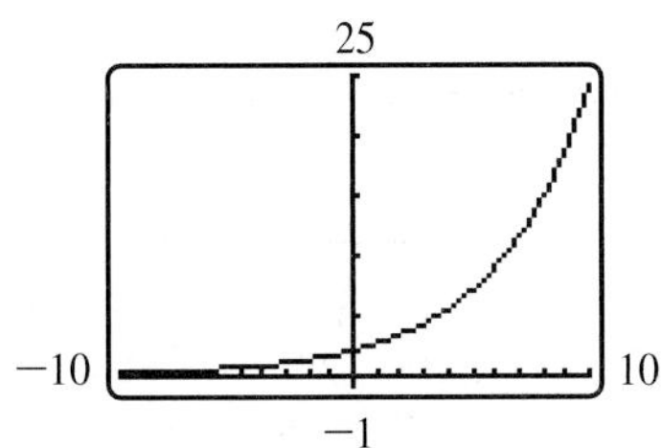

25.

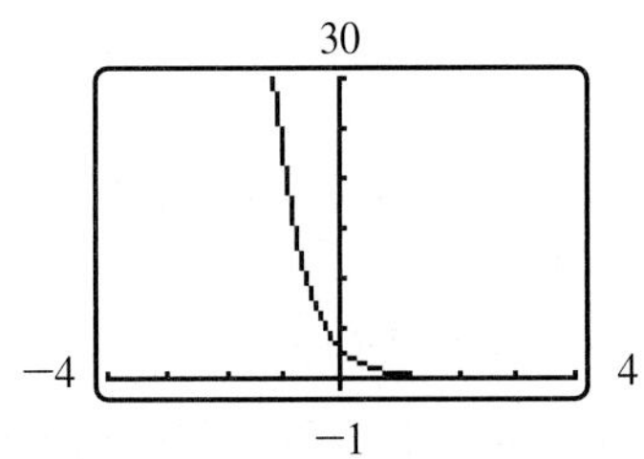

27.

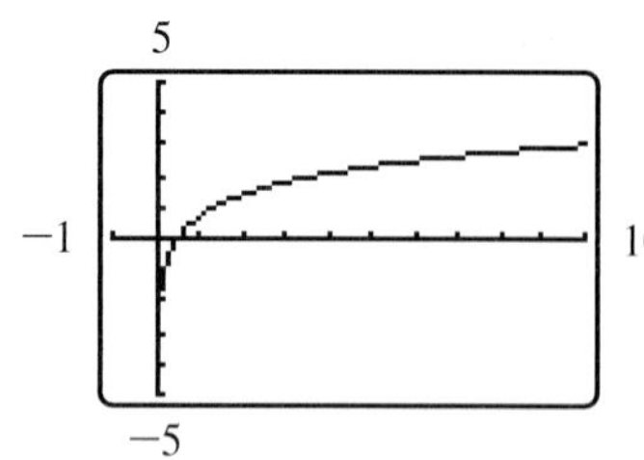

29.

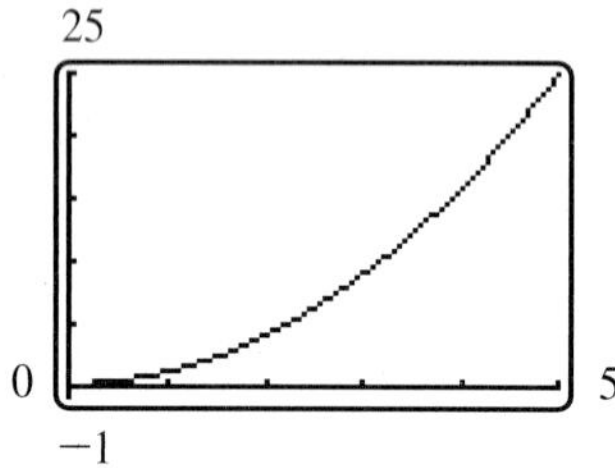

If this looks like $y = x2$, it's because it is.

31.
$$\begin{aligned} e^{2x} &= 2 \\ \ln e^{2x} &= \ln 2 \\ 2x &= \ln 2 \\ x &= \frac{\ln 2}{2} \approx 0.34657\ldots \end{aligned}$$

33. $e^x(x^2 - 1) = 0$

Either $x^2 - 1 = 0$ (hence $x = 1$ or $x = -1$), or $e^x = 0$ (never!).

35.
$$\begin{aligned} \ln 2x &= 4 \\ 2x &= e^4 \\ x &= \frac{e^4}{2} \approx 27.299\ldots \end{aligned}$$

37.
$$\begin{aligned} 4 \ln x &= -8 \\ \ln x &= -2 \\ x &= e^{-2} = \frac{1}{e^2} \approx 0.13533\ldots \end{aligned}$$

39.
$$\begin{aligned} e^{2 \ln x} &= 4 \\ 2 \ln x &= \ln 4 = \ln 2^2 = 2 \ln 2 \\ \ln x &= \ln 2 \\ x &= 2 \end{aligned}$$

41. **(a)** $\log_3 9 = \log_3(3^2) = 2$

(b) $\log_4 64 = \log_4(4^3) = 3$

(c) $\log_3 \frac{1}{27} = \log_3(3^{-3}) = -3$

43. **(a)** $\log_3 7 = \frac{\ln 7}{\ln 3} \approx 1.771$

(b) $\log_4 60 = \frac{\ln 60}{\ln 4} \approx 2.953$

(c) $\log_3 \frac{1}{24} = \frac{\ln(1/24)}{\ln 3} \approx -2.893$

45. $\ln 3 - \ln 4 = \ln\left(\frac{3}{4}\right)$

47. $\frac{1}{2} \ln 4 - \ln 2 = \ln 4^{1/2} - \ln 2 = \ln 2 - \ln 2 = 0$

49.
$$\begin{aligned} \ln \frac{3}{4} + 4 \ln 2 &= \ln \frac{3}{2^2} + \ln 2^4 \\ &= \ln\left(\frac{3}{2^2} \cdot 2^4\right) \\ &= \ln(3 \cdot 2^2) = \ln(12) \end{aligned}$$

51. Using $f(0) = 2$:
$$\begin{aligned} 2 &= ae^{b \cdot 0} = ae^0 = a \\ f(x) &= 2e^{bx} \end{aligned}$$

Using $f(2) = 6$:
$$\begin{aligned} 6 &= 2e^{2b} \\ 3 &= e^{2b} \\ \ln 3 &= \ln e^{2b} = 2b \\ b &= \frac{\ln 3}{2} \end{aligned}$$
$$f(x) = 2e^{(\frac{1}{2} \ln 3)x} = 2[e^{\ln(3)}]^{x/2} = 2 \cdot 3^{x/2}$$

53. Using $f(0) = 4$:
$$\begin{aligned} 4 &= ae^{b.0} = a \\ f(x) &= 4e^{bx} \end{aligned}$$

Using $f(2) = 2$:
$$\begin{aligned} 2 &= 4e^{2b} \\ \frac{1}{2} &= e^{2b} \\ 2 &= e^{-2b} \\ -2b &= \ln(2) \\ b &= -\frac{\ln 2}{2} \end{aligned}$$

$$f(x) = 4e^{-\left(\frac{1}{2}\ln 2\right)x} = 4[e^{\ln(2)}]^{-x/2} = \frac{4}{2^{x/2}}$$

55. $1 - \left(\frac{9}{10}\right)^{10} \approx 0.651$

57. We take on faith, whatever it may mean, that

$$\lim_{n\to\infty}\left(1+\frac{1}{n}\right)^n = e.$$

Just to take a sample starting with $n = 25$, the numbers are

$$\left(\frac{26}{25}\right)^{25}, \left(\frac{27}{26}\right)^{26}, \left(\frac{28}{27}\right)^{27},$$

and so on. If we were to try taking a similar look at the numbers in $\lim_{n\to\infty}\left(1-\frac{1}{n}\right)^n$, the numbers starting at $n = 26$ would be

$$\left(\frac{25}{26}\right)^{26}, \left(\frac{26}{27}\right)^{27}, \left(\frac{27}{28}\right)^{28},$$

and so on.
We could rewrite these as

$$\left[\left(\frac{25}{26}\right)^{25}\right]^{26/25}, \left[\left(\frac{26}{27}\right)^{26}\right]^{27/26}, \left[\left(\frac{27}{28}\right)^{27}\right]^{28/27} \cdots$$

Here, the numbers inside the square brackets are the reciprocals of the numbers in the original list, which were all pretty close to e. Therefore these must all be pretty close to $1/e$. As to the external powers, they are all close to 1 and getting closer. This limit must be $1/e$. The expression in question must approach $1 - \frac{1}{e} \approx .632$

59. We have these numbers by calculator:

u	v
.78846	2.6755
.87547	2.8495
.95551	3.0096
1.0296	3.1579
1.0986	3.2958
1.1632	3.4249

Here is the plot of the six points:

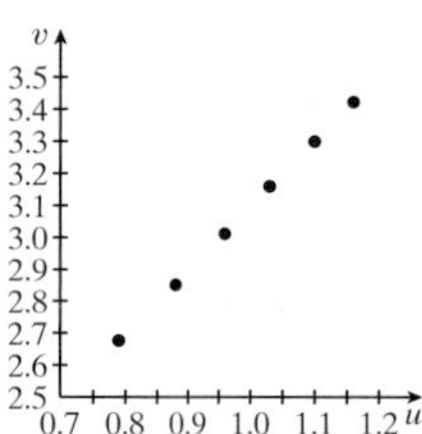

It looks like a line. If we assume it is, and try to find its slope and intercept using the first and last points, we come to

$$m = \frac{3.4249 - 2.6775}{1.1632 - .78846} \approx 2$$
$$\therefore\ v = 2u + b$$
$$2.6755 = 2.(.78846) + b$$
$$b \approx 1.099$$
$$\therefore\ a = e^b$$
$$\approx 3.001$$
$$y = 3.001x^2.$$

61. When the data are processed and the log-log points are plotted, the appearance is as follows:

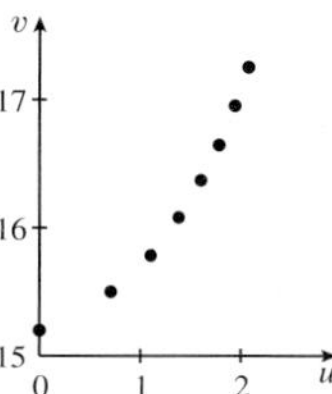

This plot does not look linear, which makes it clear that the population is *not* modeled by a power of x. The discussion in the Chapter has already strongly indicated that an exponential model is fairly good.

63. **(a)** $7 = -\log[H^+]$
$[H^+] = 10^{-7}$
(b) $[H^+] = 10^{-8}$
(c) $[H^+] = 10^{-9}$

For each increase in pH of one, $[H^+]$ is reduced to one tenth of its previous value.

65. (a)

$$\log E = 4.4 + 1.5(4) = 10.4$$
$$E = 10^{10.4}$$

(b)

$$\log E = 4.4 + 1.5(5) = 11.9$$
$$E = 10^{11.9}$$

(c)

$$\log E = 4.4 + 1.5(6) = 13.4$$
$$E = 10^{13.4}$$

For each increase in M of one, E is increased by a factor of $10^{1.5} \approx 31.6$.

67. (a) $80 = 10 \log\left(\frac{I}{10^{-12}}\right)$

$$8 = \log\left(\frac{I}{10^{-12}}\right)$$
$$10^8 = \frac{I}{10^{-12}}$$
$$I = 10^8 \cdot 10^{-12} = 10^{-4}$$

(b) $I = 10^{-3}$

(c) $I = 10^{-2}$

For each increase in dB of ten, I increases by a factor of 10.

69. $y = xe^{-x}$: maximum value $1/e$ at $x = 1$

$y = xe^{-2x}$: maximum value $1/(2e)$ at $x = \frac{1}{2}$

$y = xe^{-3x}$: maximum value $1/(3e)$ at $x = \frac{1}{3}$

Guess about $y = xe^{-kx}$:

maximum value $1/(ke)$ at $x = \frac{1}{k}$

If one believes the first of these, then the general case follows by writing

$$xe^{-kx} = \frac{kxe^{-kx}}{k} = \frac{ue^{-u}}{k} \quad (u = kx).$$

The numerator has maximum value $1/e$ when $u = 1 (x = 1/k)$. Therefore the whole expression has maximum value $(1/k)(1/e)$ when $x = 1/k$.

71. Among positive numbers u, the function $u + \frac{1}{u}$ has minimum value 2 when $u = 1$. To actually prove this, one writes

$$2 - \left(u + \frac{1}{u}\right) = \frac{2u - u^2 - 1}{u} =$$
$$\frac{(u^2 - 2u + 1)}{u} = \frac{(u-1)^2}{u} \leq 0$$

with equality only if $u = 1$. This shows that $u + \frac{1}{u} \geq 2$ with equality only if $u = 1$. Meanwhile, with $u = e^x > 0$,

$$\cosh(x) = \frac{e^x + e^{-x}}{2} = \frac{u + \frac{1}{u}}{2} \geq \frac{2}{2} = 1$$

with equality only if $u = 1$ $(x = 0)$.
This shows that the range of *cosh* includes only numbers greater than or equal to one.

As regards the hyperbolic sine, if we try to solve the equation $\sinh(x) = a$, again using $u = ex$, we can write

$$\frac{u - \frac{1}{u}}{2} = a$$
$$u^2 - 1 = 2au$$
$$u^2 - 2au - 1 = 0$$
$$u = \frac{2a \pm \sqrt{4a^2 + 4}}{2} = a + \sqrt{a^2 + 1}.$$

We simplified and chose the positive square root because $u > 0$. Because we found a unique solution no matter what a we had started with, we have actually proven that *sinh* is one-to-one by successfully finding this formula for the *inverse:*

$$\sinh^{-1}(a) = x = \ln(u) = \ln\left(a + \sqrt{a^2 + 1}\right).$$

Had we tried the same gambit with the hyperbolic cosine, we would have come to

$$u = a \pm \sqrt{a^2 - 1}.$$

We can't automatically reject the minus possibility, but we are already aware of the necessity of $a \geq 1$, so that the radicand is at least nonnegative. The two choices are reciprocals of one another (multiply them together and see), their logarithms are negatives, and so in the end we have

$$x = \ln(u) = \pm \ln\left(a + \sqrt{a^2 - 1}\right)$$

whenever $a \geq 1$.

Because it is clear that the hyperbolic cosine function is *even* $(f(-x) = f(x))$, the fact that the two solutions to the equation $\cosh(x) = a$ are negatives of one another comes as no surprise. We learn that the hyperbolic cosine *is* one-to-one on the restricted domain $[0, \infty)$ and that the range is the positive interval $[1, \infty)$.

73. The issue is purely whether or not $y = 0$ when $x = 315$, i.e., whether or not $\cosh(315/127.7) = \cosh(2.4667\ldots) = 5.9343\ldots$ is the same as $(757.7)/(127.7) = 5.9334\ldots$ We see that it's pretty close, and these numbers would be considered equal according to the level of accuracy reported in the original measurements.

75. One can look back to the discussion in Exercise 71. Because $\sinh^{-1}(0) = 0$, the equation is solved only by $x^2 - 1 = 0$, hence $x = 1$ or $x = -1$.

77. Doubling the Hz *per octave* leads to the formula $f = 220(2x)$ for $x = 0, 1, 2, 3, \ldots$ This is perfectly satisfactory, but if one wants to write in e-language, go

$$f = 220e^{x\ln(2)} = 220e^{0.693x}.$$

0.6 Transformations of Functions

1.

$$\begin{aligned}(f \circ g)(x) &= f(g(x)) \\ &= g(x) + 1 \\ &= \sqrt{x - 3} + 1\end{aligned}$$

$$D: \{x \geq | x3\}$$

We emphasize that a "domain" for a formula-type function can be any set on which the formula is properly defined. The presumption here is that we are being asked for the *largest possible* domain, which is considered the "natural" domain for a given formula.

$$\begin{aligned}(g \circ f)(x) &= g(f(x)) \\ &= \sqrt{f(x) - 1} \\ &= \sqrt{(x + 1) - 3} \\ &= \sqrt{x - 2}\end{aligned}$$

$$D: \{x | x \geq 2\}$$

3.

$$\begin{aligned}(f \circ g)(x) &= f(g(x)) \\ &= f(\ln x) \\ &= e^{\ln x} \\ &= x\end{aligned}$$

$$D: \{x | x > 0\}$$

$$\begin{aligned}(g \circ f)(x) &= g(f(x)) \\ &= g(e^x) \\ &= \ln e^x \\ &= x\end{aligned}$$

$D: (-\infty, \infty)$ (all real numbers)

5.

$$\begin{aligned}(f \circ g)(x) &= f(g(x)) \\ &= f(\sin x) \\ &= \sin^2 x + 1\end{aligned}$$

$D: (-\infty, \infty)$ (all real numbers)

$$\begin{aligned}(g \circ f)(x) &= g(f(x)) \\ &= g(x^2 + 1) \\ &= \sin(x^2 + 1)\end{aligned}$$

$D: (-\infty, \infty)$ (all real numbers)

7. Most natural answer:

$$f(x) = \sqrt{x};\ g(x) = x^4 + 1.$$

9. Most natural answer:

$$f(x) = \frac{1}{x};\ g(x) = x^2 + 1.$$

11. Most natural answer:

$$f(x) = x^2 + 3;\ g(x) = 4x + 1.$$

13. Most natural answer:

$$f(x) = x^3;\ g(x) = \sin x.$$

15. Most natural answer:

$$f(x) = e^x;\ g(x) = x^2 + 1.$$

17. One natural answer:

$$h_1(x) = (\sin x) + 2;$$
$$g_1(x) = \sqrt{x};$$
$$f_1(x) = \frac{3}{x}.$$

In doing problems like this, it helps to ask: "If I were going to calculate this number, what would I calculate first? What next? . . . What last?" (The *first* operation is known as the "Inner" function, the *last* is known as the "Outer" function. Generic *composition of functions* is Outer $\circ \ldots \circ$ Inner.) The answer is not unique. Equally reasonable in this problem would have been

$$h_2(x) = \sin x;$$
$$g_2(x) = \sqrt{x+2};$$
$$f_2(x) = \frac{3}{x}.$$

The ambiguity arises because, in a sense, the most primitive breakdown would involve *four* stages:

$$K(x) = \sin x;$$
$$H(x) = x + 2;$$
$$G(x) = \sqrt{x};$$
$$F(x) = \frac{3}{x}.$$

Observe carefully that $h1 = H \circ K$ while $g2 = G \circ H$. Therefore

$$f_1 \circ g_1 \circ h_1 = F \circ G \circ (H \circ K) \text{ while}$$
$$f_2 \circ g_2 \circ h_2 = F \circ (G \circ H) \circ K.$$

Composition is always *associative*, meaning that it doesn't matter how the parentheses are placed.

19. Applying the previous principles, to get $\cos^3(4x - 2)$ (given x) I would first need to know $4x - 2$. Then I would need its cosine. Finally I would cube that.

Thus I would take

$$h(x) = 4x - 2 \text{ (inner = first);}$$
$$g(x) = \cos(x) \text{ (intermediate);}$$
$$f(x) = x^3 \text{ (outer = last).}$$

21. Applying the previous principles, to get $4e^{x^2} - 5$ (given x) I would first need to know x^2. Then I would exponentiate. Finally I would do the linear steps more or less together. Thus I would take

$$h(x) = x^2 \text{ (inner = first);}$$
$$g(x) = e^x \text{ (intermediate);}$$
$$f(x) = 4x - 5 \text{ (outer = last).}$$

23–37. Going from separate graphs to the graph of a composition can be quite tricky. In these problems, however, although the given function f (presented by graph) may be either inner, outer, or intermediate, the other participants are *linear*. The effect in this case is a change in the scale: on the x-axis if the *inner* function is linear (Exercises 23,29,31,35), on the y-axis if the *outer* function is linear (Exercises 25,27,33,37) and of course both if f ends up intermediate (Exercise 30). Common sense can be your guide. (This has been done in the answer appendix on all but Exercise 35.)

If one wants to keep the scales unchanged, then the effect is vertical or horizontal shifting, stretching or contracting of the graph. (This has been done in Exercise 35, where some features are no longer in the visible part of the graph, and some previously unseen parts have been brought into play.)

A mirroring step is also possible, but does not appear to happen in any of these problems (until Exercise 45).

23. Shift the graph down 3 units.

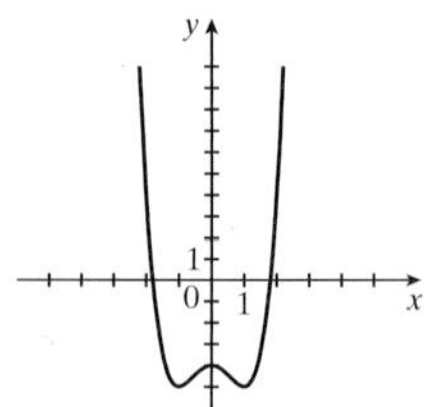

25. Shift the graph right 3 units.

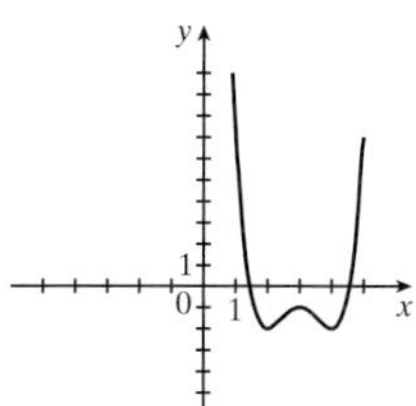

27. Multiply the scale on the x-axis by $\frac{1}{2}$.

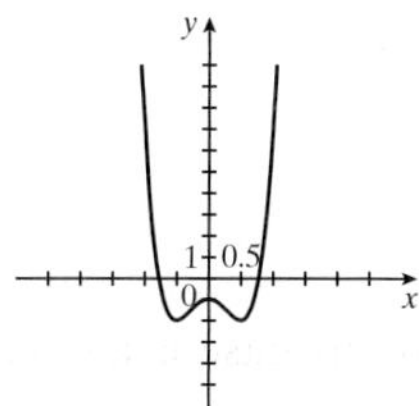

29. Multiply the scale on the y-axis by 4 and then shift the graph down 1 unit.

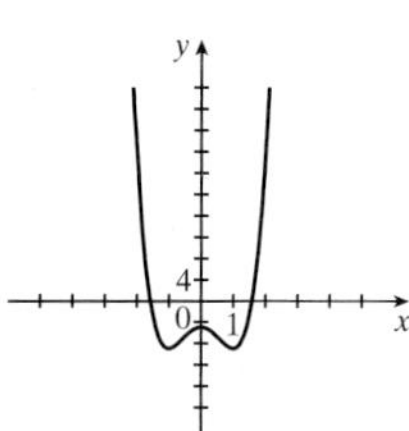

31. Shift the graph right 4 units.

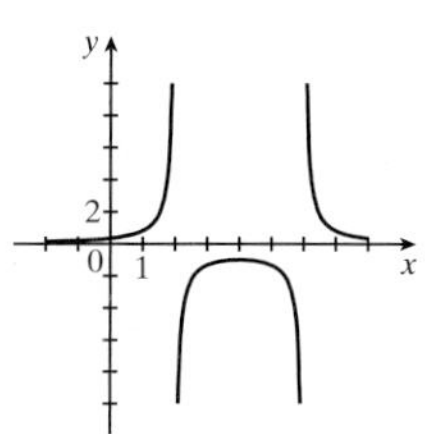

33. Multiply the scale on the x-axis by $\frac{1}{2}$.

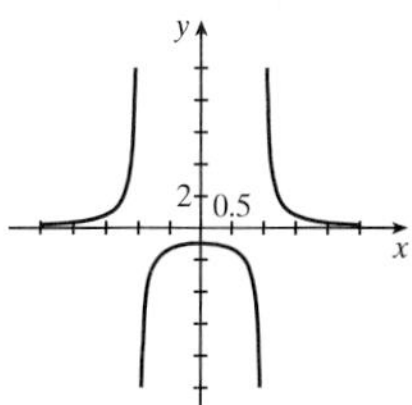

35. Shift the graph left 3 units and then multiply the scale on the x-axis by $\frac{1}{3}$.

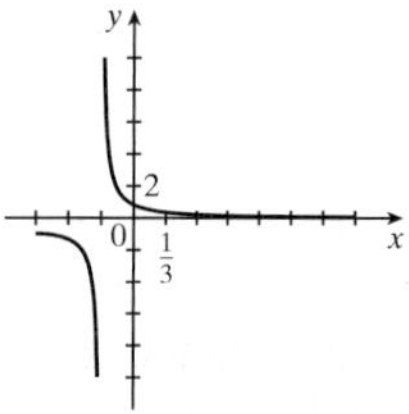

37. Multiply the scale on the y-axis by 2 and then shift the graph down 4 units.

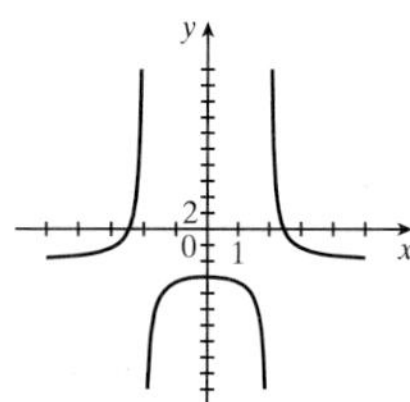

39.

$$\begin{aligned} f(x) &= x^2 + 2x + 1 \\ &= (x^2 + 2x + 1) + 1 - 1 \\ f(x) &= (x + 1)^2 \end{aligned}$$

Shift $y = x^2$ to the left 1 unit.

41.

$$\begin{aligned} f(x) &= x^2 + 2x + 4 \\ &= (x^2 + 2x + 1) + 4 - 1 \\ f(x) &= (x + 1)^2 + 3 \end{aligned}$$

Shift $y = x^2$ to the left 1 unit and up 3 units.

43.

$$\begin{aligned} f(x) &= 2x^2 + 4x + 4 \\ &= 2(x^2 + 2x + 1) + 4 - 2 \\ f(x) &= 2(x+1)^2 + 2 \end{aligned}$$

Shift $y = x^2$ to the left 1 unit, then multiply the scale on the y-axis by 2, then shift up 2 units. Alternately, using

$$f(x) = 2[(x+1)^2 + 1],$$

shift left 1 unit and up 1 unit, then double the vertical scale.

45.

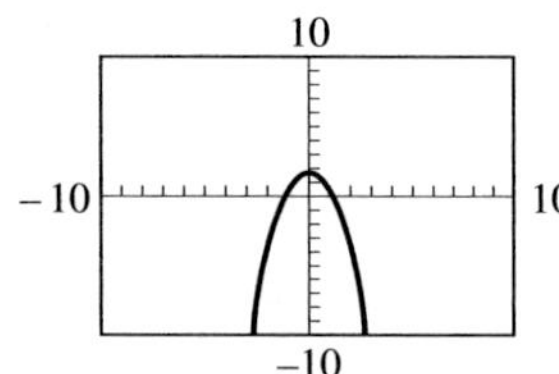

Graph is reflected across the x-axis and the scale on the y-axis is multiplied by 2.

47.

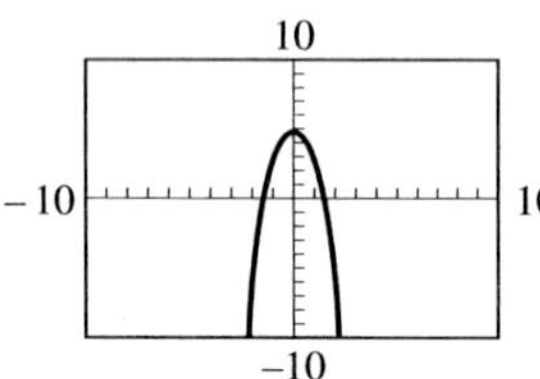

Graph is reflected across the x-axis, the scale on the y-axis is multiplied by 3, and the graph is shifted up 2 units.

49.

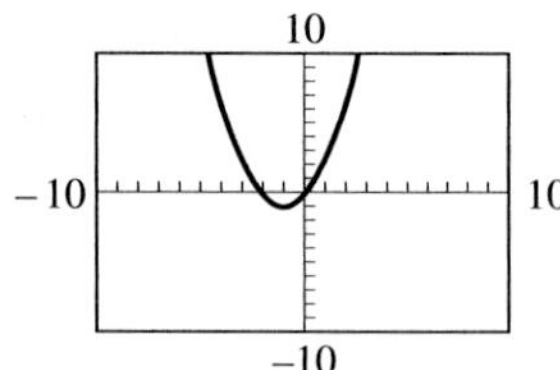

Graph is reflected across the y-axis.

51.

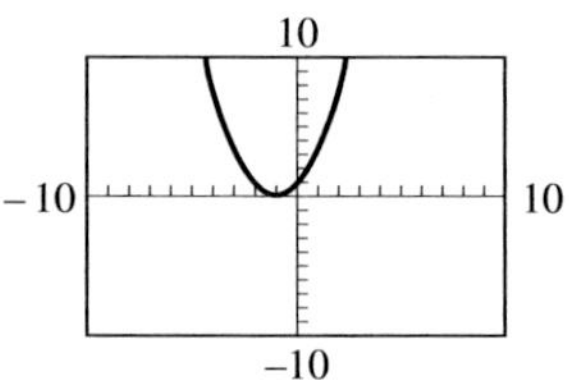

Graph is reflected across the y-axis and shifted up 1 unit.

53. The graph is reflected across the x-axis and the scale on the y-axis is multiplied by $|c|$.

55. The graph of $y = x^3$ has a part in the first quadrant, and a part in the third quadrant.

To get the graph of $y = |x|^3$, one can either reflect the first quadrant part across the y-axis, or reflect the third quadrant part across the x-axis. Both methods work only because $|x^3| = |x|^3$. In general, to graph $y = f(|x|)$ based on the graph of $y = f(x)$, the procedure is to discard the part of the graph to the left of the y-axis and replace it by a reflection in the y-axis of the part to the right of the y-axis.

57.

$$\begin{aligned} x_4 &\approx \cos .65 \approx .796 \\ x_5 &\approx \cos .796 \approx .70 \\ x_6 &\approx \cos .70 \approx .765 \\ x_7 &\approx \cos .765 \approx .721 \\ x_8 &\approx \cos .721 \approx .751 \\ x_9 &\approx \cos .751 \approx .731 \\ x_{10} &\approx \cos .731 \approx .744 \\ x_{11} &\approx \cos .744 \approx .735 \\ x_{12} &\approx \cos .735 \approx .742 \\ x_{13} &\approx \cos .742 \approx .737 \\ x_{14} &\approx \cos .737 \approx .740 \\ x_{15} &\approx \cos .740 \approx .738 \\ x_{16} &\approx \cos .738 \approx .73959 \\ x_{17} &\approx \cos .73959 \approx .73874 \\ x_{18} &\approx \cos .73874 \approx .73931 \\ x_{19} &\approx \cos .73931 \approx .73893 \\ x_{20} &\approx \cos .73893 \approx .73919 \end{aligned}$$

The question asks the student to take a different starting point and to show that the same procedure of iteration leads to the same apparent "limit." In this case "show" means "observe," it does not mean "prove."

59. They converge to 0. One of the problems in Chapter 2 asks the student to prove that $|sin(x)| < |x|$ for all but $x = 0$. This would show that 0 is the only solution to the equation $\sin(x) = x$ and offers a partial explanation (see the comments for Exercise 61) of the phenomena that the student observes.

61. If the iterates of a function f (starting from some point x_0) are going to go toward (and remain arbitrarily close to) a certain number L, this number L must be a solution of the equation $f(x) = x$. For the list of iterates $\{x_0, x_1, x_2, x_3, \ldots\}$ is, apart from the first term, the same list as the list of numbers $\{f(x_0), f(x_1), f(x_2), f(x_3), \ldots\}$. (Remember that x_{n+1} is $f(x_n)$.) If any of the numbers in the first list are close to L, then the f-values (in the second list) are close to $f(L)$. But since the lists are *identical* (apart from the first term x_0, which is not in the second list), it must be true that L and $f(L)$ are the same number.

If conditions are right (and they are in the two cases $f(x) = \cos(x)$ (Exercise 57) and $f(x) = \sin(x)$ (Exercise 59)), this "convergence" will indeed occur, and since there is in these cases only *one* solution (x about 0.739085 in Exercise 57 and $x = 0$ in Exercise 59) it won't matter where you started.

0.7 Parametric Equations and Polar Coordinates

1.

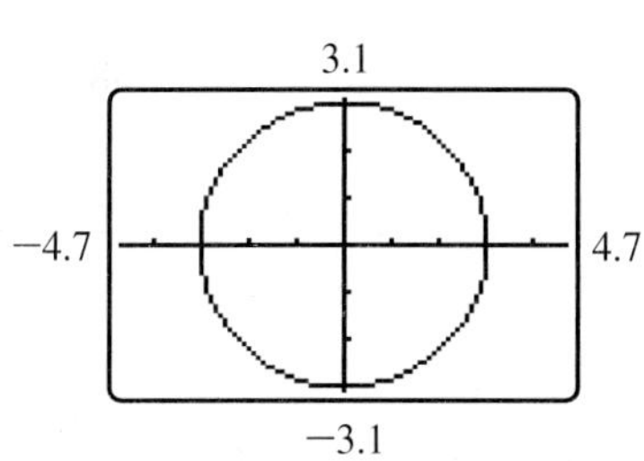

$$\begin{cases} x = 3\cos t \\ y = 3\sin t \end{cases}$$

$$\begin{cases} x^2 = 9\cos^2 t \\ y^2 = 9\sin^2 t \end{cases}$$

$$x^2 + y^2 = 9(\sin^2 t + \cos^2 t)$$

$$x^2 + y^2 = 9$$

This is a circle of radius 1, center at the origin.

3.

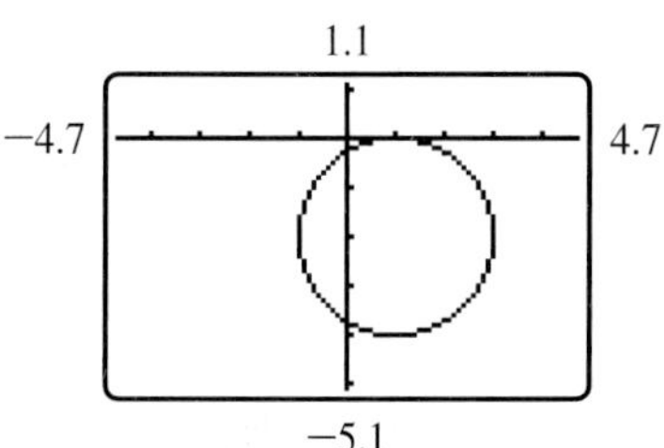

$$\begin{cases} x = 1 + 2\cos t \\ y = -2 + 2\sin t \end{cases}$$

$$\begin{cases} x - 1 = 2\cos t \\ y + 2 = 2\sin t \end{cases}$$

$$\begin{cases} (x-1)^2 = 4\cos^2 t \\ (y+2)^2 = 4\sin^2 t \end{cases}$$

$$(x-1)^2 + (y+2)^2 = 4(\sin^2 t + \cos^2 t)$$

$$(x-1)^2 + (y+2)^2 = 4$$

This is a circle of radius 2, center $(1, -2)$.

5.

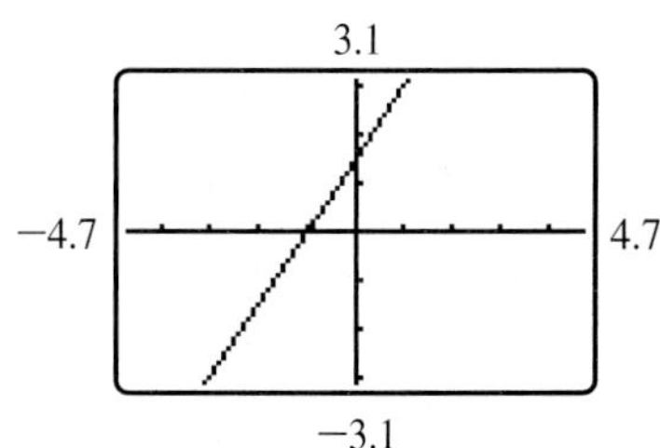

$$\begin{cases} x = -1 + 2t \\ y = 3t \end{cases}$$

$$t = \frac{y}{3}$$

$$x = -1 + 2\left(\frac{y}{3}\right)$$

$$\frac{2}{3}y = x + 1$$

$$y = \frac{3}{2}x + \frac{3}{2}$$

This is a line, slope 3/2, y-intercept 3/2.

7.

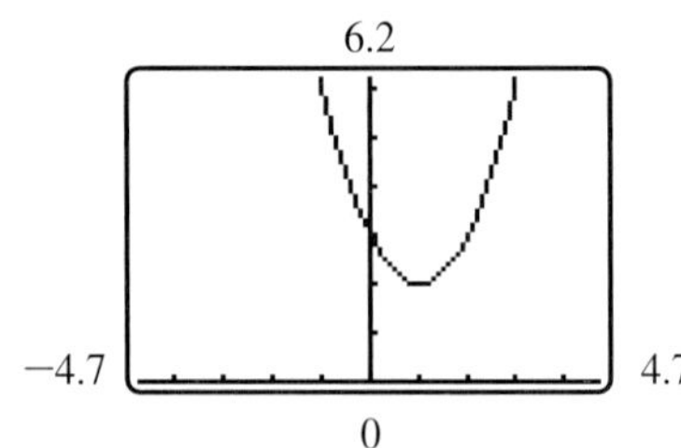

$$\begin{cases} x = 1 + t \\ y = t^2 + 2 \end{cases}$$

$$t = x - 1$$

$$y = (x - 1)^2 + 2$$

$$y = x^2 - 2x + 3$$

This is a parabola, with vertex (1, 2) opening up.

9.

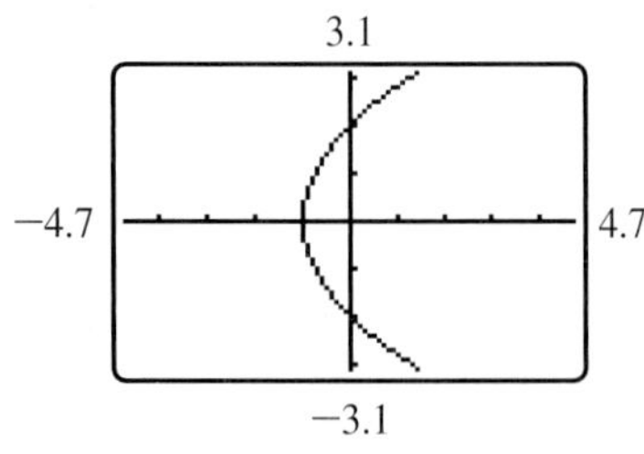

$$\begin{cases} x = t^2 - 1 \\ y = 2t \end{cases}$$

$$t = \frac{y}{2}$$

$$x = \left(\frac{y}{2}\right)^2 - 1 = \frac{1}{4}y^2 - 1$$

This is a parabola, vertex (−1, 0) opening to the right.

11.

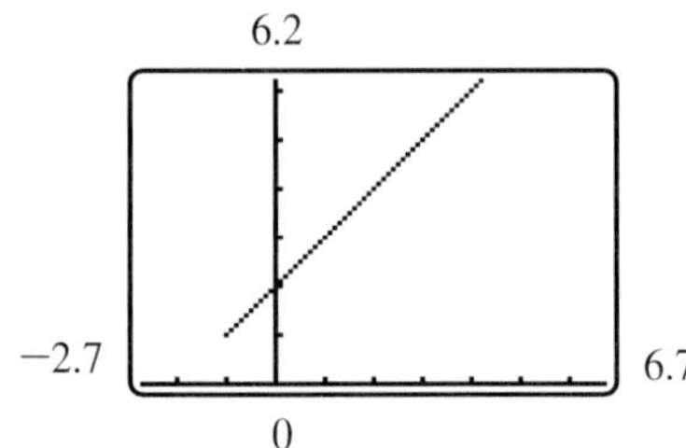

$$\begin{cases} x = t^2 - 1 \\ y = t^2 + 1 \end{cases}$$

$$y - x = t^2 + 1 - (t^2 - 1)$$

$$y = x + 2,\ x \geq -1$$

This is a ray, vertex at (−1.1), slope 1.

13.

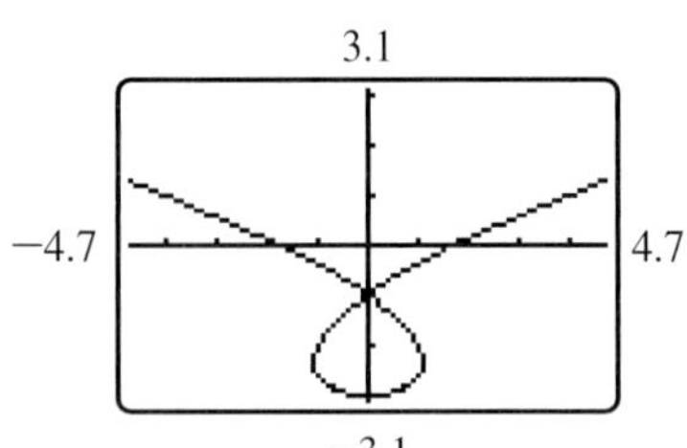

15.

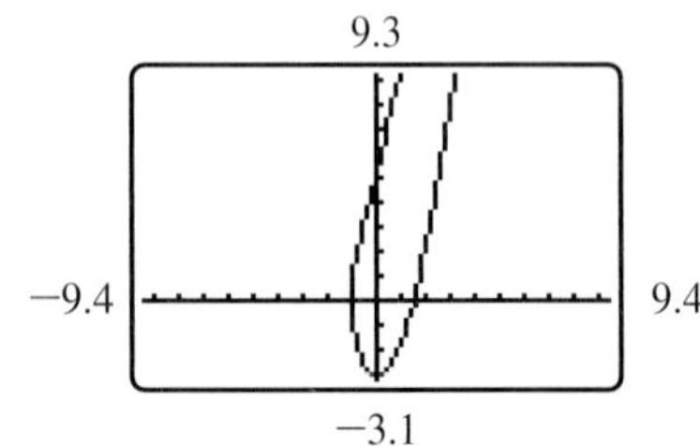

17.

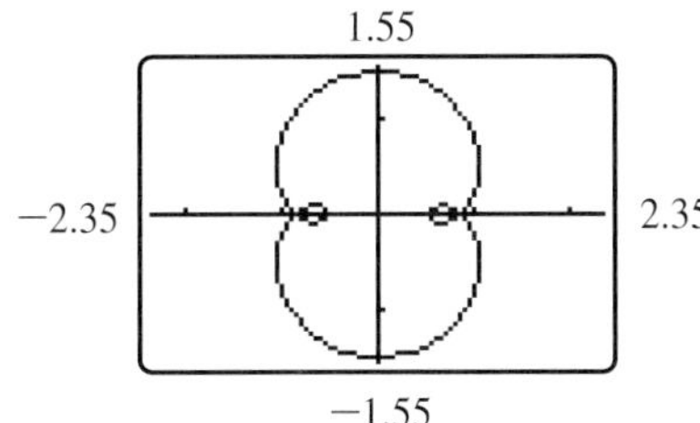

19.

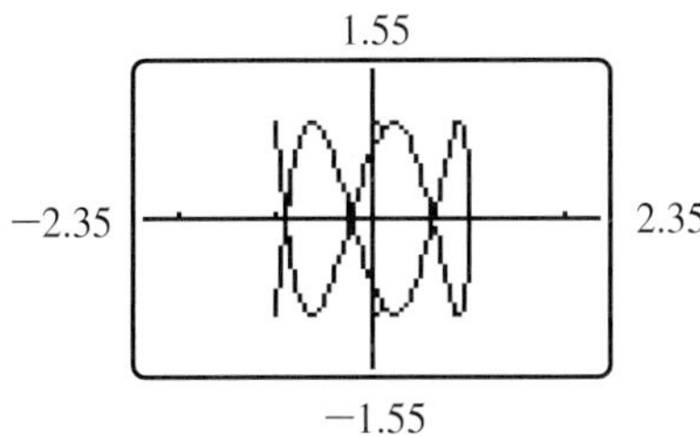

21. Since $(x+1)^2 = y$, with $x \geq -1$, this is the right half of an upward-opening parabola. It has to be Figure C.

23. x is bounded below by -1. y is bounded below by -1 and above by 1. From

$$y = \sin(t) = \sin(\pm\sqrt{x+1}) = \pm\sin(\sqrt{x+1}),$$

it has some of the features of a double sine curve, but the length of the cycles get longer as x increases. On the basis of this alone, it could be Figure B or Figure E, but the first y-intercept after $x = -1$ will be when

$$\sqrt{x+1} = \pi$$
$$x = \pi^2 - 1 \approx 8.9,$$

and this is Figure B.

25. x and y both oscillate between -1 and 1, but with different periods. This has to be Figure A.

27. Use the model $x = a + t - b$, $y = c + d - t$, $(0 \leq t \leq 1)$, with

(a) $t = 0$ corresponding to $(0, 1)$ and $t = 1$ corresponding to $(3, 4)$.

With $t = 0$, $0 = x = a$, $1 = y = c$.
With $t = 1$, $3 = x = 0 + b$, $4 = y = 1 + d$.
So $b = 3$, $d = 3$, and the equations are
$x = 3t$, $y = 1 + 3t$;

(b) $t = 0$ corresponding to $(3, 1)$ and $t = 1$ corresponding to $(1, 3)$.

With $t = 0$, $3 = x = a$, $1 = y = c$.
With $t = 1$, $1 = x = 3 + b$, $3 = y = 1 + d$.
So $b = -2$, $d = 2$, and the equations are
$x = 3 - 2t$, $y = 1 + 2t$.

29. Use the model

$$x = a + b\cos t,\ y = c + b\sin t\,(0t - 2\pi),$$

(a) The center is at $(2, 1)$, therefore $a = 2$ and $c = 1$. The radius is 3, therefore $b = 3$, and the equations are

$$x = 2 + 3\cos t,\ y = 1 + 3\sin t.$$

(b) The center is at $(-1, 3)$, therefore $a = -1$ and $c = 3$. The radius is 5, therefore $b = 5$, and the equations are

$$x = -1 + 5\cos t,\ y = 3 + 5\sin t.$$

31. (a) The graph of $y = f^{-1}(x)$ *is* the graph of $x = f(y)$. Therefore, parametrically, it is good practice to take $y = t$ and $x = f(t) = t^5 + 2t^3 + 4t - 2$.

(b) Using the same technique, come quickly to $y = t$, $x = 4t^3 + 2t$.

Comment: We have not shown and do not yet officially know that the inverse exists for either of these polynomials. After one plots the curve parametrically, one will see that it passes the *vertical* line test, which shows that the graph defines y as a function of x and in hindsight shows that f was invertible.

33. $r = 2, \theta = 0$

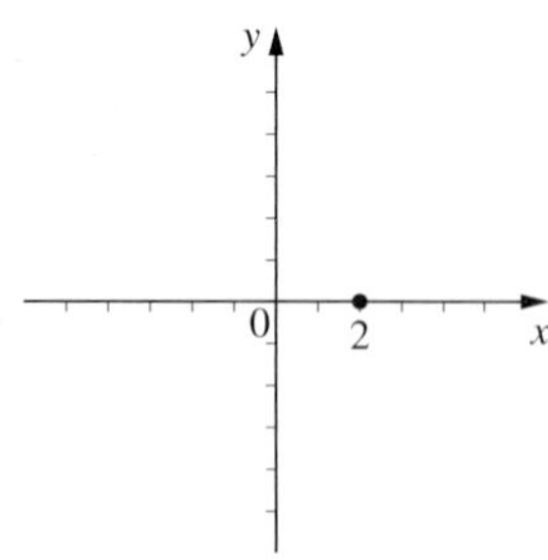

$x = 2\cos 0 = 2,\ y = 2\sin 0 = 0$
Rectangular representation: (2, 0)

35. $r = -2, \theta = \pi$

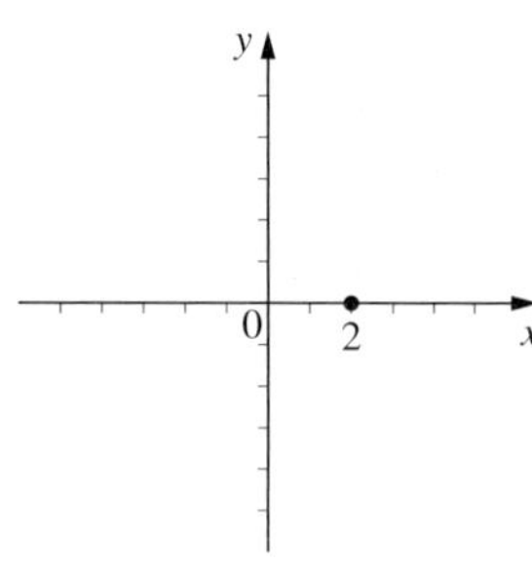

$x = -2\cos\pi = 2,\ y = -2\sin\pi = 0$
Rectangular representation: (2, 0)

37. $r = 3, \theta = -\pi$

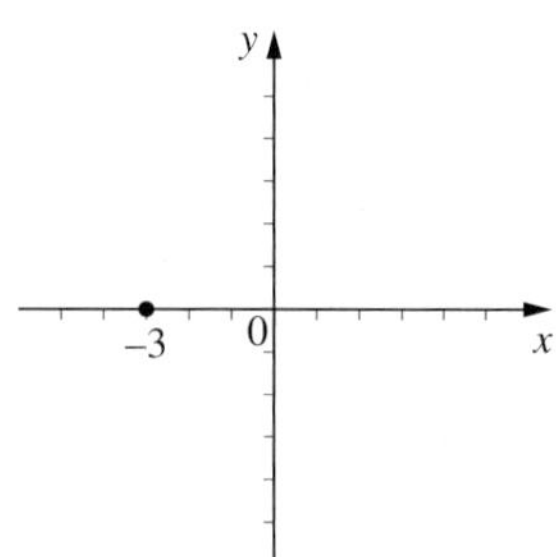

$x = 3\cos(-\pi) = -3,\ y = 3\sin(-\pi) = 0$
Rectangular representation: (−3, 0)

39.
$$r = \sqrt{2^2 + (-2)^2} = \sqrt{8} = 2\sqrt{2}$$
$$\tan\theta = \frac{-2}{2} = -1$$

The point $(x, y) = (2, -2)$ is in Quadrant IV, so θ could be $-\dfrac{\pi}{4}$.
All polar representations:

$$\left(2\sqrt{2}, -\frac{\pi}{4} \pm 2\pi n\right),\ (n = 0, 1, 2, 3, \ldots),$$

or else

$$\left(-2\sqrt{2}, \frac{3\pi}{4} \pm 2\pi n\right),\ (n = 0, 1, 2, 3, \ldots).$$

41.
$$r = \sqrt{2^2 + (-1)^2} = \sqrt{5}$$
$$\tan\theta = \frac{-1}{2} = -\frac{1}{2},$$

the point $(x, y) = (2, -1)$ is in Quadrant IV, so θ could be $-\tan^{-1}\left(\frac{1}{2}\right) \approx 0.4636$ (rad)

All polar representations:

$$\left(\sqrt{5}, -\tan^{-1}\left(\frac{1}{2}\right) \pm 2\pi n\right),$$
$$(n = 0, 1, 2, 3, \ldots),$$

or else

$$\left(-\sqrt{5}, -\tan^{-1}\left(\frac{1}{2}\right) \pm (2n+1)\pi\right),$$
$$(n = 0, 1, 2, 3, \ldots).$$

43.
$$r = \sqrt{(-\sqrt{3})^2 + 1^2} = \sqrt{3+1} = 2$$
$$\tan\theta = \frac{1}{-\sqrt{3}} = -\frac{1}{\sqrt{3}},$$

the point $(x, y) = (-\sqrt{3}, 1)$ is in Quadrant II, so θ could be $\dfrac{5\pi}{6}$.
All polar representations:

$$\left(2, \frac{5\pi}{6} \pm 2\pi n\right),$$

or else

$$\left(-2, -\frac{\pi}{6} \pm 2\pi n\right)\ (n = 0, 1, 2, 3, \ldots).$$

45.

$$r = 2; \theta = -\frac{\pi}{3}$$
$$x = 2\cos\left(-\frac{\pi}{3}\right) = 1$$
$$y = 2\sin\left(-\frac{\pi}{3}\right) = -\sqrt{3}$$

Rectangular representation: $\left(1, -\sqrt{3}\right)$.

47. $r = 3; \theta = \frac{\pi}{8}$

$$x = 3\cos\left(\frac{\pi}{8}\right) = 3(0.923879\ldots) = 2.77163\ldots$$
$$y = 3\sin\left(\frac{\pi}{8}\right) = 1.14805\ldots$$

Rectangular representation (rounded): (2.77164, 1.14805).

Trigonometry does permit exact valuation: for example,

$$\cos\left(\frac{\pi}{8}\right) = \sqrt{\frac{1+\cos\left(\frac{\pi}{4}\right)}{2}} = \sqrt{\frac{1+\frac{1}{\sqrt{2}}}{2}}.$$

This is probably not an improvement.

49.

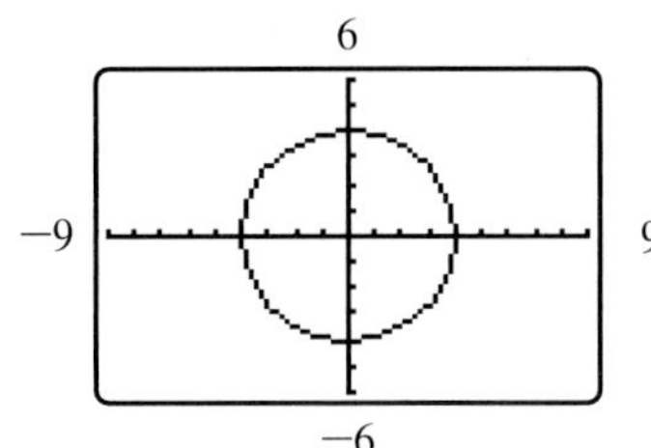

This is a circle centered at the origin with radius 4. The equation is

$$x^2 + y^2 = 4^2 = 16.$$

This fact should be known in advance, but if all else fails, go:

$$r = 4 \text{ (this case)}$$
$$x^2 + y^2 = r^2 \text{ (always)}$$
$$\therefore\ x^2 + y^2 = 16 \text{ (this case)}.$$

51.

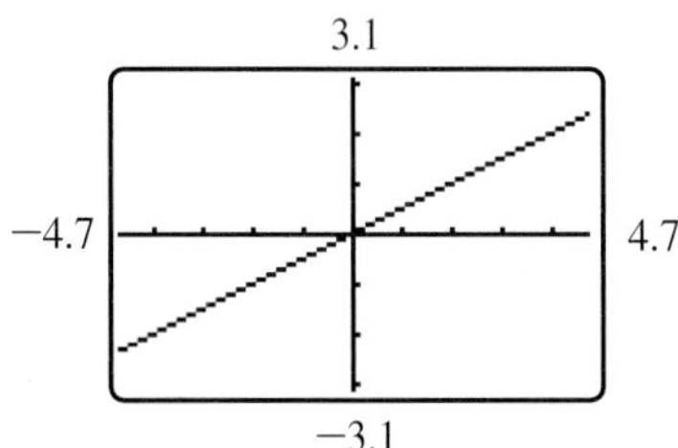

This is a line through the origin making an angle $\pi/6$ to the x-axis, i.e.,

$$\theta = \frac{\pi}{6}, \quad \frac{y}{x} = \tan\theta = \tan\frac{\pi}{6} = \frac{1}{\sqrt{3}},$$
$$y = \frac{1}{\sqrt{3}}x.$$

53.

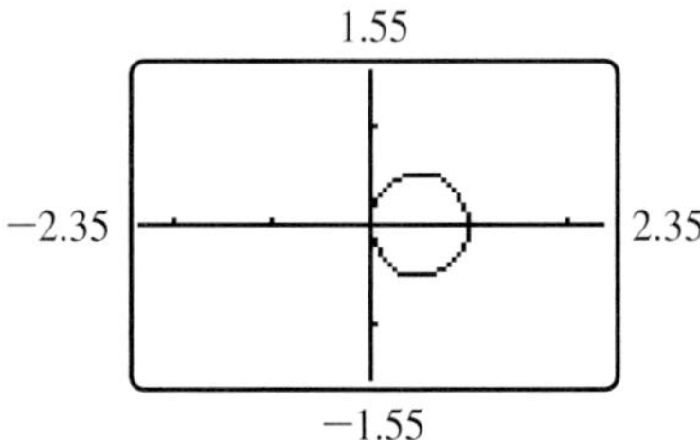

One knows from studying the examples that the figure is a circle centered at (1/2, 0) with radius 1/2. One could simply write down the equation as

$$\left(x - \frac{1}{2}\right)^2 + y^2 = \left(\frac{1}{2}\right)^2.$$

Failing the recognition, go

$$r = \cos\theta$$
$$r^2 = r\cos\theta$$
$$x^2 + y^2 = x.$$

The two equations are the same.

55.

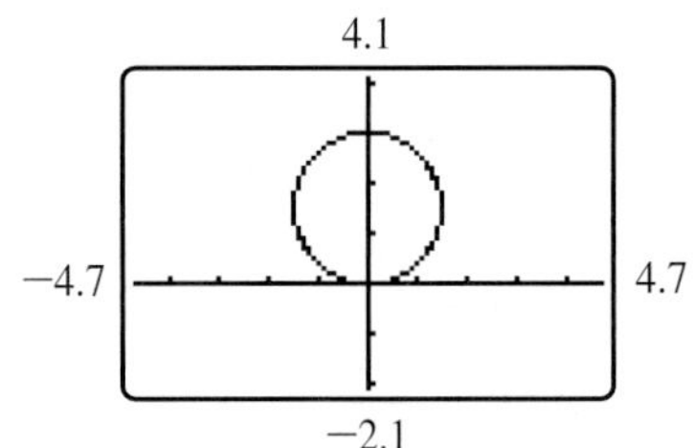

One knows from studying the examples that the figure is a circle centered at (0, 3/2) with radius 3/2. One could simply write down the equation as

$$\left(y-\frac{3}{2}\right)^2+x^2=\left(\frac{3}{2}\right)^2.$$

Failing the recognition, go

$$\begin{aligned} r &= 3\sin\theta \\ r^2 &= 3r\sin\theta \\ x^2+y^2 &= 3y. \end{aligned}$$

The two equations are the same.

57.

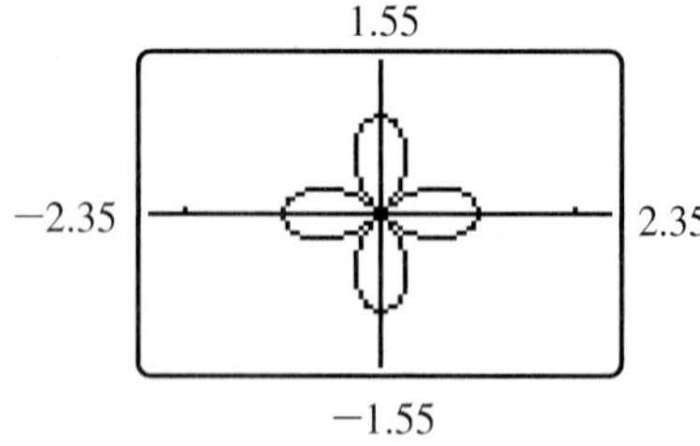

$r=0$ when $\cos(2\theta)=0$, hence,

$$\begin{aligned} 2\theta &= \frac{\pi}{2} \text{ or } \frac{3\pi}{2} \text{ or } \frac{5\pi}{2} \text{ or } \frac{7\pi}{2}, \text{ or } \ldots \\ \therefore\ \theta &= \frac{\pi}{4} \text{ or } \frac{3\pi}{4} \text{ or } \frac{5\pi}{4} \text{ or } \frac{7\pi}{4}, \text{ or } \ldots \end{aligned}$$

The pattern is clear, and this is a full list of the angles in the range $0 \le \theta \le 2\pi$.

For purposes of this kind of problem (in which 2π is a period, *whether or not the smallest possible period*), a complete list of angles within any interval of length 2π is wholly adequate. How does one know when the list is complete? By studying the pattern and anticipating all outcomes in the selected range ($[0, 2\pi)$ in this case).

59.

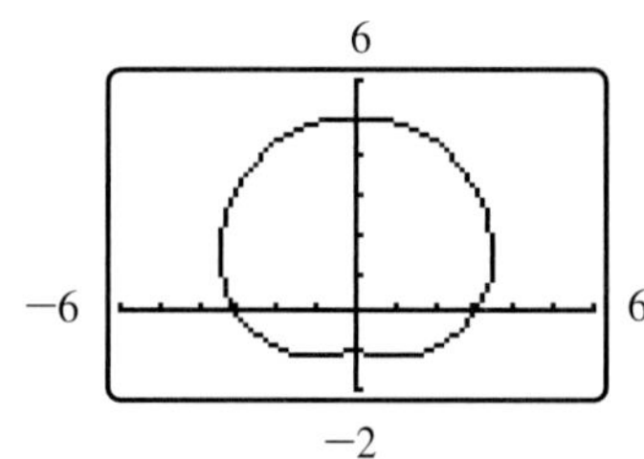

$r=0$ when $3+2\sin\theta=0$, $\sin\theta=-\dfrac{3}{2}$

Never.

61.

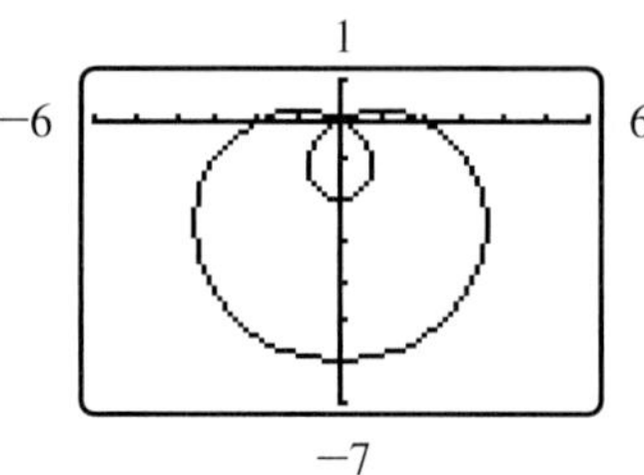

$r=0$ when $2-4\sin(\theta)=0$, hence $\sin\theta=1/2$, θ could be $\dfrac{\pi}{6}$ or $\dfrac{5\pi}{6}$, and indeed this is a full list of the angles in the range $0 \le \theta \le 2\pi$.

63.

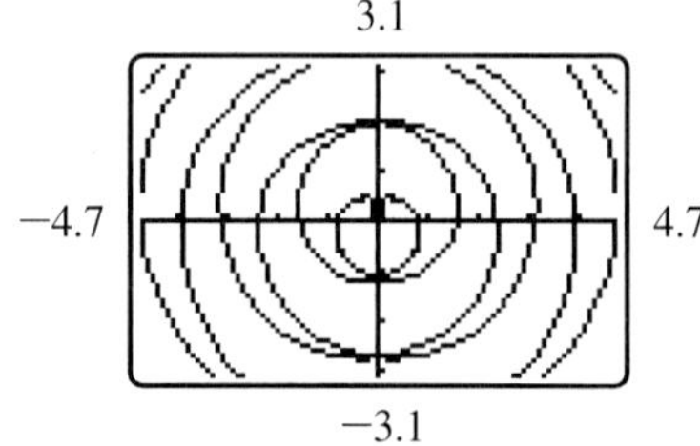

$r=0$ only when $\theta=0$, $-\infty<\theta<\infty$.

This graph $r=\theta/4$ has an unexpected right-left symmetry, for given $q>0$, both points $((q/4), q)$ and $(-(q/4), -q)$ (in polar representations (r, θ)) are on the graph. But in rectangular coordinates, these are $(a, b)=((q/4)\cos q, (q/4)\sin q)$, and $(-(q/4)\cos(-q), -(q/4)\sin(-q)=(-(q/4)\cos q, (q/4)\sin q)=(-a, b)$. We see the points have the same y-coordinate, but have oppositely signed x-coordinates. The picture shows this feature nicely.

65. One guesses, based on a number of examples in the text and the problems, that the figure is a circle of center $(a/2, 0)$ and radius $a/2$. To verify it:

$$r = a\cos\theta,\ r^2 = ar\cos\theta$$
$$x^2 + y^2 = ax,\ x^2 - ax + y^2 = 0$$
$$x^2 - ax + \frac{a^2}{4} + y^2 = \frac{a^2}{4}$$
$$\left(x\frac{a}{2}\right)^2 + y^2 = \left(\frac{a}{2}\right)^2.$$

67.
$$y^2 - x^2 = 4,$$
$$r^2\sin^2\theta - r^2\cos^2\theta = 4,$$
$$r^2(\cos^2\theta - \sin^2\theta) = -4,$$
$$r^2 = \frac{-4}{\cos(2\theta)}$$

This is an acceptable answer. Before going any farther, one should note that this will require $\cos(2\theta) < 0$, which, as a subset of $[0, 2\pi)$, puts θ in one of the two open intervals

$$\left(\frac{\pi}{4}, \frac{3\pi}{4}\right) \cup \left(\frac{5\pi}{4}, \frac{7\pi}{4}\right).$$

Subject to that quantification, one could go

$$r = \frac{2}{\sqrt{-\cos(2\theta)}} = 2\sqrt{-\sec(2\theta)},$$

eschewing the possible minus sign since the use of the minus sign merely duplicates points already "present and accounted for."

All this confirms what we might already know about the curve: It is a hyperbola, opening up and down, with asymptotes formed by the two lines $y = x$ and $y = -x$ (which correspond to the endpoints of the stated domain-intervals for θ).

69. This one is obvious: $r = 4$.

71.
$$y = 3$$
$$r\sin\theta = 3$$
$$r = \frac{3}{\sin\theta} = 3\csc\theta$$

73.
$$x = \cos 2t,\ y = \sin t$$
$$\text{From trig: } \cos 2t = 1 - 2\sin^2 t$$
$$x = 1 - 2y^2.$$

This shows that the graph is part of a parabola, with vertex $(1, 0)$ and opening to the left. However the moving point never goes left of $x = -1$, forever going back and forth on this parabola, making an about-face every time x reaches -1.

$$x = \cos t,\ y = \sin 2t$$
$$\sin 2t = 2\cos t\sin t$$
$$\sin^2 2t = 4\cos^2 t\sin^2 t$$
$$= 4\cos^2 t(1 - \cos^2 t).$$
$$\text{Therefore } y^2 = 4x^2(1 - x^2)$$
$$\text{or } y = \pm 2x\sqrt{1 - x^2}.$$

This is a considerably more complicated graph, a figure-eight in which the moving point cycles smoothly around, starting out (when $t = 0$) by moving upward from $(1, 0)$ and completing the figure every time t passes an integral multiple of 2π. The moving point passes the origin twice during each cycle (first when $t = \pi/2$ and later when $t = 3\pi/2$).

75. It is clear enough that there is a critical angle formed by the two lines from the ball, tangent to the hole. In one possible set-up, the ball is at the origin O, the hole is centered at $D = (d, 0)$ (on the x-axis, at distance d), and the critical angle A_0 is being measured from the tangent line to the center line of the hole (the x-axis). If the upper point of tangency is denoted by T, one must recall that in the triangle TOD, the right angle is at T rather than at D. (It seems to this observer that this is the only possible source of error.) This makes OD the hypotenuse,

$$\sin A_0 = \frac{\text{opposite}}{\text{hypotenuse}} = \frac{h}{d},$$

and $A_0 = sin^{-1}(h/d)$. Any acceptable angle A has to satisfy $-A0 < A < A_0$.

77. There are no more calculations to be done at this point. According to the conclusions of Exercise 75, it must be the case that $A_1 = -A_0$ while $A_2 = A_0$. From Exercise 76 one digs out

$$r_1(A) = d\cos(A) - \sqrt{d^2\cos^2 A - (d^2 - h^2)},$$

and given in this very problem is

$$r_2(A) = d + b\left(1 - \left[\frac{A}{\sin^{-1}(h/d)}\right]^2\right)$$
$$= d + b\left(1 - \left[\frac{A}{A_0}\right]^2\right).$$

One can observe, if the ball is to just drop into the side of the hole ($A \approx A_0$), that r_1 is just about $\sqrt{d^2 - h^2}$ and r_2 is just about d. There is very little margin for error on the speed. Every golfer already knows this.

79. This is a very demanding problem which will reappear in a later chapter. For now we will simply follow the instructions as given, to show what the student could possibly have done. But we must answer four questions before ever beginning: (1) *where* does the rolling circle start; (2) in which *direction* does it roll; (3) how *fast* does it roll; (4) for *which point* on the rim are we going to describe the motion.

We decide for convenience that we will have the circle rolling out on the positive x-axis with the center starting at $(0, r)$, and rolling at the rate of one revolution per minute. We consider the specific point on the rim to be the one which starts at (r, r) (directly east of the center). Now these things can be agreed: the center moves out in proportion to time (measured in minutes), i.e., $x = kt$, $y = r$ (constant). If one minute corresponds to one revolution (which amounts to one circumference of distance rolled out), we must have $2\pi r = k$. To summarize so far, the path (x_c, y_c) of the center is given by $x_c = 2\pi rt$, $y_c = r$.

Now, we have seen by example that if the circle were simply rotating in place, the motion of the given point relative to the center could be described by $x = r\cos(-2\pi t)$, $y = r\sin(-2\pi t)$. (The minus sign is there to accommodate *clockwise* rotation induced by the circle rolling *forward*, the factor of 2π is present to yield *one* revolution in the first minute.). In summary, the path (x_p, y_p) of the selected point on the rim satisfies

$$x_p - x_c = r\cos(-2\pi t) = r\cos(2\pi t),$$
$$y_p - y_c = r\sin(-2\pi t) = -r\sin(2\pi t)).$$

The final instructions are to *add* these numbers to the coordinates of the center. We can see why. It produces

$$x_p = x_c + (x_p - x_c) = 2\pi rt + r\cos(2\pi t),$$
$$y_p = y_c + (y_p - y_c) = r - r\sin(2\pi t).$$

Equally correct for the problem in its original form, and somewhat simpler looking, would have been

$$x = rt + r\cos(t), \quad y = r - r\sin(t).$$

This has the same physical set-up but an angular velocity of one *radian* per second (as opposed to one *revolution* per second that we previously used). Also equally correct for the problem would have been: $x = rt - r\sin(t)$, $y = r - r\cos(t)$. This has the same physical set-up and the same velocity as our second solution, but describes the motion of the point which started out at the origin (directly *below* the center).

81. The problem is a little hard to fully comprehend. Although individual sound waves (from different sources and originating at different times) arrive simultaneously at the point where the microphone is located, the microphone, thanks to the wonders of electronics, is able to process them and for each instant of time (t) *combine them into a single reading*, known as the *amplitude* ($r = r(t)$). However, at the given moment, the numerically recorded amplitude depends *also* on the *direction that the microphone is facing* (some angle θ from the x-axis), and in this sense the amplitude $r(t)$ becomes a function of θ, apparently known as the *polar pattern*. The polar pattern *normally* changes with time—think of the ebb and flow of an orchestral concert. However, one *could* develop a polar pattern for a particular "constant" source-pattern of sound which was *not* changing over time (a *sound-test* as opposed to a concert).

In this problem we are given a made-up polar pattern $f(\theta) = 2.5(1 + \cos(\theta))$. We do not know the circumstances, and we do not ask. If the graph of $r = f(\theta)$ is plotted treating (r, θ) as *polar coordinates*, the student observes a heart-shaped figure, with its "point" (albeit slightly rounded) toward the positive x-axis, which is the direction of maximum r. As it happens the value of r is zero when $\theta = \pi$, which would be interpreted to say that (for that exact moment) no sound whatsoever is being picked up from behind the microphone. Before I would commit to "talking behind the microphone," I would want to know (a) how much the polar pattern is likely to change over the time it would take me to make my utterance, and (b) how much (if any) of the existing sound source is currently located behind the microphone. (After all, one *possible* explanation (not the only one, and not necessarily a valid one) for the zero-amplitude reading behind the mike is that there's *no sound source* back there.) If that were the correct explanation, and if I started talking back there, I could not with certainty rule out the possibility that the echoes of my voice would strike the mike head-on and I would be heard all over the house.

Chapter 0 Review

True/False

1. False. The remark applies only to a *line*.

3. False. The remark is often true about many, but not about *all* cubics, e.g., $y = x^3$.

5. True with a qualification: the x and y axes should be scaled identically, so that the line $y = x$ is a true $45°$-line.

7. True with some qualifications.

9. False. It is the case that polar curves of the form $r = f(\theta)$ can be viewed as having the parametric form

$$x = f(\theta)\cos(\theta),\ y = f(\theta)\sin(\theta)$$

in which θ is the parameter. Therefore one could say that polar *curves* are a special case of parametric equations. But in the broader sense, polar *coordinates* are a way of describing the *whole* plane (two free variables), while a given set of parametric equations describes at most a certain point-path *within* the plane (one free variable, namely the parameter).

Review Exercises

1. $m = \dfrac{7-3}{0-2} = \dfrac{4}{-2} = -2$

3. These lines both have slope 3. They are parallel unless they are coincident. But the first line includes the point (0, 1), which does not satisfy the equation of the second line. The lines are not coincident.

5. Attach names to the points:

$$P = (1, 2),\ \ Q = (2, 4),\ \ R = (0, 6).$$

Regarding the lines:

$$PQ \text{ has slope } \frac{4-2}{2-1} = 2$$

$$QR \text{ has slope } \frac{6-4}{0-2} = -1$$

$$RP \text{ has slope } \frac{2-6}{1-0} = -4$$

Since no two of these slopes are negative reciprocals, none of the angles are right angles. The triangle is not a right triangle.

7. The line apparently goes through (1, 1) and (3, 2). If so, the slope would be

$$m = \frac{2-1}{3-1} = \frac{1}{2}.$$

The equation would be

$$y = \frac{1}{2}(x - 1) + 1 \text{ (point-slope method)}$$

or

$$y = \frac{1}{2}x + \frac{1}{2} \text{ (slope-intercept).}$$

Testing the equation when $x = 5$, we find $y = 3$. This looks to be true to the picture. Finally, using the equation with $x = 4$, we find

$$y = \frac{1}{2}(4) + \frac{1}{2} = \frac{5}{2}.$$

9. Using the point-slope method, we find

$$y + 1 = -\frac{1}{3}[x - (-1)] \text{ or}$$

$$y = -\frac{1}{3}(x + 1) - 1 \text{ or}$$

$$y = -\frac{1}{3}x - \frac{4}{3}.$$

This last is the slope-intercept form.

11. The graph passes the vertical line test, so it is a function.

13. Again we remind that we are looking for the maximal, natural domain. The radicand cannot be negative, hence we require

$$4 - x^2 \geq 0$$
$$4 \geq x^2.$$

The natural domain is the interval $\{-2 \leq x \leq 2\}$ or, in "interval-language": $[-2, 2]$.

15.

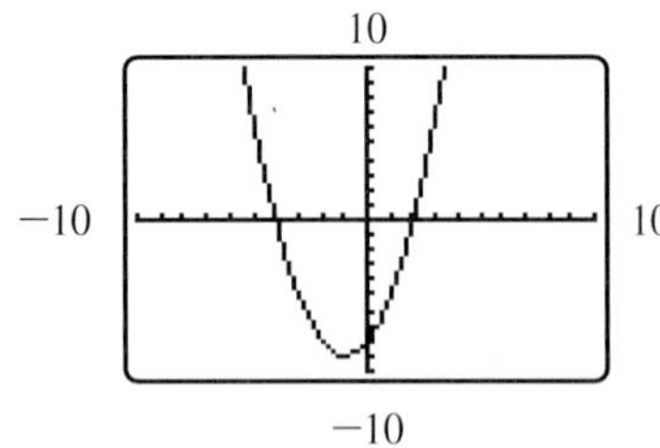

y-intercept: $(0, -8)$ (as a point).
x-intercepts: $(-4, 0)$ and $(2, 0)$ (as points, found by factoring).
Asymptotes: none.
Extrema: y takes minimum -9 when $x = 1$ (found by completing the square).

17.

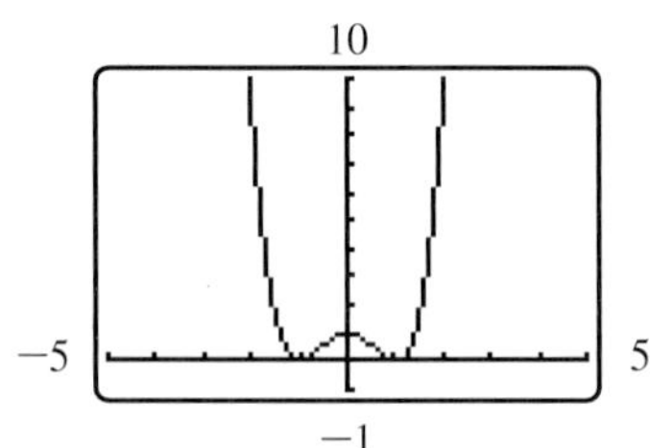

y-intercept: $(0, 1)$ (as a point).
x-intercepts: $(-1, 0)$ and $(1, 0)$ (as points, found by guessing and then factoring).
Asymptotes: none.
Extrema: $y = (x^2 - 1)^2$ (never negative) takes minimum 0 when $x = 1$ or $x = -1$. y also has a relative maximum 1 when $x = 0$ (obvious from the graph) .

19.

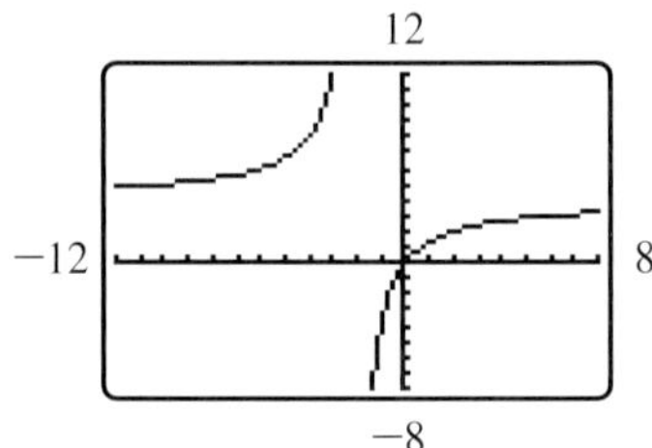

y-intercept: $(0, 0)$ (as a point).
x-intercepts: $(0, 0)$ (as a point, found by solving $y = 0$ to get $x = 0$).
Asymptotes: Vertical at $x = -2$, Horizontal at $y = 4$ (not studied yet).
Extrema: none .

21.

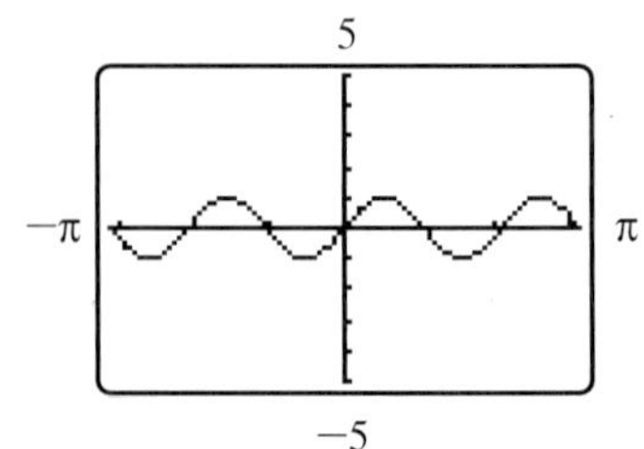

y-intercept: $(0, 0)$ (as a point).
x-intercepts: $(-n\pi/3, 0)$ and $(n\pi/3, 0)$ (as points ($n = 0, 1, 2, 3, \ldots$) found by familiarity with the sine function).
Asymptotes: none.
Extrema: y takes maximum 1 and minimum -1 with great predictability and regularity.

23.

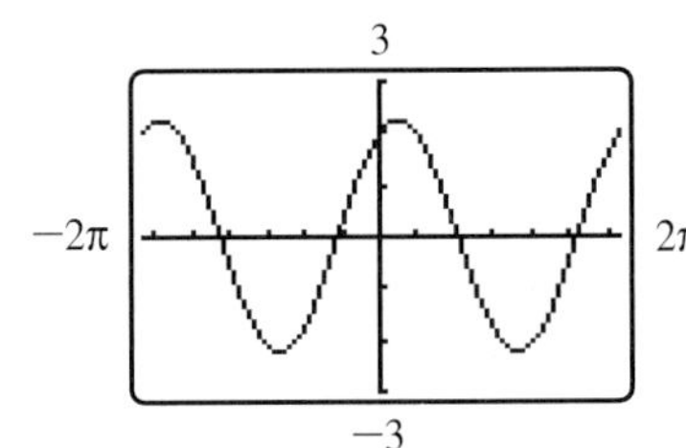

y-intercept: $(0, 2)$ (as a point).
x-intercepts: from the amplitude/phase shift form

$$f(x) = \sqrt{5}\sin\left(x + \sin^{-1}(2/\sqrt{5})\right),$$

we could write down all the intercepts only at considerable inconvenience.
Asymptotes: none.
Extrema: y takes maximum $\sqrt{5}$ and minimum $-\sqrt{5}$ with great predictability and regularity.

25.

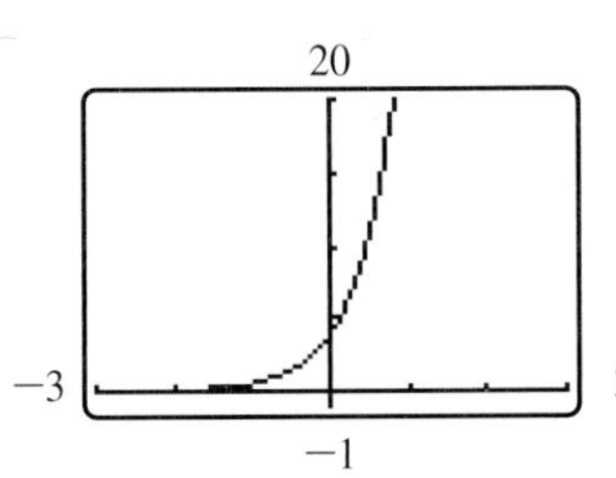

y-intercept: $(0, 4)$ (as a point).
x-intercepts: none.
Asymptotes: none vertical, Left horizontal at $y = 0$ (not studied yet).
Extrema: none.

27.

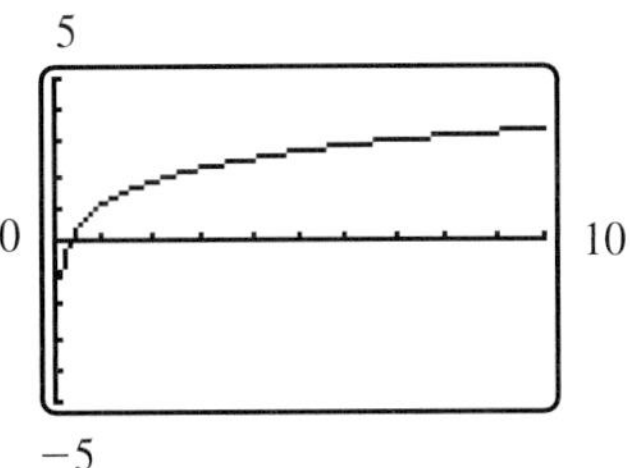

y-intercept: none.
x-intercept: $(1/3, 0)$ (as a point, found by solving $y = 0$, requiring $3x = 1$).
Asymptotes: Vertical only, at $x = 0$.
Extrema: none.

29. See Exercise 15.

31. See Exercise 19.

33. This factors:

$$x^2 - 3x - 10 = (x - 5)(x + 2).$$

The zeros are when $x = 5$ and $x = -2$.

35.

$$x^3 - 3x^2 + 2 = 0$$

Guess a root: $x = 1$.
Factor the left side:

$$(x - 1)(x^2 - 2x - 2).$$

Solve the quadratic by formula:

$$x = \frac{2 \pm \sqrt{2^2 - 4(1)(-2)}}{2} = 1 \pm \sqrt{3}.$$

Complete list of three roots:

$$x = 1,$$
$$x = 1 - \sqrt{3} \approx -.732,$$
$$x = 1 + \sqrt{3} \approx 2.732.$$

37. There are 3 solutions, one at $x = 0$ and the other two negatives of one another. The value in question is .928632 . . . , found using the function "Goal Seek" in Excel. The result can be checked, and a good graphing calculator can also find it.

39. See Example 5.8.

$$\frac{h}{50} = \tan 34°$$
$$h = 50 \tan 34° \approx 33.7 \text{ feet}$$

41. **(a)** $5^{-1/2} = \frac{1}{5^{1/2}} = \frac{1}{\sqrt{5}} = \frac{\sqrt{5}}{5}$

(b) $3^{-2} = \frac{1}{3^2} = \frac{1}{9}$

43.
$$\begin{aligned} \ln 8 - 2\ln 2 &= \ln 8 - \ln 2^2 \\ &= \ln 8 - \ln 4 \\ &= \ln\left(\frac{8}{4}\right) \\ &= \ln 2 \end{aligned}$$

45.
$$\begin{aligned} 3e^{2x} &= 8 \\ e^{2x} &= \frac{8}{3} \\ \ln e^{2x} &= \ln\left(\frac{8}{3}\right) \\ 2x &= \ln\left(\frac{8}{3}\right) \\ x &= \frac{1}{2}\ln\frac{8}{3} = \ln\left(2\sqrt{\frac{2}{3}}\right) \end{aligned}$$

It's often unclear where to stop.

47. The natural domain for f is the full real line. The natural domain for g is $\{1 \le x\}$. Because f has a universal domain, the natural domain for $f \circ g$ is the same as the domain for g, namely $\{1 \le x\}$. Because g requires its inputs be not less than 1, the domain for $g \circ f$ is the set of x for which $1 \le f(x)$, i.e., $\{1 \le x^2\} = \{1 \le |x|\}$, or in interval language $(-\infty, -1] \cup [1, \infty)$.

The formulae are easier:

$$\begin{aligned} (f \circ g)(x) &= f(g(x)) = f(\sqrt{x-1}) \\ &= (\sqrt{x-1})^2 = x - 1, \\ (g \circ f)(x) &= g(f(x)) = g(x^2) = \sqrt{x^2 - 1}. \end{aligned}$$

Caution: the *formula* for $f \circ g$ is defined for any x, but the *domain* for $f \circ g$ is restricted as stated earlier. The formula must be viewed as irrelevant outside the domain.

49. Lots of possibilities exist here. To name two, $f(x)$ could be e^x, with $g(x) = 3x^2 + 2$, or $f(x)$ could be e^{3x+2}, with $g(x) = x^2$.

51. $f(x) = x^2 - 4x + 1 = x^2 - 4x + 4 - 4 + 1 = (x-2)^2 - 3$. Starting from $y = x^2$, one needs to slide the x-axis two units left (so that old 0 is new 2), and then slide the y-axis up 3 (so that old 0 is new -3). Alternatively, moving the graph instead of moving the axes, one simply slides the graph two units right and down three. This kind of talk can be confusing and one needs to spell out very carefully whether we are moving the graph or relabeling the axes. Both are done, according to the inclination of the operator. It is known as the "alias" vs "alibi" dilemma. Let us not even ask which is which.

53. Like x^3, the function $f(x) = x^3 - 1$ passes the horizontal line test and is one-to-one. To find a formula for the inverse, start

$$\begin{aligned} y &= x^3 - 1 \\ y + 1 &= x^3 \\ (y+1)^{1/3} &= x. \end{aligned}$$

The left side of this equation is a formula for $f^{-1}(y)$, good for all y. Therefore, $f^{-1}(x) = (x+1)^{1/3}$ for all x.

55. The function is *even* $(f(-x) = f(x))$. Every horizontal line (except $y = 0$) that meets the curve at all automatically meets it at least twice. The function is not one-to-one. There is no inverse. Like many other *even* functions (including x^2), it is one-to-one on the restricted domain of positive x, and all its values are ≥ 1. With that proviso, the formula for the inverse is

$$\sqrt{\ln\sqrt{x}}(x \ge 1).$$

The student will gain if he/she can see this.

57.

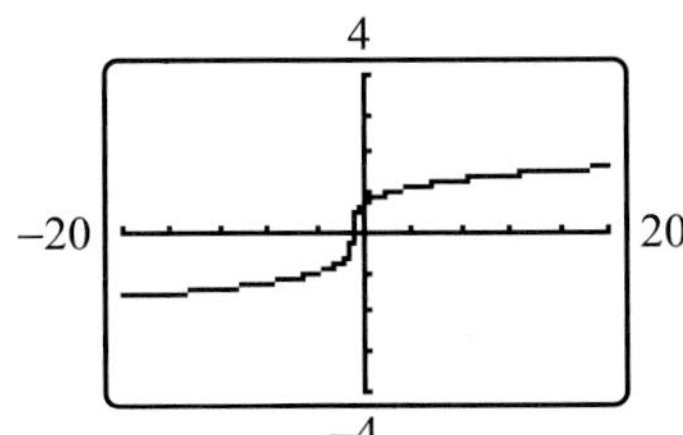

59.

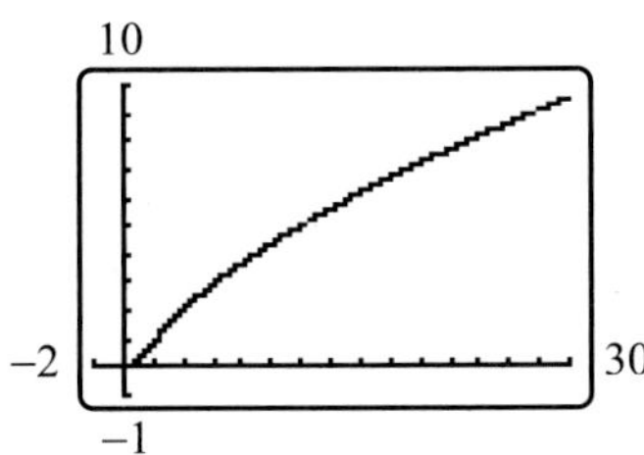

61. On the unit circle, the unique angle

$$\theta \in \left[-\frac{\pi}{2}, \frac{\pi}{2}\right]$$

for which $\sin\theta = 1$ is $\theta = \frac{\pi}{2}$. Hence, $\sin^{-1} 1 = \frac{\pi}{2}$.

63. On the unit circle, the unique angle

$$\theta \in \left[-\frac{\pi}{2}, \frac{\pi}{2}\right]$$

for which $\tan\theta = -1$ is $\theta = -\frac{\pi}{4}$. Hence, $\tan^{-1}(-1) = -\frac{\pi}{4}$.

65. If an angle θ has $\sec(\theta) = 2$, then it has $\cos(\theta) = 1/2$. Its sine could be $\pm\frac{\sqrt{3}}{2}$. But if $\theta = sec^{-1}(2)$, then in addition to all that has been stated, it is in the first quadrant, and the choice of sign (for its sine) is positive. In summary,

$$\sin(\sec^{-1} 2) = \sin\theta = \frac{\sqrt{3}}{2}.$$

67.

$$\sin^{-1}\left(\sin\left(\frac{3\pi}{4}\right)\right) = \sin^{-1}\left(\frac{\sqrt{2}}{2}\right) = \frac{\pi}{4}$$

69.

$$\begin{aligned} \sin 2x &= 1 \\ 2x &= \frac{\pi}{2} \pm 2\pi n (n = 0, 1, 2, 3 \ldots) \\ x &= \frac{\pi}{4} \pm \pi n (n = 0, 1, 2, 3 \ldots) \end{aligned}$$

71.

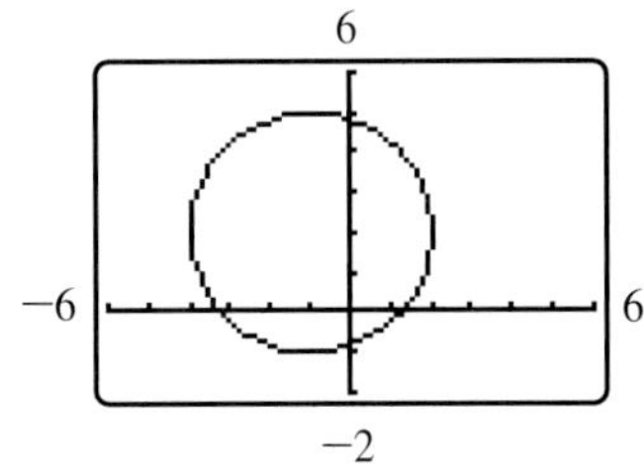

$$\begin{cases} x = -1 + 3\cos t \\ y = 2 + 3\sin t \end{cases} \quad \begin{cases} x + 1 = 3\cos t \\ y - 2 = 3\sin t \end{cases}$$

Since $(3\cos t)^2 + (3\sin t)^2 = 9$, a corresponding $x - y$ equation is $(x + 1)^2 + (y - 2)^2 = 9$.

73.

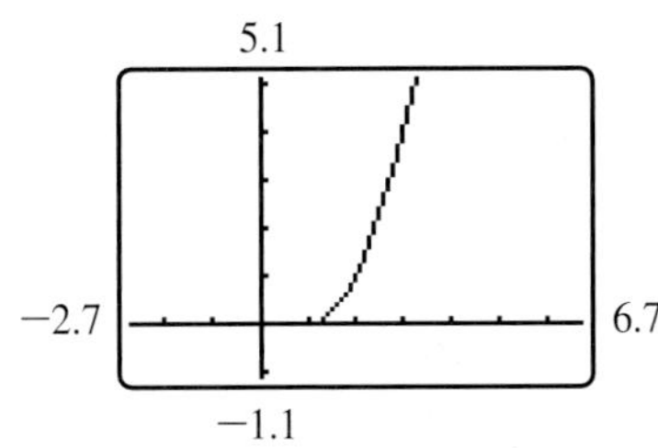

$$\begin{cases} x = t^2 + 1 \\ y = t^4 \end{cases}$$

Note that $t^2 = x - 1$; substitute this expression in the equation for y:

$$y = t^4 = (t^2)^2 = (x - 1)^2 = x^2 - 2x + 1.$$

A corresponding $x - y$ equation is

$$y = x^2 - 2x + 1 = (x - 1)^2,$$

but only for $x \geq 1$.

75.

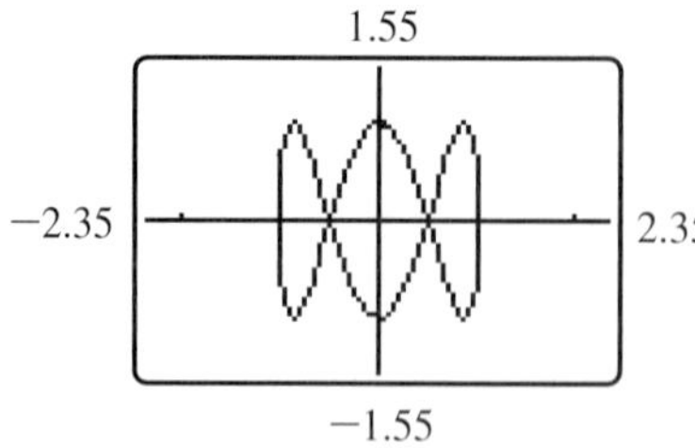

77. See Example 1.3.

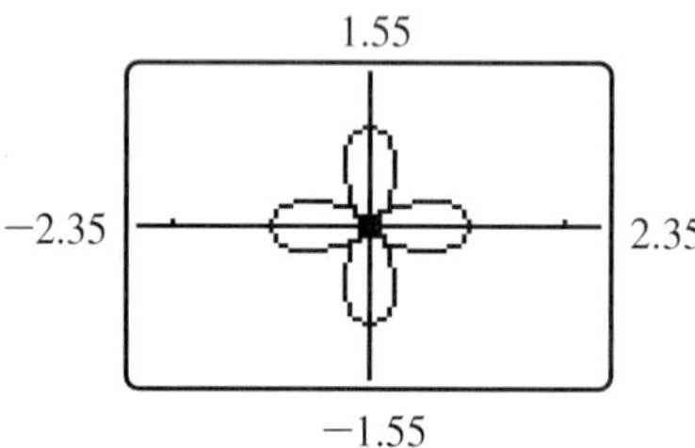

79. Eliminating t in favor of y leads to $x = y^{2/3} - 1$. This is an even function of y, reflected in north-south symmetry, among other features. Figure C

81. Having eliminated Figures C and A (given over to Exercise 80), it is easy to guess Figure B over Figure D. A brief sample of points may be needed to make the selection conclusive. Figure B

83. The ideal model is

$$\begin{cases} x = a + bt \\ y = c + dt \end{cases} \quad (0 \le t \le 1).$$

We want (2, 1) corresponding to $t = 0$, so $a = 2$ and $c = 1$. We want (4, 7) corresponding to $t = 1$, so $2 + b = 4$ and $1 + d = 7$, whence $b = 2$ and $d = 6$. In summary $\begin{cases} x = 2 + 2t \\ y = 1 + 6t \end{cases}$ is the preferred, linear parameterization of the line segment.

85.

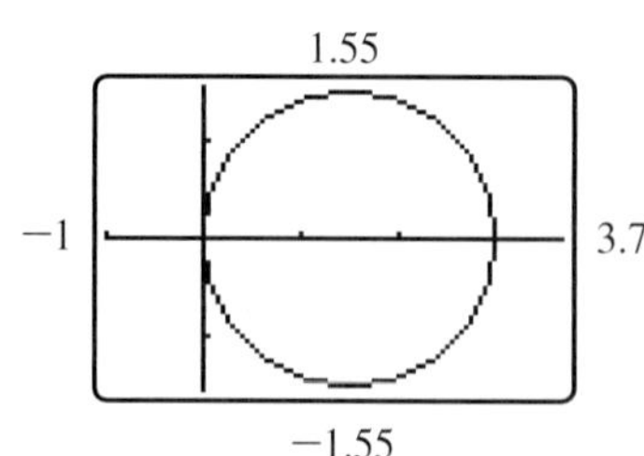

$$r = 3\cos\theta$$
$$r^2 = 3r\cos\theta$$
$$x^2 + y^2 = 3x$$

87.

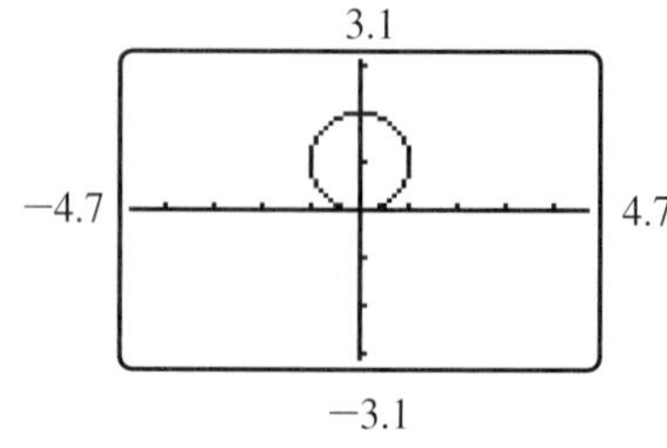

$$r = 2\sin\theta$$
$$r = 0 \text{ when } 2\sin\theta = 0$$
$$\theta = \pm n\pi\,(n = 0, 1, 2, 3, \ldots)$$

One copy of the graph is produced by the interval $0 \le \theta \le \pi$.

89.

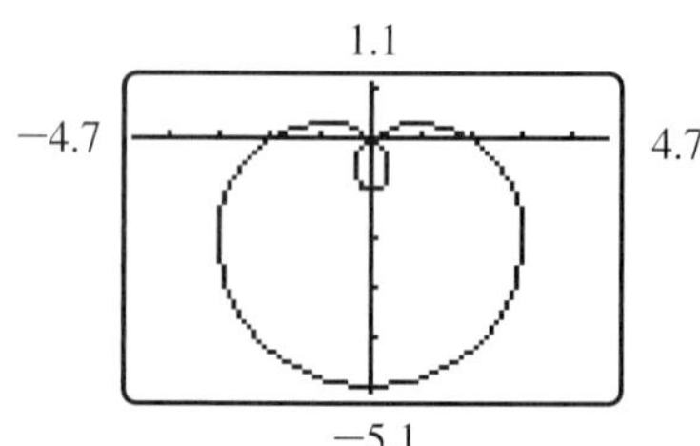

$$r = 2 - 3\sin\theta$$
$$r = 0 \text{ when } \sin\theta = \frac{2}{3}:$$
$$\theta = \sin^{-1}\left(\frac{2}{3}\right) \pm 2n\pi,$$
$$\text{or } \theta = -\sin^{-1}\left(\frac{2}{3}\right) \pm (2n + 1)\pi$$
$$(n = 0, 1, 2, 3, \ldots)$$

One copy of the graph is produced by the interval $0 \le \theta \le 2\pi$.

91.

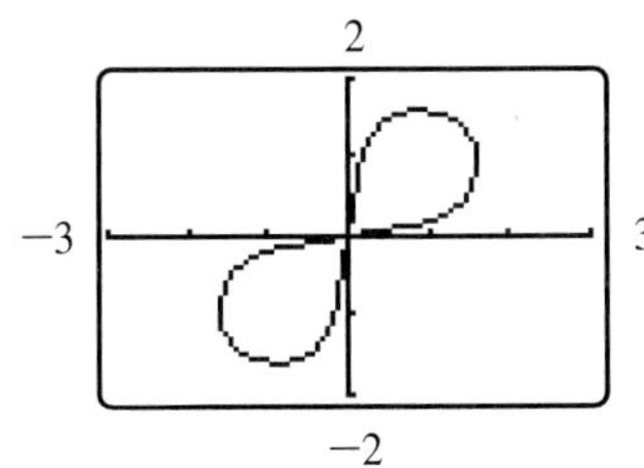

$$r^2 = 4 \sin 2\theta$$
$$r = 0 \text{ when } \sin 2\theta = 0,$$
$$2\theta = \pm n\pi,$$
$$\theta = \pm n\frac{\pi}{2}, \ (n = 0, 1, 2, 3, \ldots)$$

The northeast part of the graph is produced, not surprisingly, by the interval $0 \le \theta \le \frac{\pi}{2}$. The same interval can suffice to produce the southwest portion as well, utilizing the minus sign in $r = \pm 2\sqrt{\sin 2\theta}$. Note that there is no graph in the second or fourth quadrants, where $\sin(2\theta) < 0$. Remember: although convention permits r to be negative, r^2 is never negative.

93.

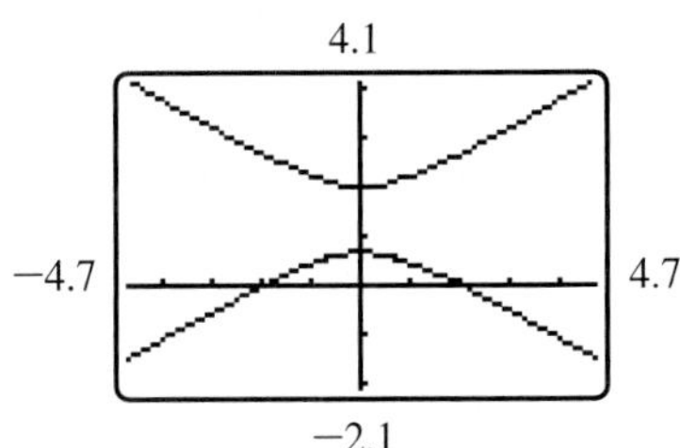

$$r = \frac{2}{1 + 2 \sin \theta}$$

There are no values of θ for which $r = 0$. A complete copy of the graph is generated by

$$0 \le \theta \le 2\pi \left(\theta \ne \frac{7\pi}{6}, \theta \ne \frac{11\pi}{6}\right).$$

95.

$$x^2 + y^2 = 9$$
$$(r \cos \theta)^2 + (r \sin \theta)^2 = 9$$
$$r^2 = 9$$
$$r = \pm 3$$

The answer may be given as simply $r = 3$, since this is sufficient to generate the entire graph.

Chapter 1

Limits and Continuity

1.1 A Brief Preview of Calculus: Tangent Lines and the Length of a Curve

1.

x coord of 2nd Point	m_{sec}	x coord of 2nd Point	m_{sec}
2	3	0	1
1.1	2.1	0.9	1.9
1.01	2.01	0.99	1.99

The slope appears to be 2.

3.

x coord of 2nd Point	m_{sec}	x coord of 2nd Point	m_{sec}
1	−0.45970	−1	0.45970
0.1	−0.04996	−0.1	0.04996
0.01	−0.00500	−0.01	0.00500

The slope appears to be 0.

5.

x coord of 2nd Point	m_{sec}	x coord of 2nd Point	m_{sec}
1	1	−1	1
0.1	0.01	−0.1	0.01
0.01	0.0001	−0.01	0.0001

The slope appears to be 0.

7.

x coord of 2nd Point	m_{sec}	x coord of 2nd Point	m_{sec}
1	0.414213562	−1	1
0.1	0.488088482	−0.1	0.513167019
0.01	0.498756211	−0.01	0.501256289

The slope appears to be $\frac{1}{2}$.

9.

x coord of 2nd Point	m_{sec}	x coord of 2nd Point	m_{sec}
1	1.718282	−1	0.632121
0.1	1.051709	−0.1	0.951626
0.01	1.005017	−0.01	0.995017

The slope appears to be 1.

11.

x coord of 2nd Point	m_{sec}	x coord of 2nd Point	m_{sec}
2	0.693147	0.1	2.558428
1.1	0.953102	0.9	1.053605
1.01	0.999500	0.99	1.005034

The slope appears to be 1. Note that we used 0.1 rather than 0 as an evaluation point because $\ln x$ is not defined at 0.

13. $n = 4$: We divide the interval [0, 2] into four subintervals of equal length to obtain the five evaluation points 0, .5, 1, 1.5, and 2. Then we evaluate the function $f(x) = x^2 + 1$ at these points to obtain five points along the curve:

$$(0, 1), (.5, 1.25), (1, 2), (1.5, 3.25), \text{ and } (2, 5).$$

Then we "connect the dots" to get four line segments. The sum of the lengths of these segments is

$$\begin{aligned}
&d\{(0, 1), (.5, 1.25)\} + d\{(.5, 1.25)(1, 2)\} \\
&+ d\{(1, 2)(1.5, 3.25)\} + d\{(1.5, 3.25)(2, 5)\} \\
&= \sqrt{(.5)^2 + (.25)^2} + \sqrt{(.5)^2 + (.75)^2} \\
&+ \sqrt{(.5)^2 + (1.25)^2} + \sqrt{(.5)^2 + (1.75)^2} \\
&\approx 0.559016994 + 0.901387819 \\
&+ 1.346291202 + 1.820027472 \\
&\approx 4.626723487.
\end{aligned}$$

$n = 8$: Now we use eight line segments:

$$
\begin{aligned}
&d\,\{(0, 1), (.25, 1.0625)\} + d\,\{(.25, 1.0625)(.5, 1.25)\} \\
&+ d\,\{(.5, 1.25)(.75, 1.5625)\} + d\,\{(.75, 1.5625)(1, 2)\} \\
&+ d\,\{(1, 2), (1.25, 2.5625)\} \\
&+ d\,\{(1.25, 2.5625)(1.5, 3.25)\} \\
&+ d\,\{(1.5, 3.25)(1.75, 4.0625)\} \\
&+ d\,\{(1.75, 4.0625)(2, 5)\} \\
&= \sqrt{(.25)^2 + (0.0625)^2} + \sqrt{(.25)^2 + (0.1875)^2} \\
&+ \sqrt{(.25)^2 + (0.3125)^2} + \sqrt{(.25)^2 + (0.4375)^2} \\
&+ \sqrt{(.25)^2 + (0.5625)^2} + \sqrt{(.25)^2 + (0.6875)^2} \\
&+ \sqrt{(.25)^2 + (0.8125)^2} + \sqrt{(.25)^2 + (0.9375)^2} \\
&\approx 0.257694102 + 0.3125 + 0.400195265 \\
&+ 0.503891109 + 0.615553613 + 0.731543744 \\
&+ 0.850091907 + 0.970260919 \\
&\approx 4.641730658.
\end{aligned}
$$

When n is large, the sum is about 4.64678.

15. $n = 4$: We get the sum of the lengths of 4 line segments:

$$
\begin{aligned}
&d\left\{(0, \cos 0), \left(\frac{\pi}{8}, \cos\frac{\pi}{8}\right)\right\} \\
&+ d\left\{\left(\frac{\pi}{8}, \cos\frac{\pi}{8}\right), \left(\frac{\pi}{4}, \cos\frac{\pi}{4}\right)\right\} \\
&+ d\left\{\left(\frac{\pi}{4}, \cos\frac{\pi}{4}\right), \left(\frac{3\pi}{8}, \cos\frac{3\pi}{8}\right)\right\} \\
&+ d\left\{\left(\frac{3\pi}{8}, \cos\frac{3\pi}{8}\right), \left(\frac{\pi}{2}, \cos\frac{\pi}{2}\right)\right\} \\
&\approx 0.400008618 + 0.448556568 + 0.509375184 \\
&+ 0.548323972 \approx 1.906264341.
\end{aligned}
$$

$n = 8$. Now we get a sum of eight terms:

$$
\begin{aligned}
&0.197287475 + 0.204429465 + 0.217008607 \\
&+ 0.232420429 + 0.248025135 + 0.261616108 \\
&+ 0.271559049 + 0.276791214 \approx 1.909137482.
\end{aligned}
$$

When n is large, the sum is about 1.91010.

17. $n = 4$:

$$
\begin{aligned}
&0.816546808 + 0.793221197 + 0.782063216 \\
&+ 0.775498194 \approx 3.167329415.
\end{aligned}
$$

$n = 8$:

$$
\begin{aligned}
&0.412816085 + 0.403988352 + 0.398513293 \\
&+ 0.394781929 + 0.392074422 + 0.390019704 \\
&+ 0.388406863 + 0.387107084 \approx 3.167707732.
\end{aligned}
$$

When n is large, the sum is about 3.16784.

19. By symmetry, the results for this curve when $n = 8$ will be twice that of the curve in Exercise 13 when $n = 4$. Hence for $n = 8$ we get

$$(2)(4.626723487) = 9.253446974.$$

For $n = 4$, we use the points

$$
\begin{aligned}
&(-2, 5), (-1, 2), (0, 1), (1, 2), \text{ and } (2, 5). \\
&d\,\{(-2, 5), (-1, 2)\} + d\,\{(-1, 2)(0, 1)\} \\
&+ d\,\{(0, 1)(1, 2)\} + d\,\{(1, 2), (2, 5)\} \\
&= (2)\,(d\,\{(0, 1)(1, 2\} + d\,\{(1, 2), (2, 5)\}) \\
&= (2)\left(\sqrt{(1)^2 + (.1)^2} + \sqrt{(1)^2 + (3)^2}\right) \\
&= (2)\left(\sqrt{2} + \sqrt{10}\right) \\
&\approx (2)(1.414213562 + 3.16227766) \\
&\approx 9.152982445.
\end{aligned}
$$

When n is large, the sum is about 9.29357.

1.2 The Concept of Limit

1. **(a)** -2 **(b)** 2
(c) does not exist **(d)** 1
(e) $\frac{1}{2}$ **(f)** -1
(g) 3 **(h)** does not exist
(i) 1.5 **(j)** 2.5

3.

$$\lim_{x\to 2^-} f(x) = \lim_{x\to 2^-} 2x = 4$$

$$\lim_{x\to 2^+} f(x) = \lim_{x\to 2^+} x^2 = 4$$

hence $\lim_{x\to 2} f(x) = 4$

$$\lim_{x\to 1} f(x) = \lim_{x\to 1} 2x = 2$$

5.

$$\lim_{x\to 1^-} f(x) = \lim_{x\to 1^-} x^2 + 1 = 2$$

$$\lim_{x\to 1^+} f(x) = \lim_{x\to 1^+} 3x + 1 = 2$$

hence $\lim_{x\to 1} f(x)$ does not exist

$$\lim_{x\to 0} f(x) = \lim_{x\to 0} 3x + 1 = 1$$

7.

x	$f(x)$	x	$f(x)$
1.5	2.22	0.5	1.71
1.1	2.05	0.9	1.95
1.01	2.01	0.99	1.99
1.001	2.00	0.999	2.00

We conjecture: $\lim_{x\to 1^+} f(x) = 2$, $\lim_{x\to 1^-} f(x) = 2$, $\lim_{x\to 1} f(x) = 2$.

9. When x is close to 1, the numerator is close to 0 and the denominator is close to 3, so the limit is 0.

11.

x	$\dfrac{x^2+x}{\sin x}$	x	$\dfrac{x^2+x}{\sin x}$
0.1	1.10	−0.1	0.90
0.01	1.01	−0.01	0.99
0.001	1.00	−0.001	1.00

We conjecture that $\lim_{x\to 0} \dfrac{x^2+x}{\sin x} = 1$.

13.

x	$\dfrac{\tan x}{\sin x}$	x	$\dfrac{\tan x}{\sin x}$
0.1	1.00502091	−0.1	1.00502091
0.01	1.00005000	−0.01	1.00005000
0.001	1.0000005	−0.001	1.0000005

We conjecture that $\lim_{x\to 0} \dfrac{\tan x}{\sin x} = 1$.

15.

x	$\dfrac{x-2}{\lvert x-2\rvert}$	x	$\dfrac{x-2}{\lvert x-2\rvert}$
2.1	1	1.9	−1
2.01	1	1.99	−1
2.001	1	1.999	−1

$\lim_{x\to 2} \dfrac{x-2}{|x-2|}$ does not exist.
There is a break in the graph at $x = 2$.

17.

x	$\dfrac{x^2-1}{x^2+2x-3}$	x	$\dfrac{x^2-1}{x^2+2x-3}$
−0.9	0.05	−1.1	−0.05
−0.99	0.005	−1.01	−0.005
−0.999	0.00	−1.001	0.00

We conjecture that $\lim_{x\to -1} \dfrac{x^2-1}{x^2+2x-3} = 0$.

19.

x	$\dfrac{x^2+1}{x-1}$	x	$\dfrac{x^2+1}{x-1}$
1.1	22.1	0.9	−18.1
1.01	202.01	0.99	−198.01
1.001	2002	0.999	−1998

$\lim_{x\to 1} \dfrac{x^2+1}{x-1}$ does not exist.

There is a vertical asymptote at $x = 1$.

21.

x	$\dfrac{x^2-1}{x^2-2x+1}$	x	$\dfrac{x^2-1}{x^2-2x+1}$
1.1	21	0.9	−19
1.01	201	0.99	−199
1.001	2001	0.999	−1999

$\lim_{x\to 1} \dfrac{x^2-1}{x^2-2x+1}$ does not exist.
There is a vertical asymptote at x = 1.

23. For $x < 0$. $\ln x$ is not defined; hence the limit does not exist. However, there is a vertical asymptote at $x = 0$. See the graph of $y = \ln x$ on page 51.

25.

x	$f(x)$	x	$f(x)$
1	1.557408	−1	1.557408
0.1	1.003347	−0.1	1.003347
0.01	1.000033	0.01	1.000033

The limit appears to be 1.

27. If $g(a) = 0$ and $f(a) \neq 0$, $\lim_{x\to a} \dfrac{f(x)}{g(x)}$ does not exist.

29. One possibility is

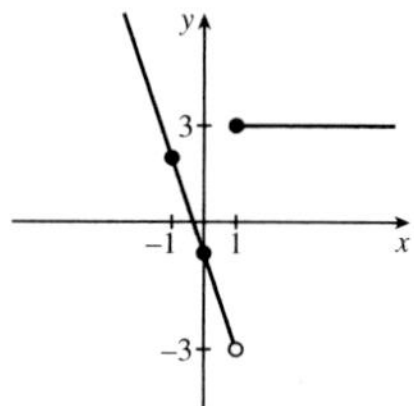

31. One possibility is

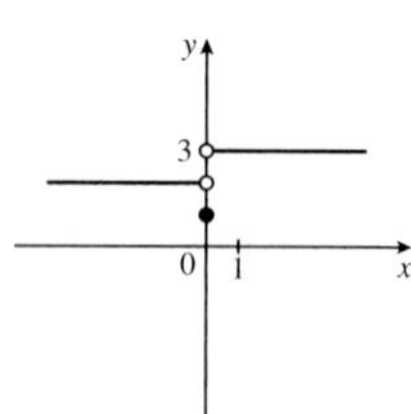

33.

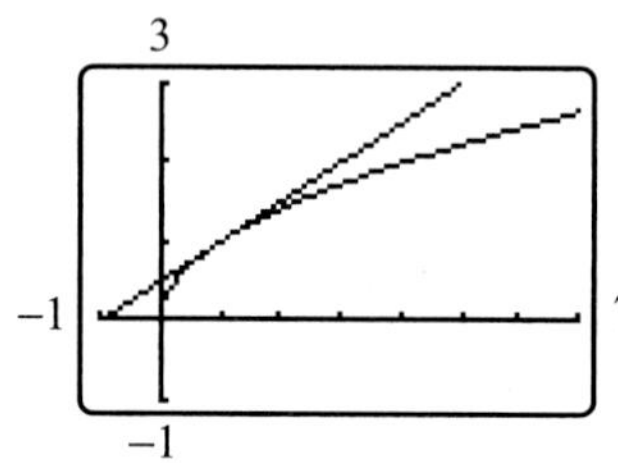

Let $f(h) = \dfrac{\sqrt{1+h}-1}{h}$

h	$\dfrac{\sqrt{1+h}-1}{h}$	h	$\dfrac{\sqrt{1+h}-1}{h}$
1	0.414214	−1	1
0.1	0.488088	−0.1	0.513167
0.01	0.498756	−0.01	0.501256

The slope m appears to be .5.

35. The first argument gives the correct value; the second argument is not valid because it looks only at certain values of x.

37.

x	$(1+x)^{1/x}$	x	$(1+x)^{1/x}$
0.1	2.59	−0.1	2.87
0.01	2.70	−0.01	2.73
0.001	2.7169	−0.001	2.7196

$\lim_{x\to 0} (1+x)^{1/x} \approx 2.7182818$

39.

x	$x^{\sec x}$
0.1	0.099
0.01	0.010
0.001	0.001

$\lim_{x\to 0^+} x^{\sec x} = 0$

For negative x the values of $x^{\sec x}$ are usually not real numbers, so $\lim_{x\to 0^-} x^{\sec x} = 0$ does not exist.

41. A possible answer is

$$f(x) = \frac{x^2}{x}$$

$$g(x) = \begin{cases} 1 & \text{if } x \le 0 \\ -1 & \text{if } x > 0 \end{cases}$$

43. As x gets arbitrarily close to a, $f(x)$ gets arbitrarily close to L.

45. For $3 \le t \le 4$, $f(t) = 8$, so $\lim_{t\to 3.5} f(t) = 8$. Also $\lim_{t\to 4^-} f(t) = 8$. On the other hand, for $4 \le t \le 5$, $f(t) = 5$, so $\lim_{t\to 4^+} f(t) = 5$. Hence $\lim_{t\to 4} f(t)$ does not exist.

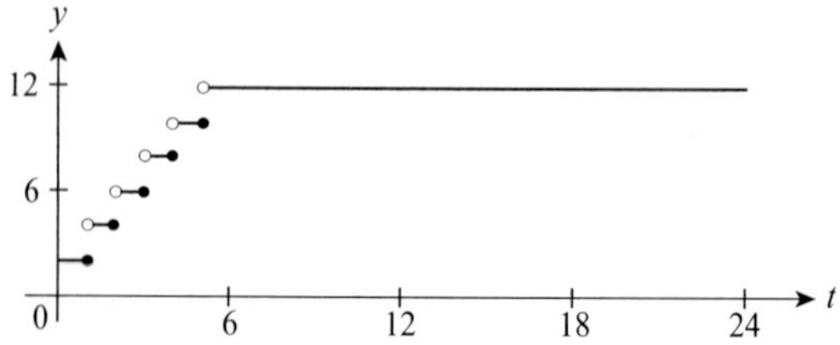

1.3 Computation of Limits

1. $\lim_{x\to 0}(x^2 - 3x + 1) = 0^2 - 3(0) + 1 = 1$

3. $\lim_{x\to 1}\sqrt{x^2+2x+4} = \sqrt{1^2+2(1)+4} = \sqrt{7}$

5. $\lim_{x\to 2}\frac{x-5}{x^2+4} = \frac{2-5}{2^2+4} = -\frac{3}{8}$

7.
$$\lim_{x\to 3}\frac{x^2-x-6}{x-3} = \lim_{x\to 3}\frac{(x-3)(x+2)}{x-3} = \lim_{x\to 3}(x+2) = 3+2 = 5$$

9.
$$\lim_{x\to 2}\frac{x^2-x-2}{x^2-4} = \lim_{x\to 2}\frac{(x-2)(x+1)}{(x+2)(x-2)} = \lim_{x\to 2}\frac{x+1}{x+2} = \frac{2+1}{2+2} = \frac{3}{4}$$

11.
$$\lim_{x\to 1}\frac{x^3-1}{x^2+2x-3} = \lim_{x\to 1}\frac{(x-1)(x^2+x+1)}{(x+3)(x-1)} = \lim_{x\to 1}\frac{x^2+x+1}{x+3} = \frac{1^2+1+1}{1+3} = \frac{3}{4}$$

13.
$$\lim_{x\to 0}\frac{\sin x}{\tan x} = \lim_{x\to 0}\frac{\sin x}{\frac{\sin x}{\cos x}} = \lim_{x\to 0}\cos x = \cos 0 = 1$$

15. $\lim_{x\to 0}\frac{xe^{-2x+1}}{x^2+x} = \lim_{x\to 0}\frac{e^{-2x+1}}{x+1} = \frac{e^{-2(0)+1}}{0+1} = e$

17.
$$\lim_{x\to 0}\frac{\sqrt{x+4}-2}{x} = \lim_{x\to 0}\frac{\sqrt{x+4}-2}{x}\left(\frac{\sqrt{x+4}+2}{\sqrt{x+4}+2}\right)$$
$$= \lim_{x\to 0}\frac{x+4-4}{x(\sqrt{x+4}+2)} = \lim_{x\to 0}\frac{x}{x(\sqrt{x+4}+2)}$$
$$= \lim_{x\to 0}\frac{1}{\sqrt{x+4}+2} = \frac{1}{\sqrt{4}+2} = \frac{1}{2+2} = \frac{1}{4}$$

19.
$$\lim_{x\to 1}\frac{x-1}{\sqrt{x}-1} = \lim_{x\to 1}\frac{(\sqrt{x}+1)(\sqrt{x}-1)}{\sqrt{x}-1} = \lim_{x\to 1}(\sqrt{x}+1) = \sqrt{1}+1 = 2$$

21.
$$\lim_{x\to 1}\left(\frac{1}{x-1}-\frac{2}{x^2-1}\right) = \lim_{x\to 1}\left(\frac{1}{x-1}-\frac{2}{(x-1)(x+1)}\right)$$
$$= \lim_{x\to 1}\left(\frac{x+1}{(x-1)(x+1)}-\frac{2}{(x-1)(x+1)}\right)$$
$$= \lim_{x\to 1}\left(\frac{x-1}{(x-1)(x+1)}\right) = \lim_{x\to 1}\left(\frac{1}{x+1}\right) = \frac{1}{2}$$

23.
$$\lim_{x\to 2^-} f(x) = \lim_{x\to 2^-} 2x = 2(2) = 4$$
$$\lim_{x\to 2^+} f(x) = \lim_{x\to 2^+} x^2 = 2^2 = 4$$
$$\lim_{x\to 2} f(x) = 4$$

25.
$$\lim_{x\to 0} f(x) = \lim_{x\to 0}(3x+1) = 3(0)+1 = 1$$

27.
$$\lim_{x\to -1^-} f(x) = \lim_{x\to -1^-}(2x+1) = 2(-1)+1 = -1$$
$$\lim_{x\to -1^+} f(x) = \lim_{x\to -1^+} 3 = 3$$
$\lim_{x\to -1} f(x)$ does not exist.

29.
$$\lim_{h\to 0}\frac{(2+h)^2-4}{h} = \lim_{h\to 0}\frac{(4+4h+h^2)-4}{h}$$
$$= \lim_{h\to 0}\frac{4h+h^2}{h} = \lim_{h\to 0} 4+h = 4$$

31. $\lim_{h\to 0} \frac{h^2}{\sqrt{h^2+h+3}-\sqrt{h+3}}$

$= \lim_{h\to 0}$

$\frac{h^2}{\sqrt{h^2+h+3}-\sqrt{h+3}} \frac{\sqrt{h^2+h+3}+\sqrt{h+3}}{\sqrt{h^2+h+3}+\sqrt{h+3}}$

$= \lim_{h\to 0} \frac{h^2(\sqrt{h^2+h+3}+\sqrt{h+3})}{(h^2+h+3)-(h+3)}$

$= \lim_{h\to 0} \frac{h^2(\sqrt{h^2+h+3}+\sqrt{h+3})}{h^2}$

$= \lim_{h\to 0} \sqrt{h^2+h+3}+\sqrt{h+3} = 2\sqrt{3}$

33. $\lim_{t\to -2} \frac{\frac{1}{2}+\frac{1}{t}}{2+t} = \lim_{t\to -2} \frac{\frac{t+2}{2t}}{2+t} = \lim_{t\to -2} \frac{1}{2t} = -\frac{1}{4}$

35.

x	$x^2 \sin\left(\frac{1}{x}\right)$	x	$x^2 \sin\left(\frac{1}{x}\right)$
−0.1	0.0054	0.1	−0.005
−0.01	5×10^{-5}	0.01	-5×10^{-5}
−0.001	-8×10^{-7}	0.001	8×10^{-7}

Conjecture: $\lim_{x\to 0} x^2 \sin\left(\frac{1}{x}\right) = 0$

Let $f(x) = -x^2$, $h(x) = x^2$.

Then $f(x) \le x^2 \sin\left(\frac{1}{x}\right) \le h(x)$.

$\lim_{x\to 0}(-x^2) = 0$, $\lim_{x\to 0}(x^2) = 0$.
Therefore, by the Squeeze Theorem,

$$\lim_{x\to 0} x^2 \sin\left(\frac{1}{x}\right) = 0.$$

37. Let $f(x) = 0$, $h(x) = \sqrt{x}$.
Since we see that

$$f(x) \le \sqrt{x}\cos^2(1/x) \le h(x),$$

$$\lim_{x\to 0^+} 0 = 0,\ \lim_{x\to 0^+} \sqrt{x} = 0.$$

Therefore, by the Squeeze Theorem,

$$\lim_{x\to 0^+} \sqrt{x}\cos^2\left(\frac{1}{x}\right) = 0.$$

39. $\lim_{x\to 4^+} \sqrt{16-x^2}$ does not exist because the domain of the function is $[-4, 4]$.

41. $\lim_{x\to -2^+} \sqrt{x^2+3x+2}$ does not exist because the domain of the function is $(-\infty, -2) \cup (-1, \infty)$.

43. $\lim_{x\to 0^+} \frac{\sqrt{1-\cos x}}{x} = \sqrt{\frac{1}{2}} = \frac{\sqrt{2}}{2}$

45. $\lim_{x\to a^-} f(x) = \lim_{x\to a^-} g(x) = g(a)$ because $g(x)$ is a polynomial. Similarly,

$$\lim_{x\to a^+} f(x) = \lim_{x\to a^+} h(x) = h(a).$$

47. (a)

$\lim_{x\to 2}(x^2 - 3x + 1)$

$= 2^2 - 3(2) + 1$ (Theorem 3.2)

$= -1$

(b)

$\lim_{x\to 0} \frac{x-2}{x^2+1}$

$= \frac{\lim_{x\to 0}(x-2)}{\lim_{x\to 0}(x^2+1)}$ (Theorem 3.1(iv))

$= \frac{\lim_{x\to 0} x - \lim_{x\to 0} 2}{\lim_{x\to 0} x^2 + \lim_{x\to 0} 1}$ (Theorem 3.1(ii))

$= \frac{0-2}{0+1}$ (Equations 3.1, 3.2, and 3.5)

$= -2$

49.

$\lim_{h\to 0} \frac{f(2+h)-f(2)}{h}$

$= \lim_{h\to 0} \frac{(2+h)^2+2-(2^2+2)}{h}$

$= \lim_{h\to 0} \frac{4h+h^2}{h}$

$= \lim_{h\to 0} 4+h$

$= 4$

51.
$$\begin{aligned}\lim_{h\to 0}\frac{f(0+h)-f(0)}{h} &= \lim_{h\to 0}\frac{(0+h)^3-(0)^3}{h}\\ &= \lim_{h\to 0}\frac{h^3}{h}\\ &= \lim_{h\to 0} h^2\\ &= 0\end{aligned}$$

53.
$$\begin{aligned}m &= \lim_{h\to 0}\frac{\sqrt{1+h}-1}{h}\cdot\frac{\sqrt{1+h}+1}{\sqrt{1+h}+1}\\ &= \lim_{h\to 0}\frac{1+h-1}{h(\sqrt{1+h}+1)}\\ &= \lim_{h\to 0}\frac{h}{h(\sqrt{1+h}+1)}\\ &= \lim_{h\to 0}\frac{1}{\sqrt{1+h}+1}\\ &= \frac{1}{\sqrt{1+0}+1}\\ &= \frac{1}{2}\end{aligned}$$

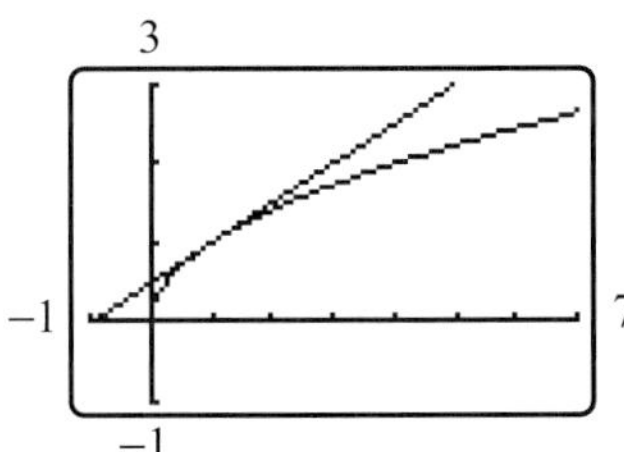

55. $\lim_{x\to 0^+}(1+x)^{1/x} = e \approx 2.71828$

57. $\lim_{x\to 0^+} x^{-x^2} = 1$

59. As x gets close to 0, $1/x$ gets larger and larger in absolute value, so $\sin(1/x)$ oscillates more and more rapidly between 1 and -1, so the limit does not exist.

61. When x is small and positive, $1/x$ is large and positive, so $\tan^{-1}(1/x)$ approaches $\pi/2$. But when x is small and negative, $1/x$ is large and negative, so $\tan^{-1}(1/x)$ approaches $-\pi/2$. See the graph on page 41. So the limit does not exist.

63.
$$\begin{aligned}\lim_{x\to a}[f(x)]^3 &= \left[\lim_{x\to a} f(x)\right]\left[\lim_{x\to a} f(x)\right]\left[\lim_{x\to a} f(x)\right]\\ &= L\cdot L\cdot L\\ &= L^3\end{aligned}$$
$$\begin{aligned}\lim_{x\to a}[f(x)]^4 &= \left[\lim_{x\to a} f(x)\right]\left[\lim_{x\to a}[f(x)]^3\right]\\ &= L\cdot L^3\\ &= L^4\end{aligned}$$

65. We can't split the limit of a product into a product of limits unless we know that both limits exist; the limit of the product of a term tending toward 0 and a term with an unknown limit is not necessarily 0 but instead is unknown.

67. One possibility is $f(x) = \frac{1}{x}$, $g(x) = -\frac{1}{x}$.

69. Yes. If $\lim_{x\to a}[f(x)+g(x)]$ exists, then, it would also be true that

$$\lim_{x\to a}[f(x)+g(x)] - \lim_{x\to a} f(x)$$

exists. But by Theorem 3.1 (ii)

$$\begin{aligned}&\lim_{x\to a}[f(x)+g(x)] - \lim_{x\to a} f(x)\\ &= \lim_{x\to a}\left[[f(x)+g(x)]-[f(x)]\right] = \lim_{x\to a} g(x),\end{aligned}$$

so $\lim_{x\to a} g(x)$ would exist, but we are given that $\lim_{x\to a} g(x)$ does not exist.

71.
$$\lim_{x\to 0^+} T(x) = \lim_{x\to 0^+}(0.14x) = 0 = T(0).$$
$$\lim_{x\to 10{,}000^-} T(x) = 0.14(10{,}000) = 1400$$
$$\lim_{x\to 10{,}000^+} T(x) = 1500 + 0.21(10{,}000) = 3600$$
$\lim_{x\to 10{,}000} T(x)$ does not exist.

A small change in income should result in a small change in tax liability. This is true near $x = 0$ but is not true near $x = 10{,}000$. As your income grows past \$10,000 your tax liability jumps enormously.

73. $\lim_{x\to 3^-}[x]=2$; $\lim_{x\to 3^+}[x]=3$
Therefore $\lim_{x\to 3}[x]$ does not exist.

1.4 Continuity and Its Consequences

1. $x=-2, x=2$

3. $x=-2, x=1, x=4$

5. $x=-2, x=2, x=4$

7. $f(1)$ is not defined and $\lim_{x\to 1} f(x)$ does not exist.

9. $f(0)$ is not defined and $\lim_{x\to 0} f(x)$ does not exist.

11.
$$\lim_{x\to 2^-} f(x)=\lim_{x\to 2^-}(x^2)=4$$
$$\lim_{x\to 2^+} f(x)=\lim_{x\to 2^+}(3x-2)=4$$
$$\lim_{x\to 2} f(x)=4;\ f(2)=3$$
$$\lim_{x\to 2} f(x)\neq f(2)$$

13. $f(x)=\dfrac{x-1}{(x+1)(x-1)}$ removable discontinuity at $x=1$, removed by $g(x)=\dfrac{1}{x+1}$; non-removable discontinuity at $x=-1$

15. no discontinuities

17. $f(x)=\dfrac{x^2\sin x}{\cos x}$ non-removable discontinuities at $x=\dfrac{\pi}{2}+n\pi, n=\ldots,-3,-2,-1,0,1,2,3,\ldots$

19. $f(x)=\dfrac{x\cos x}{\sin x}$ non-removable discontinuities at $x=n\pi$, n integer, $n\neq 0$ removable discontinuity at $x=0$, removed by

$$g(x)=\begin{cases}1, & x=0\\ \dfrac{x\cos x}{\sin x}, & x\neq 0\end{cases}$$

21. non-removable discontinuity at $x=1$

23.
$$\lim_{x\to -1^-} f(x)=\lim_{x\to -1^-}(3x-1)=-4$$
$$\lim_{x\to -1^+} f(x)=\lim_{x\to -1^+}(x^2+5x)=-4$$
$$\lim_{x\to 1^-} f(x)=\lim_{x\to 1^-}(x^2+5x)=6$$
$$\lim_{x\to 1^+} f(x)=\lim_{x\to 1^+}(3x^3)=3$$
non-removable discontinuity at $x=1$

25. $x+3\geq 0$
$x\geq -3$
$[-3,\infty)$

27. $(-\infty,\infty)$

29. $(-\infty,\infty)$

31. $x+1>0$
$x>-1$
$(-1,\infty)$

33.
$$\lim_{x\to 0^+} f(x)=\lim_{x\to 0^+} 2\frac{\sin x}{x}=2\lim_{x\to 0^+}\frac{\sin x}{x}=2.$$
Hence a must equal 2 if f is continuous.
$$\lim_{x\to 0^-} f(x)=\lim_{x\to 0^-} b\cos x=b\lim_{x\to 0^-}\cos x=b,$$
so b must equal 2 if f is continuous.

35. First note that
$$\lim_{x\to 3^+} f(x)=\lim_{x\to 3^+}\ln(x-2)+x^2$$
$$=\ln(3-2)+3^2=9.$$
Also $f(3)=2e^{3b}+1$, so if f is continuous, $2e^{3b}+1$ must equal 9; that is, $e^{3b}=4$, so $b=\dfrac{\ln 4}{3}$. Then note that $f(0)=2e^{(b)(0)}+1=3$.
Also,
$$\lim_{x\to 0^-} f(x)=\lim_{x\to 0^-} a(\tan^{-1}x+2)=a(\tan^{-1}0+2)$$
$$=a(0+2)=2a,$$
so a must equal $\dfrac{3}{2}$ if f is continuous.

37.
$$\lim_{x\to 10000^-} T(x) = \lim_{x\to 10000^-} 0.14x = 0.14(10{,}000) = 1400$$
$$\lim_{x\to 10000^+} T(x) = \lim_{x\to 10000^+} (c + 0.21x) = c + 0.21(10{,}000) = c + 2100$$
$$c + 2100 = 1400$$
$$c = -700$$

A small change in income should not result in a big change in tax, so the tax function should be continuous.

39. **(a)** For $T(x)$ to be continuous at $x = 141{,}250$ we must have

$$\lim_{x\to 141{,}250^-} T(x) = \lim_{x\to 141{,}250^+} T(x).$$

Now

$$\lim_{x\to 141{,}250^-} T(x) = \lim_{x\to 141{,}250^-} (.30)(x)a$$
$$= (.30)(141{,}250) - 5685 = 36690.$$

On the other hand,

$$\lim_{x\to 141{,}250^+} T(x) = \lim_{x\to 141{,}250^+} (.35)(x) - b$$
$$= (.35)(141{,}250) - b = 49437.50 - b.$$

Hence $b = 49437.50 - 36690 = 12{,}747.50$.

(b) For $T(x)$ to be continuous at $x = 307{,}050$ we must have

$$\lim_{x\to 307{,}050^-} T(x) = \lim_{x\to 307{,}050^+} T(x).$$

Now

$$\lim_{x\to 307{,}050^-} T(x) = \lim_{x\to 307{,}050^-} (.35)(x) - b$$
$$= (.35)(307{,}050) - 12{,}747.50 = 94{,}720.$$

On the other hand,

$$\lim_{x\to 307{,}050+} T(x) = \lim_{x\to 307{,}050^+} (.386)(x) - c$$
$$= (.386)(307{,}050) - c = 118521.3 - c.$$

Hence $c = 118{,}521.30 - 94720 = 23801.30$.

41. $f(2) = -3$, $f(3) = 2$
Since $f(2) < 0$ and $f(3) > 0$, the Intermediate Value Theorem says there is a zero in the interval [2, 3].

a	b	$f(a)$	$f(b)$	midpoint	f(midpoint)
2	3	−3	2	2.5	−0.75
2.5	3	−0.75	2	2.75	0.5625
2.5	2.75	−0.75	0.5625	2.625	−0.109375
2.625	2.75	−0.109375	0.5625	2.6875	0.223
2.625	2.6875	−0.109375	0.223	2.65625	0.557

The interval [2.625, 2.65625] contains the zero.

43. $f(-1) = 1$, $f(0) = -2$
Since $f(-1) > 0$ and $f(0) < 0$, the Intermediate Value Theorem says there is a zero in the interval [−1, 0].

a	b	$f(a)$	$f(b)$	midpoint	f(midpoint)
−1	0	1	−2	−0.5	−0.125
−1	−0.5	1	−0.125	−0.75	0.578
−0.75	−0.5	0.578	−0.125	−0.625	0.256
−0.625	−0.5	0.256	−0.125	−0.5625	0.072
−0.5625	−0.5	0.072	−0.125	−0.53125	−0.025

The interval [−0.5625, −0.53125] contains the zero.

45. $f(0) = 1$, $f(1) = \cos 1 - 1 \approx -0.46$
Since $f(0) > 0$ and $f(1) < 0$, the Intermediate Value Theorem says there is a zero in the interval [0, 1].

a	b	$f(a)$	$f(b)$	midpoint	f(midpoint)
0	1	1	−0.46	0.5	0.378
0.5	1	0.378	−0.46	0.75	−0.018
0.5	0.75	0.378	−0.018	0.625	0.186
0.625	0.75	0.186	−0.018	0.6875	0.085
0.6875	0.75	0.085	−0.018	0.71875	0.034

The interval [0.71875, 0.75] contains the zero.

47.
$$\lim_{x\to 2^+} f(x) = \lim_{x\to 2^+} (3x - 1) = 5$$
$f(2) = 3(2) - 1 = 5$
$f(x)$ is continuous from the right at $x = 2$.

49.
$$\lim_{x\to 2^+} f(x) = \lim_{x\to 2^+} (3x - 3) = 3$$
$f(2) = 2^2 = 4$
$f(x)$ is not continuous from the right at $x = 2$.

51. A function is continuous from the left at $x = a$ if $\lim_{x \to a^-} f(x) = f(a)$.

(a) $\lim_{x \to 2^-} f(x) = \lim_{x \to 2^-} x^2 = 4$

$f(2) = 5$

$f(x)$ is not continuous from the left at $x = 2$.

(b) $\lim_{x \to 2^-} f(x) = \lim_{x \to 2^-} x^2 = 4$

$f(2) = 3$

$f(x)$ is not continuous from the left at $x = 2$.

(c) $\lim_{x \to 2^-} f(x) = \lim_{x \to 2^-} x^2 = 4$

$f(2) = 4$

$f(x)$ is continuous from the left at $x = 2$.

(d) $f(x)$ is not continuous from the left at $x = 2$ because $f(2)$ is undefined.

53. Need $g(30) = 100$ and $g(34) = 0$.

$$m = \frac{0 - 100}{34 - 30} = -25$$

$$y = -25(x - 34)$$

$$g(T) = -25(T - 34)$$

55.

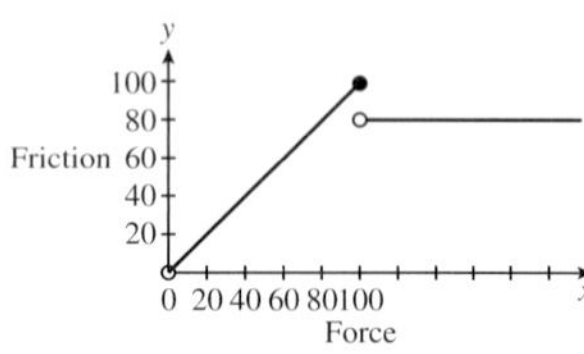

The graph is discontinuous at $x = 100$. This is when the box starts moving.

57. Let $f(t)$ be her distance from home as a function of time on Monday. Let $g(t)$ be her distance from home as a function of time on Tuesday. Let t be given in minutes, with $t = 0$ corresponding to 7:13 a.m. Then she leaves home at $t = 0$ and arrives at her destination at $t = 410$. Let $h(t) = f(t) - g(t)$. If $h(t) = 0$ for some t, then the saleswoman was at exactly the same place at the same time on both Monday and Tuesday. $h(0) = f(0) - g(0) = -g(0) < 0$ and $h(410) = f(410) - g(410) = f(410) > 0$. By the Intermediate Value Theorem, there is a t in the interval $[0, 410]$ such that $h(t) = 0$.

59.

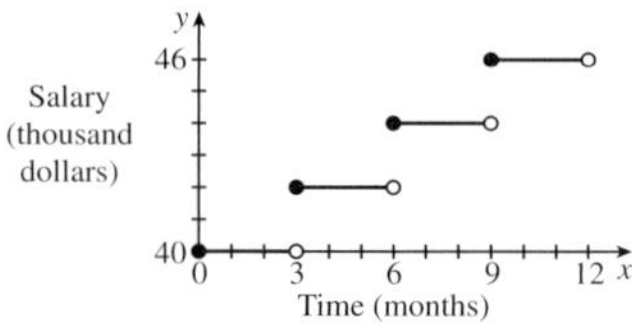

The function $s(t)$ has jump discontinuities every three months when the salary suddenly increases by \$2000. In the function $f(t)$, the \$2000 increase occurs gradually over the 3 month period, so $f(t)$ is continuous. It might be easier to do calculations with $f(t)$ because it is continuous and because it is given by a simpler formula.

61. We already know $f(x) \neq 0$ for $a < x < b$. Suppose $f(d) < 0$ for some d, $a < d < b$. Then by the Intermediate Value Theorem, there is an e in the interval $[c, d]$ such that $f(e) = 0$. But this e would also be between a and b, which is impossible. Thus, $f(x) > 0$ for all $a < x < b$.

63. $\lim_{x \to 0} x f(x) = \lim_{x \to 0} x \lim_{x \to 0} f(x) = 0 f(0) = 0$

65.

$$\lim_{x \to a} g(x) = \lim_{x \to a} |f(x)| = \left| \lim_{x \to a} f(x) \right|$$

$$= |f(a)| = g(a).$$

1.5 Limits Involving Infinity

1. $\lim_{x \to 1^-} \frac{1 - 2x}{x^2 - 1} = \infty$. As x approaches 1 from below, the numerator approaches -1 and the denominator is small but negative, so the fraction goes to ∞.

3. $\lim_{x \to 1} \frac{1 - 2x}{x^2 - 1}$ does not exist. When x is near 1, the numerator is near -1 and the denominator is small but can be positive or negative, so the limit does not exist.

5. $\lim_{x \to -1^+} \frac{1 - 2x}{x^2 - 1} = -\infty$. As x approaches -1 from above, the numerator approaches 3 and the denominator is small but negative, so the fraction goes to $-\infty$.

7. $\displaystyle\lim_{x\to 2^-} \frac{\overset{+}{4-x}}{\underset{+}{(x-2)^2}} = \infty$. As x approaches 2 from below, the numerator is near 2 and the denominator is small and positive, so the fraction goes to ∞.

9. $\displaystyle\lim_{x\to 2} \frac{4-x}{(x-2)^2} = \infty$. As x approaches 2, the numerator is near 2 and the denominator is small and positive, so the fraction goes to ∞.

11. $\displaystyle\lim_{x\to 2^-} \frac{\overset{-}{-x}}{\underset{+}{\sqrt{4-x^2}}} = -\infty$. As x approaches 2 from below, the numerator is near -2 and the denominator is small and positive, so the faction goes to $-\infty$.

13. $\displaystyle\lim_{x\to 1^-} (x^2-2x-3)^{-2/3} = \infty$. As x approaches -1, x^2-2x-3 is small, so $(x^2-2x-3)^{-2/3}$ is small and positive, so $(x^2-2x-3)^{-2/3}$ is large and positive, so the limit is ∞.

15.

$$\begin{aligned}\lim_{x\to-\infty} \frac{-x}{\sqrt{4+x^2}} &= \lim_{x\to-\infty} \frac{-x}{-x\sqrt{\frac{4}{x^2}+1}} \\ &= \lim_{x\to-\infty} \frac{1}{\sqrt{\frac{4}{x^2}+1}} \\ &= \frac{1}{\sqrt{1}} \\ &= 1\end{aligned}$$

17.

$$\begin{aligned}\lim_{x\to-\infty} \frac{x^3-2x+1}{3x^3+4x^2-1} &= \lim_{x\to-\infty} \frac{x^3\left(1-\frac{2}{x^2}+\frac{1}{x^3}\right)}{x^3\left(3+\frac{4}{x}-\frac{1}{x^3}\right)} \\ &= \lim_{x\to-\infty} \frac{1-\frac{2}{x^2}+\frac{1}{x^3}}{3+\frac{4}{x}-\frac{1}{x^3}} \\ &= \frac{1}{3}\end{aligned}$$

19.

$$\begin{aligned}\lim_{x\to\infty} \frac{x^3-2\cos x}{3x^2+4x-1} &= \lim_{x\to\infty} \frac{x^2\left(x-\frac{2\cos x}{x^2}\right)}{x^2\left(3+\frac{4}{x}-\frac{1}{x^2}\right)} \\ &= \lim_{x\to\infty} \frac{\left(x-\frac{2\cos x}{x^2}\right)}{3+\frac{4}{x}-\frac{1}{x^2}} \\ &= \infty\end{aligned}$$

21.

$$\begin{aligned}\lim_{x\to\infty} \frac{x^2-\sin x}{x^2+4x-1} &= \lim_{x\to\infty} \frac{x^2\left(1-\frac{\sin x}{x^2}\right)}{x^2\left(1+\frac{4}{x}-\frac{1}{x^2}\right)} \\ &= \lim_{x\to\infty} \frac{1-\frac{\sin x}{x^2}}{1+\frac{4}{x}-\frac{1}{x^2}} \\ &= \frac{1}{1} \\ &= 1\end{aligned}$$

23. $\displaystyle\lim_{x\to\infty} \cot^{-1} x = 0$. (Compare Example 5.8). We are looking for the angle that θ must approach as $\cot\theta$ goes to ∞. Look at the graph of $\cot\theta$ on page 36. To define the inverse cotangent, you must pick one branch of this graph, and the standard choice is the branch immediately to the right of the y-axis. Then as $\cot\theta$ goes to ∞, the angle goes to 0.

25. $\displaystyle\lim_{x\to\infty} e^{2x-1} = \infty$. As x gets large, $2x-1$ gets large, and e raised to a large positive power is large and positive.

27. $\displaystyle\lim_{x\to 0} e^{1/x^2} = \infty$. When x is small, x^2 is small and positive, so $1/x^2$ is large and positive, and e raised to a large positive power is large and positive.

29. $\displaystyle\lim_{x\to\infty} \sin 2x$ does not exist. As x gets larger and larger, the values of $\sin 2x$ oscillate between 1 and -1.

31. $\displaystyle\lim_{x\to\infty} (e^{-3x}\cos 2x) = 0$. The function $e^{-3x}\cos 2x$ is squeezed between $-e^{-3x}$ and e^{-3x}, both of which go to 0 as x goes to ∞. By the Squeeze Theorem, $\displaystyle\lim_{x\to\infty} (e^{-3x}\cos 2x) = 0$.

33. $\lim_{x\to\infty} \ln 2x = \infty$. Note that $\ln 2x = \ln 2 + \ln x$, so it is enough to show that $\ln x$ goes to ∞ as x goes to ∞. This can be seen from the graph of the function $\ln x$ on page 51.

35. $\lim_{x\to 0^+} e^{-2/x} = 0$. When x is small and positive, $-2/x$ is large and negative, and e raised to a large negative power is very small.

37. $4 + x^2 = 0$ Never
No vertical asymptotes.

$$\lim_{x\to\infty} \frac{x}{\sqrt{4+x^2}} = \lim_{x\to\infty} \frac{x}{x\sqrt{\frac{4}{x^2}+1}} = \lim_{x\to\infty} \frac{1}{\sqrt{\frac{4}{x^2}+1}} = \frac{1}{\sqrt{1}} = 1$$

$$\lim_{x\to-\infty} \frac{x}{\sqrt{4+x^2}} = \lim_{x\to-\infty} \frac{x}{-x\sqrt{\frac{4}{x^2}+1}} = \lim_{x\to-\infty} \frac{-1}{\sqrt{\frac{4}{x^2}+1}} = \frac{-1}{\sqrt{1}} = -1$$

HA: $y = 1$ and $y = -1$.

39.

$$4 - x^2 = 0$$
$$4 = x^2$$

VA: $x = \pm 2$

$$\lim_{x\to\pm\infty} \frac{x}{4-x^2} = \lim_{x\to\pm\infty} \frac{x}{x^2\left(\frac{4}{x^2}-1\right)} = \lim_{x\to\pm\infty} \frac{1}{x\left(\frac{4}{x^2}-1\right)} = 0$$

HA: $y = 0$.

41.

$$4 - x^2 = 0$$
$$4 = x^2$$

VA: $x = \pm 2$.

$$\lim_{x\to\pm\infty} \frac{x^3}{4-x^2} = \lim_{x\to\pm\infty} \frac{x^3}{x^2\left(\frac{4}{x^2}-1\right)} = \lim_{x\to\pm\infty} \frac{x}{\left(\frac{4}{x^2}-1\right)} = \mp\infty$$

No horizontal asymptote.

$$\begin{array}{r|l} & -x \\ -x^2+4 & x^3 \\ & -(+x^3-4x) \\ \hline & 4x \end{array}$$

SA: $y = -x$

43. $4 + x^2 = 0$ Never
No vertical asymptotes.

$$\lim_{x\to\pm\infty} \frac{x^3}{4+x^2} = \lim_{x\to\pm\infty} \frac{x^3}{x^2\left(\frac{4}{x^2}+1\right)} = \lim_{x\to\pm\infty} \frac{x}{\frac{4}{x^2}+1} = \pm\infty$$

No horizontal asymptote.

$$\begin{array}{r|l} & x \\ x^2+4 & x^3 \\ & -(x^3+4x) \\ \hline & -4x \end{array}$$

SA: $y = x$

45. The function is continuous for all x, so no vertical asymptotes.

$$\lim_{x\to\infty} 4\tan^{-1} x - 1 = 4(\lim_{x\to\infty} \tan^{-1} x) - 1$$
$$= 4(\pi/2) - 1 = 2\pi - 1$$
$$\lim_{x\to-\infty} 4\tan^{-1} x - 1 = 4(\lim_{x\to-\infty} \tan^{-1} x) - 1$$
$$= 4(-\pi/2) - 1 = -2\pi - 1$$

HA: $y = 2\pi - 1$ and $y = -2\pi - 1$.

47. When x is large, the value of the fraction is close to 0.

49. When x is large, the value of the fraction is very close to $\frac{1}{2}$.

51. When x is large and negative, the value of the fraction is very close to 2.

53. When x is close to -1, the value of the fraction is close to 1.

55. When x is close to 0, the value of the fraction is large and negative, so the limit appears to be $-\infty$.

57. $\lim_{x\to\infty} (\sqrt{x^2+3} - x)$

$$= \lim_{x\to\infty} (\sqrt{x^2+3} - x)\frac{\sqrt{x^2+3}+x}{\sqrt{x^2+3}+x}$$
$$= \lim_{x\to\infty} \frac{(x^2+3) - x^2}{\sqrt{x^2+3}+x} = \lim_{x\to\infty} \frac{3}{\sqrt{x^2+3}+x} = 0$$

59. $\lim_{x\to\infty} (\sqrt{5x^2+4x+7} - \sqrt{5x^2+x+3})$
If we multiply by

$$\frac{\sqrt{5x^2+4x+7} + \sqrt{5x^2+x+3}}{\sqrt{5x^2+4x+7} + \sqrt{5x^2+x+3}},$$

we get

$$\lim_{x\to\infty} \frac{(5x^2+4x+7) - (5x^2+x+3)}{\sqrt{5x^2+4x+7} + \sqrt{5x^2+x+3})}$$
$$= \lim_{x\to\infty} \frac{3x+4}{\sqrt{5x^2+4x+7} + \sqrt{5x^2+x+3}}$$
$$= \lim_{x\to\infty} \frac{3+\frac{4}{x}}{\sqrt{5+\frac{4}{x}+\frac{7}{x^2}} + \sqrt{5+\frac{1}{x}+\frac{3}{x^2}}}$$
$$= \frac{3}{2\sqrt{5}} = \frac{3\sqrt{5}}{10}.$$

61.

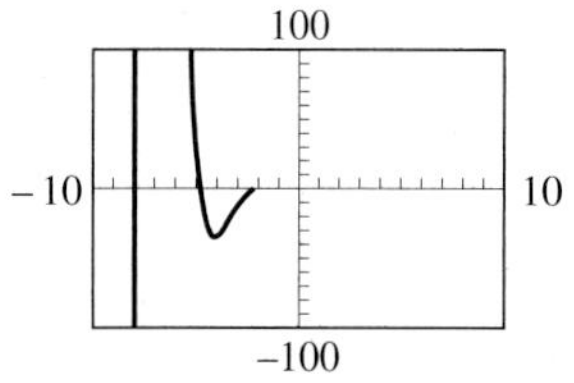

100, −10, 10, −100

on $[-10, 10]$ by $[-100, 100]$
HA: $y = 0$ approached only as $x \to \infty$.
The graph crosses the horizontal asymptote an infinite number of times.

63.
$$\lim_{x\to\infty}\left(1+\frac{1}{x}\right)^x = \lim_{x\to0^+}(1+x)^{1/x}$$
$$= \lim_{x\to0^-}(1+x)^{1/x} = \lim_{x\to-\infty}\left(1+\frac{1}{x}\right)^x$$

65.
$$h(0) = \frac{300}{1+9(.8^0)} = \frac{300}{10} = 30 \text{ mm}$$
$$\lim_{t\to\infty} \frac{300}{1+9(.8^t)} = 300 \text{ mm}$$

67.
$$\lim_{x\to0^+} \frac{80x^{-.3}+60}{9x^{-.3}+5}\left(\frac{x^{.3}}{x^{.3}}\right) = \lim_{x\to0^+} \frac{80+60x^{.3}}{9+5x^{.3}}$$
$$= \frac{80}{9}$$
$$\approx 8.89 \text{ mm}$$
$$\lim_{x\to\infty} \frac{80x^{-.3}+60}{9x^{-.3}+5} = \frac{60}{5} = 12 \text{ mm}$$

69. $f(x) = \dfrac{80x^{-0.3}+60}{10x^{-0.3}+30}$

71.
$$\lim_{t\to\infty} v_N = \lim_{t\to\infty} \frac{Ft}{m} = \infty$$
$$\begin{aligned}\lim_{t\to\infty} v_E &= \lim_{t\to\infty} \frac{Fct}{\sqrt{m^2c^2+F^2t^2}}\\ &= \lim_{t\to\infty} \frac{Fct}{t\sqrt{\frac{m^2c^2}{t^2}+F^2}}\\ &= \lim_{t\to\infty} \frac{Fc}{\sqrt{\frac{m^2c^2}{t^2}+F^2}}\\ &= \frac{Fc}{\sqrt{F^2}}\\ &= c\end{aligned}$$

73. As in Example 5.10, the terminal velocity is $-\sqrt{\frac{32}{k}}$. When $k = 0.00064$, the terminal velocity is $-\sqrt{\frac{32}{.00064}} \approx -224$. When $k = 0.00128$, the terminal velocity is $-\sqrt{\frac{32}{.00128}} \approx -158$.
Solve $\sqrt{\frac{32}{ak}} = \frac{1}{2}\sqrt{\frac{32}{k}}$. Squaring both sides,
$$\begin{aligned}\frac{32}{ak} &= \frac{1}{4}\frac{32}{k}\\ \frac{1}{a} &= \frac{1}{4}\\ a &= 4.\end{aligned}$$

75. Looking at the graph, we estimate the time to 90% of terminal velocity is about 20 seconds.

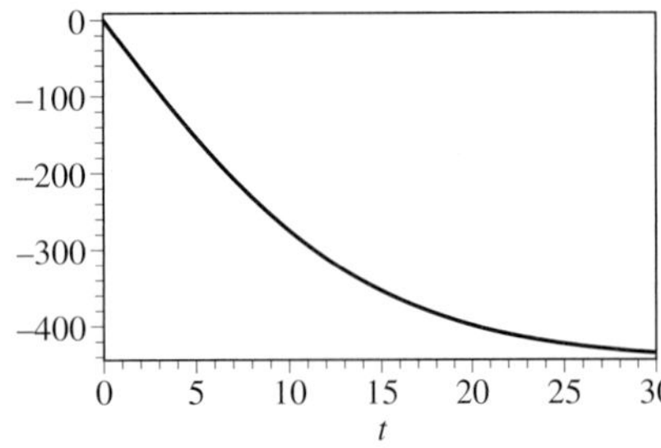

77. We must restrict the domain to $v_0 \geq 0$ because the formula makes sense only if the rocket is launched upward. To find v_e, set $19.6R - v_0^2 = 0$. Using $R \approx 6{,}378{,}000$ meters, we get $v_0 = \sqrt{19.6R} \approx 11{,}180$m/s. If the rocket is launched with initial velocity $\geq v_e$, it will never return to earth; hence v_e is called the escape velocity.

79. Suppose the degree of q is n. If we divide both $p(x)$ and $q(x)$ by x^n, then the new denominator will approach a constant while the new numerator tends to ∞, so there is no HA.

81. When we do long division, we get a remainder of $x + 2$, so the degree of p is one greater than the degree of q.

83. False if $b = 0$; otherwise true.

85. True.

87. False. For example, $f(x) = 2x$ and $g(x) = x$.

89. $x^2 - 4x = x(x-4)$, so there is a vertical asymptote at $x = 0$.
$x^2 - 7x + 10 = (x-2)(x-5)$ so there are vertical asymptotes at $x = 2$ and $x = 5$.
$$\lim_{x\to\infty} f(x) = \lim_{x\to\infty} \frac{x^2-1}{x^2-7x+10} = 1$$
$$\lim_{x\to-\infty} f(x) = \lim_{x\to-\infty} \frac{x+3}{x^2-4x} = 0$$
so HA: $y = 0$ and $y = 1$.

91.
$$\begin{aligned}\lim_{x\to-\infty} p_n(x) &= \lim_{x\to-\infty} \left(a_nx^n + a_{n-1}x^{n-1} + \cdots + a_0\right)\\ &= \lim_{x\to-\infty} \left[x^n\left(a_n + \frac{a_{n-1}}{x} + \cdots + \frac{a_0}{x}\right)\right]\end{aligned}$$
When the degree n is odd, if a_n is positive, the limit as $x \to -\infty$ is $\to -\infty$, and if a_n is negative, the limit as $x \to -\infty$ is $+\infty$.

93. $g(x) = \sin x$, $h(x) = x$

95. $\lim_{x\to0^+} x^{1/(\ln x)} = e \approx 2.71828$

97. $\lim_{x\to\infty} \left(1 - \frac{1}{x}\right)^x = \frac{1}{e} \approx 0.36788$

99. $\lim_{x\to\infty} x^{1/x} = 1$

1.6 Limits and Loss-of-Significance Errors

1. In the table below, the middle column represents values calculated using

$$f(x) = x(\sqrt{4x^2+1} - 2x).$$

x	old $f(x)$	new $f(x)$
1	0.236068	0.236068
10	0.249844	0.249844
100	0.249998	0.249998
1000	0.250000	0.250000
10000	0.250000	0.250000
100000	0.249999	0.250000
1000000	0.250060	0.250000
10000000	0.260770	0.250000
100000000	0.000000	0.250000
1000000000	0.000000	0.250000

We can rewrite $f(x)$ as

$$\begin{aligned} f(x) &= x(\sqrt{4x^2+1} - 2x)\cdot\frac{\sqrt{4x^2+1}+2x}{\sqrt{4x^2+1}+2x} \\ &= \frac{x(4x^2+1-4x^2)}{\sqrt{4x^2+1}+2x} \\ &= \frac{x}{\sqrt{4x^2+1}+2x}. \end{aligned}$$

Using this new form of the function we get the values listed in the third column of the table. We conjecture that

$$\lim_{x\to\infty} x\left(\sqrt{4x^2+1} - 2x\right) = 0.25.$$

3. As in Exercise 1, the middle column represents values calculated using the function

$$f(x) = \sqrt{x}\left(\sqrt{x+4} - \sqrt{x+2}\right).$$

We can rewrite $f(x)$ as

$$\begin{aligned} f(x) &= \sqrt{x}(\sqrt{x+4} - \sqrt{x+2})\cdot\frac{\sqrt{x+4}+\sqrt{x+2}}{\sqrt{x+4}+\sqrt{x+2}} \\ &= \frac{\sqrt{x}[(x+4)-(x+2)]}{\sqrt{x+4}+\sqrt{x+2}} \\ &= \frac{2\sqrt{x}}{\sqrt{x+4}+\sqrt{x+2}}. \end{aligned}$$

Using this new form of the function we get the values listed in the third column, and so we conjecture that

$$\lim_{x\to\infty} \sqrt{x}\left(\sqrt{x+4} - \sqrt{x+2}\right) = 1.$$

x	old $f(x)$	new $f(x)$
1	0.504017	0.504017
10	0.877708	0.877708
100	0.985341	0.985341
1000	0.998503	0.998503
10000	0.999850	0.999850
100000	0.999985	0.999985
1000000	0.999998	0.999999
10000000	1.000000	1.000000
100000000	1.000000	1.000000
1000000000	1.000000	1.000000
10000000000	1.000000	1.000000
1E+11	0.999990	1.000000
1E+12	1.000008	1.000000
1E+13	0.999862	1.000000
1E+14	0.987202	1.000000
1E+15	0.942432	1.000000
1E+16	0.000000	1.000000
1E+17	0.000000	1.000000

5. As in the previous exercises, the middle column represents values calculated using the function $f(x) = \left(\sqrt{x^2+4} - \sqrt{x^2+2}\right)$
We can rewrite $f(x)$ as

$$\begin{aligned} f(x) &= (\sqrt{x^2+4} - \sqrt{x^2+2})\cdot\frac{\sqrt{x^2+4}+\sqrt{x^2+2}}{\sqrt{x^2+4}+\sqrt{x^2+2}} \\ &= \frac{x[x^2+4-(x^2+2)]}{\sqrt{x^2+4}+\sqrt{x^2+2}} \\ &= \frac{2x}{\sqrt{x^2+4}+\sqrt{x^2+2}}. \end{aligned}$$

Using this new form of $f(x)$, we get the values listed in the third column and so we conjecture that

$$\lim_{x\to\infty} x\left(\sqrt{x^2+4}-\sqrt{x^2+2}\right)=1.$$

x	old $f(x)$	new $f(x)$
1	0.504017	0.504017
10	0.985341	0.985341
100	0.999850	0.999850
1000	0.999998	0.999999
10000	1.000000	1.000000
100000	1.000000	1.000000
1000000	1.000008	1.000000
10000000	0.987202	1.000000
100000000	0.000000	1.000000
1000000000	0.000000	1.000000

7. As in the previous exercises, the middle column represents values calculated using the function

$$f(x)=\frac{1-\cos 2x}{12x^2}.$$

We can rewrite $f(x)$ as

$$f(x)=\frac{1-\cos 2x}{12x^2}\cdot\frac{1+\cos 2x}{1+\cos 2x}=\frac{\sin^2 2x}{12x^2(1+\cos 2x)}.$$

Using this new form of the function, we get the values listed in the third column. Note that $f(x)=f(-x)$, and so we get the same values when x is negative. Hence we conjecture that $\lim_{x\to 0}\frac{1-\cos 2x}{12x^2}=\frac{1}{6}$.

x	old $f(x)$	new $f(x)$
1	0.118012	0.118012
0.1	0.166112	0.166112
0.01	0.166661	0.166661
0.001	0.166667	0.166667
0.0001	0.166667	0.166667
0.00001	0.166667	0.166667
0.000001	0.166663	0.166667
0.0000001	0.166533	0.166667
0.00000001	0.185037	0.166667
0.000000001	0	0.166667
1E−10	0	0.166667

9. As in the previous exercises, the middle column represents values calculated using the function $f(x)=\dfrac{1-\cos x^3}{x^6}$. We can rewrite $f(x)$ as

$$f(x)=\frac{1-\cos x^3}{x^6}\frac{1+\cos x^3}{1+\cos x^3}=\frac{\sin^2(x^3)}{x^6(1+\cos x^3)}.$$

Using this new form of the function, we get the values listed in the third column. Note that $f(x)=f(-x)$, and so we get the same values when x is negative. Hence we conjecture that

$$\lim_{x\to 0}\frac{1-\cos x^3}{x^6}=\frac{1}{2}.$$

x	old $f(x)$	new $f(x)$
1	0.459698	0.459698
0.1	0.500000	0.500000
0.01	0.500044	0.500000
0.001	0.000000	0.500000
0.0001	0.000000	0.500000

11.

$$\begin{aligned}\lim_{x\to 1}\frac{x^2+x-2}{x-1}&=\lim_{x\to 1}\frac{(x+2)(x-1)}{x-1}\\&=\lim_{x\to 1}(x+2)\\&=3\end{aligned}$$

$\lim_{x\to 1}\dfrac{x^2+x-2.01}{x-1}$ does not exist, since when x is close to 1, the numerator is close to $-.01$ (a small but non-zero number) and the denominator is close to 0.

13.

$f(1)=0;\ g(1)=0.00159265$
$f(10)=0;\ g(10)=-0.0159259$
$f(100)=0;\ g(100)=-0.158593$
$f(1000)=0;\ g(1000)=-0.999761$

15. $(1.000003-1.000001)\times 10^7=20$
On a computer with a 6-digit mantissa, the calculation would be

$$(1.00000-1.00000)\times 10^7=0.$$

the previous exercises

Chapter 1 Review

1. See Example 1.1.

x coord of 2nd Point	m_{sec}	x coord of 2nd Point	m_{sec}
3	3	1	1
2.1	2.1	1.9	1.9
2.01	2.01	1.99	1.99

The slope appears to be 2.

3. See Example 1.3.
With $n = 4$, we must sum the length of four line segments. We get:

$$0.276791214 + 0.271559049 + 0.261616108 + 0.248025135 \approx 1.057991506.$$

With $n = 8$, we get

$$0.138728675 + 0.138063348 + 0.136748860 + 0.134817712 + 0.132319069 + 0.129319229 + 0.125902180 + 0.122170120 \approx 1.058069194.$$

5. See Example 2.4.
Let $f(x) = \dfrac{\tan^{-1} x^2}{x^2}$.

x	$f(x)$
0.1	0.999966669
0.01	0.999999997
0.001	1.000000000
0.0001	1.000000000
0.00001	1.000000000
0.000001	1.000000000

Note that $f(x) = f(-x)$, so the results for negative x will be the same as above. The limit appears to be 1.

7.

x	$\dfrac{x+2}{\lvert x+2\rvert}$	x	$\dfrac{x+2}{\lvert x+2\rvert}$
−1.9	1	−2.1	−1
−1.99	1	−2.01	−1
−1.999	1	−2.001	−1

$\lim\limits_{x\to -2} \dfrac{x+2}{|x+2|}$ does not exist.

9.

x	$\left(1+\dfrac{2}{x}\right)^x$
10	6.1917
100	7.2446
1000	7.3743
10,000	7.3876

$$\lim_{x\to\infty}\left(1+\frac{2}{x}\right)^x = e^2 \approx 7.4$$

11. See Example 2.1.

(a) 1 **(b)** −2

(c) does not exist **(d)** 0

13. See Definition 4.1.
$x = -1, x = 1$

15. See Examples 2.2 and 3.4.

$$\lim_{x\to 2}\frac{x-2}{x^2-4} = \lim_{x\to 2}\frac{x-2}{(x+2)(x-2)} = \lim_{x\to 2}\frac{1}{x+2} = \frac{1}{2+2} = \frac{1}{4}$$

17. See page 95. $\lim\limits_{x\to -2}\dfrac{x+2}{\sqrt{x^2-4}}$ does not exist because the function is not defined to the right of $x = -2$.

19. See Theorem 3.1(iv).

$$\lim_{x\to 0}\frac{\tan x}{\sec x} = \frac{\tan 0}{\sec 0} = \frac{0}{1} = 0$$

21. See Example 3.10.

$$\lim_{x\to 1^-} f(x) = \lim_{x\to 1^-}(2x+1) = 3$$

$$\lim_{x\to 1^+} f(x) = \lim_{x\to 1^+}(x^2+1) = 2$$

$\lim\limits_{x\to 1} f(x)$ does not exist.

23. See Example 3.6.

$$\lim_{x\to 1^+} \sqrt{x^2+1} = \sqrt{2}$$

25. See Section 1.5.

$$\lim_{x\to 1^+} \frac{x+1}{x^2-1} = \lim_{x\to 1^+} \frac{x+1}{(x+1)(x-1)}$$

$$= \lim_{x\to 1^+} \frac{\overset{+}{1}}{\underset{+}{x-1}}$$

$$= \infty$$

27. See Example 5.5.

$$\lim_{x\to\infty} \frac{x^2-4}{3x^2+x+1} = \lim_{x\to\infty} \frac{x^2\left(1-\frac{4}{x^2}\right)}{x^2\left(3+\frac{1}{x}+\frac{1}{x^2}\right)}$$

$$= \lim_{x\to\infty} \frac{1-\frac{4}{x^2}}{3+\frac{1}{x}+\frac{1}{x^2}}$$

$$= \frac{1}{3}$$

29. See Section 5.7. $\lim_{x\to\infty} e^{3x} = 0$. When x is large and positive, $-3x$ is large and negative, and e raised to a large negative power is very small.

31. See Section 1.5 and the graph of $y = \ln x$ on page 51.

$$\lim_{x\to\infty} \ln 2x = \lim_{x\to\infty} (\ln 2 + \ln x)$$

$$= \ln 2 + \lim_{x\to\infty} \ln x = \infty$$

33. See Example 5.5.

$$\lim_{x\to-\infty} \frac{2x}{x^2+3x-5} = \lim_{x\to-\infty} \frac{2x}{x^2\left(1+\frac{3}{x}+\frac{5}{x^2}\right)}$$

$$= \lim_{x\to-\infty} \frac{2}{x\left(1+\frac{3}{x}+\frac{5}{x^2}\right)}$$

$$= 0$$

35. See Example 5.8.

$$\lim_{x\to-\infty} \tan^{-1} 2x = -\frac{\pi}{2}$$

37. See Example 3.9.

$$0 \le \frac{x^2}{x^2+1} < 1$$

$$-2|x| \frac{2x^3}{x^2+1} < 2|x|$$

$$\lim_{x\to 0} -2|x| = 0; \lim_{x\to 0} 2|x| = 0$$

By the Squeeze Theorem,

$$\lim_{x\to 0} \frac{2x^3}{x^2+1} = 0.$$

(This problem can also be done, perhaps more easily, without using the Squeeze Theorem . . .)

39. See Section 1.4 and Example 4.4.

$$f(x) = \frac{x-1}{x^2+2x-3} = \frac{x-1}{(x+3)(x-1)}$$

f has a non-removable discontinuity at $x = -3$ and a removable discontinuity at $x = 1$.

41. See Examples 4.2 and 4.3.

$$\lim_{x\to 0^-} f(x) = \lim_{x\to 0^-} \sin x = 0$$

$$\lim_{x\to 0^+} f(x) = \lim_{x\to 0^+} x^2 = 0$$

$$\lim_{x\to 2^-} f(x) = \lim_{x\to 2^-} x^2 = 4$$

$$\lim_{x\to 2^+} f(x) = \lim_{x\to 2^+} (4x-3) = 5$$

f has a non-removable discontinuity at $x = 2$.

43. See Examples 4.4 and 4.6.

$$f(x) = \frac{x+2}{x^2-x-6} = \frac{x+2}{(x-3)(x+2)}$$

continuous on $(-\infty, -2)$, $(-2, 3)$, $(3, \infty)$

45. See Example 4.5.

$$f(x) = \sin(1+e^x)$$

continuous $(-\infty, \infty)$

47. See Section 1.5.

$$f(x) = \frac{x+1}{(x-2)(x-1)}$$

VA: $x = 1$, $x = 2$

$$\lim_{x\to\pm\infty} \frac{x+1}{x^2-3x+2} = \lim_{x\to\pm\infty} \frac{x\left(1+\frac{1}{x}\right)}{x^2\left(1-\frac{3}{x}+\frac{2}{x^2}\right)}$$

$$= \lim_{x\to\pm\infty} \frac{1+\frac{1}{x}}{x\left(1-\frac{3}{x}+\frac{2}{x^2}\right)}$$

$$= 0$$

HA: $y = 0$

49. See Section 1.5.

$$f(x) = \frac{x^2}{x^2-1} = \frac{x^2}{(x+1)(x-1)}$$

VA: $x = -1$, $x = 1$

$$\lim_{x\to\pm\infty} \frac{x^2}{x^2-1} = \lim_{x\to\pm\infty} \frac{x^2}{x^2\left(1-\frac{1}{x^2}\right)}$$

$$= \lim_{x\to\pm\infty} \frac{1}{1-\frac{1}{x^2}}$$

$$= \frac{1}{1}$$

$$= 1$$

HA: $y = 1$

51. See Section 1.5.
$\lim_{x\to 0^+} 2e^{1/x} = \infty$, so $x = 0$ is a vertical asymptote

$$\lim_{x\to\infty} 2e^{1/x} = 2, \qquad \lim_{x\to-\infty} 2e^{1/x} = 2,$$

so $y = 2$ is HA.

53. See Section 1.5. VA when $e^x = 2$, that is, $x = \ln 2$.

$$\lim_{x\to\infty} \frac{3}{e^x-2} = 0, \qquad \lim_{x\to-\infty} \frac{3}{e^x-2} = -\frac{3}{2},$$

so $x = 0$ and $x = -3/2$ are HA.

55. See Example 6.6.
In the table below, the middle column represents values calculated using

$$f(x) = \frac{1-\cos x}{2x^2}.$$

We can rewrite $f(x)$ as

$$f(x) = \frac{1-\cos x}{2x^2} = \left(\frac{1-\cos x}{2x^2}\right)\left(\frac{1+\cos x}{1+\cos x}\right)$$

$$= \frac{1-\cos^2 x}{2x^2(1+\cos x)} = \frac{\sin^2 x}{2x^2(1+\cos x)}.$$

Using this new form of the function, we get the values listed in the third column of the table. Note that $f(x) = f(-x)$, and so we get the same values when x is negative. Hence we conjecture that

$$\lim_{x\to 0} \frac{1-\cos x}{2x^2} = \frac{1}{4}.$$

x	old $f(x)$	new $f(x)$
1	0.229849	0.229849
0.1	0.249792	0.249792
0.01	0.249998	0.249998
0.001	0.250000	0.250000
0.0001	0.250000	0.250000
0.00001	0.250000	0.250000
0.000001	0.250022	0.250000
0.0000001	0.249800	0.250000
0.00000001	0.000000	0.250000
0.000000001	0.000000	0.250000

Here are two calculator graphs with different windows:

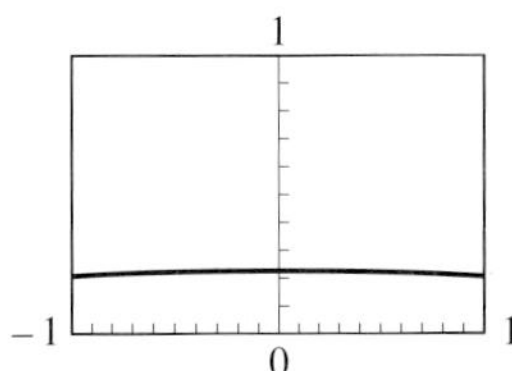

using $[-1, 1]$ by $[0, 1]$, and

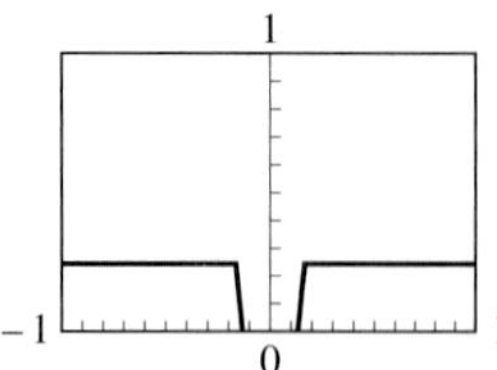

using $[-1 \times 10^{-5}, 1 \times 10^{-5}]$ by $[0, 1]$.

57. The limit of θ' as x approaches 0 is 66 radians per second, far faster than the player can maintain focus. From about 12 feet on in to the plate the player can't keep her eye on the ball.

Chapter 2

Differentiation

2.1 Tangent Lines and Velocity

1.

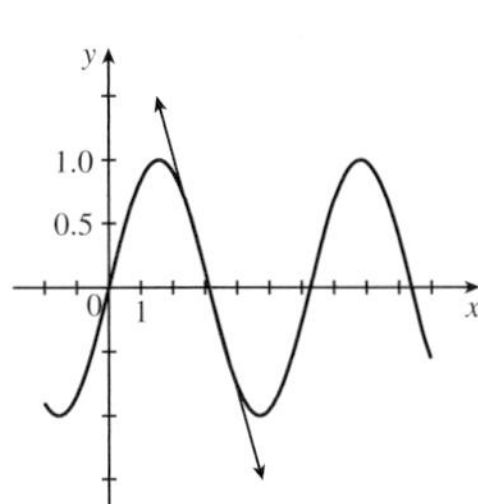

3.

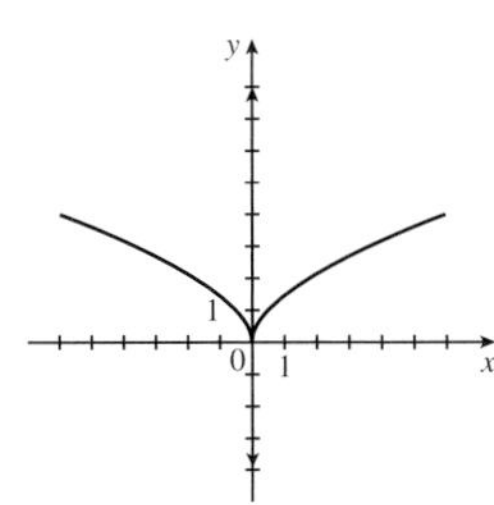

The tangent line is the y-axis.

5. At $x = 1$ the slope of the tangent line appears to be about -1.

7. C, B, A, D. At the point labeled C, the slope is very steep and negative. At point B, the slope is zero and at point A, the slope is just more than zero. The slope of the line tangent to point D is large and positive.

9. $f(x) = x^3 - x$

(a) $$m_{\text{sec}} = \frac{f(2) - f(1)}{2 - 1} = \frac{6 - 0}{1} = 6$$

(b) $$m_{\text{sec}} = \frac{f(3) - f(2)}{3 - 2} = \frac{24 - 6}{1} = 18$$

(c) $$m_{\text{sec}} = \frac{f(2) - f(1.5)}{2 - 1.5} = \frac{6 - 1.875}{.5} = 8.25$$

(d) $$m_{\text{sec}} = \frac{f(2.5) - f(2)}{2.5 - 2} = \frac{13.125 - 6}{.5} = 14.25$$

(e) $$m_{\text{sec}} = \frac{f(2) - f(1.9)}{2 - 1.9} = \frac{6 - 4.959}{.1} = 10.41$$

(f) $$m_{\text{sec}} = \frac{f(2.1) - f(2)}{2.1 - 2} = \frac{7.161 - 6}{.1} = 11.61$$

(g) tangent line slope estimate: 11 (averaging (e) and (f))

11. $f(x) = \cos x^2$

(a) $$m_{\text{sec}} = \frac{f(2) - f(1)}{2 - 1} = \frac{-.65 - .54}{1} = -1.19$$

(b) $$m_{\text{sec}} = \frac{f(3) - f(2)}{3 - 2} = \frac{-.91 - (-.65)}{1} = -.26$$

(c) $$m_{\text{sec}} = \frac{f(2) - f(1.5)}{2 - 1.5} = \frac{-.654 - (-.628)}{.5} = -.05$$

(d) $$m_{\text{sec}} = \frac{f(2.5) - f(2)}{2.5 - 2} = \frac{1.00 - (-.65)}{.5} = 3.3$$

(e) $$m_{\text{sec}} = \frac{f(2) - f(1.9)}{2 - 1.9} = \frac{-.65 - (-.89)}{.1} = 2.4$$

(f) $$m_{\text{sec}} = \frac{f(2.1) - f(2)}{2.1 - 2} = \frac{-.298 - (-.654)}{.1} = 3.6$$

(g) tangent line slope estimate: 3.0 (averaging **(e)** and **(f)**)

13.

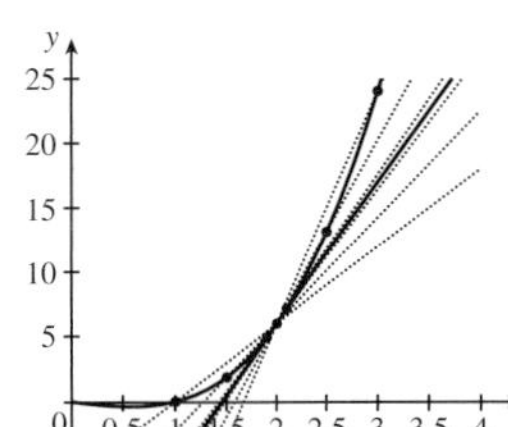

15. The answer is an animation.

17. $f(x) = x^2 - 2,\ a = 1$. We note first that $f(a) = f(1) = -1$. Then we compile the difference quotient for h not zero:

$$\frac{f(a+h)-f(a)}{h} = \frac{f(1+h)-f(1)}{h} = \frac{(1+h)^2 - 2 - (-1)}{h} = \frac{h^2+2h}{h} = h+2.$$

Now we let $h \to 0$ and get a limit of 2, which is the slope $m_{\tan}$ of the tangent line in question. The equation of the tangent line by the point-slope method is

$$y - (-1) = 2(x-1) \quad \text{or} \quad y = 2x - 3.$$

Comment: For several of the forthcoming problems, the task is to compute the slope by means of a difference quotient (DQ) and limit. To be absolutely certain about terminology, the difference quotient at *basepoint* $x = a$ and *increment* h is the ratio

$$DQ = \frac{\Delta f}{\Delta x} = \frac{f(a+h)-f(a)}{h}$$

in which the symbols in the center are code for the symbols on the right. Note that the difference quotient is not defined for $h = 0$, having the form $0/0$. However, it is typical of these problems that when properly simplified, the numerator Δf will contain a factor of h, which can cancel the denominator. The resulting simplified expression is often continuous and the limit can then be taken by substitution.

The simplest example of this phenomenon is the case in which f is initially a polynomial in x. In that case Δf is a polynomial in h, but having the value zero when $h = 0$, it must contain a factor of $h - 0 = h$. This is exactly what happened in this Exercise 17. The same principle applies in general to a rational function, and with care and "rationalization" it will also apply to problems with a square root. The advice from this corner is to avoid rushing the process. Take your time simplifying the difference quotient before ever taking any limit.

19. $f(x) = x^2 - 3x$, $a = -2$. Noting $f(-2) = 10$, and assuming h is not zero,

$$DQ = \frac{\Delta f}{\Delta x} = \frac{f(-2+h)-f(-2)}{h}$$
$$\frac{(-2+h)^2 - 3(-2+h) - (10)}{h}$$
$$\frac{4 - 4h + h^2 + 6 - 3h - 10}{h}$$
$$\frac{-7h + h^2}{h} = -7 + h.$$

At this point, we have an algebraic formula that is automatically continuous on its natural domain. The domain for the formula includes zero where the domain of the original DQ did not. The limit as $h \to 0$ is now the same as the value, clearly -7, and this is the slope. The equation of the tangent line is

$$y - 10 = -7(x+2) \quad \text{or} \quad y = -7x - 4.$$

21. $f(x) = \dfrac{2}{x+1}$, $a = 1$. Noting $f(1) = 1$, and assuming h is not zero, we have

$$DQ = \frac{f(1+h)-f(1)}{h} = \frac{\frac{2}{(1+h)+1} - \frac{2}{1+1}}{h} = \frac{\frac{2}{2+h} - 1}{h} = \frac{\left(\frac{2-(2+h)}{2+h}\right)}{h} = \frac{\left(\frac{-h}{2+h}\right)}{h} = \frac{-1}{2+h}.$$

Again we have at the end an algebraic formula whose natural domain includes $h = 0$. The limit is $-(1/2)$, which is also the slope of the tangent line. The equation of the tangent line is

$$y - 1 = \frac{1}{2}(x-1) \quad \text{or} \quad y = -\frac{x}{2} + \frac{3}{2}.$$

23. $f(x) = \sqrt{x+3}$, $a = -2$. Noting $f(-2) = 1$, and assuming h is not zero,

$$
\begin{aligned}
DQ &= \frac{f(-2+h)-f(-2)}{h} \\
&= \frac{\sqrt{(-2+h)+3}-1}{h} \\
&= \frac{\sqrt{h+1}-1}{h} \\
&= \frac{\sqrt{h+1}-1}{h}\cdot\frac{\sqrt{h+1}+1}{\sqrt{h+1}+1} \\
&= \frac{(h+1)-1}{h(\sqrt{h+1}+1)} = \frac{1}{\sqrt{h+1}+1}.
\end{aligned}
$$

The fourth step was the step of "rationalization" that we anticipated in the discussion following Exercise 17. Like the others, in this final simplified form for which $h=0$ is an allowable input, we can get the limit by substitution of $h=0$. The limit is $1/(1+1)=1/2$, which is the slope. The equation of the tangent line is

$$y-1=\frac{1}{2}(x+2) \quad \text{or} \quad y=\frac{1}{2}x+2.$$

25. $f(x)=|x-1|$, $a=1$. Here, $f(1)=0$. A quick graph makes it apparent that this function has a "corner" at $x=1$, which makes us suspect that the right and left slopes are different. To confirm this, we go

$$\lim_{h\to0^+}\frac{f(1+h)-f(1)}{h}=$$

$$\lim_{h\to0^+}\frac{|1+h-1|}{h}=\lim_{h\to0^+}\frac{|h|-0}{h}=$$

$$\lim_{h\to0^+}\left(\frac{h}{h}\right)=\lim_{h\to0^+}(1)=1$$

$$\lim_{h\to0^-}\frac{f(1+h)-f(1)}{h}=\lim_{h\to0^-}\frac{|(1+h)-1|}{h}=$$

$$\lim_{h\to0^-}\frac{|h|}{h}=\lim_{h\to0^-}\left(\frac{-h}{h}\right)=\lim_{h\to0^-}(-1)=-1.$$

Since these one-sided slopes are different, the (two-sided) limit does not exist and there is no tangent line.

27. $f(x)=\begin{cases}-2x^2 & \text{if } x<0\\ x^3 & \text{if } x\ge 0\end{cases}$ at $a=0$.

Here, $f(0)=0$, but anytime there are two formulae involved, we must compile the right and left slopes independently:

$$
\begin{aligned}
\lim_{h\to0^+}\frac{f(0+h)-f(0)}{h} &= \lim_{h\to0^+}\frac{h^3-0^3}{h} \\
&= \lim_{h\to0^+}h^2=0
\end{aligned}
$$

and

$$
\begin{aligned}
\lim_{h\to0^-}\frac{f(0+h)-f(0)}{h} &= \lim_{h\to0^-}\frac{-2h^2-0}{h} \\
&= \lim_{h\to0^-}-2h=0.
\end{aligned}
$$

These one-sided slopes are the same, and their common value (0) *is* the slope of the tangent line.

29. $f(x)=\begin{cases}x^2-1 & \text{if } x<0\\ x^3+1 & \text{if } x\ge 0\end{cases}$ at $a=0$.

We note that $f(0)=1$ from the second formula, but this does not exactly fit the first formula. We suspect that the left slope (the mismatched side) does not exist. We find

$$\lim_{h\to0^-}\frac{f(0+h)-f(1)}{h}=\lim_{h\to0^-}\frac{h^2-1-(1)}{h}=$$

$$\lim_{h\to0^-}\frac{h^2-2}{h}=\lim_{h\to0^-}\left(h-\frac{2}{h}\right).$$

This limit does not exist, and there is no tangent line.

Comment: The "mismatching" (referred to above) means that for the original function (not the DQ) the *left limit* as $x\to0^-$ is not the *value* $(+1)$. This means that in hindsight the function is not continuous at $x=0$. There is a theorem in the next section that will tell us that whenever the function is discontinuous, there can be no "slope."

31. $f(t)=16t^2+10$. Noting $f(0)=10$, $f(1)=26$, $f(2)=74$, $f(1.9)=67.76$, and $f(1.99)=73.3616$, we find

(a) $$v_{avg}=\frac{f(2)-f(0)}{2-0}=\frac{64-0}{2}=32,$$

(b) $$v_{avg}=\frac{f(2)-f(1)}{2-1}=\frac{64-26}{1}=48,$$

(c)
$$v_{avg} = \frac{f(2) - f(1.9)}{2 - 1.9} = \frac{74 - 67.76}{.1} = \frac{6.24}{.1} = 62.4,$$

(d)
$$v_{avg} = \frac{f(2) - f(1.99)}{2 - 1.99} = \frac{74 - 73.3616}{.01} = \frac{.6384}{.01} = 63.84,$$

(e) On the basis of the trend in these numbers, one might guess that they were heading to 64. Although not strictly asked, this can be confirmed:

$$\begin{aligned} v(2) &= \lim_{h\to 0} \frac{f(2+h) - f(2)}{h} \\ &= \lim_{h\to 0} \frac{16(2+h)^2 + 10 - (74)}{h} \\ &= \lim_{h\to 0} \frac{16(4 + 4h + h^2) + 10 - 74}{h} \\ &= \lim_{h\to 0} \frac{64h + 16h^2}{h} \\ &= \lim_{h\to 0} 64 + 16h = 64. \end{aligned}$$

33. $f(t) = \sqrt{t^2 + 8t}$. Noting $f(0) = 0$, $f(1) = 3$, $f(2) = \sqrt{20} = 4.47213\ldots$, $f(1.9) = \sqrt{18.81} = 4.337049\ldots$, and $f(1.99) = \sqrt{19.8801} = 4.45871\ldots$, we find

(a)
$$v_{avg} = \frac{f(2) - f(0)}{2 - 0} = \frac{\sqrt{4 + 16} - \sqrt{0}}{2} = \frac{\sqrt{20}}{2} = \sqrt{5} = 2.236068,$$

(b)
$$v_{avg} = \frac{f(2) - f(1)}{2 - 1} = \frac{\sqrt{20} - \sqrt{1 + 8}}{1} = \sqrt{20} - 3 = 1.472136,$$

(c)
$$\begin{aligned} v_{avg} &= \frac{f(2) - f(1.9)}{2 - 1.9} \\ &= \frac{\sqrt{20} - \sqrt{1.9^2 + 8(1.9)}}{.1} \\ &= \frac{\sqrt{20} - \sqrt{18.8}}{.1} = \frac{.1350862}{.1} \\ &= 1.3508627, \end{aligned}$$

(d)
$$\begin{aligned} v_{avg} &= \frac{f(2) - f(1.99)}{2 - 1.99} \\ &= \frac{\sqrt{20} - \sqrt{(1.99)^2 + 8(1.99)}}{2 - 1.99} \\ &= \frac{\sqrt{20} - \sqrt{19.8801}}{.01} = \frac{.0134253}{.01} \\ &= 1.3425375, \end{aligned}$$

(e) These numbers do not permit exact anticipation of the limit, but a guess around 1.34 is reasonable. To make an exact determination, we must rationalize a DQ as follows: assume $t < 2$, and consider v_{avg} for the interval $[t, 2]$. This would be

$$\begin{aligned} \frac{f(2) - f(t)}{2 - t} &= \frac{\sqrt{20} - \sqrt{t^2 + 8t}}{2 - t} \\ &= \frac{\sqrt{20} - \sqrt{t^2 + 8t}}{2 - t} \cdot \frac{\sqrt{20} + \sqrt{t^2 + 8t}}{\sqrt{20} + \sqrt{t^2 + 8t}} \\ &= \frac{20 - (t^2 + 8t)}{(2 - t)(\sqrt{20} + \sqrt{t^2 + 8t})} \\ &= \frac{(2 - t)(10 + t)}{(2 - t)(\sqrt{20} + \sqrt{t^2 + 8t})} \\ &= \frac{(10 + t)}{\sqrt{20} + \sqrt{t^2 + 8t}}. \end{aligned}$$

In this form, we have added $t = 2$ to the domain of the formula, and can take the limit by substitution of $t = 2$, coming to

$$v(2) = \frac{12}{2\sqrt{20}} = \frac{3}{\sqrt{5}} = 1.34164\ldots.$$

35. $f(t) = -16t^2 + 5$, velocity at $a = 1$. Noting that $f(1) = -11$, and assuming that h is not zero, we write

$$\begin{aligned} v_{avg} = DQ &= \frac{f(1+h) - f(1)}{h} \\ &= \frac{-16(1+h)^2 + 5 - (-11)}{h} \\ &= \frac{-16 - 32h - 16h^2 + 5 + 11}{h} \\ &= \frac{-32h - 16h^2}{h} \\ &= -32 - 16h. \end{aligned}$$

As usual, we can now take the limit by setting $h = 0$, finding that $v(2) = -32$.

37. $f(t) = \sqrt{t + 16}$, velocity at $a = 0$. Noting that $f(0) = 4$, and assuming h is not zero, we write

$$\begin{aligned} v_{avg} = DQ &= \frac{f(0+h) - f(0)}{h} \\ &= \frac{\sqrt{h+16} - 4}{h} \\ &= \frac{\sqrt{h+16} - 4}{h} \cdot \frac{\sqrt{h+16} + 4}{\sqrt{h+16} + 4} \\ &= \frac{(h+16) - 16}{h(\sqrt{h+16} + 4)} = \frac{1}{\sqrt{h+16} + 4}. \end{aligned}$$

As usual, we can now take the limit by setting $h = 0$, finding that $v(0) = 1/8$.

39. The hiker reached the top at the highest point on the graph (about 1.75 hours). The hiker was going the fastest on the way up at this point. The hiker was going the fastest on the way down at the point where the tangent line has the least (i.e., most negative) slope, at about 3 hours, at the end of the hike. Where the graph is level, the hiker was either resting or walking on flat ground. Q: Because the flat interval was not repeated, is it more likely he/she was resting? Perhaps, but only if we assume the descent route was the same as the ascent, an unwarranted assumption.

41.

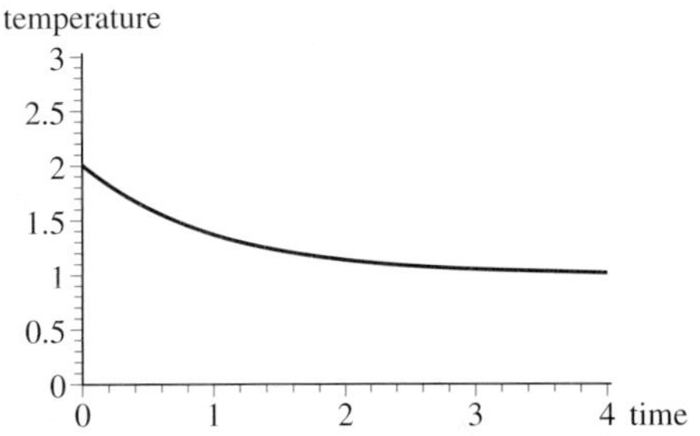

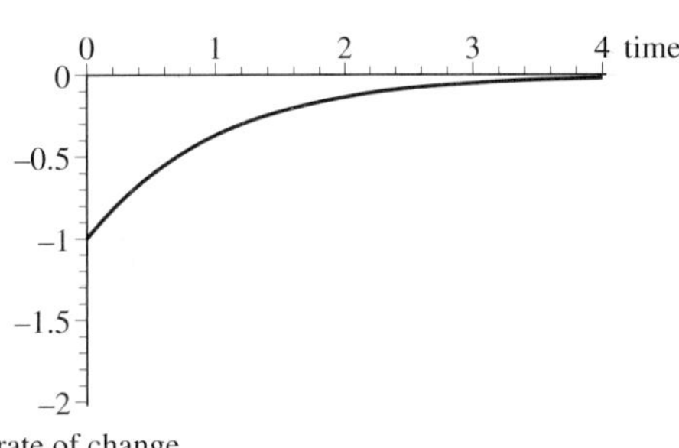

43. In general, $DQ = \dfrac{f(b) - f(a)}{b - a}$ is the average rate of change of the function $f(t)$ over the time-interval $[a, b]$.

(a) This is the case $b = 4$, $a = 2$. To say that

$$\frac{f(4) - f(2)}{2} = 21{,}034$$

per year is to say that the average rate of change in the bank balance between Jan. 1, 2002 and Jan. 1, 2004 was 21,034 ($ per year).

(b) This is the case $b = 4$, $a = 3.5$. To say that

$$2[f(4) - f(3.5)] = \frac{f(4) - f(3.5)}{1/2} = 25{,}036$$

per year is to say that the average rate of change between July 1, 2003 and Jan. 1, 2004 was 25,036 ($ per year).

(c) To say that

$$\lim_{h\to 0} \frac{f(4+h)-f(4)}{h} = \$30{,}000$$

is to say that the instantaneous rate of change in the balance on Jan. 1, 2004 was 30,000 ($ per year).

45. We are given $\theta(t) = 0.4t^2$. We are advised that θ is measured in radians, and that t is time. Let us assume that t is measured in seconds. It *has to be the case* that the factor of 0.4 has the implicit units of radians per square second. It appears to this observer that the *hint* contains a typographical error in which the symbol t appears where θ was intended. The answer to the rhetorical question in the hint is that three rotations corresponds to $\theta = 6\pi$. Proceeding, if

$$\begin{aligned} \theta(t) &= 6\pi, \text{ then} \\ 0.4t^2 &= 6\pi, \\ t^2 &= \frac{6}{.4}\pi = 15\pi, \\ t &= \sqrt{15\pi} \approx 6.865 \text{ (seconds)}. \end{aligned}$$

(No need to worry about the negative.)
At that exact moment of time (call it a) , the exact angular velocity is

$$\begin{aligned} &\lim_{h\to 0} \frac{\theta(a+h)-\theta(a)}{h} \\ &= \lim_{h\to 0} \frac{.4(\sqrt{15\pi}+h)^2 - 6\pi}{h} \\ &= \lim_{h\to 0} \frac{.4(15\pi + 2h\sqrt{15\pi} + h^2) - 6\pi)}{h} \\ &= \lim_{h\to 0} \frac{.8h\sqrt{15\pi} + .4h^2}{h} \\ &= \lim_{h\to 0} .8\sqrt{15\pi} + .4h = .8\sqrt{15\pi} \approx 5.492 \end{aligned}$$

and the units would be *radians per second.*

As regards the answer appendix, there are two things happening: (1) t rather than θ has been taken to be 6π; (2) the 15 isn't really 15, its actually $(4.8)\pi$ which is about 15.0796 . . . .

47. With $r < s$, for the time-interval $[r, s]$, we have

$$\begin{aligned} v_{avg} &= \frac{f(s)-f(r)}{s-r} \\ &= \frac{as^2 + bs + c - (ar^2 + br + c)}{s-r} \\ &= \frac{a(s^2 - r^2) + b(s-r)}{s-r} \\ &= \frac{a(s+r)(s-r) + b(s-r)}{s-r} \\ &= a(s+r) + b. \end{aligned}$$

This formula is good for all pairs (r, s) with $r < s$. Next, fixing r, taking a small positive h, and taking the case $s = r + h$, so that $s + r = 2r + h$ and $s - r = h$, we find

$$\begin{aligned} v(r) &= \lim_{h\to 0} \frac{f(r+h)-f(r)}{h} \\ &= \lim_{h\to 0} a(2r+h) + b = 2ar + b. \end{aligned}$$

This formula is good for *all* r, so it also good for s. Meanwhile, back in full generality $(r < s)$, we see that the average of the endpoint velocities is

$$\begin{aligned} \frac{v(r)+v(s)}{2} &= \frac{(2ar+b)+(2as+b)}{2} \\ \frac{2a(s+r)+2b}{2} &= a(s+r) + b = v_{avg}. \end{aligned}$$

49. Let $x = h + a$. Then $h = x - a$, and clearly

$$DQ_1 = \frac{f(a+h)-f(a)}{h}$$

is the exact same ratio as

$$DQ_2 = \frac{f(x)-f(a)}{x-a}.$$

It is also clear that $x \to a$ if and only if $h \to 0$. Therefore if one of the two limits exists, then so does the other and

$$\lim_{x\to a} DQ_2 = \lim_{h\to 0} DQ_1.$$

51. First, compute the slope of the tangent line. Using the result of Exercise 49, it is convenient to assume x is near but not exactly $1/2$ and write

$$DQ_2 = \frac{f(x) - f(1/2)}{x - (1/2)} = \frac{x^2 - (1/4)}{x - (1/2)} = \frac{(x - (1/2))(x + (1/2))}{x - (1/2)} = x + (1/2).$$

With this simplified formula for the DQ, we are free to find the limit by letting x be $1/2$, and find that the slope there is $(1/2) + (1/2) = 1$.

Next, we quickly write the equation of the tangent line in point-slope form:

$$y - (1/4) = 1(x - (1/2)) \quad \text{or} \quad y = x - (1/4).$$

The location of the tree is the point $(x, y) = (1, 3/4)$ and this point is indeed on the tangent line. The tree will be hit if the car gets that far (that being something we have no way of knowing).

2.2 The Derivative

1. $f(x) = 3x + 1$, $a = 1$, $f(a) = f(1) = 4$. Assuming h is not zero in the following steps, we find

$$\begin{aligned} f'(1) &= \lim_{h\to 0} \frac{f(1+h) - f(1)}{h} \\ &= \lim_{h\to 0} \frac{3(1+h) + 1 - (4)}{h} \\ &= \lim_{h\to 0} \frac{3h}{h} = \lim_{h\to 0} 3 = 3. \end{aligned}$$

Alternatively, assuming b is not 1,

$$\begin{aligned} \frac{f(b) - f(1)}{b - 1} &= \frac{3b + 1 - (3 + 1)}{b - 1} \\ &= \frac{3b - 3}{b - 1} = \frac{3(b - 1)}{b - 1} = 3 \end{aligned}$$

$$f'(1) = \lim_{b\to 1} \frac{f(b) - f(1)}{b - 1} = \lim_{b\to 1} 3 = 3.$$

3. $f(x) = \sqrt{3x + 1}$, $a = 1$, $f(1) = 2$. Assuming h is not zero, we write

$$\begin{aligned} \frac{f(1+h) - f(1)}{h} &= \frac{\sqrt{3(1+h) + 1} - 2}{h} \\ &= \frac{\sqrt{4 + 3h} - 2}{h} \cdot \frac{\sqrt{4 + 3h} + 2}{\sqrt{4 + 3h} + 2} \\ &= \frac{4 + 3h - 4}{h(\sqrt{4 + 3h} + 2)} = \frac{3h}{h(\sqrt{4 + 3h} + 2)} \\ &= \frac{3}{\sqrt{4 + 3h} + 2}. \end{aligned}$$

$$\begin{aligned} f'(1) &= \lim_{h\to 0} \frac{f(1+h) - f(1)}{h} \\ &= \lim_{h\to 0} \frac{3}{\sqrt{4 + 3h} + 2} = \frac{3}{\sqrt{4 + 3(0)} + 2} = \frac{3}{4}. \end{aligned}$$

Alternatively, assuming b is not 1,

$$\begin{aligned} \frac{f(b) - f(1)}{b - 1} &= \frac{\sqrt{3b + 1} - 2}{b - 1} \\ &= \frac{(\sqrt{3b + 1} - 2)(\sqrt{3b + 1} + 2)}{(b - 1)(\sqrt{3b + 1} + 2)} \\ &= \frac{(3b + 1) - 4}{(b - 1)\sqrt{3b + 1} + 2} \\ &= \frac{3(b - 1)}{(b - 1)\sqrt{3b + 1} + 2} \\ &= \frac{3}{\sqrt{3b + 1} + 2}. \end{aligned}$$

$$\begin{aligned} f'(1) &= \lim_{b\to 1} \frac{f(b) - f(1)}{b - 1} \\ &= \lim_{b\to 1} \frac{3}{\sqrt{3b + 1} + 2} = \frac{3}{\sqrt{4} + 2} = \frac{3}{4}. \end{aligned}$$

5. $f(x) = x^2 + 2x$, $a = 0$, $f(0) = 0$.

$$\begin{aligned} f'(0) &= \lim_{h\to 0} \frac{f(0+h) - f(0)}{h} \\ &= \lim_{h\to 0} \frac{f(h)}{h} \\ &= \lim_{h\to 0} \frac{h^2 + 2h}{h} = \lim_{h\to 0} (h + 2) = 2 \end{aligned}$$

7. $f(x)=x^3+4$, $a=-1$, $f(-1)=3$. Assuming b is not -1, we write

$$\frac{f(b)-f(-1)}{b-(-1)}=\frac{b^3+4-(3)}{b+1}$$
$$=\frac{b^3+1}{b+1}$$
$$=\frac{(b^2-b+1)(b+1)}{b+1}$$
$$=b^2-b+1.$$

$$f'(-1)=\lim_{b\to 1}\frac{f(b)-f(-1)}{b-(-1)}$$
$$=\lim_{b\to 1}b^2-b+1=3.$$

9. $f(x)=3x^2+1$. Assuming h is not zero,

$$\frac{f(x+h)-f(x)}{h}$$
$$=\frac{3(x+h)^2+1-(3(x)^2+1)}{h}$$
$$=\frac{3x^2+6xh+3h^2+1-(3x^2+1)}{h}$$
$$=\frac{6xh+3h^2}{h}=6x+3h.$$

$$f'(x)=\lim_{h\to 0}\frac{f(x+h)-f(x)}{h}$$
$$=\lim_{h\to 0}6x+3h=6x.$$

11. $f(x)=\dfrac{3}{x+1}$. Assuming b is not x, write

$$\frac{f(b)-f(x)}{b-x}=\frac{\frac{3}{b+1}-\frac{3}{x+1}}{b-x}$$
$$=\frac{\frac{3(x+1)-3(b+1)}{(b+1)(x+1)}}{b-x}$$
$$=\frac{-3(b-x)}{(b+1)(x+1)(b-x)}$$
$$=\frac{-3}{(b+1)(x+1)}.$$

$$f'(x)=\lim_{b\to x}\frac{f(b)-f(x)}{b-x}$$
$$=\lim_{b\to x}\frac{-3}{(b+1)(x+1)}=\frac{-3}{(x+1)^2}.$$

13. $f(x)=\sqrt{3x+1}$. Assuming b is not x, write

$$\frac{f(b)-f(x)}{b-x}=\frac{\sqrt{3b+1}-\sqrt{3x+1}}{b-x}$$
$$=\frac{(\sqrt{3b+1}-\sqrt{3x+1})(\sqrt{3b+1}+\sqrt{3x+1})}{(b-x)(\sqrt{3b+1}+\sqrt{3x+1})}$$
$$=\frac{(3b+1)-(3x+1)}{(b-x)(\sqrt{3b+1}+\sqrt{3x+1})}$$
$$=\frac{3(b-x)}{(b-x)(\sqrt{3b+1}+\sqrt{3x+1})}$$
$$=\frac{3}{(\sqrt{3b+1}+\sqrt{3x+1})}.$$

$$f'(1)=\lim_{b\to x}\frac{f(b)-f(x)}{b-x}$$
$$=\lim_{b\to x}\frac{3}{(\sqrt{3b+1}+\sqrt{3x+1})}$$
$$=\frac{3}{2\sqrt{3x+1}}.$$

15. $f(x)=x^3+2x1$. Assuming b is not x, write

$$\frac{f(b)-f(x)}{b-x}=\frac{b^3+2b-1-(x^3+2x-1)}{b-x}$$
$$=\frac{b^3-x^3+2b-2x}{b-x}$$
$$=\frac{(b-x)(b^2+bx+x^2+2)}{b-x}$$
$$=b^2+bx+x^2+2.$$
$$f'(x)=\lim_{b\to x}\frac{f(b)-f(x)}{b-x}$$
$$=\lim_{b\to x}b^2+bx+x^2+2=3x^2+2.$$

17. The function has negative slope for $x < 0$, positive slope for $x > 0$, and zero slope at $x = 0$. Its slope function (derivative) can only be **(c)**.

19. Here, moving from left to right, the slope goes from negative to positive to negative to positive. Its slope function (derivative) can only be **(a)**.

21. This function is somewhat similar to that of Exercise 19. In turn, **(b)** is vaguely like **(a)**. Such differences as can be seen between Exercise 19 and Exercise 21 and between **(b)** and **(a)** do not weaken the connection. **(b)** is right for Exercise 21.

23.

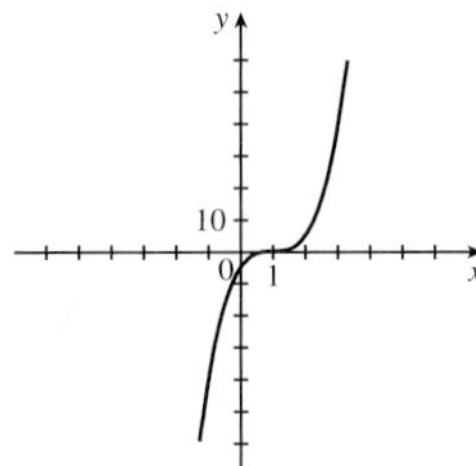

25.

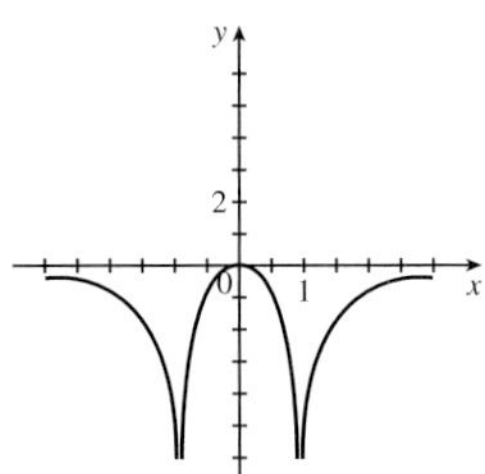

27.

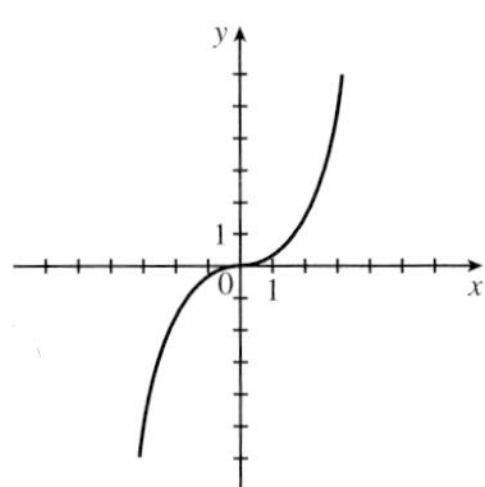

29. The function is not differentiable

at $x = -3$ (mysteriously undefined), or

at $x = -1$ (apparent vertical asymptote),

or at $x = 3$ (apparent vertical asymptote).

These are all points of discontinuity, which precludes the existence of a derivative (Theorem 2.1).

31.

x-interval	Slope of Secant Line
(0.2, 1.0)	4.25
(0.4, 1.0)	4.5
(0.6, 1.0)	4.75
(0.8, 1.0)	5.0
(1.0, 1.2)	5.0
(1.0, 1.4)	4.75
(1.0, 1.6)	4.5
(1.0, 1.8)	4.25

One expects the shorter intervals to give the better approximations. The best estimate available for the slope of the tangent line at $x = 1$ is 5.

33.

Time Interval	Average Velocity
(1.6, 2.0)	8.5
(1.7, 2.0)	9.0
(1.8, 2.0)	9.5
(1.9, 2.0)	10.0
(2.0, 2.1)	10.0
(2.0, 2.2)	9.5
(2.0, 2.3)	9.0
(2.0, 2.4)	8.5

Exactly as in Exercise 31 , our best estimate of the velocity at $t = 2$ is 10.

35. We compile the rate of change in Ton-MPG over each of the four two-year intervals for which data is given:

interval	rate of change
(1992, 1994)	$m = \dfrac{45.7 - 44.9}{2} = .4$
(1994, 1996)	$m = \dfrac{46.5 - 45.7}{2} = .4$
(1996, 1998)	$m = \dfrac{47.3 - 46.5}{2} = .4$
(1998, 2000)	$m = \dfrac{47.7 - 47.3}{2} = .2$

These rates of change are measured in Ton-MPG per year. Either the first or second (they happen to agree) could be used as an estimate for the one-year interval "1994" **(a)** while only the last is a promising estimate for the one-year interval "2000" **(b)**. The mere fact that all these numbers are positive suggests that efficiency is improving, but the last number being smaller seems to suggest that the rate of improvement is slipping.

37. We prepare a table of values for the function $f(x) = x^x$ (when x is near 1). Difference quotients based at $x = 1$ are then compiled in the last column.

n	$x = 1 + (.1)^n$	$y = x^x$	$(y-1)/(x-1)$
1	1.1000000	1.1105342	1.1053424
2	1.0100000	1.0101005	1.0100503
3	1.0010000	1.0010010	1.0010005
4	1.0001000	1.0001000	1.0001000
5	1.0000100	1.0000100	1.0000100
6	1.0000010	1.0000010	1.0000010
7	1.0000001	1.0000001	1.0000001

The evidence of this table strongly suggests that the difference quotients (essentially indistinguishable from the values themselves) are heading toward 1. If true, this would mean that $f'(1) = 1$.

39. $f(x) = \begin{cases} 2x+1 & \text{if } x < 0 \\ 3x+1 & \text{if } x \geq 0 \end{cases}$

$$D_+ f(0) = \lim_{h\to 0^+} \frac{f(h) - f(0)}{h} = \lim_{h\to 0^+} \frac{3h+1-1}{h} = 3$$

$$D_- f(0) = \lim_{h\to 0^-} \frac{f(h) - f(0)}{h} = \lim_{h\to 0^-} \frac{2h+1-1}{h} = 2$$

41. $f(x) = \begin{cases} g(x) & \text{if } x < 0 \\ k(x) & \text{if } x \geq 0 \end{cases}$

$f(x)$ is continuous at $x = 0$, and $g(x)$ and $k(x)$ are differentiable at $x = 0$.

$$D_+ f(0) = \lim_{h\to 0^+} \frac{f(h) - f(0)}{h} = \lim_{h\to 0} \frac{k(h) - k(0)}{h} = k'(0).$$

$$D_- f(0) = \lim_{h\to 0^-} \frac{f(h) - f(0)}{h} = \lim_{h\to 0} \frac{g(h) - g(0)}{h} = g'(0)$$

If $f(x)$ has a jump discontinuity at $x = 0$, it would be because its *left limit* at $x = 0$, namely $g(0)$, is not the same as the *value*, which is $k(0)$. In that case, there could be no left derivative (by Theorem 2.1) and one would have to reject the statement $D_- f(0) = g'(0)$.

43. One can say this much: If a function f has a strictly positive derivative at a point x, then the point x is surrounded by an interval (s, t) with the property that if $s < a < x < b < t$, then $f(a) < f(x) < f(b)$. To take b as an example, the ratio $\dfrac{f(b) - f(x)}{b - x}$ is fairly close to $f'(x)$ and therefore positive. The denominator being positive, the numerator must be positive as well, which is to say $f(x) < f(b)$.

This "locally increasing" property strongly suggests the stated increasing property of functions which have a positive derivative at all points of an interval, but the fairly simple formula

$$f(x) = \frac{x}{2} + x^2 \sin\left(\frac{1}{x}\right)$$

augmented by setting $f(0) = 0$, enables us to compile

$$f'(0) = \lim_{b\to 0} \frac{f(b) - f(0)}{b - 0} = \lim_{b\to 0} \frac{f(b)}{b} = \lim_{b\to 0} \frac{1}{2} + b \sin\left(\frac{1}{b}\right) = \frac{1}{2} > 0.$$

Despite this positive derivative at $x = 0$, this function has no interval of the form $(0, t)$ (no matter how small t) on which f is increasing. The interested student can compare the values at

$$x = \frac{1}{\left(2n + \frac{1}{2}\right)\pi} \text{ and } y = \frac{1}{\left(2n - \frac{1}{2}\right)\pi}$$

for large whole numbers n. Despite $x < y$, he/she will find $f(x) > f(y)$.

The example shows that there is more than single points involved in deducing the increasing nature of a function from a positive derivative. The proper tool for the inference is the so-called Mean-Value Theorem (Section 2.9), which, although not without credibility, cannot be *proved* at this introductory level.

45. $f(x) = x^{2/3}$

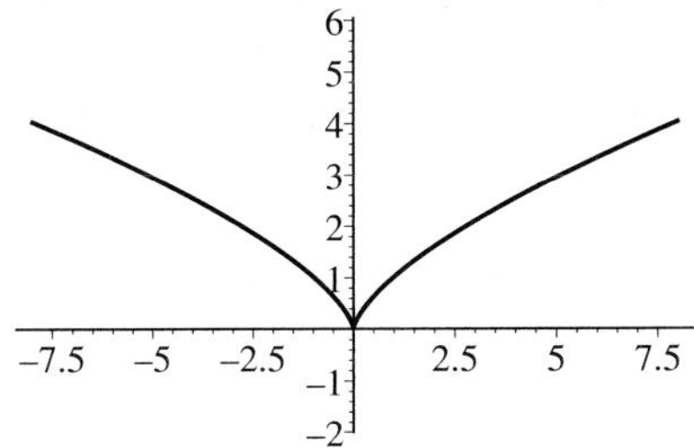

From the graph, we see that $f(x)$ appears continuous at $x = 0$, where it has both *limit* and *value* zero. However, when we try to compute its derivative at $x = 0$, we come to the difference quotient

$$\frac{f(0+h) - f(0)}{h} = \frac{f(h)}{h} = \frac{h^{2/3}}{h} = \frac{1}{h^{1/3}}.$$

Clearly this expression has no finite limit as h approaches zero. The numbers get large without bound. We do sometimes say that the vertical line $x = 0$ *is* the tangent line, but as a line it has no *slope* (just as the function has no derivative).

47. Let $f(x) = -1 - x^2$; then for all x, we have $f(x) \le x$. But at $x = -1$, we find $f(-1) = -2$ and

$$\begin{aligned} f'(-1) &= \lim_{h\to 0} \frac{f(-1+h) - f(-1)}{h} \\ &= \lim_{h\to 0} \frac{-1 - (-1+h)^2 - (-2)}{h} \\ &= \lim_{h\to 0} \frac{1 - (1 - 2h + h^2)}{h} \\ &= \lim_{h\to 0} \frac{2h - h^2}{h} = \lim_{h\to 0} 2 - h = 2. \end{aligned}$$

So, $f'(x)$ is not always less than 1.

49. **(a)** meters per second

(b) items per dollar

51. If $f'(t) < 0$, the function $f(t)$ is negatively sloped and decreasing, meaning the stock is losing value with the passing of time. This may be the basis for selling the stock if the current trend is expected to be a long-term one.

53. The following sketches are consistent with the hypotheses of infection rate rising, peaking, and returning to zero. We started with the derivative $I'(t)$ (infection rate) and had to think backwards to construct the function $I(t)$. One can see in $I(t)$ the slope increasing up to the time of peak infection rate, thereafter the *slope* decreasing but not the *values*. They merely level off.

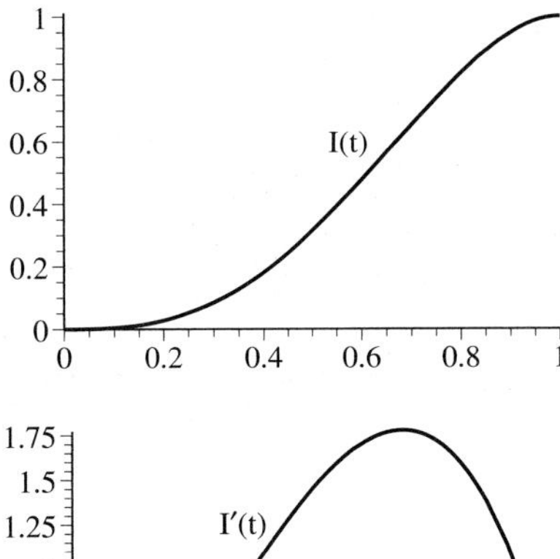

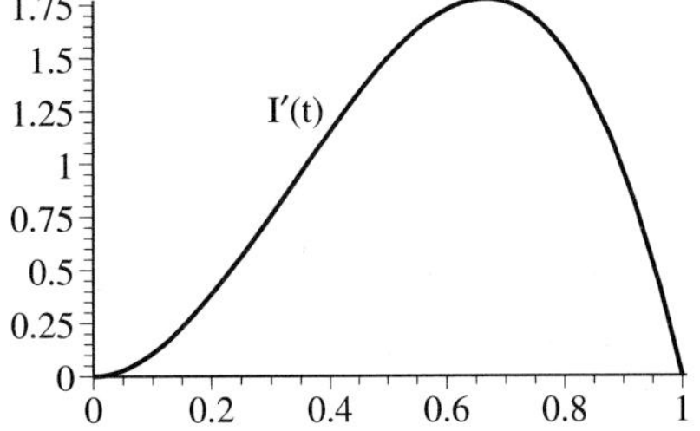

55. Because the curve appears to be bending upward, the slopes of the secant lines (based at $x = 1$ and with upper endpoint beyond 1) will increase with the upper endpoint. This has also the effect that any one of these slopes is greater than the actual derivative. Therefore,

$$f'(1) < \frac{f(1.5) - f(1)}{.5} < \frac{f(2) - f(1)}{1}.$$

As to where $f(1)$ fits in this list, it seems necessary to read the graph and come up with estimates of $f(1)$ about 4, and $f(2)$ about 7. That would put the third number in the above list at about 3, comfortably less than $f(1)$.

If one believes that the formula behind the graph is $f(x) = x^2 + 3$ (reasonable thought), one will come to the same conclusions and indeed find $f'(1) = 2$.

Comment: When we look at that graph, the tangent line that we would draw at $x = 1$ would be at just about 45 degrees to the horizontal. Before one jumps to the conclusion that its slope is 1, one has to note that the scale on the y-axis is 2 units for every unit on the x-axis. Our visual perception of 45 degrees is consistent with the numerical slope of 2.

57. This is a tricky one. It happens that for the function $f(x) = x^2 - x$, the value at $x = 1$ is *zero* ($f(1) = 0$)! Because of this fact,

$$\frac{(1+h)^2 - (1+h)}{h} = \frac{f(1+h) - f(1)}{h},$$

and the answer should be: $f(x) = x^2 - x$ and $a = 1$.

59. $\lim\limits_{h\to 0} \dfrac{\left(\frac{1}{2+h}\right) - \left(\frac{1}{2}\right)}{h}$ would be $f'(a)$ for $f(x) = \dfrac{1}{x}$ and $a = 2$.

61.

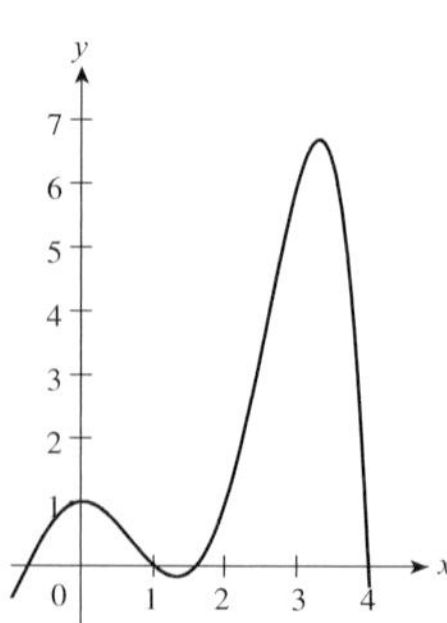

63. $f(t)$ is the price in cents of a t-minute call. There is some ambiguity, but we think the intent is that the ten-cent-per-minute-rate be applied to each of the minutes numbered 20 through 80, with the basic dollar-for-the-first-20-minutes still in effect. On this assumption we find

$$f(t) = \begin{cases} 100 & 0 < t \le 20 \\ 100 + 10(t - 20) & 20 < t \le 80 \\ 700 + 8(t - 80) & 80 < t < \infty \end{cases}.$$

This is another example of a *piecewise linear* function (this one is continuous), and although not differentiable at the transition times $t = 20$ or $t = 80$, elsewhere we have

$$f'(t) = \begin{cases} 0 & 0 < t < 20 \\ 10 & 20 < t < 80 \\ 8 & 80 < t < \infty \end{cases}.$$

The numbers 10 and 8, which are the slopes of two segments of the graph, can also be interpreted as "marginal-cost-per-minute" in their respective time intervals. Of course once the call has gone through, the caller has paid for the first 20 minutes whether he/she uses them or not, and the marginal-cost-per-minute in that interval is zero.

2.3 The Power Rule

1. $f(x) = x^3 - 2x + 1$

$$\begin{aligned} f'(x) &= \frac{d}{dx}(x^3) - \frac{d}{dx}(2x) + \frac{d}{dx}(1) \\ &= 3x^2 - 2\frac{d}{dx}(x) + 0 = 3x^2 - 2(1) \\ &= 3x^2 - 2 \end{aligned}$$

3. $f(x) = 3x^2 - 4$

$$\begin{aligned} f'(x) &= \frac{d}{dx}(3x^2) - \frac{d}{dx}(4) = 3\frac{d}{dx}(x^2) - 0 \\ &= 3(2x) = 6x \end{aligned}$$

5.

$$\begin{aligned} g(x) &= 4 \\ g'(x) &= \frac{d}{dx}(4) = 0 \end{aligned}$$

7. $f(t) = 3t^3 - 2\sqrt{t}$

$$\begin{aligned} f'(t) &= \frac{d}{dt}(3t^3) - \frac{d}{dt}\left(2\sqrt{t}\right) \\ &= 3\frac{d}{dt}(t^3) - 2\frac{d}{dt}\left(t^{1/2}\right) \\ &= 3(3t^2) - 2\left(\frac{1}{2}t^{-1/2}\right) = 9t^2 - \frac{1}{\sqrt{t}} \end{aligned}$$

9. $f(x) = \dfrac{3}{x} - 8x + 1$

$$\begin{aligned} f'(x) &= \frac{d}{dx}\left(\frac{3}{x}\right) - \frac{d}{dx}(8x) + \frac{d}{dx}(1) \\ &= 3\frac{d}{dx}(x^{-1}) - 8\frac{d}{dx}(x) + 0 \\ &= 3(-x^{-2}) - 8(1) = -\frac{3}{x^2} - 8 \end{aligned}$$

11. $h(x) = \dfrac{10}{\sqrt{x}} - 2x$

$$\begin{aligned} h'(x) &= \frac{d}{dx}\left(\frac{10}{\sqrt{x}}\right) - \frac{d}{dx}(2x) \\ &= 10\frac{d}{dx}\left(x^{-1/2}\right) - 2\frac{d}{dx}(x) \\ &= 10\left(-\frac{1}{2}x^{-3/2}\right) - 2(1) \\ &= -5x^{-3/2} - 2 = \frac{-5}{x\sqrt{x}} - 2 \end{aligned}$$

When fractional or negative exponents are not present in the statement of a problem, it is good policy to avoid them in the answer.

13. $f(s) = 2s^{3/2} - 3s^{-1/3}$

$$\begin{aligned} f'(s) &= \frac{d}{ds}\left(2s^{3/2}\right) - \frac{d}{ds}\left(3s^{-1/3}\right) \\ &= 2\frac{d}{ds}\left(s^{3/2}\right) - 3\frac{d}{ds}\left(s^{-1/3}\right) \\ &= 2\left(\frac{3}{2}s^{1/2}\right) - 3\left(-\frac{1}{3}s^{-4/3}\right) \\ &= 3s^{1/2} + s^{-4/3}\left(= 3\sqrt{s} + \frac{1}{\sqrt[3]{s^4}}\right) \end{aligned}$$

Here, fractional and negative exponents *were* present in the statement of the problem, so the last step is entirely optional.

15. $f(x) = 2\sqrt[3]{x} + 3$

$$\begin{aligned} f'(x) &= \frac{d}{dx}\left(2\sqrt[3]{x}\right) + \frac{d}{dx}(3) \\ &= 2\frac{d}{dx}\left(x^{1/3}\right) + 0 \\ &= 2\left(\frac{1}{3}x^{-2/3}\right) = \frac{2}{3}x^{-2/3} = \frac{2}{3\sqrt[3]{x^2}} \end{aligned}$$

17. $f(x) = x(3x^2 - \sqrt{x}) = 3x^3 - x^{3/2}$

$$\begin{aligned} f'(x) &= 3\frac{d}{dx}(x^3) - \frac{d}{dx}\left(x^{3/2}\right) \\ &= 3(3x^2) - \left(\frac{3}{2}x^{1/2}\right) \\ &= 9x^2 - \frac{3}{2}\sqrt{x} \end{aligned}$$

19.

$$\begin{aligned} f(x) &= \frac{3x^2 - 3x + 1}{2x} \\ &= \frac{3x^2}{2x} - \frac{3x}{2x} + \frac{1}{2x} \\ &= \frac{3}{2}x - \frac{3}{2} + \frac{1}{2}x^{-1} \end{aligned}$$

$$\begin{aligned} f'(x) &= \frac{d}{dx}\left(\frac{3}{2}x\right) - \frac{d}{dx}\left(\frac{3}{2}\right) + \frac{d}{dx}\left(\frac{1}{2}x^{-1}\right) \\ &= \frac{3}{2}\frac{d}{dx}(x) - 0 + \frac{1}{2}\frac{d}{dx}(x^{-1}) \\ &= \frac{3}{2}(1) + \frac{1}{2}(-1x^{-2}) = \frac{3}{2} - \frac{1}{2x^2} \end{aligned}$$

21. $f(x) = x^4 + 3x^2 - 2$

$$f'(x) = \frac{d}{dx}(x^4 + 3x^2 - 2) = 4x^3 + 6x$$

$$f''(x) = \frac{d}{dx}(4x^3 + 6x) = 12x^2 + 6$$

23. $f(x) = x^6 - \sqrt{x} = x^6 - x^{1/2}$

$$\frac{df}{dx} = \frac{d}{dx}\left(x^6 - x^{1/2}\right) = 6x^5 - \frac{1}{2}x^{-1/2}$$

$$\left(= 6x^5 - \frac{1}{2\sqrt{x}}\right)$$

$$\frac{d^2 f}{dx^2} = \frac{d}{dx}\left(6x^5 - \frac{1}{2}x^{-1/2}\right)$$

$$= 30x^4 - \frac{1}{2}\left(-\frac{1}{2}x^{-3/2}\right)$$

$$= 30x^4 + \frac{1}{4}x^{-3/2}$$

$$\left(= 30x^4 + \frac{1}{4x\sqrt{x}}\right)$$

In doing higher derivatives, it is wise to continue with fractional exponent notation until finished with the differentiation. Then one may wish to go back and repair the notation (in parentheses).

25. $f(t) = 4t^2 - 12 + \frac{4}{t^2} = 4t^2 - 12 + 4t^{-2}$

$$f'(t) = \frac{d}{dt}(4t^2 - 12 + 4t^{-2})$$

$$= 8t^2 - 0 + 4(-2t^{-3})$$

$$= 8t^2 - 8t^{-3}\left(= 8t^2 - \frac{8}{t^3}\right)$$

$$f''(t) = \frac{d}{dt}(8t - 8t^{-3})$$

$$= 8 - 8(-3t^{-4})$$

$$= 8 + 24t^{-4}\left(= 8 + \frac{24}{t^4}\right)$$

$$f'''(t) = \frac{d}{dt}(8 + 24t^{-4}) = 0 + 24(-4t^{-5})$$

$$= 96t^{-5}\left(= \frac{-96}{t^5}\right)$$

27.

$$f(x) = x^4 + 3x^2 - 2$$

$$f'(x) = 4x^3 + 6x$$

$$f''(x) = 12x^2 + 6$$

$$f'''(x) = 24x$$

$$f^{(4)}(x) \equiv 24 \text{ (constant)}$$

29. $f(x) = \frac{x^2 - x + 1}{\sqrt{x}} = x^{3/2} - x^{1/2} + x^{-1/2}$

$$f'(x) = \frac{d}{dx}\left(x^{3/2} - x^{1/2} + x^{-1/2}\right)$$

$$= \frac{3}{2}x^{1/2} - \frac{1}{2}x^{-1/2} - \frac{1}{2}x^{-3/2}$$

$$\left(= \frac{3x^2 - x - 1}{2x\sqrt{x}}\right)$$

$$f''(x) = \frac{d}{dx}\left(\frac{3}{2}x^{1/2} - \frac{1}{2}x^{-1/2} - \frac{1}{2}x^{-3/2}\right)$$

$$= \frac{3}{4}x^{-1/2} + \frac{1}{4}x^{-3/2} + \frac{3}{4}x^{-5/2}$$

$$\left(= \frac{3x^2 + x + 3}{4x^2\sqrt{x}}\right)$$

$$f'''(x) = \frac{d}{dx}\left(\frac{3}{4}x^{-1/2} + \frac{1}{4}x^{-3/2} + \frac{3}{4}x^{-5/2}\right)$$

$$= -\frac{3}{8}x^{-3/2} - \frac{3}{8}x^{-5/2} - \frac{15}{8}x^{-7/2}$$

$$\left(= -\frac{3(x^2 + x + 5)}{8x^3\sqrt{x}}\right)$$

31.

$$s(t) = -16t^2 + 40t + 10$$

$$v(t) = s'(t) = -32t + 40$$

$$a(t) = v'(t) = s''(t) = -32 \text{ (constant)}$$

33.

$$s(t) = \sqrt{t} + 2t = t^{1/2} + 2t$$

$$v(t) = s'(t) = \frac{1}{2}t^{-1/2} + 2(= \frac{1}{2\sqrt{t}} + 2)$$

$$a(t) = v'(t) = s''(t) = -\frac{1}{4}t^{-3/2}(= \frac{-1}{4t\sqrt{t}})$$

35.
$$h(t) = -16t^2 + 40t + 5, \ a = 1$$
$$v(t) = h'(t) = -32t + 40$$
$$v(1) = -32 + 40 = 8$$

Because $8 > 0$, the object is moving up.

$$a(t) = v'(t) = -32$$
$$a(1) = -32$$

The velocity is decreasing.

37.
$$h(t) = 10t^2 - 24t, \ a = 2$$
$$v(t) = h'(t) = 20t - 24$$
$$v(2) = 20(2) - 24 = 16$$

The object is moving up.

$$a(t) = v'(t) = 20$$
$$a(2) = 20$$

The object is gaining speed.

39.
$$f(x) = 4\sqrt{x} - 2x, \ a = 4$$
$$f(4) = 4\sqrt{4} - 2(4) = 0$$
$$f'(x) = \frac{d}{dx}\left(4x^{1/2} - 2x\right)$$
$$= 2x^{-1/2} - 2 \left(= \frac{2}{\sqrt{x}} - 2\right)$$
$$f'(4)1 = 1 - 2 = -1$$

The equation of the tangent line is

$$y = -1(x - 4) + 0 \quad \text{or} \quad y = -x + 4.$$

41.
$$f(x) = x^2 - 2, \ a = 2, \ f(2) = 2$$
$$f'(x) = 2x$$
$$f'(2) = 4$$

The equation of the tangent line is

$$y = 4(x - 2) + 2 \quad \text{or} \quad y = 4x - 6.$$

43.
$$f(x) = x^3 - 3x + 1$$
$$f'(x) = 3x^2 - 3$$

The tangent line to $y = f(x)$ is horizontal when

$$f'(x) = 0$$
$$3x^2 - 3 = 0$$
$$3(x^2 - 1) = 0$$
$$3(x + 1)(x - 1) = 0$$

$x = -1$ or $x = 1$.

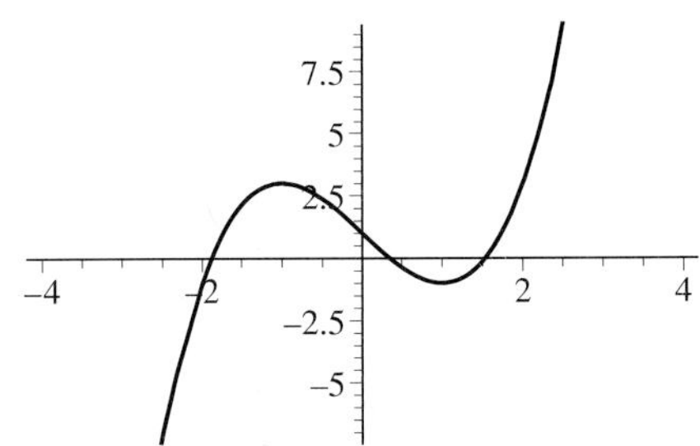

The graph shows that the first is a relative maximum, the second is a relative minimum.

45.
$$f(x) = x^{2/3}$$
$$f'(x) = \frac{2}{3}x^{-1/3} = \frac{2}{3\sqrt[3]{x}}$$

The slope of the tangent line to $y = f(x)$ does not exist where the derivative is undefined, which is only when $x = 0$.

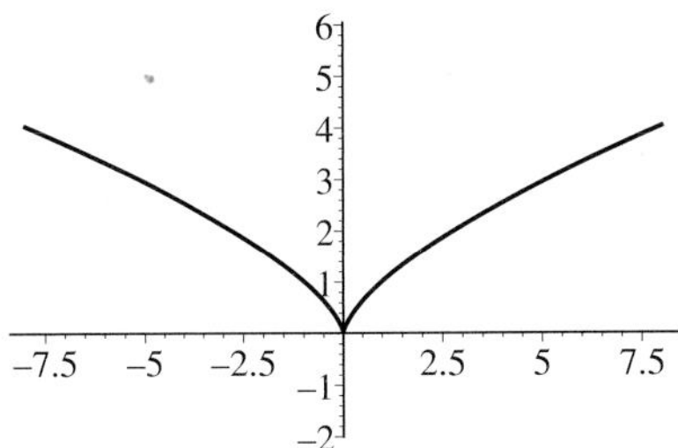

In this case, because the function is continuous, we might say that the tangent line is the vertical line $x = 0$. The feature at $x = 0$ is sometimes known as a *cusp*.

47. As regards the **(a)** function, its derivative would be negative for all negative x and positive for all positive x. Since so such function appears among the pictures, this **(a)** function has to be the one whose

derivative is absent from the list. There being no f''' in the list, **(a)** has to be f''.
This same **(a)** function is negative for a certain interval of the form $(-a, a)$, and the **(c)** function is decreasing on a similar type of interval. Thus the **(a)** function (f'') is apparently the derivative of the **(c)** function. It follows that **(c)** must be f'.

This leaves **(b)** for f itself, and our identifications are consistent in every respect.

49. $f(x) = \sqrt{x} = x^{1/2}$

$$f'(x) = \frac{1}{2}x^{-1/2}$$
$$f''(x) = \frac{1}{2}\left(-\frac{1}{2}\right)x^{-3/2}$$
$$f'''(x) = \left(\frac{1}{2}\right)\left(\frac{-1}{2}\right)\left(\frac{-3}{2}\right)x^{-5/2}$$
$$f^{(n)}(x) = (-1)^{n-1}\frac{\Pi_n}{2^n}x^{-(2n-1)/2}$$

in which Π_n is the product of the first $n-1$ *odd* integers (starting from 1 and ending at $2n-3$). Recall that the product of *all* the whole numbers from 1 to n is denoted by $n!$ If one were to multiply Π_n by product of the $n-1$ *even* numbers (from 2 to $2n-2$), one would get $(2n-2)!$ (in the numerator). Of course, one would have to do the same to the denominator, but this product of the new numbers could be written in the form $2^{n-1}(n-1)!$ A final form for an answer could be

$$f^{(n)}(x) = (-1)^{n-1}\frac{(2n-2)!}{2^{2n-1}(n-1)!}x^{(2n-1)/2}$$

Taking $n = 3$ as a test, the above expression is

$$(-1)^2\frac{4!}{2^5(2!)}x^{-5/2} = \frac{24}{2^6}x^{-5/2} = \frac{3}{8}x^{-5/2},$$

which checks.

51.
$$f(x) = ax^2 + bx + c \therefore\ f(0) = c$$
$$f'(x) = 2ax + b \quad \therefore\ f'(0) = b$$
$$f''(x) = 2a \text{ (constant)} \therefore\ f''(0) = 2a$$

Given $f''(0) = 3$, we learn $2a = 3$, $a = 3/2$. Given $f'(0) = 2$ we learn $2 = b$. And given $f(0) = -2$, we learn $c = -2$. In the end

$$f(x) = ax^2 + bx + c = \frac{3}{2}x^2 + 2x - 2.$$

53. With $f(x)$ given by cx, and h not zero, we find

$$\frac{f(x+h) - f(x)}{h} = \frac{c(x+h) - cx}{h} = \frac{ch}{h} = c.$$
$$f'(x) = \lim_{h\to 0}\frac{f(x+h) - f(x)}{h} = \lim_{h\to 0} c = c.$$

55. If $d(t)$ represents the national debt, then $d'(t)$ represents the rate of change of the national debt. The debt itself, by implication, is increasing and therefore $d'(t) > 0$.

Since the rate of increase has been reduced, this implies $d''(t)$ is being reduced. We cannot conclude anything about the size of $d(t)$.

57.
$$w(b) = cb^{3/2}$$
$$w'(b) = \frac{3c}{2}b^{1/2} = \frac{3c\sqrt{b}}{2}$$
$w'(b) > 1$ when
$$\frac{3c\sqrt{b}}{2} > 1,\ \sqrt{b} > \frac{2}{3c}$$
$$b > \frac{4}{9c^2}$$

It seems safe to assume that $b \geq w$ since the leg is attached to the body and almost by definition does not extend outside the span of the body. But under the conditions $b > \frac{4}{9c^2}$, we have

$$\frac{d}{db}(b - w) = 1 - w'(b) < 0.$$

This means that as b increases, the available space to attach the leg $(b - w)$ is decreasing, perhaps eventually to zero, after which further growth would be impossible. (Think of the massive thighs on T-Rex.)

59. We offer this table and will comment below.

time t	GDP $f(t)$	Growth during $[t-1,t]$ $\Delta f(t)$	Estimated Growth Rate est $f'(t)$	Change in Estimated Growth during $[t-1,t]$
1996	7664.8			
1997	8004.5	339.7	341.25	
1998	8347.3	342.8	343.10	1.85
1999	8690.7	343.4	334.75	−8.35
2000	9016.8	326.1	174.40	−160.35
2001	9039.5	22.7		

The numbers in the second column are differences from the first column, or growth *amounts*. Since the time period is exactly one year, these are estimators for the derivative. More precisely, as an estimator for $f'(t)$ one might use either the entry on the *same* line or the entry on the *next line down*. This works OK right up to the year 2000, when the two estimators are vastly discrepant. For that reason, we decided to take an average of the two estimators. Thus the fourth column averages the third column for the given line and the next line down. Settling on these (fourth column) as our best estimates for the derivative, we then compile in the fifth column differences from the fourth column, intending to use them as estimators for the *second* derivative. In discussing the time 2000, there is no point doing the averaging business again (column four), since we have only the one estimate anyway.

This observer finds the datum for the year 2001 to be extremely discordant with the others. It indicates a drastic and sudden drop-off in GDP (which could easily be the case in view of the historical record), but it renders the numerical estimations highly suspect.

61. Newton's Law states that force equals mass times acceleration. That is, if $F(t)$ is the driving force at time t, then $mf''(t) = ma(t) = F(t)$ in which m is the mass, appropriately unitized. The third derivative of the distance function is then $f'''(t) = a'(t) = \frac{1}{m}F'(t)$. It is both the derivative of the acceleration and directly proportional to the rate of change in force. Thus an abrupt change in acceleration or "jerk" is the direct consequence of an abrupt change in force.

63–67 Commentary: At this stage, finding a function whose derivative is given is a matter of thinking backward, or of anticipation. When the derivative is a power, one anticipates that it could have arisen from differentiating a function which was also a power, but whose exponent was one higher. That is, to get to x^p, try cx^{p+1}. After that, it is a matter of testing and adjusting the constant c. The answer is never unique, but anything offered can always be checked by differentiation.

63. $f'(x) = 4x^3$
Try $f(x) = cx^4$. Then $f'(x) = 4cx^3$, so c must be 1. One possible answer is x^4.

65. $f'(x) = 3x - 1$
Try $f(x) = cx^2 + dx$. Then $f'(x) = 2cx + d$, so $2c$ must be 3 and d must be -1.
One possible answer is $\frac{3}{2}x^2 - x$.

67. $f'(x) = \sqrt{x} = x^{1/2}$
$f(x) = \frac{2}{3}x^{3/2}$ is one possible function.

2.4 The Product and Quotient Rules

1.

$$f(x) = (x^2+3)(x^3-3x+1)$$

$$f'(x) = \left[\frac{d}{dx}(x^2+3)\right]\cdot(x^3-3x+1) + (x^2+3)\cdot\frac{d}{dx}(x^3-3x+1)$$

$$= (2x)(x^3-3x+1) + (x^2+3)(3x^2-3)$$

$$(= 5x^4 + 2x - 9)$$

Comment: The last step of simplification is a step that the student may wish to undertake, depending on what more has to be done. For example, if the problem is to find $f'(1)$, we would not simplify, but would immediately substitute the numbers, coming to $2(-1) + 4(0) = -2$. In many problems, however, it will be necessary to solve the equation $f'(x) = 0$, for which the simplified form is essential. In this section, where the emphasis is on using the product rule, we will not simplify on a regular basis, and even when we do, we will not show all the steps.

3.
$$f(x) = (3x+4)(x^3 - 2x^2 + x)$$
$$f'(x) = \left[\frac{d}{dx}(3x+4)\right](x^3 - 2x^2 + x) + (3x+4)\frac{d}{dx}(x^3 - 2x^2 + x)$$
$$= 3(x^3 - 2x^2 + x) + (3x+4)(3x^2 - 4x + 1)$$

5.
$$f(x) = (\sqrt{x} + 3x)\left(5x^2 - \frac{3}{x}\right) = (x^{1/2} + 3x)(5x^2 - 3x^{-1})$$
$$f'(x) = \left(\frac{1}{2}x^{-1/2} + 3\right)(5x^2 - 3x^{-1}) + (x^{1/2} + 3x)(10x + 3x^{-2})$$

7.
$$f(x) = \frac{3x-2}{5x+1}$$
$$f'(x) = \frac{\left(\begin{array}{l}(5x+1)\frac{d}{dx}(3x-2)\\ -(3x-2)\frac{d}{dx}(5x+1)\end{array}\right)}{(5x+1)^2}$$
$$= \frac{3(5x+1) - (3x-2)5}{(5x+1)^2}$$
$$= \frac{15x + 3 - 15x + 10}{(5x+1)^2} = \frac{13}{(5x+1)^2}$$

9.
$$f(x) = \frac{x-2}{x^2+x+1}$$
$$f'(x) = \frac{\left(\begin{array}{l}(x^2+x+1)\frac{d}{dx}(x-2)\\ -(x-2)\frac{d}{dx}(x^2+x+1)\end{array}\right)}{(x^2+x+1)^2}$$
$$= \frac{(x^2+x+1)1 - (x-2)(2x+1)}{(x^2+x+1)^2}$$
$$= \frac{x^2 + x + 1 - (2x^2 - 3x - 2)}{(x^2+x+1)^2}$$
$$= \frac{-x^2 + 4x + 3}{(x^2+x+1)^2}$$

11.
$$f(x) = \frac{3x - 6\sqrt{x}}{5x^2 - 2} = \frac{3(x - 2x^{1/2})}{5x^2 - 2}$$
$$f'(x) = 3\frac{\left(\begin{array}{l}(5x^2-2)\frac{d}{dx}(x - 2x^{1/2})\\ -(x - 2x^{1/2})\frac{d}{dx}(5x^2-2)\end{array}\right)}{(5x^2-2)^2}$$
$$= 3\frac{\left(\begin{array}{l}(5x^2-2)(1 - x^{-1/2})\\ -(x - 2x^{1/2})(10x)\end{array}\right)}{(5x^2-2)^2}$$
$$= 3\frac{\left(\begin{array}{l}(5x^2 - 2 - 5x^{3/2} + 2x^{-1/2})\\ -(10x^2 - 20x^{3/2})\end{array}\right)}{(5x^2-2)^2}$$
$$= \frac{3(-5x^2 + 15x^{3/2} + 2x^{-1/2} - 2)}{(5x^2-2)^2}$$

13.
$$f(x) = \frac{(x+1)(x-2)}{x^2 - 5x + 1} = \frac{x^2 - x - 2}{x^2 - 5x + 1}$$
$$f'(x) = \frac{\left(\begin{array}{l}(x^2 - 5x + 1)\frac{d}{dx}(x^2 - x - 2)\\ -(x^2 - x - 2)\frac{d}{dx}(x^2 - 5x + 1)\end{array}\right)}{(x^2 - 5x + 1)^2}$$
$$= \frac{\left(\begin{array}{l}(x^2 - 5x + 1)(2x - 1)\\ -(x^2 - x - 2)(2x - 5)\end{array}\right)}{(x^2 - 5x + 1)^2}$$
$$\left(= \frac{-4x^2 + 6x - 11}{(x^2 - 5x + 1)^2}\right)$$

We omitted many steps in the simplification.

15. We do not recommend treating this one as a quotient, but advise preliminary simplification.
$$f(x) = \frac{x^2 + 3x - 2}{\sqrt{x}} = \frac{x^2}{\sqrt{x}} + \frac{3x}{\sqrt{x}} - \frac{2}{\sqrt{x}}$$
$$= x^{3/2} + 3x^{1/2} - 2x^{-1/2}$$
$$f'(x) = \frac{3}{2}x^{1/2} + \frac{3}{2}x^{-1/2} + x^{-3/2}$$

17. Preliminary simplification is superior to using the product rule.

$$f(x) = x\left(\sqrt[3]{x} + 3\right) = x^{4/3} + 3x$$
$$f'(x) = \frac{4}{3}x^{1/3} + 3$$

19.

$$f(x) = (x^2 - 1)\frac{x^3 + 3x^2}{x^2 + 2} = \frac{x^5 + 3x^4 - x^3 - 3x^2}{x^2 + 2}$$

$$\frac{df}{dx} = \frac{\begin{pmatrix} (x^2+2)\frac{d}{dx}(x^5 + 3x^4 - x^3 - 3x^2) \\ -(x^5 + 3x^4 - x^3 - 3x^2)\frac{d}{dx}(x^2+2) \end{pmatrix}}{(x^2+2)^2}$$

$$= \frac{\begin{pmatrix} (x^2+2)(5x^4 + 12x^3 - 3x^2 - 6x) \\ -(x^5 + 3x^4 - x^3 - 3x^2)(2x) \end{pmatrix}}{(x^2+2)^2}$$

$$= \frac{3x^6 + 6x^5 + 9x^4 + 24x^3 - 6x^2 - 12x}{(x^2+2)^2}$$

21.

$$\frac{d}{dx}\left[f(x)g(x)h(x)\right] = \frac{d}{dx}\left[(f(x)g(x))\,h(x)\right]$$
$$= (f(x)g(x))\frac{dh}{dx} + h(x)\frac{d}{dx}(f(x)g(x))$$
$$= (f(x)g(x))\frac{dh}{dx} + h(x)\left(f(x)\frac{dg}{dx} + g(x)\frac{df}{dx}\right)$$
$$= \frac{df}{dx}g(x)h(x) + f(x)\frac{dg}{dx}h(x) + f(x)g(x)\frac{dh}{dx}$$

In the general case of a product of n functions, the derivative will have n terms to be added, each term a product of all but one of the functions multiplied by the derivative of the missing function.

23.

$$f'(x) = \left[\frac{d}{dx}(x^{2/3})\right](x^2 - 2)(x^3 - x + 1)$$
$$+ x^{2/3}\left[\frac{d}{dx}(x^2 - 2)\right](x^3 - x + 1)$$
$$+ x^{2/3}(x^2 - 2)\frac{d}{dx}(x^3 - x + 1)$$
$$= \frac{2}{3}x^{-1/3}(x^2 - 2)(x^3 - x + 1)$$
$$+ x^{2/3}(2x)(x^3 - x + 1)$$
$$+ x^{2/3}(x^2 - 2)(3x^2 - 1)$$

25.

$$f(x) = (x+1)(x^3 + 4x)(x^5 - 3x^2 + 1)$$
$$\frac{df}{dx} = \left[\frac{d}{dx}(x+1)\right](x^3 + 4x)(x^5 - 3x^2 + 1)$$
$$+ (x+1)\left[\frac{d}{dx}(x^3 + 4x)\right](x^5 - 3x^2 + 1)$$
$$+ (x+1)(x^3 + 4x)\left[\frac{d}{dx}(x^5 - 3x^2 + 1)\right]$$
$$= (x^3 + 4x)(x^5 - 3x^2 + 1)$$
$$+ (x+1)(3x^2 + 4)(x^5 - 3x^2 + 1)$$
$$+ (x+1)(x^3 + 4x)(5x^4 - 6x)$$

27. It is essential in this problem to make estimates of the values and slopes of the two curves. We estimated $f(1)$ to be -2 and $g(1)$ to be 1. After that, we drew in tangent lines and estimated their intercepts: For f we got an x-intercept of 1.8 and a y-intercept of -4 (a line with those intercepts has slope $4/1.8 = 20/9$); for g we got an x-intercept of 1.25 and a y-intercept of 4.5 (a line with those intercepts has slope $-4.5/1.25 = -18/5$). On the basis of those estimates, we conclude that $f(1)g(1) = -2$, while

$$(fg)'(1) = f'(1)g(1) + f(1)g'(1)$$
$$= \left(\frac{20}{9}\right)1 - 2\left(\frac{-18}{5}\right) = \frac{424}{45} \approx 9.$$

We'll call it 9 since there is so much estimation involved. The equation of the tangent line would then be

$$y - (-2) = 9(x - 1) \quad \text{or} \quad y = 9x - 11.$$

Anything close to this is reasonable.

29. Here we have the same problems all over again from Exercise 27. We made these estimates: $f(0) = -1$,

$g(0) = 3,\ f'(0) = -3/2$, and $g'(0) = -6/5$. Accepting these figures, we come to

$$h(0) = \frac{f(0)}{g(0)} = -1/3$$
$$h'(0) = \frac{f'(0)g(0) - f(0)g'(0)}{[g(0)]^2} = $$
$$= \frac{-\left(\frac{3}{2}\right)3 - (-1)\left(-\frac{6}{5}\right)}{3^2} = \frac{-57}{90} \approx -\frac{2}{3}.$$

The equation of the line tangent line at $x = 0$ is

$$y = -\left(\frac{2}{3}\right)x - \frac{1}{3} = \frac{-(2x+1)}{3}.$$

31. Using our estimates from Exercise 27,

$$(f(1) = -2 \quad \text{and} \quad f'(1) = 20/9),$$

with

$$h(x) = x^2 f(x) \quad (\therefore\ h(1) = 1 \cdot f(1) = -2),$$
$$h'(x) = x^2 f'(x) + 2x \cdot f(x)$$
$$h'(1) = 1 \cdot f'(1) + 2 \cdot f(1)$$
$$= (20/9) - 2 \cdot 2 \approx -2.$$

The equation of the line tangent to $y = h(x)$ at $x = 1$ is

$$y + 2 = -2(x - 1) \quad \text{or} \quad y = -2x.$$

33. The rate at which the quantity Q changes is Q'. Since the amount is said to be "decreasing at a rate of 4%," we have to ask "4% of *what*?" The answer in this type of context is usually 4% of *itself*. In other words, $Q' = -.04Q$. As for P, the 3% rate of increase would translate as $P' = .03P$. By the product rule, with $R = PQ$, we have:

$$R' = (PQ)' = P'Q + PQ' = (.03P)Q + P(-.04Q)$$
$$= -(.01)PQ = (-.01)R.$$

In other words, revenue is decreasing at a rate of 1%. We have no comment on what is or is not obvious, especially after Exercise 53, Section 0.3.

35. $R' = Q'P + QP'$
(derivatives with respect to time)

At a certain moment of time (call it t_0) we are given

$$P(t_0) = 20 \text{ (\$/item)}$$
$$Q(t_0) = 20{,}000 \text{ (items)}$$
$$P'(t_0) = 1.25 \text{ (\$/item/year)}$$
$$Q'(t_0) = 2{,}000 \text{ (items/year)}$$
$$\therefore\ R'(t_0) = 2{,}000(20) + (20{,}000)1.25$$
$$= 65{,}000 \text{ (\$/year)}.$$

So revenue is increasing by \$65,000/year at the moment t_0.

37. If

$$u(m) = \frac{82.5m - 6.75}{m + .15},$$

then using the quotient rule,

$$\frac{du}{dm} = \frac{(m + .15)(82.5) - (82.5m - 6.75)1}{(m + .15)^2}$$
$$= \frac{19.125}{(m + .15)^2} \text{ (clearly positive)}.$$

It seems to be saying that initial ball speed is an increasing function of the mass of the bat. Meanwhile,

$$u'(1) = \frac{19.125}{1.15^2} \approx 14.46$$

and

$$u'(1.2) = \frac{19.125}{1.35^2} \approx 10.49.$$

The units here are "meters per second per kilogram," not too easy to interpret.

39. If

$$u(m) = \frac{14.11}{m + .05} = \frac{282.2}{20m + 1},$$

then

$$\frac{du}{dm} = \frac{(20m + 1) \cdot 0 - 282.2(20)}{(20m + 1)^2}$$
$$= \frac{-5644}{(20m + 1)^2}.$$

This is clearly negative, which means that impact speed of the ball is a decreasing function of the weight of the club. It appears that the explanation may have to do with the stated fact that the speed of the club is inversely proportional to its mass. Although the lesson of Example 4.6 was that a heavier club makes for greater ball velocity, that was assuming a fixed club speed, quite a different assumption from this problem.

41. Our CAS gave

$$f'(x) = \frac{2x}{x^2+1} - \frac{2x(x^2-2)}{(x^2+1)^2},$$

which has the form of the expression in Exercise 22.

43. Our CAS gave

$$f'(x) = \frac{-(4x^2-6x+11)}{(x^2-5x+1)^2},$$

which has the form of the expression obtained when using the quotient rule.

We cannot speculate about what other CAS's might have given.

45. Using the quotient rule, we got a derivative in the form

$$\frac{3x}{2\sqrt{3x^3+x^2}},$$

which could be written

$$\frac{3x}{2\sqrt{x^2(3x+1)}}.$$

One *could* then factor $\sqrt{x^2}$ out of the denominator as $|x|$ and use

$$\frac{x}{|x|} = \begin{cases} 1 & \text{if } x > 0 \\ -1 & \text{if } x < 0 \end{cases}.$$

This last is a dubious improvement. We favor leaving the expression as it first appeared, since that is the language in which the problem was presented. It is difficult to program into a CAS the intangible "good judgment" about how far to go in simplifying.

47. If $F(x) = f(x)g(x)$, then

$$F'(x) = f'(x)g(x) + f(x)g'(x) \quad \text{and}$$

$$\begin{aligned} F''(x) &= f''(x)g(x) + f'(x)g'(x) \\ &\quad + f'(x)g'(x) + f(x)g''(x) \\ &= f''(x)g(x) + 2f'(x)g'(x) + f(x)g''(x). \end{aligned}$$

$$\begin{aligned} F'''(x) &= f'''(x)g(x) + f''(x)g'(x) \\ &\quad + 2f''(x)g'(x) + 2f'(x)g''(x) \\ &\quad + f'(x)g''(x) + f(x)g'''(x) \\ &= f'''(x)g(x) + 3f''(x)g'(x) \\ &\quad + 3f'(x)g''(x) + f(x)g'''(x). \end{aligned}$$

One can see obvious parallels to the binomial coefficients as they come from Pascal's Triangle:

$$(a+b)^2 = a^2 + 2ab + b^2$$
$$(a+b)^3 = a^3 + 3a^2b + 3ab^2 + b^3.$$

On this basis, one could correctly predict the pattern of the fourth or any higher derivative.

49. If $g(x) = [f(x)]^2 = f(x)f(x)$, then

$$\begin{aligned} g'(x) &= f'(x)f(x) + f(x)f'(x) \\ &= 2f(x)f'(x). \end{aligned}$$

51.

$$\left(P + \frac{n^2a}{V^2}\right)(V - nb) = nRT$$

$$P + \frac{n^2a}{V^2} = \frac{nRT}{V-nb}$$

$$P = \frac{nRT}{V-nb} - \frac{n^2a}{V^2}$$

From this, we find with some difficulty

$$P'(V) = \frac{dP}{dV} = \frac{-nRT}{(V-nb)^2} + \frac{2n^2a}{V^3}$$

$$P''(V) = \frac{d}{dV}\left(\frac{dP}{dV}\right) = \frac{2nRT}{(V-nb)^3} - \frac{6n^2a}{V^4}.$$

Obviously, if $P'(V) = 0$, then

$$\frac{2na}{V^3} = \frac{RT}{(V-nb)^2}(= X),$$

in which X is a temporary name. If $P''(V)$ is *also* zero, then

$$0 = P''(V) = \frac{2nX}{(V - nb)} - \frac{3nX}{V}$$
$$= nX\left[\frac{2}{V - nb} - \frac{3}{V}\right] = \frac{nX(3nb - V)}{V(V - nb)},$$
$$\therefore\ V = 3nb,\ V - nb = 2nb,$$
$$X = \frac{2na}{V^3} = \frac{2a}{27n^2b^3},$$
$$RT = (V - nb)^2X = 4n^2b^2X = \frac{8a}{27b},$$
$$T = \frac{8a}{27bR},\quad P = \frac{nRT}{V - nb} - \frac{n^2a}{V^2}$$
$$P = \frac{8an}{27b(2nb)} - \frac{n^2a}{9n^2b^2} = \frac{a}{27b^2}.$$

In summary,

$$(T_c, P_c, V_c) = \left(\frac{8a}{27bR}, \frac{a}{27b^2}, 3nb\right).$$

Substituting the given numbers, in particular $Tc = 647°$ (Kelvin).

53.
$$f(x) = \frac{x^{2.7}}{1 + x^{2.7}}$$
$$f'(x) = \frac{\left(1 + x^{2.7}\right)\cdot 2.7x^{1.7} - 2.7x^{1.7}\cdot\left(x^{2.7}\right)}{\left(1 + x^{2.7}\right)^2}$$
$$= \frac{2.7x^{1.7}}{\left(1 + x^{2.7}\right)^2}$$

The fact that $0 < f(x) < 1$ when $x > 0$ suggest to us that f may be some kind of concentration ratio or percentage-of-presence of the allosteric enzyme in some system. If so, the derivative would be interpreted as the rate of change in the concentration per unit of activator. Inasmuch as this rate goes to zero as x tends to either zero or infinity, the activation would apparently be most effective at some intermediate value of x.

55. $\frac{d}{dx}\left[x^3 f(x)\right] = 3x^2 \cdot f(x) + x^3 f'(x)$

57. Utilizing $\frac{d}{dx}(\sqrt{x}) = \frac{1}{2\sqrt{x}}$ (which is a special case of the power rule), we find

$$\frac{d}{dx}\left(\frac{\sqrt{x}}{f(x)}\right) = \frac{f(x)\frac{1}{2\sqrt{x}} - \sqrt{x}f'(x)}{[f(x)]^2}$$
$$= \frac{f(x) - 2xf'(x)}{2\sqrt{x}[f(x)]^2}.$$

2.5 The Chain Rule

1. $f(x) = (x^3 - 1)^2$

Using the chain rule:

$$f'(x) = 2(x^3 - 1)(3x^2) = 6x^2(x^3 - 1).$$

Using the product rule:

$$f(x) = (x^3 - 1)(x^3 - 1)$$
$$f'(x) = (3x^2)(x^3 - 1) + (x^3 - 1)(3x^2)$$
$$= 2(3x^2)(x^3 - 1)$$
$$= 6x^2(x^3 - 1).$$

Using preliminary multiplication:

$$f(x) = x^6 + 2x^3 + 1$$
$$f'(x) = 6x^5 + 6x^2$$
$$= 6x^2(x^3 - 1).$$

3. $f(x) = (x^2 + 1)^3$

Chain rule:

$$f'(x) = 3(x^2 + 1)^2 \cdot 2x.$$

Using preliminary multiplication:

$$f(x) = x^6 + 3x^4 + 3x^2 + 1$$
$$f'(x) = 6x^5 + 12x^3 + 6x.$$

Advisory: The square root function and the reciprocal function come up so often in the following problems and in real life that we do not go to fractional or negative exponents, but use the formulae

$$\frac{d}{dx}\sqrt{x} = \frac{1}{2\sqrt{x}} \quad \text{and} \quad \frac{d}{dx}\left(\frac{1}{x}\right) = \frac{-1}{x^2},$$

often in conjunction with the chain rule. The student is advised to have these two formulae in his/her active memory. Helpful and necessary as fractional and negative exponents are, their appearances can be deceiving and their use requires careful final presentation.

5.
$$f(x) = \sqrt{x^2+4}$$
$$f'(x) = \frac{1}{2\sqrt{x^2+4}} \cdot 2x = \frac{x}{\sqrt{x^2+4}}$$

7.
$$f(x) = (x^3 + x - 1)^3$$
$$f'(x) = 3(x^3 + x - 1)^2(3x^2 + 1)$$

9.
$$f(x) = x^5\sqrt{x^3+2}$$
$$f'(x) = \left(\begin{array}{c} x^5\frac{1}{2\sqrt{x^3+2}}3x^2 \\ +5x^4\sqrt{x^3+2} \end{array}\right)$$
$$= \frac{3x^7 + 10x^4(x^3+2)}{2\sqrt{x^3+2}}$$
$$= \frac{13x^7 + 20x^4}{2\sqrt{x^3+2}}$$

11.
$$f(x) = \frac{x^3}{(x^2+4)^2}$$
$$f'(x) = \frac{3x^2(x^2+4)^2 - x^3 2(x^2+4)(2x)}{(x^2+4)^4}$$
$$= \frac{3x^4 + 12x^2 - 4x^4}{(x^2+4)^3} = \frac{x^2(12 - x^2)}{(x^2+4)^3}$$

13.
$$f(x) = \frac{6}{\sqrt{x^2+4}} = 6(x^2+4)^{-1/2}$$
$$f'(x) = -3(x^2+4)^{-3/2} \cdot 2x$$
$$= \frac{-6x}{(x^2+4)^{3/2}}$$

15.
$$f(x) = (\sqrt{x} + 3)^{4/3}$$
$$f'(x) = \frac{4(\sqrt{x}+3)^{1/3}}{3} \cdot \frac{1}{2\sqrt{x}}$$
$$= \frac{2(\sqrt{x}+3)^{1/3}}{3\sqrt{x}}$$

17.
$$f(x) = \left(\sqrt{x^3+2} + 2x\right)^{-2}$$
$$f'(x) = -2\left(\sqrt{x^3+2} + 2x\right)^{-3}\left[\frac{3x^2}{2\sqrt{x^3+2}} + 2\right]$$
$$= -\frac{3x^2 + 4\sqrt{x^3+2}}{(\sqrt{x^3+2} + 2x)^3 \cdot \sqrt{x^3+2})}$$

19.
$$f(x) = \frac{x}{\sqrt{x^2+1}}$$
$$f'(x) = \frac{\sqrt{x^2+1} - x\frac{1}{2\sqrt{x^2+1}}2x}{x^2+1}$$
$$= \frac{1}{(x^2+1)\sqrt{x^2+1}}$$

21.
$$f(x) = \sqrt{\frac{x}{x^2+1}}$$
$$f'(x) = \frac{1}{2\sqrt{\frac{x}{x^2+1}}} \cdot \frac{(x^2+1) - 2x^2}{(x^2+1)^2}$$
$$= \frac{1 - x^2}{2\sqrt{x}(x^2+1)^{3/2}}$$

23.
$$f(x) = \sqrt{x^2+16}, \ a = 3, \ f(3) = 5$$
$$f'(x) = \frac{1}{2\sqrt{x^2+16}}(2x) = \frac{x}{\sqrt{x^2+16}}$$
$$f'(3) = \frac{3}{\sqrt{3^2+16}} = \frac{3}{5}$$
$$f(3) = \sqrt{3^2+16} = 5$$

So the tangent line is $y = \frac{3}{5}(x-3) + 5$ or $y = \frac{3}{5}x + \frac{16}{5}$.

25.

$$s(t) = \sqrt{t^2 + 8},$$
$$v(t) = s'(t) = \frac{2t}{2\sqrt{t^2+8}} = \frac{t}{\sqrt{t^2+8}}$$
$$v(2) = \frac{2}{\sqrt{12}} = \frac{1}{\sqrt{3}} = \frac{\sqrt{3}}{3}$$

27. For higher derivatives, fractional exponents will be required.

$$f(x) = \sqrt{2x+1} = (2x+1)^{1/2}$$
$$f'(x) = \frac{1}{2}(2x+1)^{-1/2} \cdot 2 = (2x+1)^{-1/2}$$
$$f''(x) = \frac{1}{2}(2x+1)^{-3/2}(2)$$
$$= -(2x+1)^{-3/2}$$
$$f'''(x) = -\left(-\frac{3}{2}\right)(2x+1)^{-5/2} \cdot 2$$
$$= 3(2x+1)^{-5/2}$$
$$f^{(4)}(x) = 3\left(-\frac{5}{2}\right)(2x+1)^{-7/2} \cdot 2$$
$$= -15(2x+1)^{-7/2}$$
$$f^{(n)}(x) = (-1)^{n+1}(1)(3)\ldots(2n-3)(2x+1)^{-(2n-1)/2}$$

29. Slope $2/1 = 2$ at $a(2)$ b looks like a line of slope -3, $\therefore\ f'(2) = 2(-3) = -6$.

31. Slope of -2 at $a(-1)$, $c'(0) = -3$ giving an answer of 6.

33.

$$h(x) = f(g(x))g(x)$$
$$h'(1) = f'(g(1))g'(1)$$
$$= f'(4) \cdot g'(1)$$

Utilizing values given in the table we know that the intervals (3, 4) and (4, 5) are our best estimators for the slope of f at $x = 4$.

x-interval	slope of secant line (f)
(3, 4)	$m = \dfrac{0-(-2)}{4-3} = 2$
(4, 5)	$m = \dfrac{2-0}{5-4} = 2$

They agree, so our guess is that $f'(4) = 2$.

Likewise, the intervals (0, 1) and (1, 2) are the best we have for estimating the slope of g at $x = 1$.

x-interval	slope of secant line (g)
(0, 1)	$m = \dfrac{4-2}{1-0} = 2$
(1, 2)	$m = \dfrac{6-4}{2-1} = 2$

our guess is that $g'(1) = 2$.
So $h'(1) = f'(4)g'(1) = 2 \cdot 2 = 4$.

35.

$$k(x) = g(f(x))f(x)$$
$$k'(x) = g'(f(x))f'(x)$$
$$k'(3) = g'(f(3)) \cdot f'(3) = g'(-2) \cdot f'(3)$$

Following the procedures used in Exercise 33,

x-interval	slope of secant line (g)
$(-3, -2)$	$m = \dfrac{4-6}{-2-(-3)} = -2$
$(-2, -1)$	$m = \dfrac{2-4}{-1-(-2)} = -2$

our guess is that $g'(-2) = -2$.

x-interval	slope of secant line (f)
(2, 3)	$m = \dfrac{-2-(-3)}{3-2} = 1$
(3, 4)	$m = \dfrac{0-(-2)}{4-3} = 2$

Our estimate is that $f'(3) = 3/2$ (an average). So we finally estimate

$$k'(3) = g'(-2)f'(3) = -2 \cdot (3/2) = 3.$$

37.

$$h'(x) = f'(g(x))g'(x)$$
$$h'(1) = f'(g(1))g'(1)$$
$$= f'(2) \cdot (-2) = -6$$

39. $f(x) = x^3 + 4x - 1$ is a one-to-one function with $f(0) = -1$ and $f'(0) = 4$. Therefore, $g(1) = 0$ and

$$g'(-1) = \frac{1}{f'(g(-1))} = \frac{1}{f'(0)} = \frac{1}{4}.$$

41. $f(x) = x^5 + 3x^3 + x$ is a one-to-one function with $f(1) = 5$ and $f'(1) = 5 + 9 + 1 = 15$. Therefore, $g(5) = 1$ and

$$g'(5) = \frac{1}{f'(g(5))} = \frac{1}{f'(1)} = \frac{1}{15}.$$

43. $f(x) = \sqrt{x^3 + 2x + 4}$ is a one-to-one function and $f(0) = 2$ so $g(2) = 0$. Meanwhile,

$$f'(x) = \frac{1}{2\sqrt{x^3 + 2x + 4}}(3x^2 + 2)$$

$$f'(0) = 1/2$$

$$g'(2) = \frac{1}{f'(g(2))} = \frac{1}{f'(0)} = 2.$$

45. $\begin{cases} x = t^2 - 2 \\ y = t^3 - t \end{cases}$

(a) At $t = -1$: $x = -1$, $y = 0$, and

$$m = \frac{y'(-1)}{x'(-1)} = \left.\frac{3t^2 - 1}{2t}\right|_{t=-1} = \frac{2}{-2} = -1.$$

The tangent line is $y = -(x + 1)$.

(b) At $t = 1$: $x = -1$, $y = 0$, and

$$m = \frac{y'(1)}{x'(1)} = \left.\frac{3t^2 - 1}{2t}\right|_{t=1} = \frac{2}{2} = 1.$$

The tangent line is $y = x + 1$.

47. $\begin{cases} x = (t^2 + 1)^2 \\ y = t^4 - 1 \end{cases}$

(a) At $t = 2$: $x = 25$, $y = 15$, and

$$m = \frac{y'(2)}{x'(2)} = \left.\frac{4t^3}{2(t^2 + 1)2t}\right|_{t=2} = \frac{32}{40} = \frac{4}{5}.$$

The tangent line is

$$y - 15 = \frac{4}{5}(x - 25) \text{ or}$$

$$y = \frac{4x}{5} - 5.$$

(b) At $t = -2$: $x = 25$, $y = 15$, and

$$m = \frac{y'(-2)}{x'(-2)} = \left.\frac{4t^3}{2(t^2 + 1)2t}\right|_{t=-2} = \frac{-32}{-40} = \frac{4}{5}.$$

The tangent line is again $y = \frac{4}{5}x - 5$.

It's the same tangent line. The path is an even function of t, and everything is the same at $t = -2$ as at $t = 2$ except for the direction of motion. However, the tangent line does not recognize that as grounds for a difference.

49. $\begin{cases} x = t^2 - 2t \\ y = t^2 - t \end{cases}$

(a) At $t = 1$: $x = 3$, $y = 2$, and

$$m = \frac{y'(-1)}{x'(-1)} = \left.\frac{2t - 1}{2t - 2}\right|_{t=-1} = \frac{-3}{-4} = \frac{3}{4}.$$

The tangent line is

$$y - 2 = \frac{3}{4}(x - 3) \text{ or}$$

$$y = \frac{3}{4}x - \frac{1}{4}.$$

(b) At $t = 1$: $x = -1$, $y = 0$, and

$$m = \frac{y'(1)}{x'(1)} = \left.\frac{2t - 1}{2t - 2}\right|_{t=1}$$

is an undefined expression. However, we usually consider this situation ($x' = 0$, y' not zero) to indicate a "vertical tangent," and would say that the tangent line is $x = -1$.

51. $f(x) = (x^2 + 3)^2 \cdot 2x$

Recognizing the "$2x$" as the derivative of $x^2 + 3$, we guess $g(x) = c(x^2 + 3)^3$.

$$g'(x) = 3c(x^2 + 3)^2 \cdot 2x,$$

which will be $f(x)$ only if $3c = 1$, $\therefore$ $c = 1/3$, and

$$g(x) = \frac{(x^2 + 3)^3}{3}.$$

53. $f(x) = \dfrac{x}{\sqrt{x^2 + 1}}$. Recognizing the "$x$" as half the derivative of $x^2 + 3$, and knowing that differentiation throws the square root into the denominator, we guess

$g(x) = c\sqrt{x^2 + 1}$ and find that

$g'(x) = \dfrac{c}{2\sqrt{x^2+1}}(2x)$ will match $f(x)$

if $c = 1$, $\therefore$ $g(x) = \sqrt{x^2 + 1}$.

55. As a temporary device, given *any* f, set $g(x) = f(-x)$. Then by the chain rule,

$$g'(x) = f'(-x)(-1) = -f'(-x).$$

In the even case ($g = f$), this reads $f'(x) = f'(x)$ and shows f' is odd. In the odd case ($g = -f$ and therefore $g' = -f'$), this reads $-f'(x) = -f'(-x)$ or $f'(x) = f'(-x)$ and shows f' is even.

57.

$$\begin{aligned}\frac{d}{dx} f(\sqrt{x}) &= f'(\sqrt{x}) \cdot \frac{d}{dx}\sqrt{x} \\ &= f'(\sqrt{x}) \cdot \frac{1}{2\sqrt{x}}\end{aligned}$$

59.

$$\begin{aligned}&\frac{d}{dx}\left(\frac{1}{1+[f(x)]^2}\right) \\ &= -\left(\frac{1}{1+[f(x)]^2}\right)^2 \cdot \frac{d}{dx}\left(1+[f(x)]^2\right) \\ &= -\frac{1}{\left(1+[f(x)]^2\right)^2} \cdot 2f(x) \cdot f'(x)\end{aligned}$$

61. In order to tackle this challenging problem, the student is first referred to Exercise 79 in Section 0.7, where we pointed out the issues concerning rates, initial set-up, and selection of the point to be tracked. For this problem, we set up with the fixed circle centered at the origin, the rolling circle initially centered on the positive x-axis (at $(R + r, 0)$), and we plan to track the point of the rolling circle which is initially farthest from the origin, i.e., at the location $(R + 2r, 0)$. The location of this point at time t will be $P = P(t) = (x_p, y_p)$.

As before, we denote the position of the center of the rolling circle by $C = C(t) = (x_c, y_c)$, and it is clear that at any time t, there is an angle θ such that $x_c = (R + r)\cos\theta$, $y_c = (R + r)\sin\theta$. We can assume that $\theta = \omega_1 t$ for some frequency parameter ω_1 yet to be determined.

There is another angle ϕ that measures the angle made by the radius CP to the horizontal. This brings about the relations $x_p - x_c = r\cos\phi$, $y_p - y_c = r\sin\phi$. We can assume that $\phi = \omega_2 t$ for some other frequency parameter ω_2.

Let $Q = Q(t)$ be the position at time t of the point on the rolling circle that started out *in contact* with the fixed circle, that is, at $(R, 0)$. There is a third angle that measures the angle formed from the center line CO to the radius CQ. This angle is also directly proportional to time, and by way of normalization of the speed, we assume *this* angle is t itself. This brings us to the picture, in which (although not labeled) C is perfectly obvious and Q is the point slightly below and to the right, connected to C by a marked radial segment. P is not shown, but it would be diametrically opposite Q.

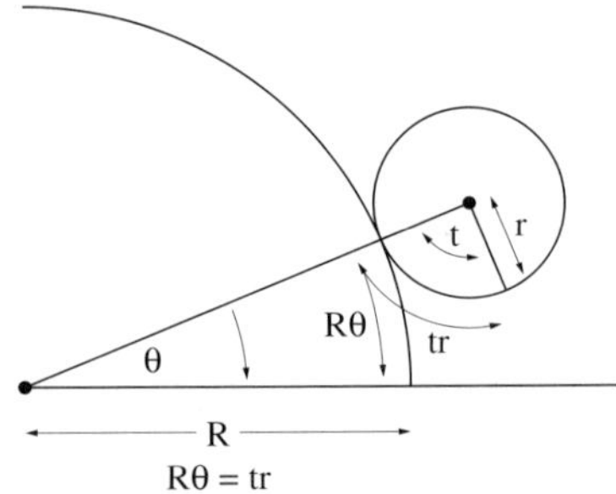

The distances of circumference on the two circles which must match are $R\theta$ and rt (radius times radians), which gives us

$$R(t\omega_1) = R\theta = rt \quad \text{or} \quad \omega_1 = \frac{r}{R}.$$

Now let us advance in time to the first moment after $t = 0$ that Q is directly opposite from the point of contact of the two circles. This would be at time $t = \pi$, and it seems clear that at that moment $\phi = \pi + \theta$. This would translate to

$$\begin{aligned}&\omega_2\pi = \phi = \pi + \omega_1\pi \quad \text{or} \\ &\omega_2 = 1 + \omega_1 = 1 + \frac{r}{R} = \frac{R+r}{R}.\end{aligned}$$

Collecting all the parts, we now find

$$\begin{aligned} x_p &= x_c + (x_p x_c) \\ &= (R+r)\cos\theta + r\cos\phi \\ &= (R+r)\cos(\omega_1 t) + r\cos(\omega_2 t) \\ &= (R+r)\cos\left(\frac{rt}{R}\right) + r\cos\left(\frac{(R+r)t}{R}\right) \\ y_p &= (R+r)\sin\left(\frac{rt}{R}\right) + r\sin\left(\frac{(R+r)t}{R}\right). \end{aligned}$$

The tangent line is vertical at every point where $dx/dt = 0$. In particular, the tangent line is vertical when $t = 0$. Finding other values of t for which this happens (in the absence of special values of r/R) would require numerical methods. Only if the ratio r/R is *rational* does the point P *ever* return to its starting point. (Comment: we find the answer in the text appendix to have some merit but also some flaws.)

2.6 Derivatives of Trigonometric Functions

1. The peaks and valleys of $\cos(x)$ (e.g., $0, \pi, 2\pi$, etc.) are matched with the zeros of $\sin(x)$, and the decreasing intervals for $\cos(x)$ (e.g., $[0, \pi]$) correspond to the intervals where $\sin(x)$ is positive, hence where $-\sin(x)$ is negative. These features lend credibility to the notion that $-\sin(x)$ might be the derivative of $\cos(x)$.

3.
$$\begin{aligned} f(x) &= 4\sin x - x \\ f'(x) &= 4\cos x - 1 \end{aligned}$$

5.
$$\begin{aligned} f(x) &= \tan 3x - \csc 4x \\ f'(x) &= 3\sec^2 3x + 4\csc 4x \cot 4x \end{aligned}$$

7.
$$\begin{aligned} f(x) &= x\cos 5x^2 \\ f'(x) &= (1)\cos 5x^2 + x(-\sin 5x^2)\cdot 10x \\ &= \cos 5x^2 - 10x^2 \sin 5x^2 \end{aligned}$$

9.
$$\begin{aligned} f(x) &= \sin(\tan(x^2)) \\ f'(x) &= \cos(\tan(x^2))\cdot \sec^2(x^2)\cdot 2x \end{aligned}$$

11.
$$\begin{aligned} f(x) &= \frac{\sin(x^2)}{x^2} \\ f'(x) &= \frac{x^2\cos(x^2)\cdot 2x - \sin(x^2)\cdot 2x}{x^4} \\ &= \frac{2x[x^2\cos(x^2)\cdot 2x - \sin(x^2)]}{x^4} \\ &= \frac{2[x^2\cos(x^2) - \sin(x^2)]}{x^3} \end{aligned}$$

13.
$$\begin{aligned} f(t) &= \sin t \sec t = \tan t \\ f'(t) &= \sec^2 t \end{aligned}$$

15.
$$\begin{aligned} f(x) &= \frac{1}{\sin(4x)} = \csc(4x) \\ f'(x) &= -\csc(4x)\cot(4x)\cdot(4) \\ &= -4\csc(4x)\cot(4x) \\ &= \frac{-4\cos(4x)}{\sin^2(4x)} \end{aligned}$$

The matter of the simplification, or at any rate the "rewriting" of trigonometric expressions, can go on indefinitely, and there is little consensus on what constitutes an ideal final form. This writer has one guideline that the student will have heard before: Express the answer in the language in which the problem was originally stated. For examples, don't drag in double angle formulae where none were originally involved, and don't drag secants into a problem which has only sines and cosines.

17.
$$\begin{aligned} f(x) &= 2\sin x\cos x \\ f'(x) &= 2\cos x\cdot\cos x + 2\sin x(-\sin x) \\ &= 2\cos^2 x - 2\sin^2 x (= 2\cos 2x) \end{aligned}$$

19.
$$\begin{aligned} f(x) &= 4x^2\tan x \\ f'(x) &= 8x\tan x + 4x^2\sec^2 x \end{aligned}$$

21.
$$\begin{aligned} f(x) &= 4\sin^2 x + 4\cos^2 x \\ &= 4(\sin^2 x + \cos^2 x) \equiv 4 \\ f'(x) &\equiv 0 \end{aligned}$$

Failing to observe that the function is a constant, the student who is willing to simplify the answer will still get a derivative of zero, having started out

$$f'(x) = 4\,(2\sin x\cdot\cos x + 2\cos x\cdot(-\sin x)).$$

23. $f(x) = 2\sin x \cos x$

Our CAS gave $f'(x) = 4\cos^2 x - 2$. This is correct, possibly motivated by a guideline to "use cosines over sines whenever possible." See our comment following Exercise 15.

25. $f(x) = 2\sin^2 x + \cos 2x$

Our CAS gave

$$f'(x) = 4\sin x \cos x - 2\sin 2x,$$

which is a result obtainable by straightforward differentiation. The CAS does not recognize this as zero, which, as in Exercise 21, can be found either by altering the function before differentiation (it's the constant 1) or diligent simplification afterward.

27.

$$f(x) = \sin 4x,\ a = \frac{\pi}{8},\ f\left(\frac{\pi}{8}\right) = \sin\frac{\pi}{2} = 1$$

$$f'(x) = 4\cos 4x$$

$$f'\left(\frac{\pi}{8}\right) = 4\cos\frac{\pi}{2} = 0$$

So the equation of the tangent line is

$$y - 1 = 0\left(x - \frac{\pi}{8}\right) \quad \text{or} \quad y = 1.$$

29.

$$f(x) = \cos x,\ a = \frac{\pi}{2},\ f\left(\frac{\pi}{2}\right) = \cos\frac{\pi}{2} = 0$$

$$f'(x) = -\sin x$$

$$f'\left(\frac{\pi}{2}\right) = -\sin\frac{\pi}{2} = -1$$

So the equation of the tangent line is

$$y - 0 = -1\left(x - \frac{\pi}{2}\right) \quad \text{or} \quad y = -x + \frac{\pi}{2}.$$

31.

$$s(t) = t^2 - \sin(\sqrt{t}),\ a = 0,\ s(0) = 0$$

$$v(t) = s'(t) = 2t - \cos(\sqrt{t}) \cdot \frac{1}{2\sqrt{t}}$$

$$v(0) \quad \text{wants to be} \quad 2 \cdot 0 - \frac{\cos 0}{0}$$

but is undefined.

33.

$$s(t) = \frac{\cos t}{t},\ a = \pi,\ s(\pi) = -1/\pi$$

$$v(t) = s'(t)$$

$$= \frac{-1}{t^2}\cos t + \frac{1}{t}(-\sin t)$$

$$v(\pi) = -\frac{\cos\pi}{\pi^2} - \frac{\sin\pi}{\pi}$$

$$= \frac{1}{\pi^2} - \frac{1}{\pi}(0) = \frac{1}{\pi^2}$$

35. Let Q be the starting position of our moving point. Let the position of the point at time t be known as $P = (x, y)$, and let θ be the angle between the radius OP and the radius OQ.

If Q is on the positive x-axis, then $x = \cos(\theta)$ and $y = \sin(\theta)$. This is the virtual definition of sine and cosine.

If $Q = (1, 0)$ and one revolution takes 2π seconds, then

$$\frac{t}{2\pi} = \frac{\text{elapsed time}}{\text{total time}} = \frac{\text{elapsed angle}}{\text{total angle}} = \frac{\theta}{2\pi}.$$

Thus $\theta = t$ and the motion is $x = \cos(t)$, $y = \sin(t)$.

If $Q = (3, 0)$, the only difference is $x = 3\cos(t)$, $y = 3\sin(t)$.

If $Q = (1, 0)$ and one revolution takes π seconds (this is twice as fast), then the angle θ at time t is $2t$, and the conclusion is $x = \cos(2t)$, $y = \sin(2t)$.

If $Q = (1, 0)$ and one revolution takes 2 seconds, then the angle θ at time t satisfies

$$\frac{t}{2} = \frac{\text{elapsed time}}{\text{total time}} = \frac{\text{elapsed distance}}{\text{total distance}} = \frac{\theta}{2\pi},$$

then $\theta = \pi t$ and the conclusion is $x = \cos(\pi t)$, $y = \sin(\pi t)$.

If the motion is clockwise, then t is replaced by $-t$, resulting in no change in x, and in y being replaced by $-y$.

If $Q = (0, 1)$, then reviewing the definition of θ, we see that it is unchanged in its relation to t, but the angle from OP to the positive x-axis (which is the key quantity) is now $\theta + \pi/2$. Thanks to the trig laws

$$\sin(\theta + \pi/2) = \cos(\theta)$$
$$\cos(\theta + \pi/2) = -\sin(\theta)$$

we can easily adjust our formulae by replacing sine by cosine, and replacing cosine by minus sine. The effect is to replace x by $-y$ and y by x.

37.

$$f(t) = 4\sin 3t$$
$$f'(t) = 12\cos 3t$$

The maximum speed of 12 occurs when the vertical position is zero.

39.

$$Q(t) = 3\sin 2t + t + 4$$
$$I(t) = \frac{dQ}{dt} = 6\cos 2t + 1$$

At time $t = 0$, $I(0) = 7$ amps. At time $t = 1$, $I(1) = 6\cos 2 + 1 \approx -1.497$ amps.

41.

$$f(x) = \sin 2x = 2\sin x\cos x$$
$$f'(x) = 2\cos x \cdot \cos x + 2\sin x(-\sin x)$$
$$= 2(\cos^2 x - \sin^2 x)$$
$$= 2\cos 2x$$

43.

$$f(x) = \sin x$$
$$f'(x) = \cos x$$
$$f''(x) = -\sin x$$
$$f'''(x) = -\cos x$$
$$f^{(4)}(x) = \sin x = f(x) \text{ (starting over)}$$
$$\therefore\ f^{(75)}(x) = (f^{(72)})^{(3)}(x) = f''' = -\cos x$$
$$f^{(150)}(x) = (f^{(148)})^{(2)}(x) = f'' = -\sin x$$

45. If $f(x) = \cos(x)$, then

$$\frac{f(x+h) - f(x)}{h} =$$
$$\frac{\cos(x+h) - \cos(x)}{h} =$$
$$\frac{\cos x\cos h - \sin x\sin h - \cos x}{h} =$$
$$(\cos x)\frac{(\cos h - 1)}{h} - (\sin x)\left(\frac{\sin h}{h}\right).$$

Taking the limit according to Lemma 6.1,

$$f'(x) = \lim_{h\to 0}\frac{f(x+h) - f(x)}{h}$$
$$= \begin{pmatrix} (\cos x)\cdot\lim_{h\to 0}\frac{\cos(h)-1}{h} \\ -(\sin x)\cdot\lim_{h\to 0}\frac{\sin(h)}{h}\end{pmatrix}$$
$$= \cos x\cdot 0 - \sin x\cdot 1 = -\sin x.$$

47. (a)

$$\lim_{x\to 0}\frac{\sin 3x}{x} = \lim_{x\to 0}\frac{3\sin 3x}{3x}$$
$$= 3\cdot\lim_{x\to 0}\frac{\sin(3x)}{(3x)} = 3\cdot 1 = 3$$

(b)

$$\lim_{t\to 0}\frac{\sin t}{4t} = \frac{1}{4}\lim_{t\to 0}\frac{\sin t}{t} = \frac{1}{4}\cdot 1 = \frac{1}{4}$$

(c)

$$\lim_{x\to 0}\frac{\cos x - 1}{5x} = \frac{1}{5}\lim_{x\to 0}\frac{\cos x - 1}{x} = 0$$

(d) Let $u = x^2$: then $u \to 0$ as $x \to 0$, and

$$\lim_{x\to 0}\frac{\sin x^2}{x^2} = \lim_{u\to 0}\frac{\sin u}{u} = 1.$$

49. The sketch: $y = x$ and $y = \sin(x)$

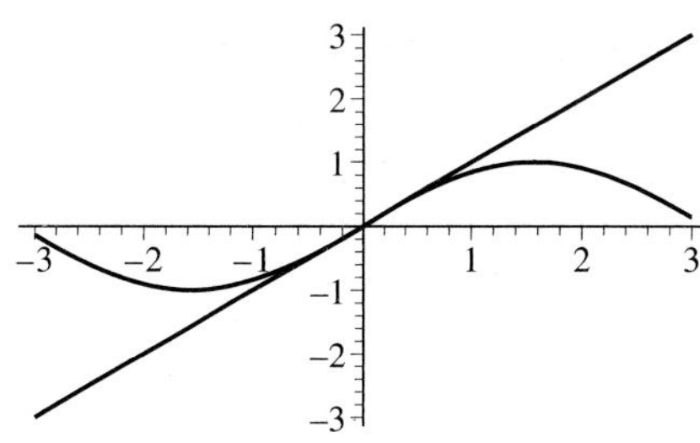

It is not possible visually to either detect or rule out intersections near $x = 0$ (other than zero itself), but since both functions are *odd*, any intersections on the negative side will also appear on the positive side. Therefore we need only worry about positive x. We have to disagree with the suggested line of reasoning for showing that there are no other intersections. We see it as a task for the Mean Value Theorem (Section 2.9). There is however a simple argument based on the assumption that the shortest path from a point to a line is the perpendicular line. For $\pi \geq \theta > 0$, the point $(\cos(\theta), \sin(\theta))$ has perpendicular distance to the x-axis measured by $\sin(\theta)$. But θ itself is the arc length (angle in radians times the radius) to the x-axis as measured *along the unit circle* from the same point. This being a longer route, $\sin(\theta) < \theta$. For $\theta > \pi$, there is simply *no issue* $(\pi > 1 \geq \sin(\theta))$.

51. As seen from the graphs, changing the scale on the x-axis increases the number of oscillations or periods on the display. As the number of periods on the display increase, the graph looks more and more like a bunch of line segments. Its inflection points and concavity are no longer detectable.

2.7 Derivatives of Exponential and Logarithmic Functions

1.
$$f(x) = 4e^x - x$$
$$f'(x) = 4e^x - 1$$

3.
$$f(x) = x^3 e^x$$
$$f'(x) = 3x^2 \cdot e^x + x^3 \cdot e^x = e^x x^2 (x+3)$$

5.
$$f(x) = x + 2^x$$
$$f'(x) = 1 + 2^x \ln 2$$

7.
$$f(x) = 2e^{4x+1}$$
$$f'(x) = 2e^{4x+1} \cdot 4 = 8e^{4x+1}$$

9.
$$f(x) = \left(\frac{1}{3}\right)^{x^2}$$
$$f'(x) = \left(\frac{1}{3}\right)^{x^2} \cdot \ln\left(\frac{1}{3}\right) \cdot 2x$$
$$= -2x \ln(3) \left(\frac{1}{3}\right)^{x^2}$$

11.
$$f(x) = 4^{-3x+1}$$
$$f'(x) = 4^{-3x+1} \cdot \ln 4 \cdot (-3)$$
$$= -6 \ln(2) 4^{-3x+1}$$

13.
$$f(x) = \frac{e^{4x}}{x}$$
$$f'(x) = \frac{x \cdot 4e^{4x} - e^{4x} \cdot 1}{x^2} = \frac{e^{4x}(4x-1)}{x^2}$$

15.
$$f(x) = \ln 2x$$
$$f'(x) = \frac{1}{2x} \cdot (2) = \frac{1}{x}$$

17.
$$f(x) = \ln(x^3 + 3x)$$
$$f'(x) = \frac{3x^2+3}{x^3+3x} = \frac{3(x^2+1)}{x(x^2+3)}$$

19.
$$f(x) = \ln(\cos x)$$
$$f'(x) = \frac{1}{\cos x} \cdot - \sin x$$
$$= -\tan x$$

21. If $f(x) = \sin\left[\ln(\cos[x^3])\right]$, then
$$f'(x) = \cos\left[\ln(\cos[x^3])\right] \cdot \frac{1}{\cos[x^3]} \cdot \sin[x^3] \cdot 3x^2$$
$$= -3x^2 \cdot \cos\left[\ln(\cos[x^3])\right] \cdot \tan[x^3].$$

23.
$$f(x) = \frac{\sqrt{\ln(x^2)}}{x} = \frac{\sqrt{2\ln(x)}}{x}$$
$$f'(x) = \frac{x \cdot \frac{1}{2\sqrt{2\ln(x)}} \cdot \frac{2}{x} - \sqrt{2\ln(x)} \cdot 1}{x^2}$$
$$= \frac{1 - 2\ln(x)}{x^2\sqrt{2\ln(x)}} = \frac{1 - \ln(x^2)}{x^2\sqrt{\ln(x^2)}}$$

25.
$$f(x) = \ln(\sec x + \tan x)$$
$$f'(x) = \frac{\sec x \tan x + \sec^2 x}{\sec x + \tan x}$$
$$= \sec x$$

27.
$$f(x) = 3e^x, \ a = 1, \ f(1) = 3e^1 = 3e$$
$$f'(x) = 3e^x$$
$$f'(1) = 3e^1 = 3e$$

So the equation of the tangent line is

$$y - 3e = 3e(x-1) \quad \text{or} \quad y = 3ex.$$

29.
$$f(x) = 3^x, \ a = 1, \ f(1) = 3$$
$$f'(x) = 3^x \ln 3$$
$$f'(1) = 3 \cdot \ln 3$$

So the equation of the tangent line is

$$y = (3 \cdot \ln 3)(x-1) + 3.$$

31.
$$f(x) = x^2 \ln x, \ a = 1, \ f(1) = 0$$
$$f'(x) = 2x \ln x + x^2 \cdot \frac{1}{x} = 2x \ln x + x$$
$$f'(1) = 2 \cdot 1 \ln 1 + 1 = 2 \cdot 0 + 1 = 1$$

So the equation of the tangent line is

$$y = 1(x-1) + 0 \quad \text{or} \quad y = x - 1.$$

33.
$$v(t) = 100 \cdot 3^t$$
$$v'(t) = 100 \cdot 3^t \ln 3$$
$$\frac{v'(t)}{v(t)} = \frac{100 \cdot 3^t \ln 3}{100 \cdot 3^t} = \ln 3 \approx 1.10$$

So the percentage change is about 110%.

35.
$$v(t) = 100e^t$$
$$v'(t) = 100e^t$$
$$\frac{v'(t)}{v(t)} = \frac{100e^t}{100e^t} = 1$$

So the percentage change is 100%.

37.
$$p(t) = 200 \cdot 3^t$$
$$\ln(p(t)) = \ln(300) + t \ln(3)$$
$$\frac{p'(t)}{p(t)} = \frac{d}{dt}[\ln(p(t)] = \ln 3 \approx 1.099$$

So the rate of change of population is about 110% per unit of time.

39.
$$f(t) = Ae^{rt}$$
$$APY = \frac{f(1) - A}{A} = \frac{Ae^r - A}{A} = e^r - 1$$

(a) $APY = e^{0.05} - 1 \approx .05127\ldots$ (5.1%).

(b) $APY = e^{0.1} - 1 \approx .10517\ldots$ (10.5%)

(c) $APY = e^{0.2} - 1 \approx .22140\ldots$ (22.1%)

(d) $APY = e^{\ln 2} - 1 = 1$ (exact) (100%)

(e) $APY = e^1 - 1 \approx 1.71828\ldots$ (171.8%)

Comment: The acronym APY stands for "Annual Percentage Yield," which, in the Theory of Interest, is the thing actually known as *the interest rate* and usually denoted by i. The r-parameter, known here as the "continuous compounding rate," is known there as the *force of interest*. These things surface elsewhere from time to time (Chapter 0 and Chapter 6).

41.
$$f(x) = x^{\sin x}$$
$$\ln f(x) = \sin x \cdot \ln x$$
$$\frac{f'(x)}{f(x)} = \frac{d}{dx}(\sin x \cdot \ln x)$$
$$= \cos x \cdot \ln x + \frac{\sin x}{x}$$
$$f'(x) = x^{\sin x}\left(\frac{x \cos x \cdot \ln x + \sin x}{x}\right)$$

43.
$$f(x) = (\sin x)^x$$
$$\ln f(x) = x \cdot \ln(\sin x)$$
$$\frac{f'(x)}{f(x)} = \frac{d}{dx}(x \cdot \ln(\sin x))$$
$$= \frac{x \cos x}{\sin x} + \ln(\sin x)$$
$$= x \cot x + \ln(\sin x)$$
$$f'(x) = (\sin x)^x \cdot (x \cot x + \ln(\sin x))$$

45.
$$f(x) = x^{\ln x}$$
$$\ln f(x) = \ln x \cdot \ln x = \ln^2 x$$
$$\frac{f'(x)}{f(x)} = \frac{d}{dx}\left(\ln^2 x\right) = \frac{2 \ln x}{x}$$
$$f'(x) = x^{\ln x}\left[\frac{2 \ln x}{x}\right] = 2x^{[(\ln x) - 1]} \ln x$$

47.
$$f(t) = e^{-t} \cos t$$
$$v(t) = f'(t) = -e^{-t} \cos t + e^{-t}(-\sin t)$$
$$= -e^{-t}(\cos t + \sin t)$$

If the velocity is zero, it is because

$$\cos t = \sin t$$
$$t = \frac{3\pi}{4}, \frac{7\pi}{4}, \ldots, \frac{(3+4n)\pi}{4}, \ldots$$
$$\cos(t) = \frac{-1}{\sqrt{2}}, \frac{1}{\sqrt{2}}, \ldots \text{ (alternating signs).}$$

Position when velocity is zero:

$$f\left(\frac{3\pi}{4}\right) = e^{\frac{-3\pi}{4}} \cos\frac{3\pi}{4}$$
$$= e^{\frac{-3\pi}{4}}\left(-\frac{1}{\sqrt{2}}\right) \approx -.067020$$
$$f\left(\frac{7\pi}{4}\right) = e^{\frac{-7\pi}{4}} \cos\frac{7\pi}{4}$$
$$= e^{\frac{-7\pi}{4}}\left(\frac{1}{\sqrt{2}}\right) \approx .002896.$$

In general,

$$f(\frac{(3+4n)\pi}{4}) = \frac{(-1)^{n+1}}{\sqrt{2}} e^{-(3+4n)\pi/4}.$$

There is also an operative identity that the student can investigate:

$$f'(t) = -\sqrt{2}e^{-\pi/4} f\left(t - \frac{\pi}{4}\right)$$
$$\approx -0.6448 f\left(t - \frac{\pi}{4}\right).$$

In the following picture, we see $f(t)$ (solid), zero at $\pi/2$ and $3\pi/2$. The derivative $f'(t)$ is the dotted curve, zero at $3\pi/4$ and $7\pi/4$, which are relative extreme points for f. We are observing what are known as "damped oscillations." The damping is so strong that only the first minimum is easily detected. (A maximum at $7\pi/4$ is unrecognizable as such. We cannot show t near zero or it would swamp what we do see, but the *value* would be 1 and the *derivative* would be -1 at time zero .)

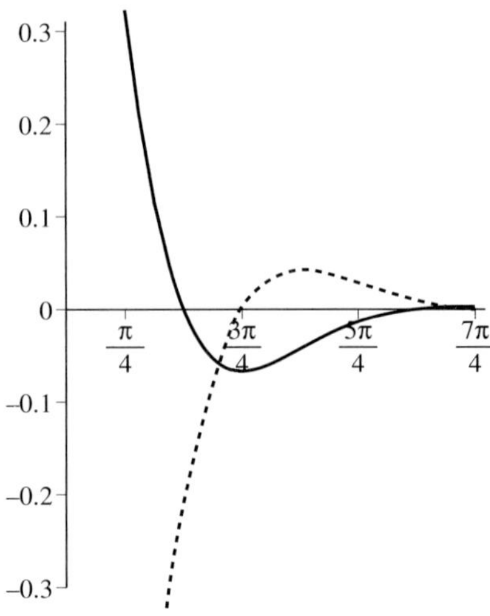

49. With a certain amount of effort, one finds that the acceleration is

$$a(t) = v'(t) = 2e^{-t}[\sin(t)].$$

This is zero for $t = m\pi (m = 0, 1, 2, 3 \ldots)$; the corresponding positions are $(-1)^m e^{-m\pi}$, and the corresponding extremal velocities are the same except for sign (and units).

This phenomenon can be observed at $t = \pi$ on the graph.

The initial impetus is down (time zero, position 1 (say one foot) , velocity -1) and that is the moment of greatest speed (not shown). The moment of maximum *positive* velocity is $t = \pi$, position $-e^{-\pi} = -.0432$ (feet), and velocity .0432 (feet per second).

51.

$$f(x) = \sinh x = \frac{e^x - e^{-x}}{2}$$
$$f'(x) = \frac{e^x + e^{-x}}{2} = \cosh x$$

53.
$$f(x) = \tanh x = \frac{\sinh x}{\cosh x}$$
$$f'(x) = \frac{\cosh x \cdot \cosh x - \sinh x \cdot \sinh x}{\cosh^2 x}$$
$$= \frac{\cosh^2 x - \sinh^2 x}{\cosh^2 x} = \frac{1}{\cosh^2 x}$$
$$= \text{sec h}^2 x \left(= \frac{4}{(e^x + e^{-x})^2}\right)$$

55.
$$f(x) = \sinh(\cos x)$$
$$f'(x) = -\cosh(\cos x)\sin x$$

57.
$$f(x) = e^{\ln x^2}$$
$$f'(x) = e^{\ln x^2} \cdot \frac{d}{dx} \ln x^2 = e^{\ln x^2} \cdot \frac{2}{x} = 2x$$

Much easier if one noticed at the outset that $f(x) = x^2$.

59.
$$f(x) = \ln\sqrt{4e^{3x}} = \frac{1}{2}\left[\ln\left(4 \cdot e^{3x}\right)\right]$$
$$= \frac{1}{2}\left[\ln 4 + \ln e^{3x}\right] = \frac{\ln 4 + 3x}{2}$$
$$f'(x) = \frac{3}{2}$$

61. Taking h to be .001 and $-.001$ as values numerically close to zero:

$$\frac{3^{.001} - 1}{0.001} \approx 1.09922 \quad \text{and}$$
$$\frac{3^{-.001} - 1}{-.001} \approx 1.09801.$$

Both approximate $\ln 3 \approx 1.09861$ reasonably well, and their average (1.098615) does so even better.

63.
$$x(t) = \frac{6}{2e^{-8t} + 1} = 6(2e^{-8t} + 1)^{-1}$$
$$x'(t) = -6(2e^{-8t} + 1)^{-2} \cdot -16e^{-8t}$$
$$= \frac{96e^{-8t}}{(2e^{-8t} + 1)^2}$$

Since $e^{-8t} > 0$ for any t, both numerator and denominator are positive, so that $x'(t) > 0$. Then, since $x(t)$ is an increasing function with a limiting value of 6 (as t goes to infinity), the concentration never exceeds (indeed, never reaches) the value of 6.

65. If $g(x) = e^x$, then
$$g'(x) = e^x$$
$$g''(x) = e^x$$
$$g(0) = g'(0) = g''(0) = e^0 = 1.$$

If $f(x) = \dfrac{a + bx}{1 + cx}$, then $f(0) = a$,
$$f'(x) = \frac{b(1 + cx) - (a + bx)(c)}{(1 + cx)^2}$$
$$= \frac{b - ac}{(1 + cx)^2} = (b - ac)(1 + cx)^{-2}$$
$$f'(0) = b - ac$$
$$f''(x) = (b - ac)(-2)(1 + cx)^{-3}c$$
$$= \frac{-2c(b - ac)}{(1 + cx)^3}$$
$$f''(0) = -2c(b - ac)$$
$$1 = g(0) = f(0) = a \quad \text{so} \quad a = 1.$$
$$1 = g'(0) = f'(0) = b - ac = b - c$$
$$1 = g''(0) = f''(0) = -2c(b - ac) = -2c$$
$$\text{so} \quad c = -\frac{1}{2} \quad \text{and} \quad b = 1 + c = 1 - \frac{1}{2} = \frac{1}{2}$$
$$\text{so} \quad b = \frac{1}{2}.$$

In summary, $a = 1$, $b = \frac{1}{2}$, $c = -\frac{1}{2}$, and
$$g(x) = \frac{1 + (x/2)}{1 - (x/2)} = \frac{2 + x}{2 - x}.$$

67.
$$f(x) = e^{-x^2/2}$$
$$f'(x) = e^{-x^2/2} \cdot -\frac{2x}{2}$$
$$= -xe^{-x^2/2}(= -xf(x))$$
$$f''(x) = -\left[x(-xe^{-x^2/2}) + 1 \cdot e^{-x^2/2}\right]$$
$$= e^{-x^2/2}(x^2 - 1)(= (x^2 - 1)f(x))$$

This will be zero only when $x = \pm 1$.

69. It helps immensely to leave the name f as it was in Exercise 67, and give a new name g to the new function here, so that

$$g(x) = e^{-\left[\frac{(x-m)^2}{2c^2}\right]} = f(u)$$

in which $u = \dfrac{x-m}{c}$. Then

$$g'(x) = f'(u)\frac{du}{dx} = \frac{f'(u)}{c} = \frac{-uf(u)}{c}$$
$$= \frac{-(x-m)e^{-\left[\frac{(x-m)^2}{2c^2}\right]}}{c^2},$$
$$g''(x) = \frac{d}{dx}\left(\frac{f'(u)}{c}\right) = \frac{f''(u)\frac{du}{dx}}{c}$$
$$= \frac{f''(u)}{c^2} = \frac{(u^2-1)f(u)}{c^2}$$
$$= \frac{((x-m)^2 - c^2)e^{-\left[\frac{(x-m)^2}{2c^2}\right]}}{c^4}.$$

This will be zero only when $x = m \pm c$.

2.8 Implicit Differentiation and Inverse Trigonometric Functions

1. $x^2 + 4y^2 = 8$ at $(2, 1)$

Explicitly:

$$4y^2 = 8 - x^2$$
$$y^2 = \frac{8-x^2}{4}$$
$$y = \pm\frac{\sqrt{8-x^2}}{2} \text{ (choose plus to fit (2,1)).}$$

For $y = \dfrac{\sqrt{8-x^2}}{2}$,

$$y' = \frac{1}{2}\frac{(-2x)}{2\sqrt{8-x^2}} = \frac{-x}{2\sqrt{8-x^2}},$$
$$y'(2) = -1/2.$$

Implicitly:

$$\frac{d}{dx}(x^2+4y^2) = \frac{d}{dx}(8)$$
$$2x + 8y \cdot y' = 0$$
$$y' = \frac{-2x}{8y} = \frac{-x}{4y}$$
$$\text{at } (2, 1): y' = \frac{-2}{4\cdot 1} = -\frac{1}{2}.$$

3. $y - 3x^2y = \cos x$ at $(0, 1)$

Explicitly:

$$y(1-3x^2) = \cos x$$
$$y = \frac{\cos x}{1-3x^2}$$
$$y'(x) = \frac{(1-3x^2)(-\sin x) - \cos x(-6x)}{(1-3x^2)^2}$$
$$= \frac{-\sin x + 3x^2\sin x + 6x\cos x}{(1-3x^2)^2}$$
$$y'(0) = 0.$$

Implicitly:

$$\frac{d}{dx}(y - 3x^2y) = \frac{d}{dx}(\cos x)$$
$$y' - (6xy + 3x^2y') = -\sin x$$
$$y'(1-3x^2) = 6xy - \sin x$$
$$y' = \frac{6xy - \sin x}{1-3x^2}$$
$$\text{at } (0, 1): y' = 0 \text{ (again).}$$

5.

$$x^2y^2 + 3y = 4x$$
$$\frac{d}{dx}(x^2y^2 + 3y) = \frac{d}{dx}(4x)$$
$$2xy^2 + x^2 2y \cdot y' + 3y' = 4$$
$$y'(2x^2y + 3) = 4 - 2xy^2$$
$$y' = \frac{4 - 2xy^2}{2x^2y + 3}$$

7.

$$\sqrt{xy} - 4y^2 = 12$$
$$\frac{d}{dx}(\sqrt{xy} - 4y^2) = \frac{d}{dx}(12)$$
$$\frac{1}{2\sqrt{xy}} \cdot \frac{d}{dx}(xy) - 8y \cdot y' = 0$$
$$\frac{1}{2\sqrt{xy}} \cdot (xy' + y) - 8y \cdot y' = 0$$
$$(xy' + y) - 16y \cdot y'\sqrt{xy} = 0$$
$$y'\left(x - 16y\sqrt{xy}\right) = -y$$
$$y' = \frac{-y}{(x - 16y\sqrt{xy})} = \frac{y}{16y\sqrt{xy} - x}$$

9.

$$\frac{x+3}{y} = 4x + y^2$$
$$x + 3 = 4xy + y^3$$
$$1 = \frac{d}{dx}\left(4xy + y^3\right) = 4(xy' + y) + 3y^2y'$$
$$1 - 4y = y'(3y^2 + 4x)$$
$$y' = \frac{1-4y}{3y^2 + 4x}$$

Differentiating the quotient leads eventually to

$$y' = \frac{y - 4y^2}{2y^3 + x + 3}\left(= \frac{1-4y}{2y^2 + \frac{x+3}{y}}\right).$$

The expressions are the same because the fraction in the denominator of the second form is already $4x + y^2$ (has been from the beginning). It's an arena where equally correct answers can look quite different.

11.

$$e^{x^2y} - e^y = x$$
$$\frac{d}{dx}(e^{x^2y} - e^y) = \frac{d}{dx}(x)$$
$$e^{x^2y}\frac{d}{dx}(x^2y) - e^y y' = 1$$
$$e^{x^2y}(2xy + x^2y') - e^y y' = 1$$
$$y'(x^2e^{x^2y} - e^y) = 1 - 2xye^{x^2y}$$
$$y' = \frac{1 - 2xye^{x^2y}}{x^2e^{x^2y} - e^y}$$

13.

$$\sqrt{x+y} - 4x^2 = y$$
$$\frac{d}{dx}\left(\sqrt{x+y} - 4x^2\right) = \frac{d}{dx}(y)$$
$$\frac{1}{2\sqrt{x+y}} \cdot (1 + y') - 8x = y'$$
$$y'\left(\frac{1}{2\sqrt{x+y}} - 1\right) = \frac{-1}{2\sqrt{x+y}} + 8x$$
$$y'\left(\frac{1 - 2\sqrt{x+y}}{2\sqrt{x+y}}\right) = \frac{16x\sqrt{x+y} - 1}{2\sqrt{x+y}}$$
$$y' = \frac{16x\sqrt{x+y} - 1}{1 - 2\sqrt{x+y}}$$

15.

$$e^{4y} - \ln y = 2x$$
$$\frac{d}{dx}\left(e^{4y} - \ln y\right) = \frac{d}{dx}(2x)$$
$$e^{4y} \cdot 4y' - \frac{1}{y} \cdot y' = 2$$
$$y'\left(4e^{4y} - \frac{1}{y}\right) = 2$$
$$y'\left(\frac{4ye^{4y} - 1}{y}\right) = 2$$
$$y' = \frac{2y}{4ye^{4y} - 1}$$

17. $x^2 - 4y^2 = 0$ at $(2, 1)$

Rewrite: $x^2 = 4y^2$
Differentiate by x:

$$2x = 8y \cdot y'$$
$$y' = \frac{2x}{8y} = \frac{x}{4y}$$
$$\text{at } (2, 1) : y' = \frac{2}{4 \cdot 1} = \frac{1}{2}$$

The equation of the tangent line is

$$y - 1 = \frac{1}{2}(x - 2) \quad \text{or} \quad y = \frac{1}{2}x.$$

19. $x^2 - 4y^3 = 0$ at $(2, 1)$

It's almost like the previous, the difference being

$$y' = \frac{2x}{12y^2} = \frac{x}{6y^2}$$

$$\text{at } (2, 1) : y' = \frac{2}{6 \cdot 1^2} = \frac{1}{3}.$$

The equation of the tangent line is

$$y - 1 = \frac{1}{3}(x - 2) \quad \text{or} \quad y = \frac{1}{3}(x + 1).$$

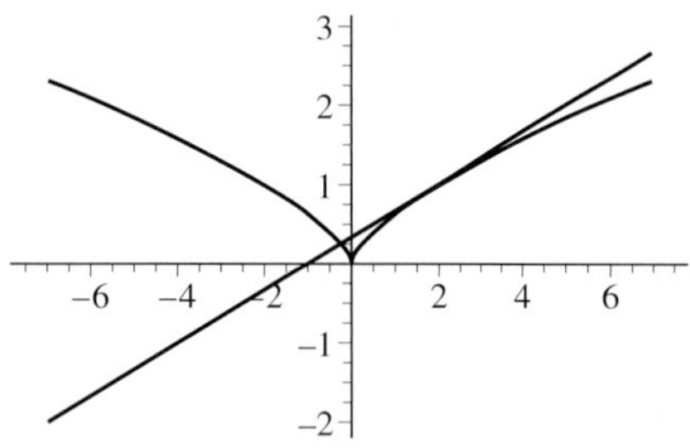

21. $x^2y^2 = 4y$ at $(2, 1)$

This one has $y = 0$ as part of the curve(s) , but our point of reference is not on that part, so we can assume y is not zero, cancel it, and come to

$$x^2y = 4 \qquad \left(y = \frac{4}{x^2}\right)$$

$$\frac{d}{dx}(x^2y) = \frac{d}{dx}(4)$$

$$2xy + x^2 \cdot y' = 0$$

$$y' = \frac{-2y}{x} \left(= \frac{-8}{x^3}\right)$$

$$\text{at } (2, 1) : y' = -2/2 = -1.$$

The equation of the tangent line is

$$y - 1 = (-1)(x - 2) \quad \text{or} \quad y = -x + 3.$$

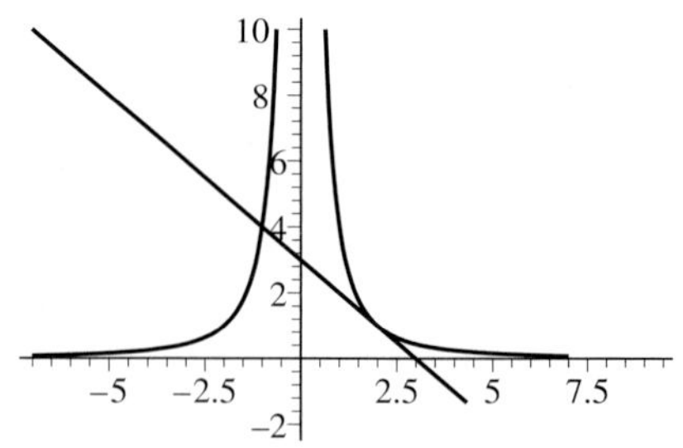

23. $x^3y^3 = 9y$ at $(1, 3)$. This is much like Exercise 21:

$$x^3y^2 = 9$$

$$\frac{d}{dx}(x^3y^2) = \frac{d}{dx}(9)$$

$$3x^2y^2 + x^3 \cdot 2y \cdot y' = 0$$

$$y' = \frac{-3x^2y^2}{x^3 \cdot 2y} = \frac{-3y}{2x}$$

At $(1, 3)$: $y' = -\frac{9}{2}$.
The equation of the tangent line is

$$y - 3 = \frac{9}{2}(x - 1) \quad \text{or} \quad y = -\frac{9}{2}x + \frac{15}{2}.$$

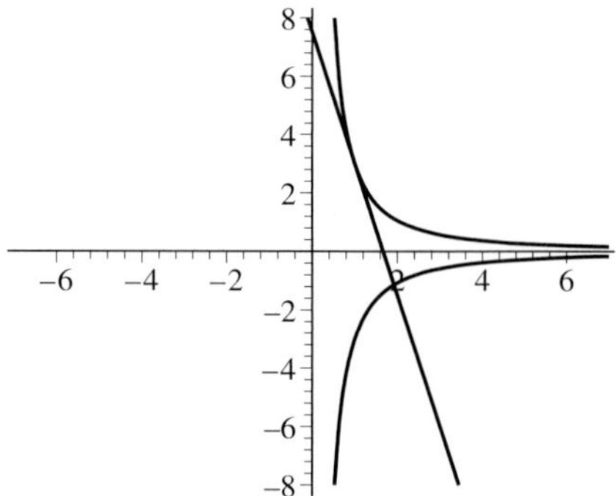

25. $x^2 + y^3 - 3y = 4$

$$\frac{d}{dx}(x^2 + y^3 - 3y) = \frac{d}{dx}(4)$$

$$2x + 3y^2y' - 3y' = 0$$

$$y'(3y^2 - 3) = -2x$$

$$y' = \frac{2x}{3 - 3y^2}$$

Horizontal tangents: from the formula, $y' = 0$ only when $x = 0$.

When $x = 0$, we have $0^2 + y^3 - 3y = 4$. Using a CAS to solve this, we find that

$$y = \left(2 - \sqrt{3}\right)^{1/3} + \left(2 + \sqrt{3}\right)^{1/3} \approx 2.2$$

is a horizontal tangent line, tangent to the curve at the (approximate) point $(0, 2.2)$.

Vertical tangents: The denominator in y' must be zero.

$$\begin{aligned} 3 - 3y^2 &= 0 \\ y^2 &= 1 \quad \text{or} \quad y = \pm 1. \end{aligned}$$

When $y = 1$ we have

$$\begin{aligned} x^2 + (1)^3 - 3(1) &= 4 \\ x^2 &= 6 \quad \text{or} \quad x = \pm\sqrt{6} \approx \pm 2.4. \end{aligned}$$

Also, when $y = -1$, we have

$$\begin{aligned} x^2 + (-1)^3 - 3(-1) &= 4 \\ x^2 &= 2 \\ x &= \pm\sqrt{2} \approx \pm 1.4. \end{aligned}$$

Thus, we find 4 vertical tangent lines:

$$x = -\sqrt{6},\ x = -\sqrt{2},\ x = \sqrt{2},\ x = \sqrt{6},$$

tangent to the curve (respectively) at the points

$$\left(-\sqrt{6}, 1\right),\ \left(-\sqrt{2}, -1\right),\ \left(\sqrt{2}, -1\right),\ \left(\sqrt{6}, 1\right).$$

27.

$$\begin{aligned} x^2 + y^2 &= 4 \\ \frac{d}{dx}(x^2 + y^2) &= \frac{d}{dx}(4) \\ 2x + 2yy' &= 0 \\ y' &= \frac{-2x}{2y} = \frac{-x}{y} \end{aligned}$$

$$\begin{aligned} y'' &= -\frac{d}{dx}\left(\frac{x}{y}\right) = -\frac{y \cdot 1 - xy'}{y^2} \\ &= \frac{y \cdot 1 - x\left(-\frac{x}{y}\right)}{y^2} = -\frac{y^2 + x^2}{y^3} = -\frac{4}{y^3} \end{aligned}$$

29.

$$\begin{aligned} x^2y^2 + 3x - 4y &= 5 \\ \frac{d}{dx}(x^2y^2 + 3x - 4y) &= \frac{d}{dx}5 \\ x^2 2yy' + 2xy^2 + 3 - 4y' &= 0 \\ (2x^2y - 4)y' &= -2xy^2 - 3 \\ y' &= \frac{2xy^2 + 3}{4 - 2x^2y} \end{aligned}$$

This is about to get nasty. Let us try something different: Differentiate the third line with respect to x. It will go like this:

$$\begin{aligned} &2(2xyy' + x^2(y')^2 + x^2yy'') \\ &\quad + 2(2xyy' + y^2) - 4y'' = 0. \end{aligned}$$

Cancel a common factor of 2.

$$\begin{aligned} y(2 - x^2y) &= 4xyy' + x^2(y')^2 + y^2 \\ y &= \frac{4xyy' + x^2(y')^2 + y^2}{2 - x^2y}. \end{aligned}$$

One could go ahead and substitute the known formula for y', but our interest is likely to be either (a) at a critical point (where $y' = 0$) or (b) at some other specific point, in which case we find y' numerically and use it as a third input to the formula.

31.

$$\begin{aligned} y^2 &= x^3 - 6x + 4\cos y \\ \frac{d}{dx}(y^2) &= \frac{d}{dx}(x^3 - 6x + 4\cos y) \\ 2yy' &= 3x^2 - 6 - 4\sin y \cdot y' \\ (2y + 4\sin y)y' &= 3x^2 - 6 \\ y' &= \frac{3x^2 - 6}{2y + 4\sin y} \end{aligned}$$

At the third line, differentiating in x:

$$\begin{aligned}
&2[yy'' + (y')^2] \\
&= 6x - 4[\sin y \cdot y'' + \cos y \cdot (y')^2], \\
&yy'' + (y')^2 = \\
&3x - 2 \sin y \cdot y'' - 2 \cos y \cdot (y')^2, \\
&y''(y + 2 \sin y) = 3x - [2 \cos y + 1](y')^2 \\
&y'' = \frac{3x - [2 \cos y + 1](y')^2}{y + 2 \sin y}.
\end{aligned}$$

33.

$$\begin{aligned}
x^2 + y^3 - 2y &= 3 \\
y' &= \frac{-2x}{3y^2 - 2}
\end{aligned}$$

If $x = 1.9$, solving for y requires solving the equation $y^3 - 2y + 0.61 = 0$. Using the equation of the tangent line found in Example 8.1, $y = -4x + 9$, $y(1.9) \approx 1.4$.
If $x = 2.1$, solving for y requires solving the equation $y^3 - 2y + 1.41 = 0$. Using the equation of the tangent line found in Example 8.1, $y = -4x + 9$, $y(2.1) \approx 0.6$.

Note from Figure 2.43 that for $x = 2.1$ the tangent takes one far away from the curve. The only y corresponding to $x = 2.1$ on the curve is at approximately -1.68. The approximation at $x = 2.1$ is not a good one.

35. Both of the points $(-3, 0)$ and $(0, 3)$ are on the curve:

$$\begin{aligned}
0^2 &= (-3)^3 - 6(-3) + 9 = -27 + 18 + 9 \\
3^2 &= (0)^3 - 6(0) + 9 = 9.
\end{aligned}$$

The equation of the line through these points has slope

$$\frac{0 - (-3)}{-3 - 0} = \frac{-3}{-3} = 1$$

and y-intercept 3, so $y = x + 3$. This line intersects the curve at:

$$\begin{aligned}
y^2 &= x^3 - 6x + 9 \\
(x + 3)^2 &= x^3 - 6x + 9 \\
x^2 + 6x + 9 &= x^3 - 6x + 9 \\
x^3 - 12x - x^2 &= 0 \\
x(x^2 - x - 12) &= 0.
\end{aligned}$$

Therefore, $x = 0, -3,$ or 4 and so the third point is $(4, 7)$.

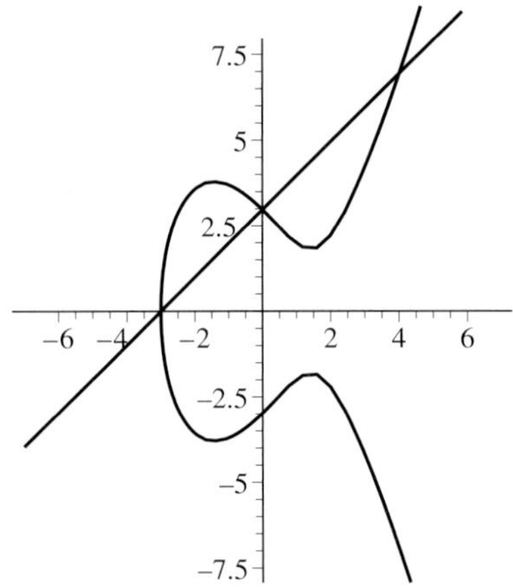

37. $x^2y - 2y = 4$

$$\begin{aligned}
\frac{d}{dx}(x^2y - 2y) &= \frac{d}{dx}(4) \\
2xy + x^2y' - 2y' &= 0 \\
y'(x^2 - 2) &= -2xy \\
y' &= \frac{-2xy}{x^2 - 2} = \frac{2xy}{2 - x^2}
\end{aligned}$$

In this case, we can solve for y rather easily, finding

$$y = \frac{4}{x^2 - 2}.$$

In this explicit form,

$$y' = \frac{-8x}{(x^2 - 2)^2} = \frac{-2xy}{x^2 - 2}.$$

Contrary to the suggestions, we did not expect to find any vertical *tangents*, and we don't. Vertical *asymptotes* may occur at points where the function is not defined (i.e., when the denominator is zero), in this case, when $x = \pm\sqrt{2}$, and the numerator not being zero, they are indeed present.

Horizontal *tangents* occur where the derivative is zero, in this case only when $x = 0$ and $y = -2$. A horizontal *asymptote* arises if there is a limit for y as $x \to \pm\infty$. In this case the limit being zero, the line $y = 0$ is a horizontal asymptote.

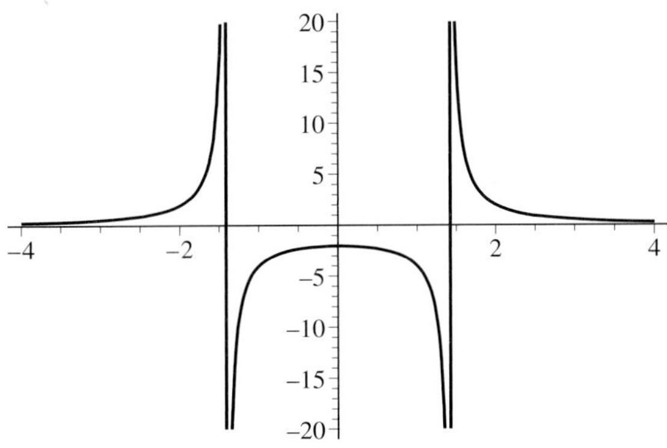

39. If $y_1 = \dfrac{c}{x}$, then $y_1' = \dfrac{-c}{x^2} = -\dfrac{y_1}{x}$. If $y_2^2 = x^2 + k$, then $2y_2(y_2') = 2x$ and $y_2' = \dfrac{x}{y_2}$. If we are at a particular point (x_0, y_0) on both graphs, this means $y_1(x_0) = y_0 = y_2(x_0)$ and

$$y_1' \cdot y_2' = \left(\frac{-y_0}{x_0}\right) \cdot \left(\frac{x_0}{y_0}\right) = -1.$$

This means that the slopes are negative reciprocals and the curves are orthogonal.

41. This one will be covered by Exercise 43, so we'll pass directly on.

43. Conjecture: The family of functions $\{y_1 = cx^n\}$ is orthogonal to the family of functions $\{x^2 + ny_2^2 = k\}$ whenever $n \neq 0$.
If $y_1 = cx^n$, then $y_1' = cnx^{n-1} = \dfrac{ny_1}{x}$. If $ny_2^2 = -x^2 + k$, then $2ny_2(y_2') = -2x$ and $y_2' = -\dfrac{x}{ny_2}$. If we are at a particular point (x_0, y_0) on both graphs, this means $y_1(x_0) = y_0 = y_2(x_0)$ and

$$y_1' \cdot y_2' = \left(\frac{ny_0}{x_0}\right) \cdot \left(-\frac{x_0}{ny_0}\right) = -1.$$

This means that the slopes are negative reciprocals and the curves are orthogonal.

45.
$$f(x) = \tan^{-1}\sqrt{x}$$
$$f'(x) = \frac{1}{1+(\sqrt{x})^2} \cdot \frac{d}{dx}\sqrt{x} = \frac{1}{2(1+x)\sqrt{x}}$$

47.
$$f(x) = \tan^{-1}(\cos x)$$
$$f'(x) = \frac{1}{1+(\cos x)^2} \cdot \frac{d}{dx}\cos x = \frac{-\sin x}{1+(\cos x)^2}$$

49. If, as written, $f(x) = 4\sec(x^4)$, then

$$f'(x) = 4\sec(x^4)\tan(x^4) \cdot 4x^3.$$

Because of the nature of all the companion problems, some think the original intent was

$$f(x) = 4\sec^{-1}(x^4),$$

in which case

$$f'(x) = \frac{4}{|x^4|\sqrt{(x^4)^2 - 1}} \cdot \frac{d}{dx}x^4 = \frac{16}{x\sqrt{(x^4)^2 - 1}} \quad (|x| > 1)$$

(The absolute value symbol around the x^4 is completely unnecessary. As an even power, that object is positive.)

51.
$$f(x) = e^{\tan^{-1} x}$$
$$f'(x) = e^{\tan^{-1} x}\frac{d}{dx}\tan^{-1} x = \frac{e^{\tan^{-1} x}}{1+x^2}$$

53.
$$f(x) = \frac{x^2}{\cot^{-1} x}$$
$$f'(x) = \frac{\cot^{-1} x \cdot 2x - x^2 \cdot \frac{-1}{1+x^2}}{(\cot^{-1} x)^2} = \frac{x\left(\cot^{-1} x \cdot 2(1+x^2) + x\right)}{(\cot^{-1} x)^2(1+x^2)}$$

55. $f(x) = \tan^{-1}(\ln x^2)$

$$f'(x) = \frac{1}{1+(\ln x^2)^2} \cdot \frac{d}{dx}(\ln x^2)$$

$$= \frac{1}{1+(\ln x^2)^2} \cdot \frac{1}{x^2} \cdot 2x = \frac{2}{x\left(1+(\ln x^2)^2\right)}$$

57. From Example 8.6, the rate of change of the angle is

$$\theta'(t) = \frac{1}{1+\left[\frac{d(t)}{2}\right]^2} \cdot \frac{d'(t)}{2}.$$

Given a maximum rotational rate of $\theta'(t) = -3$ (radians/second), the distance from the plate at which a player can track the ball can be obtained by solving the equation

$$-3 = \frac{4}{4+[d(t)]^2} \cdot \frac{d'(t)}{2}$$

for $d(t)$ in terms of $d'(t)$. This leads to

$$d(t) = \frac{\sqrt{-6 \cdot d'(t) - 36}}{3},$$

if $d'(t) \leq -6$, which may be reasonable since the distance is decreasing as the ball approaches the plate. We get $d(t) = 4$ for $d'(t) = -30$ ft/sec and $d(t) = 9.45$ for $d'(t) = -140$ ft/sec. This would mean a player can track the ball to within 4 feet from the plate in slowpitch, but only to within 9.45 feet from the plate in the major leagues.

59. We've seen this one before (Section 0.4 Exercise 73), and this time we'll simply assume that "six feet tall" is intended to mean "eye level at six feet." If A is the viewing angle formed between the rays from the person's eye to the top of the frame and to the bottom of the frame, and if x is the distance between the person and the wall, then since the frame extends from 6 to 8 feet, we have $\tan A = \frac{2}{x}$, or $A = \arctan \frac{2}{x}$. Then,

$$\frac{dA}{dx} = \frac{1}{1+\left(\frac{2}{x}\right)^2} \cdot \left(\frac{-2}{x^2}\right) = \frac{-2}{x^2+4}.$$

Since the derivative is negative, the angle is a decreasing function of x. Strictly speaking, $\arctan \frac{2}{x}$ is undefined at $x = 0$, but $\arctan \frac{2}{x} \to \frac{\pi}{2}$ as $x \to 0$. The angle A continues to enlarge (up to a right angle) as x decreases to zero. In this case the maximal viewing angle is not a feasible one.

2.9 The Mean Value Theorem

1.

$$f(x) = x^2 + 1, \ [-2, 2]$$
$$f(-2) = 5 = f(2)$$

As a polynomial, $f(x)$ is continuous on $[-2, 2]$, differentiable on $(-2, 2)$, and the conditions of Rolle's Theorem hold. There exists $c \in (-2, 2)$ such that $f'(c) = 0$. But $f'(c) = 2c$, $\therefore\ c = 0$.

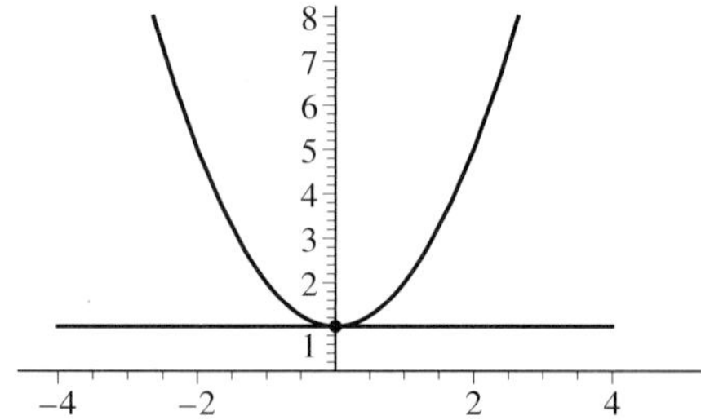

3. $f(x) = x^3 + x^2$, on $[0, 1]$, with $f(0) = 0$. $f(1) = 2$. As a polynomial, $f(x)$ is continuous on $[0, 1]$ and differentiable on $(0, 1)$. Since the conditions of the Mean Value Theorem hold, there exists a number $c \in (0, 1)$ such that

$$f'(c) = \frac{f(1) - f(0)}{1-0} = \frac{2-0}{1-0} = 2.$$

But $f'(c) = 3c^2 + 2c$.

$$\therefore\ 3c^2 + 2c = 2,$$
$$3c^2 + 2c - 2 = 0.$$

By the quadratic formula,

$$c = \frac{-2 \pm \sqrt{2^2 - 4(3)(-2)}}{2(3)} = \frac{-2 \pm \sqrt{28}}{6}$$
$$= \frac{-2 \pm 2\sqrt{7}}{6} = \frac{-1 \pm \sqrt{7}}{3},$$
$$\therefore\ c \approx -1.22 \quad \text{or} \quad c \approx 0.55.$$

But since $-1.22 \notin (0, 1)$, we accept only the other alternative: $c = \dfrac{-1 + \sqrt{7}}{3} \approx 0.55$.

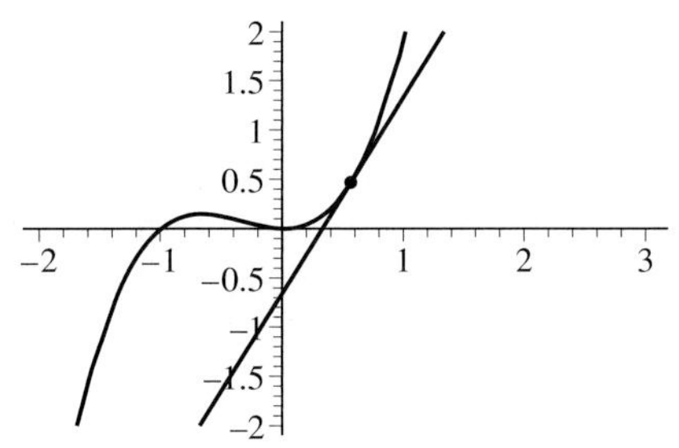

5. $f(x) = \sin x$, $\left[0, \dfrac{\pi}{2}\right]$, $f(0) = 0$, $f(\pi/2) = 1$. As a trig function, $f(x)$ is continuous on $\left[0, \dfrac{\pi}{2}\right]$ and differentiable on $\left(0, \dfrac{\pi}{2}\right)$. The conditions of the Mean Value Theorem hold, and there exists $c \in \left(0, \dfrac{\pi}{2}\right)$ such that

$$f'(c) = \frac{f\left(\frac{\pi}{2}\right) - f(0)}{\frac{\pi}{2} - 0} = \frac{1 - 0}{\frac{\pi}{2} - 0} = \frac{2}{\pi}.$$

But $f'(c) = \cos(c)$ and c is to be in the first quadrant, therefore,

$$c = \cos^{-1}\left(\frac{2}{\pi}\right) \approx .88.$$

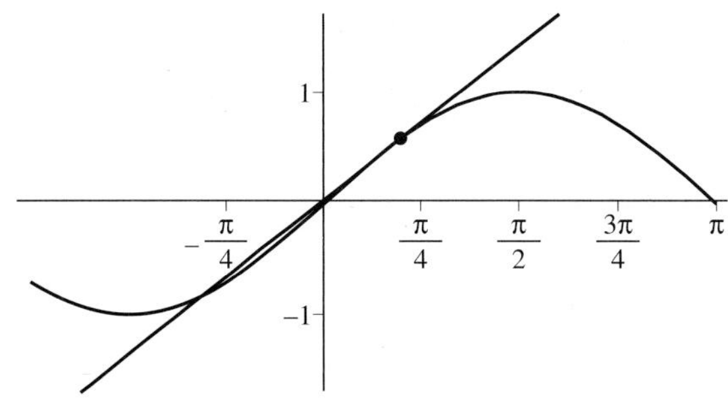

7. If $f'(x) > 0$ for all x, then for each (a, b) with $a < b$ we know there exists a $c \in (a, b)$ such that

$$\frac{f(b) - f(a)}{b - a} = f'(c) > 0.$$

$a < b$ makes the denominator positive, and so we must have the numerator also positive, $\therefore\ f(a) < f(b)$.

9. $f(x) = x^3 + 5x + 1$
$f'(x) = 3x^2 + 5$. This is positive for all x, so $f(x)$ is increasing.

11. $f(x) = -x^3 - 3x + 1$
$f'(x) = -3x^2 - 3$. This is negative for all x, so $f(x)$ is decreasing.

13. $f(x) = e^x$
$f'(x) = e^x$. This is positive for all x, so $f(x)$ is increasing.

15. $f(x) = \ln x$
$f'(x) = \frac{1}{x}$
$f'(x) > 0$ for $x > 0$, that is, for all x in the domain of f. So $f(x)$ is increasing.

17. $f(x) = x^3 + 5x + 1$

As a polynomial, $f(x)$ is continuous and differentiable for all x, with $f'(x) = 3x^2 + 5$ (the answer in the text appendix is in error on this), which is positive for all x.

Simple checking around gives $f(1) = 7$ and $f(-1) = -5$ $(-5 < 0 < 7)$, so the Intermediate Value Theorem gives at least one number $c \in (-1, 1)$ such that $f(c) = 0$.

By Theorem 9.2 , if the equation $f(x)=0$ had *two* (or more) solutions, then the equation $f'(x)=0$ would have *at least one* solution, whereas in fact it has *none.* Thus, the solution (c) to $f(x)=0$ is unique.

19. $f(x)=x^4+3x^2-2$. A fourth degree polynomial like this, with leading coefficient positive (1 in this case) will have limit $+\infty$ as $x\to+\infty$ and also as $x\to-\infty$. Since the value in this case is -2 at $x=0$, it follows by the Intermediate Value Theorem that f has at least two roots, one positive and one negative. The derivative

$$f'(x)=4x^3+6x=x(4x^2+6)$$

is evidently zero only once, at $x=0$. Therefore there can be at most two roots and hence exactly two roots.

An aside: Because the function is *even*, the roots are negatives of one another, and in this case they can even be found by radicals. By the quadratic formula,

$$0=f(x)=(x^2)^2+3x^2-2 \text{ makes}$$
$$x^2=\frac{-3\pm\sqrt{9+8}}{2}=\frac{-3+\sqrt{17}}{2}$$

(we must take the "plus" option because $x^2>0$). Thus, the two roots are

$$\pm\sqrt{\frac{-3+\sqrt{17}}{2}}\approx\pm0.74937.$$

21. $f(x)=x^3+ax+b$, $a>0$. Any cubic (actually any *odd degree*) polynomial heads in opposite directions ($\pm\infty$) as x goes to the oppositely signed infinities, and therefore by the Intermediate Value Theorem has at least one root. For the uniqueness, we look at the derivative, in this case $3x^2+a$. Because $a>0$ by assumption, this expression is strictly positive. The function is strictly increasing and can have at most one root.

23. $f(x)=x^5+ax^3+bx+c$, $a>0$, $b>0$. Here is another odd degree polynomial (see Exercise 21) with at least one root. $f'(x)=5x^4+3ax^2+b$ is evidently strictly positive because of our assumptions about a, b. Exactly as in Exercise 21, there can be at most one root.

25. As a fourth degree polynomial,

$$f(x)=ax^4+bx^3+cx^2+dx+e$$

has a cubic derivative and a quadratic second derivative. The quadratic, as is well known, can have at most two roots. It follows that the derivative (the cubic) can have at most three roots, and that the original can have at most four. We are using over and over the principle by which the roots of a function are separated by roots of the derivative (Theorem 9.2).

27. We have learned in previous sections how to get at least one antiderivative (g_o) by raising the power up one and adjusting a constant multiplier. In this section we have gained the additional information that any two antiderivatives on a common interval differ from one another by a constant. Therefore, from one, we can get all others by adding an "arbitrary" constant. In this case,

$$f(x)=x^2.$$

One candidate: $g_0(x)=kx^3$
Because we require $x^2=g_0'(x)=3kx^2$, we must have $3k=1$, $k=1/3$.
Most general solution:

$$g(x)=g_0(x)+c=\frac{x^3}{3}+c.$$

29.

$$f(x)=x^4$$
$$g(x)=\frac{x^5}{5}+c$$

We have done as in Exercise 27 without showing so many steps.

31. This one has a wrinkle. See our comments for Exercise 27. Although the obvious first candidate is

$$g_0(x)=\frac{-1}{x},$$

due to the disconnection of the domain by the discontinuity at $x = 0$, we could add *different* constants, one for negative x, another for positive x. Thus, the answer *could* appear as

$$g(x) = \begin{cases} \frac{-1}{x} + a & \text{for } x > 0 \\ \frac{-1}{x} + b & \text{for } x < 0. \end{cases}$$

33. $f(x) = \sin x$
$g(x) = -\cos x + c$

35. $f(x) = \frac{1}{x}$ on $[-1, 1]$. We easily see that $f(1) = 1$, $f(-1) = -1$, and $f'(x) = \frac{-1}{x^2}$. If we try to find the c in the interval $(-1, 1)$ for which

$$f'(c) = \frac{f(1) - f(1)}{1 - (-1)} = \frac{1 - (-1)}{1 - (-1)} = \frac{2}{2} = 1,$$

the equation would be $\frac{-1}{c^2} = 1$ or $c^2 = -1$. There is of course no such c, and the explanation is that the function is not defined for $x = 0 \in (-1, 1)$. The hypotheses for the Mean Value Theorem are not fulfilled.

37. $f(x) = \tan x$ on $[0, \pi]$, $f'(x) = \sec^2(x)$. We know the tangent has a massive discontinuity at $x = \frac{\pi}{2}$, so, as in Exercise 35, we should not be surprised if the Mean Value Theorem does not apply. As applied to the interval $[0, \pi]$, it would say

$$\sec^2(c) = f'(c) = \frac{f(\pi) - f(0)}{\pi - 0} = \frac{\tan \pi - \tan 0}{\pi - 0} = 0.$$

But secant = 1/cosine is never in the interval $(-1, 1)$, so no such c exists.

39. If the student will look back through our solutions to Exercise 43, Section 2.2 , he/she will find that there we showed that if a derivative g' is positive at a single point $x = b$, then $g(x) > g(b)$ for $x > b$ but sufficiently near b. In this problem, we will apply that remark to $g = f'$ at $x = 0$, and conclude from $f''(0) > 0$ that $f'(x) > f'(0) = 0$ for $x > 0$ but sufficiently small. This being true about the derivative f', it tells that f itself is increasing on some interval $(0, a)$, and in particular that $f(x) > f(0) = 0$ for $0 < x < a$. On the other side (the negative side), f' is negative, f is decreasing (to zero) and therefore likewise positive. In summary, $x = 0$ is a genuine relative minimum.

41. This problem has a lot in common with Exercise 49 in Section 2.6, and we advised at that time that a proper solution would utilize the Mean Value Theorem. In our solution to Exercise 49 (Section 2.6), we proved and utilized the inequality $\sin(a) < a$, valid whenever $0 < a$. We are still not certain how the weaker inequality $|\sin(a)| \leq |a|$ (Example 9.4) was intended to be used, nor do we care to repeat our thorough solution to the earlier problem, but we *will* show how the *stronger* inequality follows from the Mean Value Theorem.

Consider the function $g(x) = x - \sin(x)$, obviously with $g(0) = 0$ and $g'(x) = 1 - \cos(x)$. If there was ever a point $a > 0$ with $\sin(a) \geq a$, $(g(a) \leq 0)$, then by the MVT applied to g on the interval $[0, a]$ there would be a point $c(0 < c < a)$ with

$$g'(c) = \frac{g(a) - g(0)}{a - 0} = \frac{g(a)}{a} \leq 0.$$

This would read $1 - \cos(c) = g'(c) \leq 0$ or $\cos(c) \geq 1$. The latter condition is possible only if $\cos(c) = 1$ and $\sin(c) = 0$, in which case c (being positive) would be *at minimum* π. But even in this unlikely case we still would have $\sin(a) \leq 1 < \pi \leq c < a$.

43. Since the inverse sine function is increasing on the interval $[0, 1)$ (it has a positive derivative), we start from the previously proven inequality $\sin(x) < x$ for $0 < x$. If indeed $0 < x < 1$, we can apply the inverse sine and conclude

$$x = \sin^{-1}(\sin(x)) < \sin^{-1}(x).$$

45. $f(x) = \begin{cases} 2x & x \leq 0 \\ 2x - 4 & x > 0 \end{cases}$

$f(x) = 2x - 4$ is continuous and differentiable on $(0, 2)$. Also, $f(0) = 0 = f(2)$. But $f'(x) \equiv 2$ on $(0, 2)$, so there is no c such that $f'(c) = 0$. Rolle's Theorem requires that $f(x)$ be continuous on the closed interval, but we have a jump discontinuity at $x = 0$, which is enough to preclude the applicability of Rolle's.

Chapter 2 Review

True/False

1. false ("corners")

3. false (the derivative gives the slope of tangent lines to the graph of a function)

5. true

7. false (the order does matter)

9. false ($f'(x) = 4\cos(4x)$, $f'(0) = 4$)

11. true ($\ln(ax) = \ln(a) + \ln(x)$)

13. Not exactly. We would consider Rolle's to be the special case, essential in proving the general case.

Regular Exercises

1.

Interval	Average
(0, 1)	$\dfrac{3.0 - 2.0}{10} = 1.0$
(.5, 1)	$\dfrac{3.0 - 2.6}{.5} = 0.8$
(1, 1.5)	$\dfrac{3.4 - 3.0}{.5} = 0.8$
(1, 2)	$\dfrac{4.0 - 3.0}{10} = 1.0$

Using the shortest intervals, (0.5, 1.0) and (1.0, 1.5), gives the most promising estimates for $f'(1)$, namely 0.8.

3. Find $f'(2)$ for $f(x) = x^2 - 2x$. Noting $f(2) = 0$, and assuming $h \neq 0$,

$$\begin{aligned} \frac{f(2+h) - f(2)}{h} &= \frac{(2+h)^2 - 2(2+h) - (0)}{h} \\ &= \frac{4 + 4h + h^2 - 4 - 2h}{h} \\ &= \frac{2h + h^2}{h} = 2 + h \\ f'(2) &= \lim_{h\to 0} \frac{f(2+h) - f(2)}{h} \\ &= \lim_{h\to 0} 2 + h = 2. \end{aligned}$$

5. Find $f'(1)$ for $f(x) = \sqrt{x}$.
Noting $f(1) = 1$, and assuming $h \neq 0$,

$$\begin{aligned} \frac{f(1+h) - f(1)}{h} &= \frac{\sqrt{1+h} - 1}{h} \\ &= \frac{\sqrt{1+h} - 1}{h} \cdot \frac{\sqrt{1+h} + 1}{\sqrt{1+h} + 1} \\ &= \frac{1 + h - 1}{h(\sqrt{1+h} + 1)} = \frac{1}{\sqrt{1+h} + 1} \\ f'(1) &= \lim_{h\to 0} \frac{f(1+h) - f(1)}{h} \\ &= \lim_{h\to 0} \frac{1}{\sqrt{1+h} + 1} = \frac{1}{2}. \end{aligned}$$

7. Find $f'(x)$for $f(x) = x^3 + x$.
Assuming h not zero,

$$\begin{aligned} \frac{f(x+h) - f(x)}{h} &= \frac{(x+h)^3 + (x+h) - (x^3 + x)}{h} \\ &= \frac{\begin{pmatrix} x^3 + 3x^2h + 3xh^2 + h^3 \\ +x + h - x^3 - x \end{pmatrix}}{h} \end{aligned}$$

$$= \frac{3x^2h + 3xh^2 + h^3 + h}{h}$$
$$= 3x^2 + 3xh + h^2 + 1.$$
$$f'(x) = \lim_{h\to 0} \frac{f(x+h) - f(x)}{h}$$
$$= \lim_{h\to 0} 3x^2 + 3xh + h^2 + 1$$
$$= 3x^2 + 1.$$

9. Find the slope of the tangent line to

$$y = f(x) = x^4 - 2x + 1 \quad \text{at} \quad x = 1.$$

Noting $f(1) = 0$ and assuming h not zero,

$$\frac{f(1+h) - f(1)}{h}$$
$$= \frac{(1+h)^4 - 2(1+h) + 1 - (0)}{h}$$
$$= \frac{1 + 4h + 6h^2 + 4h^3 + h^4 - 2 - 2h + 1}{h}$$
$$= \frac{2h + 6h^2 + 4h^3 + h^4}{h}$$
$$= 2 + 6h + 4h^2 + h^3$$
$$m_{\tan} = \lim_{h\to 0} \frac{f(1+h) - f(1)}{h}$$
$$= \lim_{h\to 0} 2 + 6h + 4h^2 + h^3 = 2.$$

The equation of the tangent line is

$$y - 0 = 2(x - 1) \quad \text{or} \quad y = 2x - 2.$$

11. Find the tangent line to

$$y = f(x) = 3e^{2x} \quad \text{at} \quad x = 0.$$

Clearly $f(0) = 3$. Difference quotient method not feasible here.

$$\text{Know: } f'(x) = 6e^{2x}$$
$$\therefore\ f'(0) = 6e^{2\cdot 0} = 6.$$

So the equation of the tangent line is

$$y - 3 = 6(x - 0) \quad \text{or} \quad y = 6x + 3.$$

13. Find the slope to $y - x^2y^2 = x - 1$ at $(1, 1)$.

$$\frac{d}{dx}(y - x^2y^2) = \frac{d}{dx}(x - 1)$$
$$y' - 2xy^2 - x^2 2y \cdot y' = 1$$
$$y'(1 - x^2 2y) = 1 + 2xy^2$$
$$y' = \frac{1 + 2xy^2}{1 - 2x^2y}$$

At $(1, 1)$: $y' = \dfrac{1 + 2(1)(1)^2}{1 - 2(1)^2(1)} = \dfrac{3}{-1} = -3$

The equation of the tangent line is

$$y - 1 = -3(x - 1) \quad \text{or} \quad y = -3x + 4.$$

15.
$$s(t) = -16t^2 + 40t + 10$$
$$v(t) = s'(t) = -32t + 40$$
$$a(t) = v'(t) = -32$$

17.
$$s(t) = 10e^{-2t} \sin 4t$$
$$v(t) = s'(t) = 10\begin{pmatrix} -2e^{-2t} \sin 4t \\ +4e^{-2t} \cos 4t \end{pmatrix}$$
$$a(t) = v'(t) = 10 \cdot (-2)\begin{bmatrix} -2e^{-2t} \sin 4t \\ +e^{-2t} 4 \cos 4t \end{bmatrix}$$
$$+ 10(4) \cdot \begin{bmatrix} -2e^{-2t} \cos 4t \\ -e^{-2t} 4 \sin 4t \end{bmatrix}$$
$$= 160e^{-2t} \cos 4t - 120e^{-2t} \sin 4t$$

19.
$$s(t) = -16t^2 + 40t + 10$$
$$v(t) = -32t + 40$$
$$v(1) = -32(1) + 40 = 8.$$

The ball is rising.

$$v(2) = -32(2) + 40 = 24.$$

The ball is falling.

21. $f(x) = \sqrt{x + 1}$

(a)
$$m_{\sec} = \frac{f(2) - f(1)}{2 - 1}$$
$$= \frac{\sqrt{3} - \sqrt{2}}{1} \approx .318$$

(b)
$$m_{\sec} = \frac{f(1.5) - f(1)}{1.5 - 1} = \frac{\sqrt{2.5} - \sqrt{2}}{.5} \approx .334$$

(c)
$$m_{\sec} = \frac{f(1.1) - f(1)}{1.1 - 1} = \frac{\sqrt{2.1} - \sqrt{2}}{.1} \approx .349$$

The best estimate for the slope of the tangent line is **(c)** (approximately .349).

23.
$$f(x) = x^4 - 3x^3 + 2x - 1$$
$$f'(x) = 4x^3 - 9x^2 + 2$$

25.
$$f(x) = \frac{3}{\sqrt{x}} + \frac{5}{x^2} = 3x^{-1/2} + 5x^{-2}$$
$$f'(x) = -\frac{3}{2}x^{-3/2} - 10x^{-3} = \frac{-3}{2x\sqrt{x}} - \frac{10}{x^3}$$

27.
$$f(t) = t^2(t+2)^3$$
$$f'(t) = 2t(t+2)^3 + t^2 \cdot 3(t+2)^2 \cdot 1 = 2t(t+2)^3 + 3t^2(t+2)^2 = t(t+2)^2(5t+4)$$

29.
$$g(x) = \frac{x}{3x^2 - 1}$$
$$g'(x) = \frac{(3x^2 - 1) \cdot 1 - x(6x)}{(3x^2 - 1)^2} = \frac{3x^2 - 1 - 6x^2}{(3x^2 - 1)^2} = -\frac{3x^2 + 1}{(3x^2 - 1)^2}$$

31.
$$f(x) = x^2 \sin x$$
$$f'(x) = 2x \sin x + x^2 \cos x$$

33.
$$f(x) = \tan \sqrt{x}$$
$$f'(x) = \sec^2 \sqrt{x} \cdot \frac{1}{2\sqrt{x}}$$

35.
$$f(t) = t \csc t$$
$$f'(t) = \csc t \cdot 1 + t \cdot (-\csc t \cdot \cot t) = \csc t - t \csc t \cot t$$
$$\csc t(1 - t \cot t)$$

37.
$$u(x) = 2e^{-x^2}$$
$$u'(x) = 2e^{-x^2}(-2x) = -4xe^{-x^2}$$

39.
$$f(x) = x \ln x^2$$
$$f'(x) = 1 \cdot \ln x^2 + x \cdot \frac{1}{x^2} \cdot 2x = \ln x^2 + 2$$

41.
$$f(x) = \sqrt{\sin 4x}$$
$$f'(x) = \frac{1}{2\sqrt{\sin 4x}} \cdot \cos 4x \cdot 4 = \frac{2 \cos 4x}{\sqrt{\sin 4x}}$$

43.
$$f(x) = \left(\frac{x+1}{x-1}\right)^2$$
$$f'(x) = 2\left(\frac{x+1}{x-1}\right)\frac{d}{dx}\left(\frac{x+1}{x-1}\right) = 2\left(\frac{x+1}{x-1}\right)\frac{1 \cdot (x-1) - (x+1) \cdot 1}{(x-1)^2} = 2\left(\frac{x+1}{x-1}\right)\frac{-2}{(x-1)^2} = \frac{-4(x+1)}{(x-1)^3}$$

45.
$$f(t) = te^{4t}$$
$$f'(t) = e^{4t} \cdot 1 + te^{4t} \cdot 4 = (1 + 4t)e^{4t}$$

47.
$$f(x) = \sin^{-1} 2x$$
$$f'(x) = \frac{1}{\sqrt{1 - (2x)^2}} \cdot 2 = \frac{2}{\sqrt{1 - 4x^2}}$$

49.
$$f(x) = \tan^{-1}(\cos 2x)$$
$$f'(x) = \frac{1}{1 + (\cos 2x)^2} \cdot (-2 \sin 2x) = -\frac{2 \sin 2x}{1 + \cos^2 2x}$$

51. We provide in the picture simultaneous graphs of the function (bold) and its derivative (dotted).

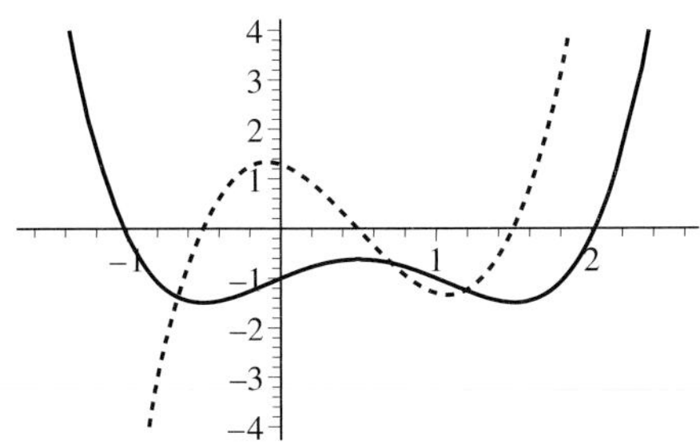

We did this by guessing the formula, which the student was not expected to do. The student's success is evidenced by any reasonably similar sketch. The features to be noted are that the function is decreasing wherever the derivative is negative, and increasing wherever the derivative is positive.

53.
$$f(x) = x^4 - 3x^3 + 2x^2 - x - 1$$
$$f'(x) = 4x^3 - 9x^2 + 4x - 1$$
$$f''(x) = 12x^2 - 18x + 4$$

55.
$$f(x) = xe^{2x}$$
$$f'(x) = 1 \cdot e^{2x} + xe^{2x} \cdot 2 = e^{2x} + 2xe^{2x}$$
$$f''(x) = e^{2x} \cdot 2 + 2 \cdot \left(e^{2x} + 2xe^{2x}\right)$$
$$= 4e^{2x} + 4xe^{2x} \left(= 4(1 + x)e^{2x}\right)$$
$$f'''(x) = 4e^{2x} \cdot 2 + 4\left(e^{2x} + 2xe^{2x}\right)$$
$$= 12e^{2x} + 8xe^{2x} = 4(3 + 2x)e^{2x}$$

57.
$$f(x) = \tan x$$
$$f'(x) = \sec^2 x$$
$$f''(x) = 2 \sec x \cdot \sec x \tan x$$
$$= 2 \sec^2 x \tan x$$

59.
$$f(x) = \sin 3x$$
$$f'(x) = \cos 3x \cdot 3 = 3 \cos 3x$$
$$f''(x) = 3(-\sin 3x \cdot 3) = -9 \sin 3x$$
$$f'''(x) = -9 \cos 3x \cdot 3 = -27 \cos 3x$$
$$f^{(26)}(x) = -3^{26} \sin 3x$$

61.
$$R(t) = P(t)Q(t)$$
$$R'(t) = Q'(t) \cdot P(t) + Q(t) \cdot P'(t)$$
$$P(0) = 2.4(\$)$$
$$Q(0) = 12 \text{ (thousands)}$$
$$Q'(t) = -1.5 \text{ (thousands per year)}$$
$$P'(t) = 0.1 \text{ (\$ per year)}$$
$$R'(0) = (-1.5) \cdot (2.4) + 12 \cdot (0.1)$$
$$= -2.4 \text{ (thousand \$ per year)}$$

Revenue is decreasing at a rate of \$2400 per year.

63.
$$f(t) = 4 \cos 2t$$
$$v(t) = f'(t) = 4(-\sin 2t) \cdot 2 = -8 \sin 2t$$

(a) The velocity is zero when $v(t) = -8 \sin 2t = 0$.

$$2t = 0,\ \pi,\ 2\pi,\ \ldots \text{ so } t = 0,\ \frac{\pi}{2},\ \pi\frac{3\pi}{2},\ \ldots$$
$$f(t) = 4cos2t = 4 \text{ for } t = 0,\ ?,\ 2?,\ \ldots$$
$$f(t) = 4 \cos 2t = -4 \text{ for } t = \frac{\pi}{2},\ \frac{3\pi}{2},\ \ldots$$

The position of the spring when the velocity is zero is 4 or −4.

(b) The velocity is a maximum when $v(t) = -8 \sin 2t = 8$.

$$2t = \frac{3\pi}{2},\ \frac{7\pi}{2},\ \ldots \text{ so } t = \frac{3\pi}{4},\ \frac{7\pi}{4},\ \ldots$$
$$f(t) = 4 \cos 2t = 0 \text{ for } t = \frac{3\pi}{4},\ \frac{7\pi}{4},\ \ldots$$

The position of the spring when the velocity is at a maximum is zero.

(c) Velocity is at a minimum when $v(t) = -8 \sin 2t = -8$.

$$2t = \frac{\pi}{2},\ \frac{5\pi}{2},\ \ldots \text{ so } t = \frac{\pi}{4},\ \frac{5\pi}{4},\ \ldots$$
$$f(t) = 4 \cos 2t = 0 \text{ for } t = \frac{\pi}{4},\ \frac{5\pi}{4},\ \ldots$$

The position of the spring when the velocity is at a minimum is also zero.

65.
$$x^2y - 3y^3 = x^2 + 1$$
$$\frac{d}{dx}(x^2y - 3y^3) = \frac{d}{dx}(x^2 + 1)$$
$$2xy + x^2y' - 3 \cdot 3y^2 \cdot y' = 2x$$
$$y'(x^2 - 9y^2) = 2x - 2xy$$
$$y' = \frac{2x(1-y)}{x^2 - 9y^2}$$

67.
$$\frac{y}{x+1} - 3y = \tan x$$
$$\frac{d}{dx}\left(\frac{y}{x+1} - 3y\right) = \frac{d}{dx}\tan x$$
$$\frac{(x+1)y' - y \cdot (1)}{(x+1)^2} - 3y' = \sec^2 x$$
$$y'(x+1) - y = (x+1)^2(3y' + \sec^2 x)$$
$$y' = \frac{\sec^2 x(x+1)^2 + y}{(x+1)[1 - 3(x+1)]}$$

69. When $x = 0$, $-3y^3 = 1$, $y = \frac{-1}{\sqrt[3]{3}}$ (call this a).
From our formula (Exercise 65) , we find $y' = 0$. Meanwhile, back at the third line of Exercise 65, we differentiate, learning

$$2(xy' + y) + (2xy' + x^2y'') - 9\left[2y(y')^2 + y^2y''\right] = 2.$$

At $(0, a)$ with $y' = 0$, we find $2a - 9a^2y'' = 2$,

$$y'' = \frac{-2\sqrt[3]{3}}{9}\left(\sqrt[3]{3} + 1\right)$$
$$x^2y - 3y^3 = x^2 + 1.$$

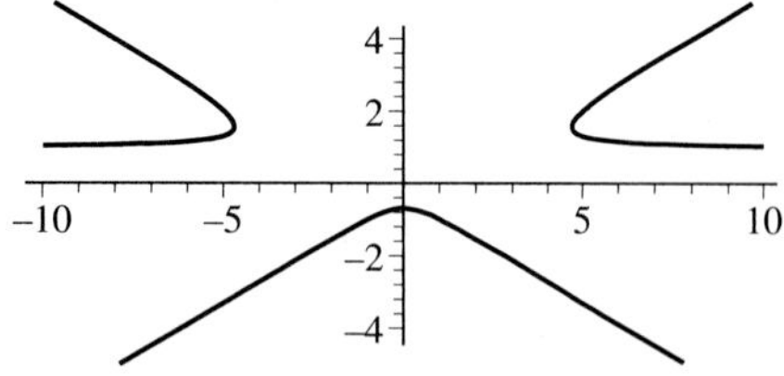

71.
$$y = x^3 - 6x^2 + 1$$
$$y' = 3x^2 - 12x = 3x(x - 4)$$

(a) $y' = 0$ for $x = 0$ $(y = 1)$, and $x = 4(y = -31)$, so there are horizontal tangent lines at $(0, 1)$ and $(4, -31)$.

(b) y' is defined for all x, so there are no vertical tangent lines.

73.
$$x^2y - 4y = x^2$$
$$\frac{d}{dx}(x^2y - 4y) = \frac{d}{dx}x^2$$
$$2xy + x^2y' - 4y' = 2x$$
$$y'(x^2 - 4) = 2x - 2xy$$
$$y' = \frac{2x - 2xy}{x^2 - 4} = \frac{2x(1-y)}{x^2 - 4}$$

(a) $y' = 0$ when $x = 0$ or $y = 1$.

At $y = 1$, $x^2 \cdot 1 - 4 \cdot 1 = x^2$

$x^2 - 4 = x^2$ (*never*).

There is no x for which $y = 1$.

At $x = 0$, $0^2 \cdot y - 4y = 0^2$, so $y = 0$.

Therefore, there is a horizontal tangent line at $(0, 0)$.

(b) y' is not defined when $x^2 - 4 = 0$, or $x = \pm 2$. At $x = \pm 2$, $4y - 4y = 4$ (*never*), so the function is not defined at $x = \pm 2$. There are no vertical tangent lines.

75. $f(x) = x^3 + 7x - 1$
$f(x)$ is continuous and differentiable for all x, and $f'(x) = 3x^2 + 7$, which is positive for all x. By Theorem 9.2, if the equation $f(x) = 0$ has two solutions, then $f'(x) = 0$ would have at least one solution, but it has none. We discussed at length (Section 2.9) why every odd degree polynomial has at least one root, so in this case there is exactly one root.

77. $f(x) = x^5 + 2x^3 - 1$ is a one-to-one function with $f(1) = 2$, $f'(1) = 11$. If g is the name of the inverse, then $g(2) = 1$ and

$$g'(-2) = \frac{1}{f'(g(2))} = \frac{1}{f'(1)} = \frac{1}{11}.$$

79. The inverse sine function picks out for any $x \in [-1, 1]$ the angle θ in the interval $\left[\frac{-\pi}{2}, \frac{\pi}{2}\right]$ whose sine is x. There can be absolutely no question that among all the possible angles whose sine is 1, $sin^{-1}(1) = \pi/2$.

81. The inverse tangent function picks out for any x the angle θ in the interval

$$\left(\frac{-\pi}{2}, \frac{\pi}{2}\right)$$

whose tangent is x. There can be absolutely no question that among all the possible angles whose tangent is -1, $tan^{-1}(-1) = -\pi/4$.

83. $\sin(\sec^{-1} 2)$? Call $\sec^{-1}(2)$ by the temporary name a. Then $\sec(a) = 2$, $\cos(a) = 1/2$, a itself is $\pi/3$ (60°), and its sine is $\sqrt{3}/2$.

85. $\begin{cases} x = t^3 - 3t \\ y = t^2 - t + 1 \end{cases}$

(a) At $t = 0$, $x = 0$, $y = 1$, and

$$m = \frac{y'(0)}{x'(0)} = \left.\frac{2t-1}{3t^2-3}\right|_{t=0} = \frac{1}{3}.$$

(b) At $t = 1$, $x = -2$, $y = 1$, and

$$m = \frac{y'(1)}{x'(1)} = \left.\frac{2t-1}{3t^2-3}\right|_{t=1}.$$

The denominator is zero; the slope is undefined but geometrically vertical.

(c) At $(x, y) = (2, 3)$, $3 = y = t^2 - t + 1$, $t^2 - t - 2 = 0$, $(t-2)(t+1) = 0$, t is 2 or -1. But -1 does not work for $x = -1$ (it gives $x = 0$), so indeed $t = 2$ (which does give $x = 2$).

$$m = \frac{y'(2)}{x'(2)} = \left.\frac{2t-1}{3t^2-3}\right|_{t=2} = \frac{3}{9} = \frac{1}{3}.$$

87. $\begin{cases} x = t^3 - 3t \\ y = t^2 + 2t \end{cases}$

The horizontal component of velocity at $t = 0$ is

$$\frac{dx}{dt} = 3t^2 - 3\Big|_{t=0} = -3$$

and the vertical component of velocity at $t = 0$ is

$$\frac{dy}{dt} = 2t + 2|_{t=0} = 2.$$

The speed, therefore, at $t = 0$ is

$$\sqrt{\left(\frac{dx}{dt}\right)^2 + \left(\frac{dy}{dt}\right)^2} = \sqrt{(-3)^2 + (2)^2} = \sqrt{13}.$$

At $t = 0$ the position is $(0, 0)$, after which the object moves to the left and up.

89.

$$f(x) = x^2 - 2x \text{ on } [0, 2]$$

$$f(2) = 0 = f(0)$$

$$\text{If } f'(c) = \frac{f(2) - f(0)}{2 - 0} = \frac{0-0}{2} = 0,$$

then $2c - 2 = f'(c) = 0$, so $c = 1$.

91.

$$f(x) = 3x^2 - \cos x$$

One trial: $g_o(x) = kx^3 - \sin x$

$$g_o'(x) = 3kx^2 - \cos x.$$

Need $3k = 3$, $k = 1$, and the general solution is

$$g(x) = g_o(x) + c = x^3 - \sin x + c.$$

93. $x = 1$ is to be double root of

$$f(x) = (x^3 + 1) - [m(x-1) + 2].$$

Rewriting,

$$f(x) = (x^3 - 1) - m(x - 1)$$
$$= (x - 1)\left[x^2 + x + 1 - m\right].$$

Let $g(x) = x^2 + x + 1 - m$. Then $x = 1$ is a *double* root of f only if $(x - 1)$ is a *factor* of g, in which case $g(1) = 0$. Therefore we require $0 = g(1) = 3 - m$ or $m = 3$. Now,

$$g(x) = x^2 + x - 2 = (x - 1)(x + 2),$$
$$f(x) = (x - 1)g(x) = (x - 1)^2(x + 2),$$

and $x = 1$ is a double root.

The line tangent to the curve $y = x^3 + 1$ at the point $(1, 2)$ has slope $y' = 3x^2 = 3(1) = 3(= m)$. The equation of the tangent line is $y - 2 = 3(x - 1)$ or $y = 3x - 1(= m(x - 1) + 2)$.

Chapter 3

Applications of Differentiation

3.1 Linear Approximation and Newton's Method

1.

$$f(x_0) = f(1) = \sqrt{1} = 1$$
$$f'(x) = \frac{1}{2}x^{-1/2}$$
$$f'(x_0) = f'(1) = \frac{1}{2}$$

So

$$L(x) = f(x_0) + f'(x_0)(x - x_0)$$
$$= 1 + \frac{1}{2}(x - 1)$$
$$= \frac{1}{2} + \frac{1}{2}x$$

3.

$$f(x) = \sqrt{2x + 9}, \ x_0 = 0$$
$$f(x_0) = f(0) = \sqrt{2 \cdot 0 + 9} = 3$$
$$f'(x) = \frac{1}{2}(2x + 9)^{-1/2} \cdot 2 = (2x + 9)^{-1/2}$$
$$f'(x_0) = f'(0) = (2 \cdot 0 + 9)^{-1/2} = \frac{1}{3}$$

So

$$L(x) = f(x_0) + f'(x_0)(x - x_0)$$
$$= 3 + \frac{1}{3}(x - 0)$$
$$= 3 + \frac{1}{3}x$$

5.

$$f(x) = \sin 3x, \ x_0 = 0$$
$$f(x_0) = f(0) = \sin(3 \cdot 0) = \sin 0 = 0$$
$$f'(x) = 3 \cos 3x$$
$$f'(x_0) = f'(0) = 3 \cos 3 \cdot 0 = 3$$
$$L(x) = f(x_0) + f'(x_0)(x - x_0)$$
$$= 0 + 3(x - 0)$$
$$= 3x$$

7. In each case, $L(x) = f(0) + f'(0)x$.
In each case, $f(0) = 1$ and $f'(0) = 2$, so in all three cases $L(x) = 1 + 2x$.

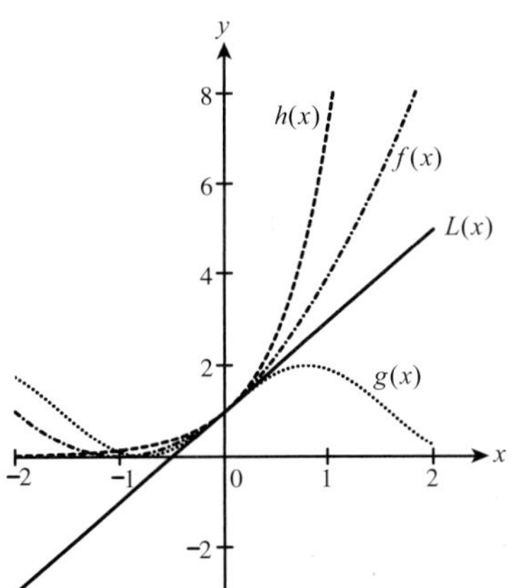

9.

$$f(x) = \sin x, \ x_0 = \frac{\pi}{3}$$
$$f\left(\frac{\pi}{3}\right) = \frac{\sqrt{3}}{2}$$
$$f'(x) = \cos x$$
$$f'\left(\frac{\pi}{3}\right) = \cos\frac{\pi}{3} = \frac{1}{2}$$

$$L(x) = f\left(\frac{\pi}{3}\right) + f'\left(\frac{\pi}{3}\right)\left(x - \frac{\pi}{3}\right)$$
$$= \frac{\sqrt{3}}{2} + \frac{1}{2}\left(x - \frac{\pi}{3}\right)$$
$$L(1) = \frac{\sqrt{3}}{2} + \frac{1}{2}\left(1 - \frac{\pi}{3}\right) \approx .842$$

11. $f(x) = \sqrt[4]{16 + x},\ x_0 = 0$

$$f(0) = \sqrt[4]{16 + 0} = 2$$
$$f'(x) = \frac{1}{4}(16 + x)^{-3/4}$$
$$f'(0) = \frac{1}{4}(16 + 0)^{-3/4} = \frac{1}{32}$$
$$L(x) = f(0) + f'(0)(x - 0) = 2 + \frac{1}{32}x$$
$$L(0.04) = 2 + \frac{1}{32}(0.04) = 2.00125$$

13. $f(x) = \sqrt[4]{16 + x},\ x_0 = 0$

$$L(x) = 2 + \frac{1}{32}x \text{ (from Exercise 11)}$$
$$L(.16) = 2 + \frac{1}{32}(.16) = 2.005$$

15.

$$\sqrt[4]{16.04} = 2.0012488$$
$$L(0.04) = 2.00125$$
$$|2.0012488 - 2.00125| = .00000117$$

$$\sqrt[4]{16.08} = 2.0024953$$
$$L(.08) = 2.0025$$
$$|2.0024953 - 2.0025| = .00000467$$

$$\sqrt[4]{16.16} = 2.0049814$$
$$L(.16) = 2.005$$
$$|2.0049814 - 2.005| = .0000186$$

17. (a)

$$f(24) \approx f(20) + \frac{4}{10}(f(30) - f(20))$$
$$= 18 + 0.4(14 - 18)$$
$$= 16.4$$

(b)

$$f(36) \approx f(30) + \frac{6}{10}(f(40) - f(30))$$
$$= 14 + 0.6(12 - 14)$$
$$= 12.8$$

19. (a)

$$f(208) \approx f(200) + \frac{8}{20}(f(220) - f(200))$$
$$= 128 + 0.4(142 - 128)$$
$$= 133.6$$

(b)

$$f(232) \approx f(220) + \frac{12}{20}(f(240) - f(220))$$
$$= 142 + 0.6(136 - 142)$$
$$= 138.4$$

21. The first tangent line intersects the x-axis at a point a little to the right of 1. So x_1 is about 1.25 (very roughly). The second tangent line intersects the x-axis at a point between 1 and x_1, so x_2 is about 1.1 (very roughly). Newton's Method will converge to the zero at $x = 1$.

23. It wouldn't work because $f'(0) = 0$.

25.

$$f(x) = x^3 + 3x^2 - 1 = 0,\ x_0 = 1$$
$$f'(x) = 3x^2 + 6x$$

(a)

$$x_1 = x_0 - \frac{f(x_0)}{f'(x_0)}$$
$$= 1 - \frac{1^3 + 3 \cdot 1^2 - 1}{3 \cdot 1^2 + 6 \cdot 1}$$
$$= 1 - \frac{3}{9}$$
$$= \frac{2}{3}$$
$$x_2 = x_1 - \frac{f(x_1)}{f'(x_1)}$$
$$= \frac{2}{3} - \frac{\left(\frac{2}{3}\right)^3 + 3\left(\frac{2}{3}\right)^2 - 1}{3\left(\frac{2}{3}\right)^2 + 6\left(\frac{2}{3}\right)}$$
$$= \frac{79}{144}$$
$$\approx .5486$$

(b) .53209

27. $$f(x) = x^4 - 3x^2 + 1 = 0,\ x_0 = 1$$
$$f'(x) = 4x^3 - 6x$$

(a) $$x_1 = x_0 - \frac{f(x_0)}{f'(x_0)} = 1 - \left(\frac{1^4 - 3 \cdot 1^2 + 1}{4 \cdot 1^3 - 6 \cdot 1}\right) = \frac{1}{2}$$
$$x_2 = x_1 - \frac{f(x_1)}{f'(x_1)}$$
$$= \frac{1}{2} - \left(\frac{\left(\frac{1}{2}\right)^4 - 3\left(\frac{1}{2}\right)^2 + 1}{4\left(\frac{1}{2}\right)^3 - 6\left(\frac{1}{2}\right)}\right)$$
$$= \frac{5}{8}$$

(b) .61803

29. $$f(x) = x^3 + 4x^2 - 3x + 1$$
$$f'(x) = 3x^2 + 8x - 3$$

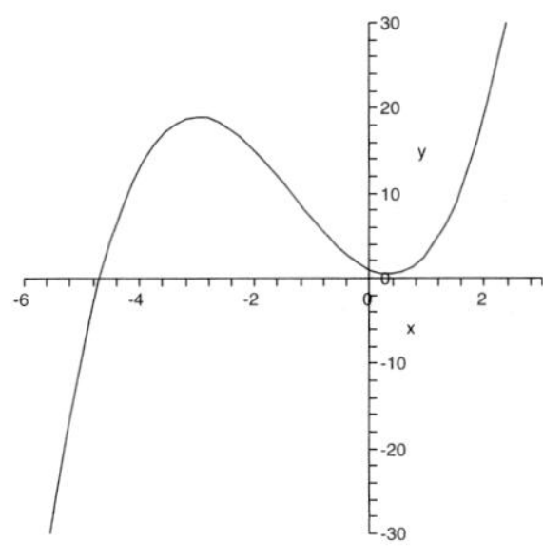

$$x_0 = -5$$
$$x_{n+1} = x_n - \frac{f(x_n)}{f'(x_n)} \quad \text{for} \quad n = 0, 1, 2, 3, \ldots$$

n	x_n
1	−4.718750
2	−4.686202
3	−4.6857796
4	−4.6857795

$x \approx -4.685780$

31. $$f(x) = x^5 + 3x^3 + x - 1$$
$$f'(x) = 5x^4 + 9x^2 + 1$$

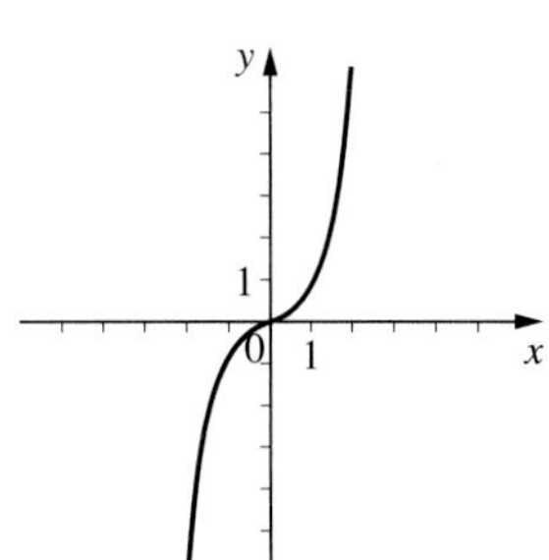

$$x_0 = .5$$
$$x_{n+1} = x_n - \frac{f(x_n)}{f'(x_n)} \quad \text{for} \quad n = 0, 1, 2, 3, \ldots$$

n	x_n
1	.526316
2	.525262
3	.525261
4	.525261

$x \approx .525260$

33. $$f(x) = \cos x - x$$
$$f'(x) = -\sin x - 1$$

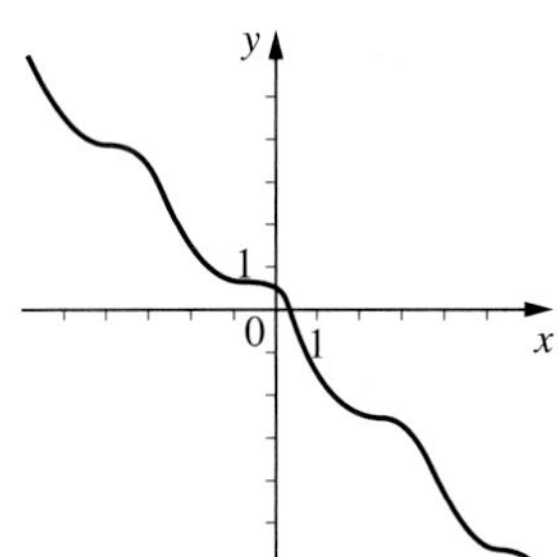

$$x_0 = 1$$
$$x_{n+1} = x_n - \frac{f(x_n)}{f'(x_n)} \quad \text{for} \quad n = 0, 1, 2, 3, \ldots$$

n	x_n
1	.750364
2	.739113
3	.739085
4	.739085

$x \approx .739085$

35.

$$\sin x = x^2 - 1$$
$$f(x) = \sin x - x^2 + 1$$
$$f'(x) = \cos x - 2x$$

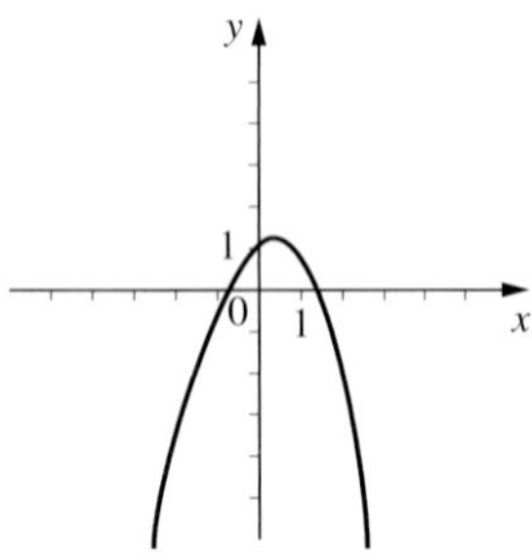

$x_0 = -.5$ or $x = 1.5$

$$x_{n+1} = x_n - \frac{f(x_n)}{f'(x_n)} \quad \text{for} \quad n = 0, 1, 2, 3, \ldots$$

n	x_n
1	−.644108
2	−.636751
3	−.636733
4	−.636733

$x \approx -.636733$
or

n	x_n
1	1.413799
2	1.409634
3	1.409624
4	1.409624

$x \approx 1.409624$

37.

$$e^x = -x$$
$$f(x) = e^x + x$$
$$f'(x) = e^x + 1$$

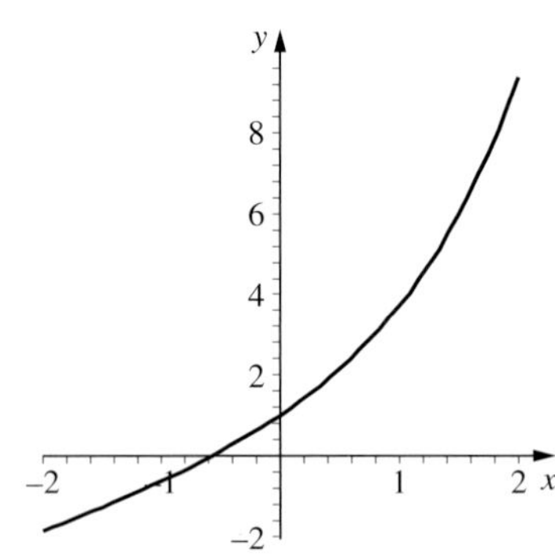

$x_0 = -.5$

$$x_{n+1} = x_n - \frac{f(x_n)}{f'(x_n)} \quad \text{for} \quad n = 0, 1, 2, 3, \ldots$$

n	x_n
1	−.566311
2	−.567143
3	−.567143
4	−.567143

$x \approx -.567143$

39.

$$\begin{aligned} x_{n+1} &= x_n - \frac{f(x_n)}{f'(x_n)} \\ &= x_n - \left(\frac{x_n^2 - c}{2x_n}\right) \\ &= x_n - \frac{x_n^2}{2x_n} + \frac{c}{2x_n} \\ &= \frac{x_n}{2} + \frac{c}{2x_n} \\ &= \frac{1}{2}\left(x_n + \frac{c}{x_n}\right) \end{aligned}$$

If $x_0 < \sqrt{a}$, then $\frac{a}{x_0} > \sqrt{a}$, so $x_0 < \sqrt{a} < \frac{a}{x_0}$.

41. $f(x) = x^2 - 11$; $x_0 = 3$; $\sqrt{11} \approx 3.316625$

43. $f(x) = x^3 - 11$; $x_0 = 2$; $\sqrt[3]{11} \approx 2.22398$

45. $f(x) = x^4 - 24$; $x_0 = 2$; $\sqrt[4]{24} \approx 2.213364$

47. $f(x) = x^{4.4} - 24$; $x_0 = 2$; $\sqrt[4.4]{24} \approx 2.059133$

49.

$$f(x) = 4x^3 - 7x^2 + 1 = 0,\ x_0 = 0$$
$$f'(x) = 12x^2 - 14x$$
$$x_1 = x_0 - \frac{f(x_0)}{f'(x_0)} = 0 - \frac{1}{0}$$

The method fails because $f'(x_0) = 0$.
Roots are .3454, .4362, 1.659.

51.

$$f(x) = x^2 + 1,\ x_0 = 0$$
$$f'(x) = 2x$$
$$x_1 = x_0 - \frac{f(x_0)}{f'(x_0)} = 0 - \frac{1}{0}$$

The method fails because $f'(x_0) = 0$.
There are no roots.

53. $f(x) = \dfrac{4x^2 - 8x + 1}{4x^2 - 3x - 7} = 0,\ x_0 = -1$

Note: $f(x_0) = f(-1)$ is undefined, so Newton's Method fails because x_0 is not in the domain of f. Notice that $f(x) = 0$ only when $4x^2 - 8x + 1 = 0$. So using Newton's Method on $g(x) = 4x^2 - 8x + 1$ with $x_0 = -1$ leads to $x \approx .1339$. The other root is $x \approx 1.8660$.

55. With $x_0 = 1.2$

n	x_n
1	0.800000000
2	0.950000000
3	0.995652174
4	0.999962680
5	0.999999997
6	1.000000000
7	1.000000000

With $x_0 = 2.2$

n	x_n
0	2.200000
1	2.107692
2	2.056342
3	2.028903
4	2.014652
5	2.007378
6	2.003703
7	2.001855
8	2.000928
9	2.000464
10	2.000232
11	2.000116
12	2.000058
13	2.000029
14	2.000015
15	2.000007
16	2.000004
17	2.000002
18	2.000001
19	2.000000
20	2.000000

The convergence is much faster with $x_0 = 1.2$. Indeed, we did these calculations with an Excel spreadsheet. When we did the part (b) calculation displayed to nine decimal places, the answers stabilized at the wrong answer $x_n = 2.000000065$ for $n \geq 22$. This is another example of why you cannot trust your calculator or computer.

57. With $x_0 = -1.1$

n	x_n
1	−1.0507937
2	−1.0256065
3	−1.0128572
4	−1.0064423
5	−1.0032246
6	−1.0016132
7	−1.0008068
8	−1.0004035
9	−1.0002017
10	−1.0001009
11	−1.0000504
12	−1.0000252
13	−1.0000126
14	−1.0000063
15	−1.0000032
16	−1.0000016
17	−1.0000008
18	−1.0000004
19	−1.0000002
20	−1.0000001
21	−1.0000000
22	−1.0000000

With $x_0 = 2.1$

n	x_n
0	2.100000000
1	2.006060606
2	2.000024340
3	2.000000000
4	2.000000000

As with Exercise 55, not only did one calculation (in this case, part (a)) converge more slowly, but when displayed to more decimal places part (a) does not converge to -1.0

59.
$$f(x) = \tan x$$
$$f(0) = \tan 0 = 0$$
$$f'(x) = \sec^2 x$$
$$f'(0) = \sec^2 0 = 1$$
$$L(x) = f(0) + f'(0)(x-0) = 0 + 1(x-0) = x$$
$$L(0.01) = .01$$
$$f(0.01) = \tan 0.01 \approx .0100003$$
$$L(0.1) = 0.1$$
$$f(0.1) = \tan(0.1) \approx .1003$$
$$L(1) = 1$$
$$f(1) = \tan 1 \approx 1.557$$

61.
$$f(x) = \sqrt{4+x}$$
$$f(0) = \sqrt{4+0} = 2$$
$$f'(x) = \frac{1}{2}(4+x)^{-1/2}$$
$$f'(0) = \frac{1}{2}(4+0)^{-1/2} = \frac{1}{4}$$
$$L(x) = f(0) + f'(0)(x-0) = 2 + \frac{1}{4}x$$
$$L(0.01) = 2 + \frac{1}{4}(0.01) = 2.0025$$
$$f(0.01) = \sqrt{4+0.01} \approx 2.002498$$
$$L(0.1) = 2 + \frac{1}{4}(0.1) = 2.025$$
$$f(0.1) = \sqrt{4+0.1} \approx 2.0248$$
$$L(1) = 2 + \frac{1}{4}(1) = 2.25$$
$$f(1) = \sqrt{4+1} \approx 2.2361$$

63. If you graph $|\tan x - x|$, you see that the difference is less than .01 on the interval $-.306 < x < .306$ (In fact, a slightly larger interval would work as well.)

65. For small x we approximate e^x by $x+1$.

$$\frac{Le^{2\pi d/L} - e^{-2\pi d/L}}{e^{2\pi d/L} + e^{-2\pi d/L}} \approx \frac{L\left[\left(1+\frac{2\pi d}{L}\right) - \left(1-\frac{2\pi d}{L}\right)\right]}{\left(1+\frac{2\pi d}{L}\right) + \left(1-\frac{2\pi d}{L}\right)} \approx \frac{L\left(\frac{4\pi d}{L}\right)}{2} = 2\pi d$$
$$f(d) \approx \frac{4.9}{\pi} \cdot 2\pi d = 9.8d$$

67. The smallest positive solution of the first equation is 0.132782, and for the second equation the smallest positive solution is 1, so the species modeled by the second equation is certain to go extinct. This is consistent with the models, since the expected number of offspring for the population modeled by the first equation is 2.2, while for the second equation it is only 1.3.

69. The only positive solution is 0.6407.

71.
$$W(x) = \frac{PR^2}{(R+x)^2}, \; x_0 = 0$$
$$W'(x) = \frac{-2PR^2}{(R+x)^3}$$
$$L(x) = W(x_0) + W'(x_0)(x-x_0)$$
$$= \frac{PR^2}{(R+0)^2} + \left(\frac{-2PR^2}{(R+0)^3}\right)(x-0)$$
$$= P - \frac{2Px}{R}$$

$$L(x) = 120 - .01(120) = P - \frac{2Px}{R}$$
$$= 120 - \frac{2 \cdot 120x}{R}$$
$$.01 = \frac{2x}{R}$$
$$x = .005R = .005(20{,}900{,}000) = 104{,}500 \text{ ft}$$

73. To find the smallest positive solution of $\tan\left(\sqrt{x}\right) = \sqrt{x}$, plot $f(x) = \tan\left(\sqrt{x}\right) - \sqrt{x}$ to see that it crosses the x-axis at approximately $x = 20$. Newton's Method (3 iterations) leads to $L \approx 20.19$.

$$y = \sqrt{L} - \sqrt{L}x - \sqrt{L}\cos\sqrt{L}x + \sin\sqrt{L}x$$
$$= 4.493 - 4.493x - 4.493\cos 4.493x + \sin 4.493x$$

75.
$$0^3 - 3 \cdot 0^2 + 2 \cdot 0 = 0$$
$$1^3 - 3 \cdot 1^2 + 2 \cdot 1 = 0$$
$$2^3 - 3 \cdot 2^2 + 2 \cdot 2 = 0$$

77. **(a)** 0 **(b)** 2 **(c)** 1

3.2 Indeterminate Forms and L'Hôpital's Rule

Note: If there is an easier method than L'Hôpital's Rule, we use the easier method.

1.
$$\lim_{x \to -2} \frac{x+2}{x^2-4}$$
$$= \lim_{x \to -2} \frac{x+2}{(x+2)(x-2)}$$
$$= \lim_{x \to -2} \frac{1}{x-2} = -\frac{1}{4}$$

3.
$$\lim_{x \to \infty} \frac{3x^2+2}{x^2-4} = \lim_{x \to \infty} \frac{3 + \frac{2}{x^2}}{1 - \frac{4}{x^2}} = \frac{3}{1} = 3$$

5. $\lim_{x \to 0} \frac{\sin x}{x} = 1$, as we learned in sections 1.2 and 2.6. It is also possible to use L'Hôpital's Rule:

$$\lim_{x \to 0} \frac{\sin x}{x} = \lim_{x \to 0} \frac{\cos x}{1} = 1,$$

but this argument involves circular reasoning; in order to use L'Hôpital's Rule, you need to know the derivative of $\sin x$, and in order to find the derivative of $\sin x$, you need to know that $\lim_{x \to 0} \frac{\sin x}{x} = 1$.

7.
$$\lim_{x \to 0} \frac{\tan^{-1} x}{\sin x} = \lim_{x \to 0} \frac{1/\left(1+x^2\right)}{\cos x} = \lim_{x \to 0} \frac{1}{1} = 1$$

9.
$$\lim_{x \to \pi} \frac{\sin 3x}{\sin x} = \lim_{x \to \pi} \frac{3\cos 3x}{\cos x} = \frac{3(-1)}{-1} = 3$$

11.
$$\lim_{x \to 0} \frac{\sin x - x}{x^3} = \lim_{x \to 0} \frac{\cos x - 1}{3x^2}$$
$$= \lim_{x \to 0} \frac{-\sin x}{6x} = \frac{-1}{6}$$

13.
$$\lim_{x \to 1} \frac{\sqrt{x}-1}{x-1} = \lim_{x \to 1} \frac{\sqrt{x}-1}{x-1} \frac{\sqrt{x}+1}{\sqrt{x}+1}$$
$$= \lim_{x \to 1} \frac{x-1}{(x-1)\left(\sqrt{x}+1\right)}$$
$$= \lim_{x \to 1} \frac{1}{\sqrt{x}+1} = \frac{1}{2}$$

15.
$$\lim_{x \to \infty} \frac{x^3}{e^x} = \lim_{x \to \infty} \frac{3x^2}{e^x} = \lim_{x \to \infty} \frac{6x}{e^x} = \lim_{x \to \infty} \frac{6}{e^x} = 0$$

17.
$$\lim_{x \to 0} \frac{e^{2x}-1}{x} = \lim_{x \to 0} \frac{2e^{2x}}{1} = \frac{2}{1} = 2$$

19.
$$\lim_{x \to 1} \frac{\sin \pi x}{x-1} = \lim_{x \to 1} \frac{\pi \cos \pi x}{1} = \frac{\pi(-1)}{1} = -\pi$$

21.
$$\lim_{x \to \infty} \frac{\ln x}{x^2} = \lim_{x \to \infty} \frac{1/x}{2x} = \lim_{x \to \infty} \frac{1/x}{2x} = \lim_{x \to \infty} \frac{1}{2x^2} = 0$$

23.
$$\lim_{x \to \infty} xe^{-x} = \lim_{x \to \infty} \frac{x}{e^x} = \lim_{x \to \infty} \frac{1}{e^x} = 0$$

25. As x approaches 1 from below, $\ln x$ is a small negative number. Hence, $\ln(\ln x)$ is undefined, so the limit is undefined.

27.
$$\lim_{x \to 0^+} x \ln x = \lim_{x \to 0^+} \frac{\ln x}{1/x} = \lim_{x \to 0^+} \frac{1/x}{-1/x^2}$$
$$= \lim_{x \to 0^+} -x = 0$$

29.
$$\lim_{x \to 0^+} \frac{\ln x}{\cot x} = \lim_{x \to 0^+} \frac{1/x}{-\csc^2 x} = \lim_{x \to 0^+} \frac{-\sin^2 x}{x}$$
$$= \lim_{x \to 0^+} \left(-\sin x \frac{\sin x}{x}\right) = (0)(1) = 0$$

31. $\lim_{x\to\infty}\left(\sqrt{x^2+1}-x\right)$

$$= \lim_{x\to\infty}\left(\left(\sqrt{x^2+1}-x\right)\frac{\sqrt{x^2+1}+x}{\sqrt{x^2+1}+x}\right)$$

$$= \lim_{x\to\infty}\left(\frac{x^2+1-x^2}{\sqrt{x^2+1}+x}\right) = \lim_{x\to\infty}\frac{1}{\sqrt{x^2+1}+x} = 0$$

33. Let $y = \left(1+\frac{1}{x}\right)^x$. Then $\ln y = x\ln\left(1+\frac{1}{x}\right)$. Then

$$\lim_{x\to\infty}\ln y = \lim_{x\to\infty} x\ln\left(1+\frac{1}{x}\right) = \lim_{x\to\infty}\frac{\ln\left(1+\frac{1}{x}\right)}{1/x}$$

$$= \lim_{x\to\infty}\frac{\frac{1}{1+\frac{1}{x}}\left(\frac{-1}{x^2}\right)}{-1/x^2} = \lim_{x\to\infty}\frac{1}{1+\frac{1}{x}} = 1$$

Hence, $\lim_{x\to\infty} y = \lim_{x\to\infty} e^{\ln y} = e$

35. $$\lim_{x\to 0^+}\left(\ln x + \frac{1}{x}\right) = \lim_{x\to 0^+}\frac{x\ln x + 1}{x}.$$

As $x\to 0^+$, the numerator get close to 1 (see Exercise 27), while the denominator is small and positive. Hence, the limit is ∞.

37. Let $y = (1/x)^x$. Then $\ln y = x\ln(1/x)$. Then $\lim_{x\to 0^+}\ln y = \lim_{x\to 0^+} x\ln(1/x) = 0$, by Exercise 27. Thus, $\lim_{x\to 0^+} y = \lim_{x\to 0^+} e^{\ln y} = 1$.

39. L'Hopital's Rule does not apply. As $x\to 0$, the numerator gets close to 1 and the denominator is small and positive. Hence the limit is ∞.

41. L'Hopital's Rule does not apply. As $x\to 0$, the numerator is small and positive while the denominator goes to $-\infty$. Hence, the limit is 0.

43. The first step is plausible, since for x close to 0, $\sin x\approx x$. The second step (dividing out the x's) is actually correct.

45. **(a)** $\lim_{x\to 0}\frac{\sin x^2}{x^2} = \lim_{x\to 0}\frac{2x\cos x^2}{2x} = \lim_{x\to 0}\cos x^2 = 1$, which is the same as $\lim_{x\to 0}\frac{\sin x}{x}$.

(b)
$$\lim_{x\to 0}\frac{1-\cos x^2}{x^4} = \lim_{x\to 0}\frac{2x\sin x^2}{4x^3}$$
$$= \lim_{x\to 0}\frac{\sin x^2}{2x^2} = \frac{1}{2}\lim_{x\to 0}\frac{\sin x^2}{x^2}$$
$$= (1/2)\,(1)\text{ (by part (a))}$$
$$= 1/2,$$

while

$$\lim_{x\to 0}\frac{1-\cos x}{x^2} = \lim_{x\to 0}\frac{\sin x}{2x}$$
$$= \frac{1}{2}\lim_{x\to 0}\frac{\sin x}{x} = \frac{1}{2}(1) = \frac{1}{2},$$

so both of these limits are the same.

47. $$\lim_{x\to 0}\frac{\sin kx^2}{x^2} = \lim_{x\to 0}\frac{2kx\cos kx^2}{2x}$$
$$= \lim_{x\to 0} k\cos kx^2 = k(1) = k$$

49. $\lim_{x\to\infty}\frac{e^x}{x^n} = \infty$ since n applications of L'Hopital's Rule yields $\lim_{x\to\infty}\frac{e^x}{n!} = \infty$. Hence, e^x dominates x^n.

51. $$\lim_{x\to 0}\frac{e^{cx}-1}{x} = \lim_{x\to 0}\frac{ce^{cx}}{1} = c$$

53. If $x\to 0$, then $x^2\to 0$, so if $\lim_{x\to 0}\frac{f(x)}{g(x)} = L$, then $\lim_{x\to 0}\frac{f(x^2)}{g(x^2)} = L$ (but not conversely). If $a\neq 0$ or 1, then $\lim_{x\to a}\frac{f(x)}{g(x)}$ involves the behavior of the quotient near a, while $\lim_{x\to a}\frac{f(x^2)}{g(x^2)}$ involves the behavior of the quotient near the different point a^2.

55. $$\lim_{\omega\to 0}\frac{2.5(4\omega t - \sin 4\omega t)}{4\omega^2}$$
$$= \lim_{\omega\to 0}\frac{2.5(4t - 4t\cos 4\omega t)}{8\omega}$$
$$= \lim_{\omega\to 0}\frac{2.5(16t^2\sin 4\omega t)}{8} = 0$$

57. **(a)** $\frac{(x+1)(2+\sin x)}{x(2+\cos x)}$

(b) $\frac{x}{e^x}$

(c) $\dfrac{3x+1}{x-7}$

(d) $\dfrac{3-8x}{1+2x}$

59. The area of triangular region 1 is

$$(1/2)(\text{base})(\text{height}) = (1/2)(1-\cos\theta)(\sin\theta).$$

Let P be the center of the circle. The area of region 2 equals the area of sector APC minus the area of triangle APB. The area of the sector is $\theta/2$, while the area of triangle APB is

$$(1/2)(\text{base})(\text{height}) = (1/2)(\cos\theta)(\sin\theta).$$

Hence, the area of region 1 divided by the area of region 2 is

$$\frac{(1/2)(1-\cos\theta)(\sin\theta)}{\theta/2-(1/2)(\cos\theta)(\sin\theta)} = \frac{(1-\cos\theta)(\sin\theta)}{\theta-\cos\theta\sin\theta}$$
$$= \frac{\sin\theta-\cos\theta\sin\theta}{\theta-\cos\theta\sin\theta} = \frac{\sin\theta-(1/2)\sin 2\theta}{\theta-(1/2)\sin 2\theta}.$$

Then

$$\lim_{\theta 0}\frac{\sin\theta-(1/2)\sin 2\theta}{\theta-(1/2)\sin 2\theta} = \lim_{\theta 0}\frac{\cos\theta-\cos 2\theta}{1-\cos 2\theta}$$
$$= \lim_{\theta 0}\frac{-\sin\theta+2\sin 2\theta}{2\sin 2\theta} = \lim_{\theta 0}\frac{-\cos\theta+4\cos 2\theta}{4\cos 2\theta}$$
$$= \frac{-1+4(1)}{4(1)} = \frac{3}{4}.$$

3.3 Maximum and Minimum Values

1. **(a)** no absolute extrema
(b) $f(0) = -1$ is absolute max
(c) no absolute extrema

3. **(a)** $f\left(\dfrac{\pi}{2}+2n\pi\right) = 1$ is abs max;
$f\left(\dfrac{3\pi}{2}+2n\pi\right) = -1$ is abs min

(b) $f(0) = 0$ is abs min; $f(\pi/4) = \dfrac{\sqrt{2}}{2}$ is abs max

(c) $f(\pi/2) = 1$ is abs max

5.
$$f(x) = x^2+5x-1$$
$$f'(x) = 2x+5$$
$$2x+5 = 0$$

$x = \dfrac{-5}{2}$ is a critical number. This is a parabola opening upward, so we have a minimum.

7.
$$f(x) = x^3-3x+1$$
$$f'(x) = 3x^2-3$$
$$3x^2-3 = 3(x^2-1) = 3(x+1)(x-1) = 0$$

$x = \pm 1$ are critical numbers.
$x = -1$ is a local max; $x = 1$ is a local min.

9.
$$f(x) = x^3-3x^2+3x$$
$$f'(x) = 3x^2-6x+3$$
$$3x^2-6x+3 = 3(x^2-2x+1) = 3(x-1)^2 = 0$$

$x = 1$ is a critical number, but it is neither a local min nor max.

11.
$$f(x) = x^4-3x^3+2$$
$$f'(x) = 4x^3-9x^2$$
$$4x^3-9x^2 = x^2(4x-9) = 0$$

$x = 0, \dfrac{9}{4}$ are critical numbers.
$x = \dfrac{9}{4}$ is a local min.
$x = 0$ is neither a local max nor min.

13.
$$f(x) = x^{3/4}-4x^{1/4}$$
$$f'(x) = \frac{3}{4x^{1/4}} - \frac{1}{x^{3/4}}$$

If $x \neq 0$, $f'(x) = 0$ when $3x^{3/4} = 4x^{1/4}$
$x = 0, \dfrac{16}{9}$ are critical numbers.
$x = \dfrac{16}{9}$ is a local min.
$x = 0$ is neither a local max nor min.

15.
$$f(x) = x^3-2x^2-4x$$
$$f'(x) = 3x^2-4x-4$$
$$3x^2-4x-4 = (x-2)(3x+2) = 0$$

$x = 2, -\frac{2}{3}$ are critical numbers.

$x = -\frac{2}{3}$ is a local max; $x = 2$ is local min.

17.
$$f(x) = \sin x \cos x \text{on } [0, 2\pi]$$
$$f'(x) = \cos x \cos x + \sin x(-\sin x)$$
$$= \cos^2 x - \sin^2 x$$
$$\cos^2 x - \sin^2 x = 0$$
$$\cos^2 x = \sin^2 x$$
$$\cos x = \pm \sin x$$

$x = \frac{\pi}{4}, \frac{3\pi}{4}, \frac{5\pi}{4}, \frac{7\pi}{4}$ are critical numbers.

$x = \frac{\pi}{4}, \frac{5\pi}{4}$ are local max;

$x = \frac{3\pi}{4}, \frac{7\pi}{4}$ are local min.

19. $f(x) = \dfrac{x^2 - 2}{x + 2}$

Note that $x = -2$ is not in the domain of f.

$$f'(x) = \frac{(2x)(x + 2) - (x^2 - 2)(1)}{(x + 2)^2}$$
$$= \frac{2x^2 + 4x - x^2 + 2}{(x + 2)^2}$$
$$= \frac{x^2 + 4x + 2}{(x + 2)}$$

$f'(x) = 0$ when $x^2 + 4x + 2 = 0$, so the critical numbers are $x = -2 \pm \sqrt{2}$.
$x = -2 + \sqrt{2}$ is a local min; $x = -2 - \sqrt{2}$ is a local max.

21.
$$f(x) = \frac{x}{x^2 + 1}$$
$$f'(x) = \frac{1(x^2 + 1) - x(2x)}{(x^2 + 1)^2}$$
$$= \frac{x^2 + 1 - 2x^2}{(x^2 + 1)^2}$$
$$= \frac{1 - x^2}{(x^2 + 1)^2}$$

$f'(x) = 0$ for $1 - x^2 = 0$, $x = 1, -1$

$f'(x)$ is defined for all x,
so $x = 1, -1$ are the critical numbers.
$x = -1$ is local min; $x = 1$ is local max.

23.
$$f(x) = \frac{e^x + e^{-x}}{2}$$
$$f'(x) = \frac{e^x - e^{-x}}{2}$$
$$\frac{e^x - e^{-x}}{2} = 0 \text{ when } e^x = e^{-x}, \text{ that is, } x = 0.$$

$f'(x)$ is defined for all x.
So $x = 0$ is a critical number.
$x = 0$ is a local min.

25.
$$f(x) = x^{4/3} + 4x^{1/3} + 4x^{-2/3}$$
f is not defined at $x = 0$.
$$f'(x) = \frac{4}{3}x^{1/3} + \frac{4}{3}x^{-2/3} - \frac{8}{3}x^{-5/3}$$
$$= \frac{4}{3}x^{-5/3}(x^2 + x - 2)$$
$$= \frac{4}{3}x^{-5/3}(x - 1)(x + 2)$$

$x = -2, 1$ are critical numbers.
$x = -2$ and $x = 1$ are local minima.

27. $f(x) = 2x\sqrt{x + 1} = 2x(x + 1)^{1/2}$
Domain of f is all $x \geq -1$.

$$f'(x) = 2(x + 1)^{1/2} + 2x\left(\frac{1}{2}(x + 1)^{-1/2}\right) \cdot 1$$
$$= \frac{2(x + 1) + x}{\sqrt{x + 1}}$$
$$= \frac{3x + 2}{\sqrt{x + 1}}$$
$$f'(x) = 0 \quad \text{for} \quad 3x + 2 = 0, \ x = -\frac{2}{3}$$

$f'(x)$ is undefined for $\sqrt{x + 1} = 0$, $x = -1$ so $x = -\frac{2}{3}, -1$ are critical numbers.

$x = -\frac{2}{3}$ is a local min.

$x = -1$ is an endpoint, so it is neither a local min nor a local max, though it is a maximum on the interval $[-1, 0)$.

29. $f(x) = \sin^{-1} x^2$
Domain of f is $[-1, 1]$

$$f'(x) = \frac{2x}{\sqrt{1 - x^4}}$$

Critical numbers are $x = 0$ (since derivative is zero) and $x = \pm 1$ (since derivative is undefined). $x = 0$ is a local minimum. $x = \pm 1$ are endpoints, so they are not local extrema.

31. Because of the absolute value sign, there may be critical numbers where the function $x^2 - 1$ changes sign, that is, at $x = \pm 1$. For $x > 1$ and for $x < -1$, $f(x) = x^2 - 1$ and $f'(x) = 2x$, so there are no critical numbers on these intervals. For $-1 < x < 1$, $f(x) = 1 - x^2$ and $f'(x) = -2x$, so 0 is a critical number.

33. First, let's find the critical numbers for $x < 0$. In this case,

$$f(x) = x^2 + 2x - 1$$
$$f'(x) = 2x + 2 = 2(x + 1),$$

so the only critical number in this interval is $x = -1$, and it is a local minimum. Now for $x > 0$,

$$f(x) = x^2 - 4x + 3$$
$$f'(x) = 2x - 4 = 2(x - 2),$$

so the only critical number is $x = 2$, and it is a local minimum. Finally, $x = 0$ is also a critical number, since f is not continuos and hence not differentiable at $x = 0$. Indeed, $x = 0$ is a local maximum.

35.
$$f(x) = x^3 - 3x + 1$$
$$f'(x) = 3x^2 - 3 = 3(x^2 - 1)$$
$$f'(x) = 0 \quad \text{for} \quad x = 1, -1$$

On $[0, 2]$, 1 is the only critical number. We calculate:
$f(0) = 1$;
$f(1) = -1$ is the absolute minimum;
$f(2) = 3$ is the absolute maximum.
On the interval $[-3, 2]$, we have both 1 and -1 as critical numbers. We calculate:
$f(-3) = -17$ is the absolute minimum;
$f(-1) = 3$ is the absolute maximum;
$f(1) = -1$.
$f(2) = 3$ is also the absolute maximum.

37.
$$f(x) = x^{2/3}$$
$$f'(x) = \frac{2}{3}x^{-1/3} = \frac{2}{3\sqrt[3]{x}}$$

$f'(x) \neq 0$ for any x, but $f'(x)$ undefined for $x = 0$, so $x = 0$ is critical number.
On $[-4, -2]$, $0 \in [-4, -2]$, so we only look at endpoints.

$$f(-4) = \sqrt[3]{16} \approx 2.52$$
$$f(-2) = \sqrt[3]{4} \approx 1.59$$

So $f(-4) = \sqrt[3]{16}$ is the absolute maximum and $f(-2) = \sqrt[3]{4}$ is the absolute minimum.
On $[-1, 3]$, we have 0 as a critical number.
$f(-1) = 1$
$f(0) = 0$ is the absolute minimum
$f(3) = 3^{2/3}$ is the absolute maximum.

39.
$$f(x) = e^{-x^2}$$
$$f'(x) = -2xe^{-x^2}$$

Hence, $x = 0$ is the only critical number.
On $[0, 2]$,
$f(0) = 1$ is the absolute maximum and
$f(2) = e^{-4}$ is the absolute minimum.
On $[-3, 2]$,
$f(-3) = e^{-9}$ is the absolute minimum and
$f(0) = 1$ is the absolute maximum.
$f(2) = e^{-4}$.

41.
$$f(x) = \frac{3x^2}{x-3}$$

Note that $x = 3$ is not in the domain of f.

$$f'(x) = \frac{6x(x-3) - 3x^2(1)}{(x-3)^2}$$
$$= \frac{6x^2 - 18x - 3x^2}{(x-3)^2}$$
$$= \frac{3x^2 - 18x}{(x-3)^2}$$
$$= \frac{3x(x-6)}{(x-3)^2}$$

The critical points are $x = 0$, $x = 6$. On $[-2, 2]$,
$f(-2) = -12/5$
$f(-2) = -12$
$f(0) = 0$.
Hence, absolute max is $f(0) = 0$ and absolute min is $f(2) = -12$.
On $[2, 8]$, the function is not continuos and in fact has no absolute max or min

43. **(a)** absolute min at $(-1, 3)$;
absolute max at $(0.3660, 1.3481)$

(b) absolute min at $(-1.3660, -3.8481)$;
absolute max at $(-3, 49)$

45. **(a)** absolute min at $(0.6371, -1.1305)$;
absolute max at $(-1.2269, 2.7463)$

(b) absolute min at $(-2.8051, -0.0748)$;
absolute max at $(-5, 29.2549)$

47. **(a)** absolute min at $(0, 0)$;
absolute max at $\left(\pm\frac{\pi}{2}, 3 + \frac{\pi}{2}\right)$.

(b) absolute min at $(4.913, -1.814)$;
absolute max at $(2.029, 4.820)$

49. If an absolute max or min occurs only at the endpoint of a closed interval, then there will be no absolute max or min on the open interval.

35) on $(0, 2)$, $f(1) = -1$ is min.; no max.
on $(-3, 2)$, $f(-1) = 3$ is max, no min

37) on $(-4, -2)$, no max or min.
on $(-1, 3)$, $f(0) = 0$ is min.; no max

39) on $(0, 2)$, no max or min
on $(-3, 2)$, $f(0) = 1$ is max; no min

41) on $(-2, 2)$, $f(0) = 0$ is max; no min
on $(2, 8)$, no max or min

51.

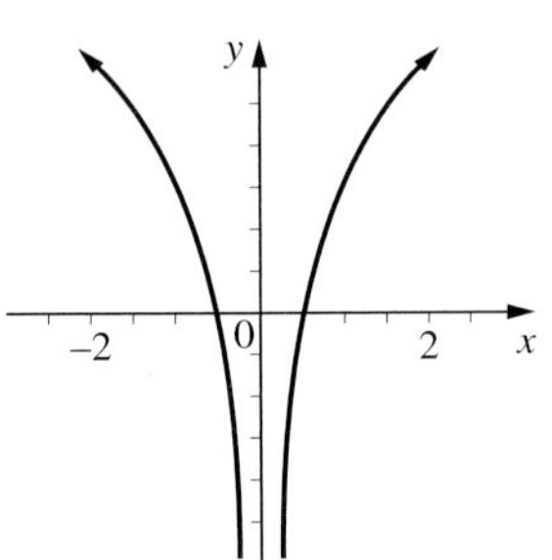

53.

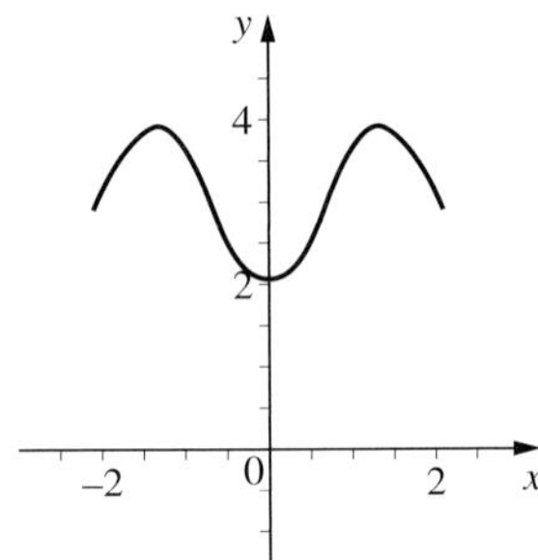

55. You will not be able to construct an example with a continuous function: There are many examples using a function with a discontinuity, for example $f(x) = \sec^2 x$.

57.
$$f(x) = x^3 + cx + 1$$
$$f'(x) = 3x^2 + c$$

We know (perhaps from a pre-calculus course) that for any cubic polynomial with positive leading coefficient, when x is large and positive, the value of the polynomial is very large and positive, and when x is large and negative, the value of the polynomial is very large and negative.

Type 1: $c > 0$. There are no critical numbers, As you move from left to right, the graph of f is always rising.

Type 2: $c < 0$.There are two critical numbers, $x = \pm\sqrt{-c/3}$. As you move from left to right, the graph rises until we get to the first critical number, then the graph must fall until we get to the second critical number, and then the graph rises again. So the critical number on the left is a local maximum and the critical number on the right is a local minimum.

Type 3: $c = 0$. There is only one critical number, which is neither a local max nor a local min.

59.
$$f(x) = x^3 + bx^2 + cx + d$$
$$f'(x) = 3x^2 + 2bx + c$$

The quadratic formula says that the critical numbers are

$$x = \frac{-2b \pm \sqrt{4b^2 - 12c}}{6} = \frac{-b \pm \sqrt{b^2 - 3c}}{3}.$$

So if $c < 0$, the quantity under the square root is positive and there are two critical numbers. This is like the Type 2 polynomials in Exercise 57. We know that as x goes to infinity, the polynomial $x3 + bx2 + cx + d$ gets very large and positive, and when x goes to minus infinity, the polynomial is very large but negative. Therefore, the critical number on the left must be a local max, and the critical number on the right must be a local min.

61.
$$f(x) = x^4 + cx^2 + 1$$
$$f'(x) = 4x^3 + 2cx = 2x(2x^2 + c)$$

So $x = 0$ is always a critical number.
Case 1: $c \geq 0$. The only solution to $2x(2x^2 + c) = 0$ is $x = 0$, so $x = 0$ is the only critical number. This must be a minimum, since we know that the function $x^4 + cx^2 + 1$ is large and positive when $|x|$ is large (so the graph is roughly U-shaped). We could also note that $f(0) = 1$, and 1 is clearly the absolute minimum of this function if $c \geq 0$.
Case 2: $c < 0$. Then there are two other critical numbers, $x = \pm\sqrt{\frac{-c}{2}}$. Now $f(0)$ is still equal to 1, but the value of f at both new critical numbers is less than 1. Hence, $f(0)$ is a local max, and both new critical numbers are local minimums.

63.

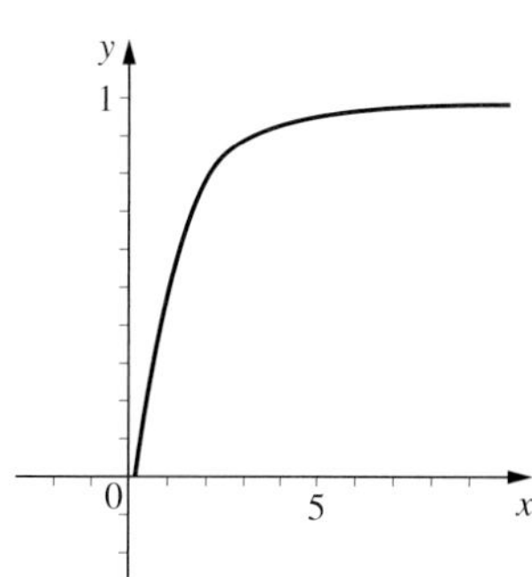

$$f(x) = \frac{x^2}{x^2 + 1}, \ x > 0$$
$$f'(x) = \frac{2x(x^2 + 1)^2 - 2x \cdot 2(x^2 + 1) \cdot 2x}{(x^2 + 1)^2}$$
$$= \frac{2x}{(x^2 + 1)^2}$$
$$f''(x) = \frac{2 \cdot (x^2 + 1)^2 - 2x \cdot 2(x^2 + 1) \cdot 2x}{(x^2 + 1)^4}$$
$$= \frac{2(x^2 + 1)\left[(x^2 + 1) - 4x^2\right]}{(x^2 + 1)^4}$$
$$= \frac{2\left[1 - 3x^2\right]}{(x^2 + 1)^3}$$
$$f''(x) = 0 \quad \text{for} \quad x = \pm\frac{1}{\sqrt{3}}, \ x = -\frac{1}{\sqrt{3}} \notin (0, \infty)$$

$x = \frac{1}{\sqrt{3}}$ is steepest point.

65.

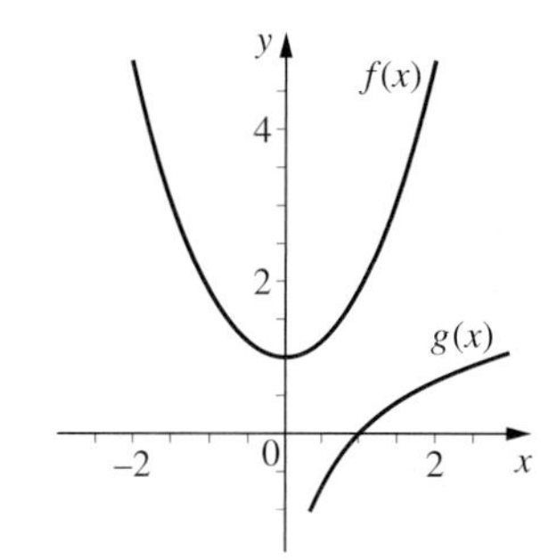

$$f(x) = x^2 + 1$$
$$g(x) = \ln x$$
$$h(x) = f(x) - g(x) = x^2 + 1 - \ln x$$
$$h'(x) = 2x - \frac{1}{x} = 0$$
$$2x^2 = 1$$
$$x = \pm\sqrt{\frac{1}{2}}$$
$$x = \sqrt{\frac{1}{2}} \text{ is min}$$
$$f'(x) = 2x$$
$$g'(x) = \frac{1}{x}$$
$$f'\left(\sqrt{\frac{1}{2}}\right) = 2\sqrt{\frac{1}{2}} = \sqrt{2}$$
$$g'\left(\sqrt{\frac{1}{2}}\right) = \frac{1}{\sqrt{\frac{1}{2}}} = \sqrt{2}$$

So the tangents are parallel.

67.
$$y = x^5 - 4x^3 - x + 10,\ x \in [-2, 2]$$
$$y' = 5x^4 - 12x^2 - 1$$

$x = -1.575, 1.575$ are critical numbers of y. There is a local max at $x = -1.575$, local min at $x = 1.575$.

$x = -1.575$ represents the top, and $x = 1.575$ represents the bottom of the roller coaster.

$$y''(x) = 20x^3 - 24x = 4x(5x^2 - 6) = 0$$

$x = 0, \pm\sqrt{\frac{6}{5}}$ are critical numbers of y'. We calculate y' at the critical numbers and at the endpoints $x = \pm 2$:

$$y'(0) = -1$$
$$y'\left(\pm\sqrt{\frac{6}{5}}\right) = -\frac{41}{5}$$
$$y'(\pm 2) = 31.$$

So the points where the roller coaster is making the steepest descent are $x = \pm\sqrt{\frac{6}{5}}$.

69.
$$W(t) = a \cdot e^{-be^{-t}}$$
as $t \to \infty$, $-be^{-t} \to 0$, so $a \cdot W(t) \to a$.
$$W'(t) = a \cdot e^{-be^{t}} \cdot be^{-t}$$
as $t \to \infty$, $be^{-t} \to 0$, so $W'(t) \to 0$.
$$W''(t) = (a \cdot e^{-be^{-t}} \cdot be^{-t}) \cdot be^{-t} + (a \cdot e^{-be^{-t}}) \cdot (-be^{-t})$$
$$= a \cdot e^{-be^{-t}} \cdot be^{-t}\left[be^{-t} - 1\right]$$
$$W''(t) = 0 \text{ when } be^{-t} = 1$$
$$e^{-t} = b^{-1}$$
$$-t = \ln b^{-1}$$
$$t = \ln b$$
$$W'(\ln b) = a \cdot e^{-be^{-\ln b}} \cdot be^{-\ln b}$$
$$= a \cdot e^{-b\left(\frac{1}{b}\right)} \cdot b \cdot \frac{1}{b} = ae^{-1}$$

Maximum growth rate is ae^{-1} when $t = \ln b$.

71. Label the triangles as illustrated.

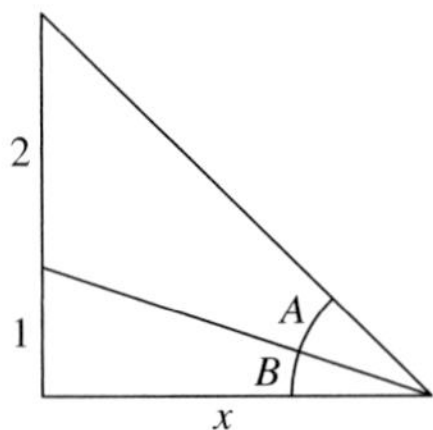

$$\tan(A + B) = \frac{3}{x}$$
$$A + B = \tan^{-1}\left(\frac{3}{x}\right)$$
$$\tan B = \frac{1}{x}$$
$$B = \tan^{-1}\frac{1}{x}$$

Therefore,

$$A = (A + B) - B,$$

$$A = \tan^{-1}\left(\frac{3}{x}\right) - \tan^{-1}\left(\frac{1}{x}\right),$$

$$\frac{dA}{dx} = \frac{-\frac{3}{x^2}}{1 + \left(\frac{3}{x}\right)^2} - \frac{-\frac{1}{x^2}}{1 + \left(\frac{1}{x}\right)^2}$$

$$= \frac{1}{x^2 + 1} - \frac{3}{x^2 + 9}.$$

The maximum viewing angle will occur at a critical value.

$$\frac{dA}{dx} = 0$$

$$\frac{1}{x^2 + 1} = \frac{3}{x^2 + 9}$$

$$x^2 + 9 = 3x^2 + 3$$

$$2x^2 = 6$$

$$x^2 = 3$$

$$x = \sqrt{3}\text{ft} \approx 1.73\text{ft}$$

This is a maximum because when x is large and when x is a little bigger than 0, the angle is small.

3.4 Increasing and Decreasing Functions

1. $y = x^3 - 3x + 2$

$y' = 3x^2 - 3 = 3(x^2 - 1) = 3(x + 1)(x - 1)$

$x = 1, 1$ are critical numbers.
$(x + 1) > 0$ on $(-1, \infty)$, $(x + 1) < 0$ on $(-\infty, -1)$;
$(x - 1) > 0$ on $(1, \infty)$, $(x - 1) < 0$ on $(-\infty, -1)$;
$3(x + 1)(x - 1) > 0$ on $(1, \infty) \cup (-\infty, -1)$, so y is increasing on $(1, \infty)$ and on $(-\infty, -1)$.
$3(x + 1)(x - 1) < 0$ on $(-1, 1)$, so y is decreasing on $(-1, 1)$.

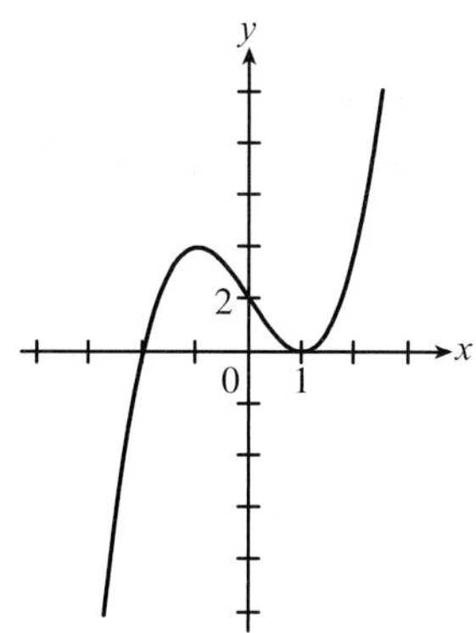

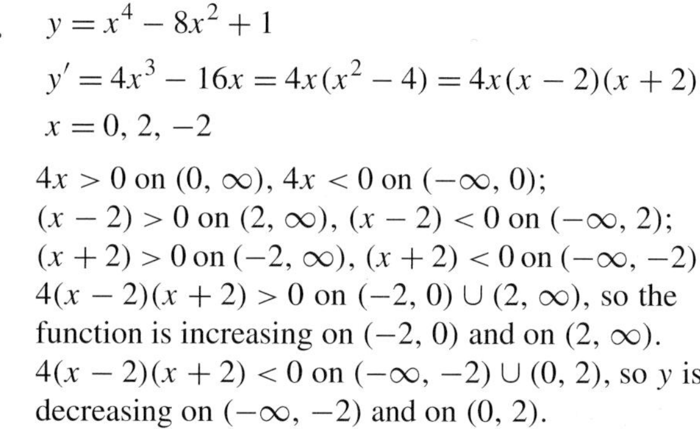

3. $y = x^4 - 8x^2 + 1$

$y' = 4x^3 - 16x = 4x(x^2 - 4) = 4x(x - 2)(x + 2)$

$x = 0, 2, -2$

$4x > 0$ on $(0, \infty)$, $4x < 0$ on $(-\infty, 0)$;
$(x - 2) > 0$ on $(2, \infty)$, $(x - 2) < 0$ on $(-\infty, 2)$;
$(x + 2) > 0$ on $(-2, \infty)$, $(x + 2) < 0$ on $(-\infty, -2)$.
$4(x - 2)(x + 2) > 0$ on $(-2, 0) \cup (2, \infty)$, so the function is increasing on $(-2, 0)$ and on $(2, \infty)$.
$4(x - 2)(x + 2) < 0$ on $(-\infty, -2) \cup (0, 2)$, so y is decreasing on $(-\infty, -2)$ and on $(0, 2)$.

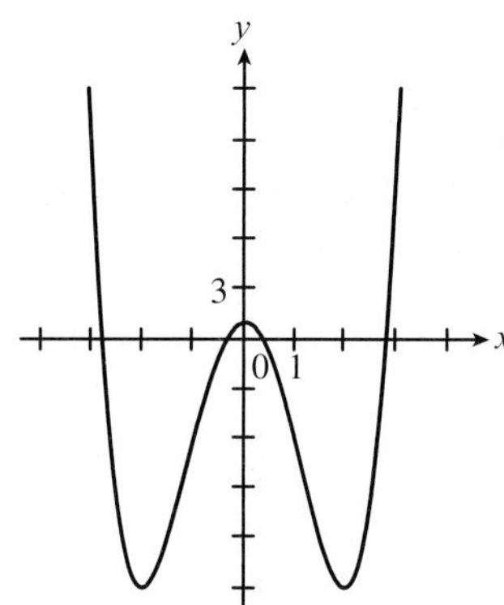

5. $y = (x + 1)^{2/3}$

$$y' = \frac{2}{3}(x + 1)^{-1/3} = \frac{2}{3\sqrt[3]{x + 1}}$$

y' is not defined for $x = -1$.

$\frac{2}{3\sqrt[3]{x + 1}} > 0$ on $(-1, \infty)$, y is increasing

$\frac{2}{3\sqrt[3]{x + 1}} < 0$ on $(-\infty, 1)$, y is decreasing

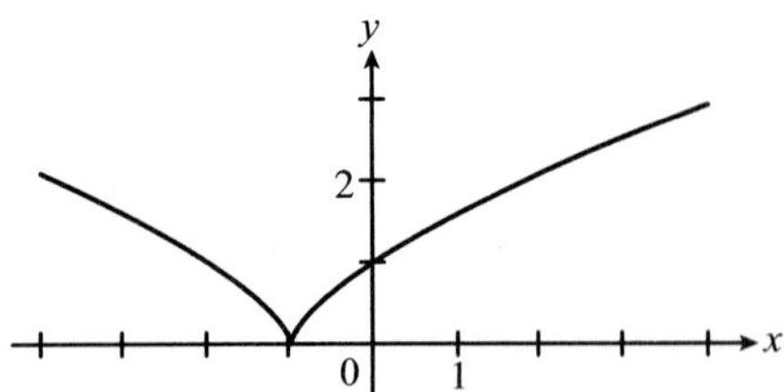

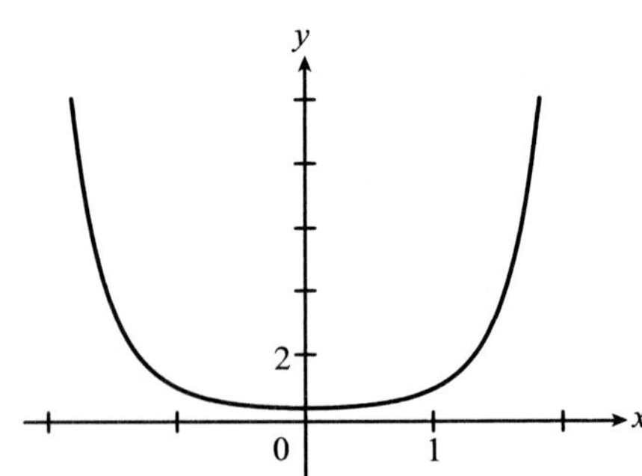

7.

$$\begin{aligned} y &= \sin x + \cos x \\ y' &= \cos x - \sin x = 0 \\ \cos x &= \sin x \\ x &= \frac{\pi}{4}, \frac{5\pi}{4}, \frac{9\pi}{4}, \text{ etc.} \end{aligned}$$

$$\cos x - \sin x \begin{cases} > 0 \text{ on } \left(-\frac{3\pi}{4}, \frac{\pi}{4}\right) \cup \left(\frac{5\pi}{4}, \frac{9\pi}{4}\right) \cup \ldots \\ < 0 \text{ on } \left(\frac{\pi}{4}, \frac{5\pi}{4}\right) \cup \left(\frac{9\pi}{4}, \frac{13\pi}{4}\right) \cup \ldots \end{cases}$$

So $y = \sin x + \cos x$ is decreasing on

$$\left(\frac{\pi}{4}, \frac{5\pi}{4}\right), \left(\frac{9\pi}{4}, \frac{13\pi}{4}\right),$$

etc. and is increasing on

$$\left(\frac{-3\pi}{4}, \frac{\pi}{4}\right), \left(\frac{5\pi}{4}, \frac{9\pi}{4}\right), \text{ etc.}$$

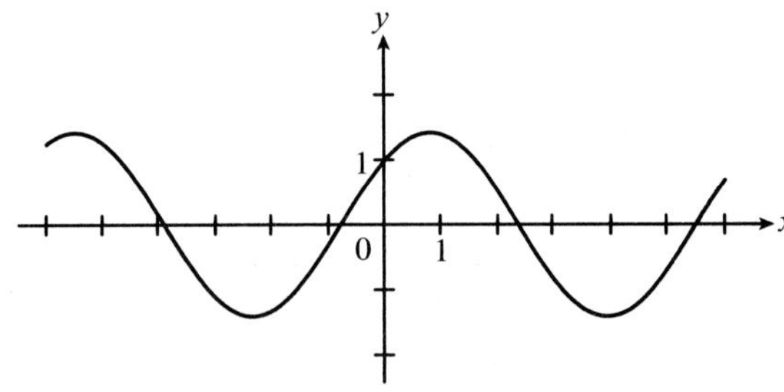

9.

$$\begin{aligned} y &= e^{x^2-1} \\ y' &= e^{x^2-1} \cdot 2x = 2xe^{x^2-1} \\ x &= 0 \end{aligned}$$

$2xe^{x^2-1} > 0$ on $(0, \infty)$, y is increasing
$2xe^{x^2-1} < 0$ on $(-\infty, 0)$, y is decreasing

11.

$$\begin{aligned} y &= x^4 + 4x^3 - 2 \\ y' &= 4x^3 + 12x^2 = 4x^2(x+3) \end{aligned}$$

Critical numbers are $x = 0$, $x = -3$.
$4x2(x+3) > 0$ on $(-3, 0) \cup (0, \infty)$;
$4x2(x+3) < 0$ on $(-\infty, -3)$.
Hence, $x = -3$ is a local minimum and $x = 0$ is not an extremum.

13.

$$\begin{aligned} y &= xe^{-2x} \\ y' &= 1 \cdot e^{-2x} + x \cdot e^{-2x}(-2) \\ &= e^{-2x} - 2xe^{-2x} \\ &= e^{-2x}(1-2x) \\ x &= \frac{1}{2} \end{aligned}$$

$e^{-2x}(1-2x) > 0$ on $\left(-\infty, \frac{1}{2}\right)$;
$e^{-2x}(1-2x) < 0$ on $\left(\frac{1}{2}, \infty\right)$.
So $y = xe^{-2x}$ has a local maximum at $x = \frac{1}{2}$.

15.

$$\begin{aligned} y &= \tan^{-1}(x^2) \\ y' &= \frac{2x}{1+x^4} \end{aligned}$$

Critical number is $x = 0$.
$\frac{2x}{1+x^4} > 0$ for $x > 0$;
$\frac{2x}{1+x^4} < 0$ for $x < 0$.
Hence, $x = 0$ is a local minimum.

17. $y = \dfrac{x}{1+x^3}$

Note that the function is not defined for $x = -1$.

$$\begin{aligned} y' &= \frac{1(1+x^3) - x(3x^2)}{(1+x^3)} \\ &= \frac{1+x^3-3x^3}{(1+x^3)^2} \\ &= \frac{1-2x^3}{(1+x^3)^2}. \end{aligned}$$

Critical number is $x = \sqrt[3]{1/2}$.

$y' > 0$ on $(-1, -\sqrt[3]{1/2})$;

$y' < 0$ on $(\sqrt[3]{1/2}, \infty)$.

Hence, $x = \sqrt[3]{1/2}$ is a local max.

19. $y = \sqrt{x^3+3x^2} = (x^3+3x^2)^{1/2}$

Domain is all $x \geq -3$.

$$\begin{aligned} y' &= \frac{1}{2}(x^3+3x^2)^{-1/2}(3x^2+6x) \\ &= \frac{3x^2+6x}{2\sqrt{x^3+3x^2}} \\ &= \frac{3x(x+2)}{2\sqrt{x^3+3x^2}}. \end{aligned}$$

$x = 0, -2, -3$ are critical numbers.

y' undefined at $x = 0, -3$.

$$y' \begin{cases} > 0 \text{ on } (-3, -2) \cup (0,) \\ < 0 \text{ on } (-2, 0) \end{cases}.$$

So $y = \sqrt{x^3+3x^2}$ has local max at $x = -2$, local min at $x = 0$. $x = -3$ is an endpoint, and so is not a local extremum.

21.

$$\begin{aligned} y &= \frac{2x}{x^2-1} \\ y' &= \frac{2(x^2-1) - 2x(2x)}{(x^2-1)^2} \\ &= \frac{2x^2-2-4x^2}{(x^2-1)^2} \\ &= \frac{2(-x^2-1)}{(x^2-1)^2}. \end{aligned}$$

$y' \neq 0$ for any x; y' not defined for $x = -1, 1$. But $x = -1, 1$ are not in the domain of y, so there are no critical numbers and thus no extrema. The vertical asymptotes are $x = -1$ and $x = 1$. When $|x|$ is large, the function approaches 0, so $y = 0$ is the horizontal asymptote.

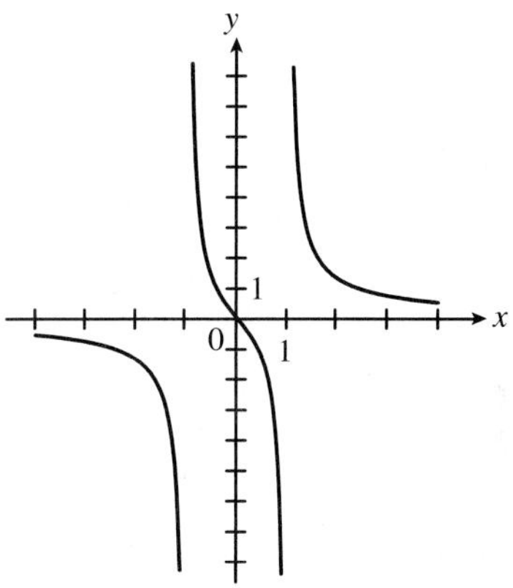

23. $y = \dfrac{x^2}{x^2-4x+3} = \dfrac{x^2}{(x-1)(x-3)}$

Vertical asymptotes are $x = 1$, $x = 3$. When $|x|$ is large, the function approaches the value 1, so $y = 1$ is a horizontal asymptote.

$$\begin{aligned} y' &= \frac{2x(x^2-4x+3) - x^2(2x-4)}{(x^2-4x+3)^2} \\ &= \frac{2x^3-8x^2+6x-2x^3+4x^2}{(x^2-4x+3)^2} \\ &= \frac{-4x^2+6x}{(x^2-4x+3)^2} \\ &= \frac{2x(-2x+3)}{(x^2-4x+3)^2} \\ &= \frac{2x(-2x+3)}{[(x-3)(x-1)]^2}. \end{aligned}$$

Critical numbers are $x = 0$ (local min) and $x = 3/2$ (local max).

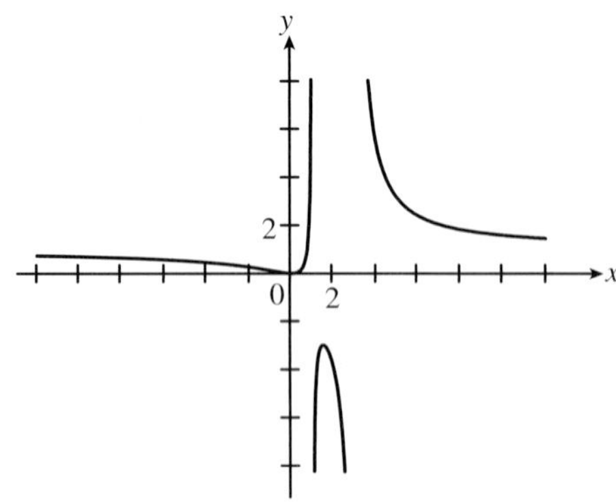

25. $y = \dfrac{3x}{x^2 - 2x - 3} = \dfrac{3x}{(x-3)(x+1)}$
Vertical asymptotes are $x = 3$, $x = -1$. When $|x|$ is large, the function approaches the value 0, so $y = 0$ is horizontal asymptote.

$$\begin{aligned} y' &= \frac{3(x^2 - 2x - 3) - 3x(2x - 2)}{(x^2 - 2x - 3)^2} \\ &= \frac{3x^2 - 6x - 9 - 6x^2 + 6x}{(x^2 - 2x - 3)^2} \\ &= \frac{-3x^2 - 9}{(x^2 - 2x - 3)^2} = \frac{-3(x^2 + 9)}{(x^2 - 2x - 3)^2}, \end{aligned}$$

so there are no critical points.

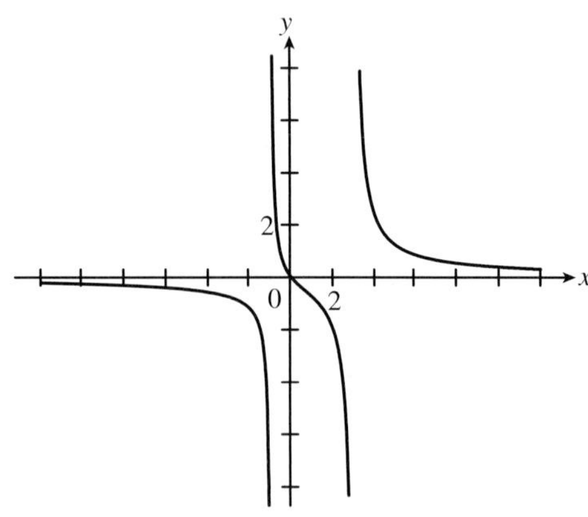

27. $y = \dfrac{4x}{x^2 - x + 1}$
The denominator has no real roots, so there are no vertical asymptotes. When $|x|$ is large, the function approaches 0, so $y = 0$ is a horizontal asymptote.

$$\begin{aligned} y' &= \frac{4(x^2 - x + 1) - 4x(2x - 1)}{(x^2 - x + 1)^2} \\ &= \frac{4x^2 - 4x + 4 - 8x^2 + 4x)}{(x^2 - x + 1)^2} \\ &= \frac{-4x^2 + 4}{(x^2 - x + 1)^2} = \frac{-4(x^2 - 1)}{(x^2 - x + 1)^2}, \end{aligned}$$

so critical numbers are $x = -1$ (a local min) and $x = 1$ (a local max).

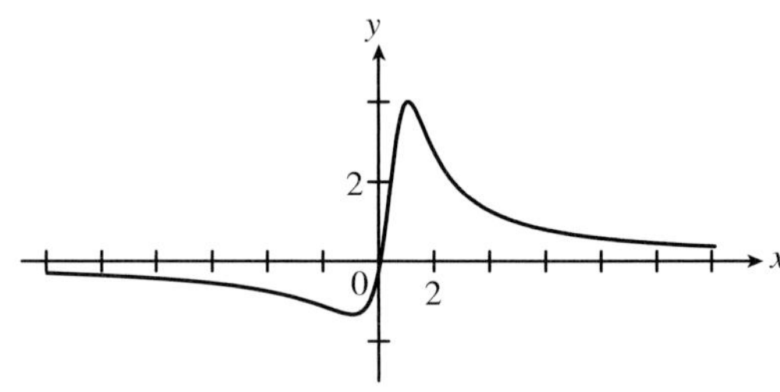

29. $y = \dfrac{x}{\sqrt{x^2 + 1}}$
There are no vertical asymptotes, since the function is defined and continuos for all x. As x approaches ∞, the function approaches the value 1, and as x approaches $-\infty$, the function approaches the value -1, so horizontal asymptotes are $y = 1$ and $y = -1$.

$$\begin{aligned} y' &= \frac{1\sqrt{x^2 + 1} - x\frac{d}{dx}\sqrt{x^2 + 1}}{x^2 + 1} \\ &= \frac{1\sqrt{x^2 + 1} - x(x)(x^2 + 1)^{-1/2}}{x^2 + 1} \\ &= \frac{\sqrt{x^2 + 1} - x^2(x^2 + 1)^{-1/2}}{x^2 + 1}. \end{aligned}$$

The derivative = 0 when

$$\begin{aligned} &\sqrt{x^2 + 1} - x^2(x^2 + 1)^{-1/2} = 0. \\ &\sqrt{x^2 + 1} = x^2(x^2 + 1)^{-1/2} \\ &x^2 + 1 = x^2. \end{aligned}$$

This never happens, so there are no critical numbers.

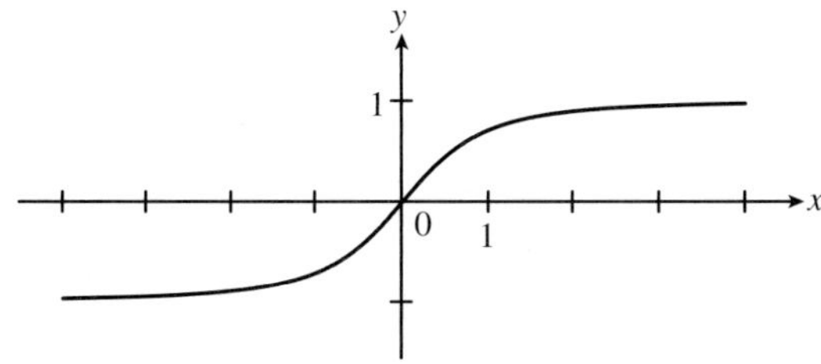

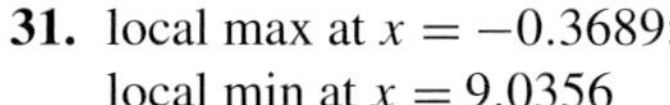

31. local max at $x = -0.3689$;
local min at $x = 9.0356$

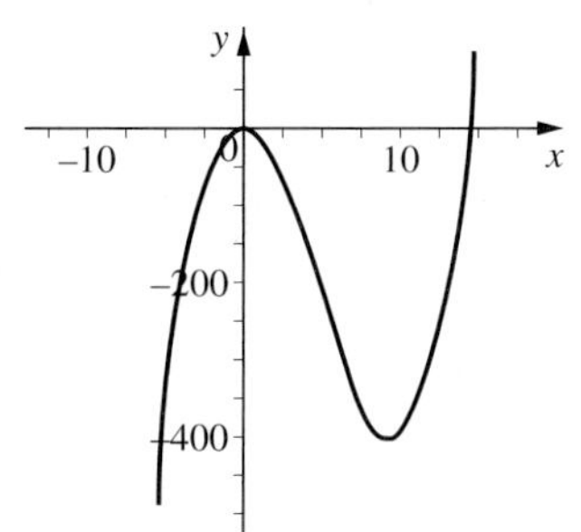

33. local minima at $x = -0.9474$, 11.2599;
local max at 0.9374

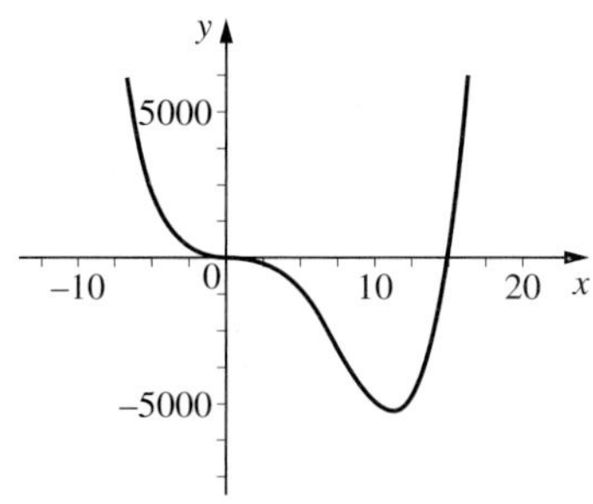

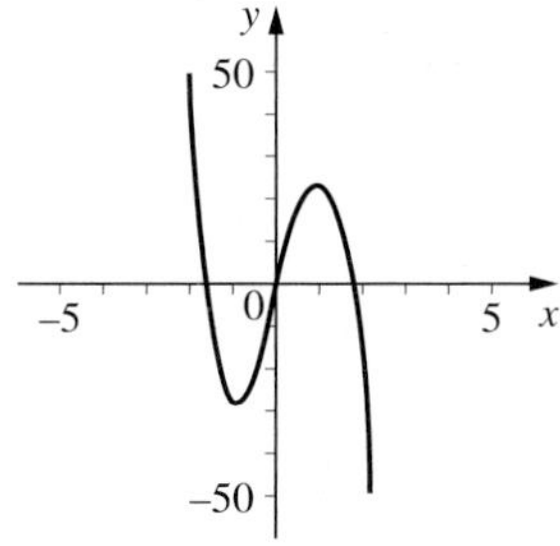

35. local minima at $x = -1.0084$, 10.9079;
local maxima at $x = -10.9079$, 1.0084

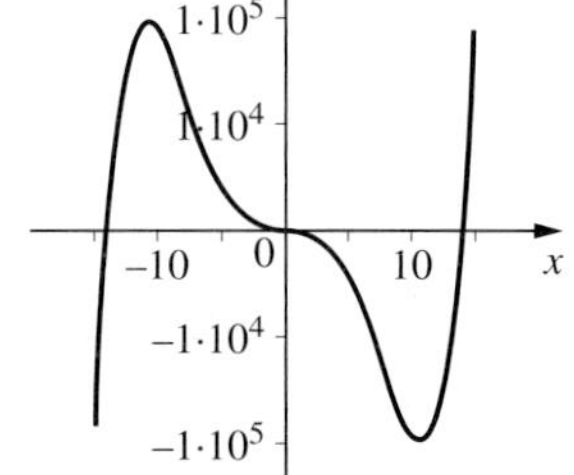

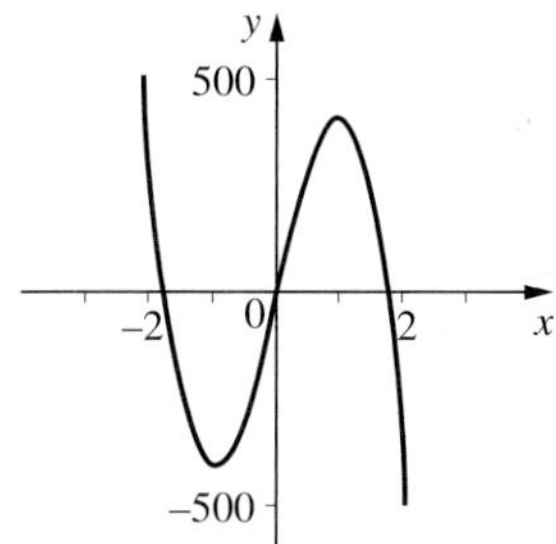

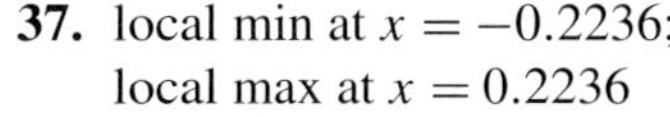

37. local min at $x = -0.2236$;
local max at $x = 0.2236$

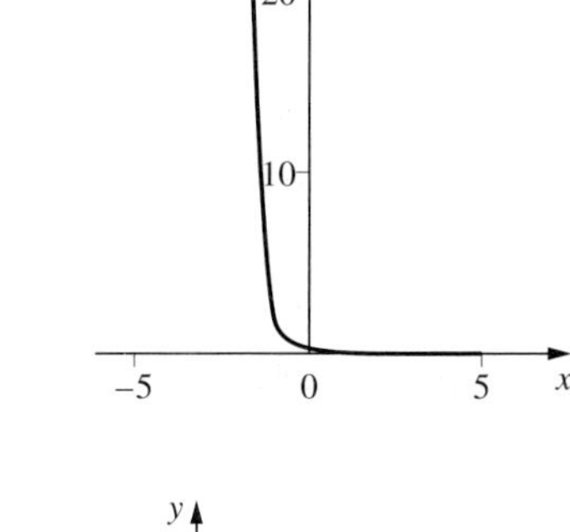

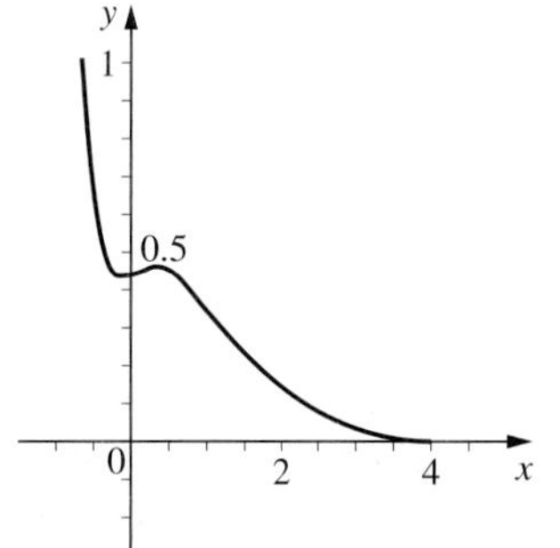

39.

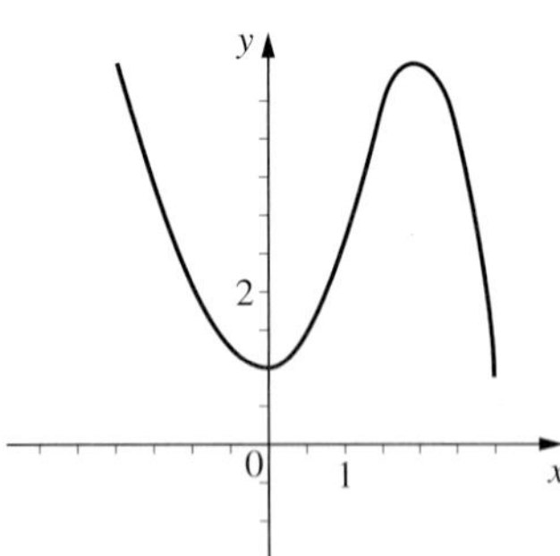

41.

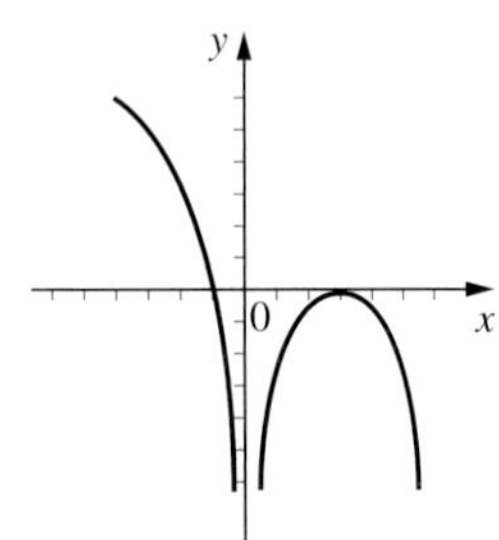

43. Answers will vary.

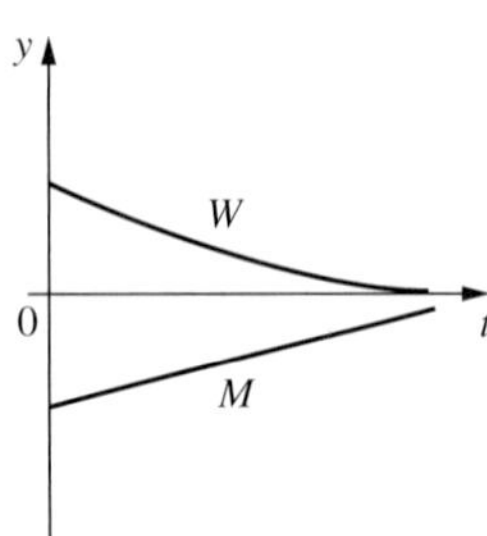

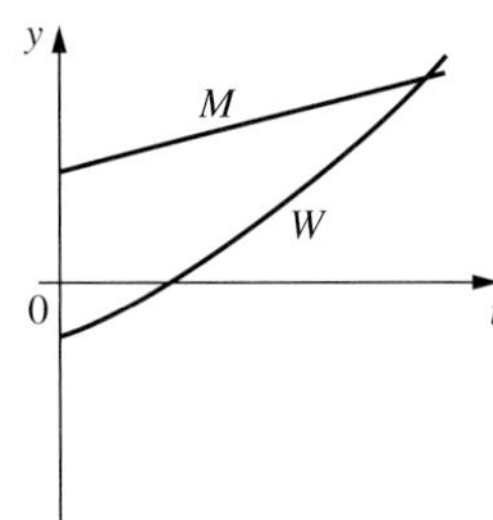

45. f is continuous on $[a, b]$, and $c \in (a, b)$ is a critical number.

(i) If $f'(x) > 0$ for all $x \in (a, c)$ and $f'(x) < 0$ for all $x \in (c, b)$, by Theorem 3.1, f is increasing on (a, c) and decreasing on (c, b), so $f(c) > f(x)$ for all $x \in (a, c)$ and $x \in (cb)$. Thus $f(c)$ is a local max.

(ii) If $f'(x) < 0$ for all $x \in (a, c)$ and $f'(x) > 0$ for all $x \in (c, b)$, by Theorem 3.1, f is decreasing on (a, c) and increasing on (c, b). So $f(c) < f(x)$ for all $x \in (a, c)$ and $x \in (c, b)$. Thus $f(c)$ is a local min.

(iii) If $f'(x) > 0$ on (a, c) and (c, b), then $f(c) > f(x)$ for all $x \in (a, c)$ and $f(c) < f(x)$ for all $x \in (c, b)$, so c is not a local extremum. If $f'(x) < 0$ on (a, c) and (c, b), then $f(c) < f(x)$ for all $x \in (a, c)$ and $f(c) > f(x)$ for all $x \in (c, b)$, so c is not a local extremum.

47. Let $f(x) = 2\sqrt{x}$, $g(x) = 3 - \dfrac{1}{x}$.

Then $f(1) = 2\sqrt{1} = 2$, and $g(1) = 3 - \dfrac{1}{1} = 2$, so $f(1) = g(1)$.

$$f'(x) = \frac{1}{\sqrt{x}};$$

$$g'(x) = \frac{1}{x^2}.$$

Therefore, $f'(x) > g'(x)$ for all $x > 1$, and

$$f(x) = 2\sqrt{x} > 3 - \frac{1}{x} = g(x)$$

for all $x > 1$.

49. Let $f(x) = e^x$, $g(x) = x + 1$.
Then $f(0) = e^0 = 1$, $g(0) = 0 + 1 = 1$, so $f(0) = g(0)$.

$$f'(x) = e^x; \; g'(x) = 1.$$

So $f'(x) > g'(x)$ for $x > 0$.
Thus, $f(x) = e^x > x + 1 = g(x)$ for $x > 0$.

51. Let $f(x) = 3 + e^{-x}$; then $f(0) = 4$, $f'(x) = -e^{-x} < 0$, so f is decreasing. But $f(x) = 3 + e^{-x} = 0$ has no solution.

53. Let y_1 and y_2 be two points in the domain of f^1 with $y_1 < y_2$. Let $x_1 = f^1(y_1)$ and $x_2 = f^1(y_2)$. We want to show $x_1 < x_2$. Suppose not. Then $x_2 < x_1$. But then, since f is increasing, $f(x_2) \leq f(x_1)$. That is, $y_2 \leq y_1$, which contradicts our choice of y_1 and y_2.

55. The domain of $\sin^{-1} x$ is the interval $[-1, 1]$. The function is increasing on the entire domain.

57. True. If $x_1 < x_2$, then $g(x_1) < g(x_2)$ since g is increasing, and then $f(g(x_1)) < f(g(x_2))$ since i is increasing.

59.
$$s(t) = \sqrt{t+4} = (t+4)^{1/2}$$
$$s'(t) = \frac{1}{2}(t+4)^{-1/2} = \frac{1}{2\sqrt{t+4}} > 0$$

So total sales are always increasing at the rate of $\frac{1}{2\sqrt{t+4}}$ thousand dollars per month.

61. True. $(f \circ g)'(c) = f'(g(c))g'(c) = 0$, since c is a critical number of g.

3.5 Concavity and Overview of Curve Sketching

1.
$$f(x) = x^3 - 3x^2 + 4x - 1$$
$$f'(x) = 3x^2 - 6x + 4$$
$$f''(x) = 6x - 6 = 6(x-1)$$
$$f''(x) \begin{cases} > 0 \text{ on } (1, \infty) \\ < 0 \text{ on } (-\infty, 1) \end{cases}$$

So f is concave down on $(-\infty, 1)$ and concave up on $(1, \infty)$.

3.
$$f(x) = x + \frac{1}{x} = x + x^{-1}$$
$$f'(x) = 1 - x^2$$
$$f''(x) = 2x^{-3}$$
$$f''(x) \begin{cases} > 0 \text{ on } (0, \infty) \\ < 0 \text{ on } (-\infty, 0) \end{cases}$$

So f is concave up on $(0, \infty)$ and concave down on $(-\infty, 0)$.

5.
$$f(x) = \sin x - \cos x$$
$$f'(x) = \cos x + \sin x$$
$$f''(x) = -\sin x + \cos x$$

$$f''(x) \begin{cases} < 0 \ldots \left(\frac{\pi}{4}, \frac{5\pi}{4}\right) \cup \left(\frac{9\pi}{4}, \frac{13\pi}{4}\right) \ldots \\ > 0 \ldots \left(\frac{3\pi}{4}, \frac{\pi}{4}\right) \cup \left(\frac{5\pi}{4}, \frac{9\pi}{4}\right) \ldots \end{cases}$$

f is concave down on

$$\ldots \left(\frac{\pi}{4}, \frac{5\pi}{4}\right) \cup \left(\frac{9\pi}{4}, \frac{13\pi}{4}\right) \ldots, \text{ concave up on}$$
$$-\ldots \left(-\frac{3\pi}{4}, \frac{\pi}{4}\right) \cup \left(\frac{5\pi}{4}, \frac{9\pi}{4}\right) \ldots.$$

7.
$$f(x) = x^{4/3} + 4x^{1/3}$$
$$f'(x) = \frac{4}{3}x^{1/3} + \frac{4}{3}x^{-2/3}$$
$$f''(x) = \frac{4}{9}x^{-2/3} - \frac{8}{9}x^{-5/3}$$
$$= \frac{4}{9x^{2/3}}\left(1 - \frac{2}{x}\right)$$

The quantity $\frac{4}{9x^{2/3}}$ is never negative, so the sign of the second derivative is the same as the sign of $1 - \frac{2}{x}$. Hence, the function is concave up for $x > 2$ and $x < 0$, and is concave down for $0 < x < 2$.

9.
$$f(x) = x^3 - 3x^2 9x$$
$$f'(x) = 3x^2 - 6x - 9$$
$$= 3(x^2 - 2x - 3)$$
$$= 3(x-3)(x+1)$$

Critical numbers are $x = -1, 3$. $f''(x) = 6x - 6$. Hence, $f''(-1) = -12 < 0$ and $f''(3) = 12 > 0$. So $f(-1)$ is a local maximum and $f(3)$ is a local minimum.

11.
$$f(x) = x^2 - \frac{16}{x}$$
$$f'(x) = 2x + \frac{16}{x^2}$$

So the only critical number is $x = -2$.

$$f''(x) = 2 - \frac{32}{x^3}, \text{ so } f''(-2) = 6 > 0,$$

so $f(-2)$ is a local minimum.

13.

$$f(x) = x \ln x$$
$$f'(x) = 1 + \ln x$$

So the only critical number is $x = 1/e$.

$$f''(x) = \frac{1}{x}, \quad \text{so} \quad f''(1/e) = e > 0,$$

so $f(1/e)$ is a local minimum.

15.

$$f(x) = x^3 - 3x^2 + 4$$
$$f'(x) = 3x^2 - 6x = 3x(x - 2)$$
$$f''(x) = 6x - 6 = 6(x - 1)$$
$$f'(x) > 0 \text{ on } (-\infty, 0) \cup (2, \infty)$$
$$f'(x) < 0 \text{ on } (0, 2)$$
$$f''(x) > 0 \text{ on } (1, \infty)$$
$$f''(x) < 0 \text{ on } (-\infty, 1)$$
$$f''(0) = -6$$
$$f''(2) = 6$$

Therefore, f is increasing on $(-\infty, 0)$ and on $(2, \infty)$, decreasing on $(0, 2)$. $x = 0$ is a local maximum, and $x = 2$ is a local minimum. f is concave up on $(1, \infty)$, concave down on $(-\infty, 1)$, and $x = 1$ is an inflection point.

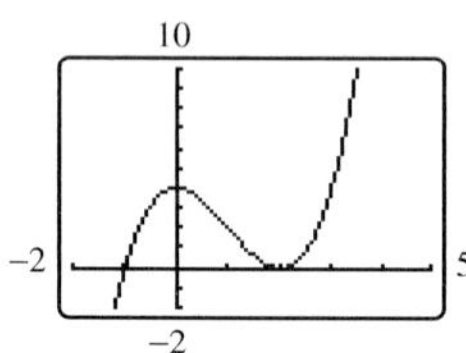

17.

$$y = x^3 + 3x - 1$$
$$y' = 3x^2 + 3 = 3(x^2 + 1)$$
$$y'' = 6x$$

y' is always > 0. So the function is always increasing.

$$y'' \begin{cases} > 0 \text{ on } (0, \infty) \\ < 0 \text{ on } (-\infty, 0). \end{cases}$$

So the function is concave up on $(0, \infty)$ and concave down on $(-\infty, 0)$; $x = 0$ is an inflection point.

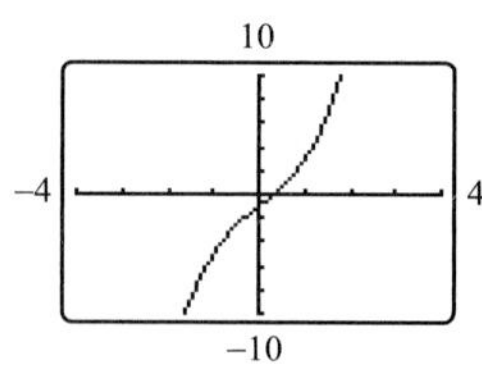

19.

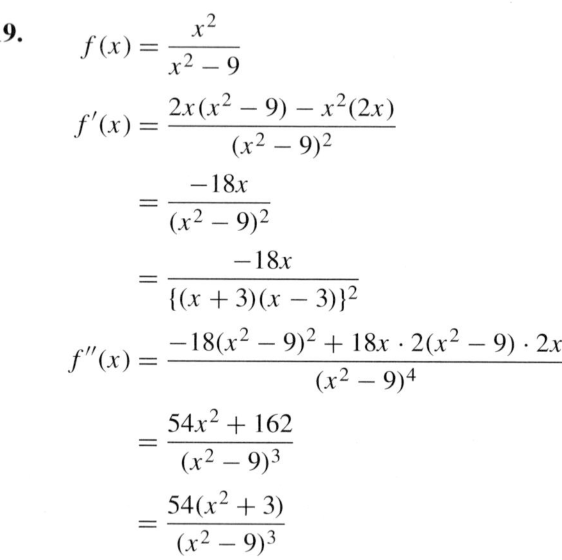

$$f(x) = \frac{x^2}{x^2 - 9}$$

$$f'(x) = \frac{2x(x^2 - 9) - x^2(2x)}{(x^2 - 9)^2}$$
$$= \frac{-18x}{(x^2 - 9)^2}$$
$$= \frac{-18x}{\{(x + 3)(x - 3)\}^2}$$

$$f''(x) = \frac{-18(x^2 - 9)^2 + 18x \cdot 2(x^2 - 9) \cdot 2x}{(x^2 - 9)^4}$$
$$= \frac{54x^2 + 162}{(x^2 - 9)^3}$$
$$= \frac{54(x^2 + 3)}{(x^2 - 9)^3}$$

$$f'(x) \begin{cases} > 0 \text{ on } (-\infty, -3) \cup (-3, 0) \\ < 0 \text{ on } (0, 3) \cup (3, \infty) \end{cases}$$

$$f''(x) \begin{cases} > 0 \text{ on } (-\infty, -3) \cup (3, \infty) \\ < 0 \text{ on } (-3, 3) \end{cases}$$

$$f''(0) = \frac{162}{(-9)^3}$$

f is increasing on $(-\infty, -3) \cup (-3, 0)$, decreasing on $(0, 3) \cup (3, \infty)$, concave up on $(-\infty, -3) \cup (3, \infty)$, concave down on $(-3, 3)$; $x = 0$ is a local max.

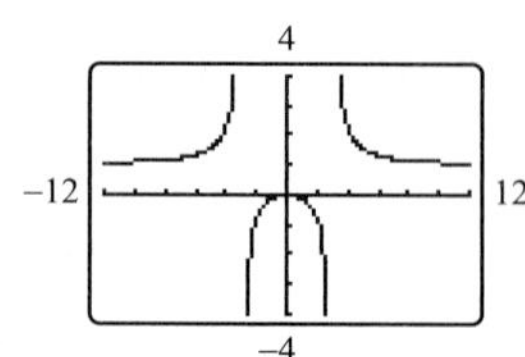

21.

$$y = x^{1/5}(x+1) = x^{6/5} + x^{1/5}$$

$$y' = \frac{6}{5}x^{1/5} + \frac{1}{5}x^{-4/5}$$

$$y'' = \frac{6}{25}x^{-4/5}\frac{4}{25}x^{-9/5}$$

$x = 0$ is one critical point, since y' is not defined at 0. To find other critical points, set

$$\frac{6}{5}x^{1/5} + \frac{1}{5}x^{-4/5} = 0.$$

Multiplying by $5x^{4/5}$, this simplifies to $6x + 1 = 0$, so $x = -1/6$ is the other critical point.

$$y' \begin{cases} > 0 \text{ on } (-1/6, 0) \cup (0, \infty) \\ < 0 \text{ on } (-\infty, -1/6) \end{cases}.$$

Also

$$y'' \begin{cases} > 0 \text{ on } (-\infty, 0) \cup (2/3, \infty) \\ < 0 \text{ on } (0, 2/3) \end{cases}.$$

So the function is increasing on $(-1/6, \infty)$; decreasing on $(-\infty, -1/6)$; concave up on $(-\infty, 0)$ and on $(2/3, \infty)$; concave down on $(0, 2/3)$. Also $x = -1/6$ is a local min and $x = 0$ and $x = 2/3$ are inflection points.

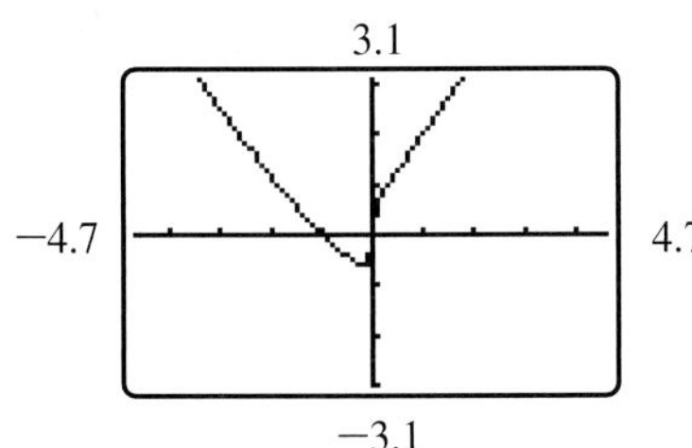

23.

$$f(x) = e^{-2/x}$$

$$f'(x) = e^{-2/x}\left(\frac{2}{x^2}\right) = \frac{2}{x^2}e^{-2/x}$$

$$f''(x) = \frac{-4}{x^3}e^{-2/x} + \frac{2}{x^2}e^{-2/x}\left(\frac{2}{x^2}\right)$$

$$= \frac{4}{x^4}e^{-2/x} - \frac{4}{x^3}e^{-2/x}$$

$f'(x) > 0$ on $(-\infty, 0) \cup (0, \infty)$

$$f''(x) \begin{cases} > 0 \text{ on } (-\infty, 0) \cup (0, 1) \\ < 0 \text{ on } (1, \infty) \end{cases}$$

f increasing on $(-\infty, 0)$ and on $(0, \infty)$, concave up on $(-\infty, 0) \cup (0, 1)$, concave down on $(1, \infty)$, inflection point at $x = 1$. f is undefined at $x = 0$.

$$\lim_{x\to 0^+} e^{-2/x} = \lim_{x\to 0^+} \frac{1}{e^{2/x}} = 0, \text{ and}$$

$$\lim_{x\to 0^-} e^{-2/x} = \infty.$$

So f has a vertical asymptote at $x = 0$.

$$\lim_{x\to\infty} e^{-2/x} = \lim_{x\to-\infty} e^{-2/x} = 1.$$

So f has a horizontal asymptote at $y = 1$.

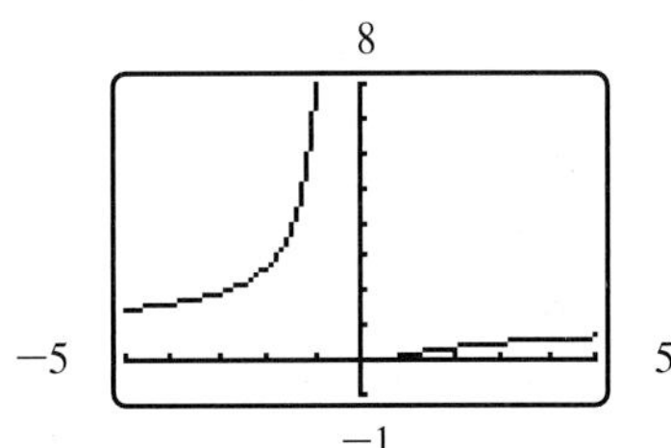

25.

$$f(x) = x + \sin x$$

$$f'(x) = 1 + \cos x$$

$$f''(x) = -\sin x$$

f is increasing on $(-\infty, \infty)$.
There are inflection points at $n\pi$.
f is concave up between each odd multiple of π and the following even multiple of π.
f is concave down between each even multiple of π and the following odd multiple of π.

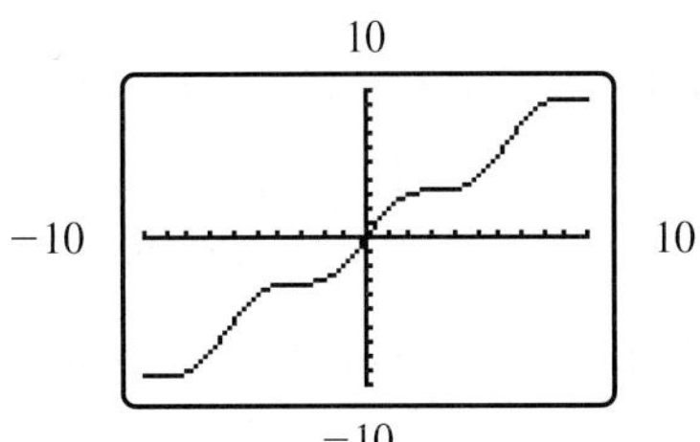

27.

$$y = (x^2 + 1)^{2/3}$$
$$y' = \frac{2}{3}(x^2 + 1)^{-1/3}(2x)$$
$$= \frac{4x(x^2 + 1)^{-1/3}}{3}$$

So the only critical number is $x = 0$.

$$y'' = \frac{4}{3}\left[(x^2 + 1)^{-1/3} + x\left(\frac{-1}{3}\right)(x^2 + 1)^{-4/3}(2x)\right]$$
$$= \frac{4}{3}\frac{(x^2 + 1 - \frac{2x^2}{3})}{(x^2 + 1)^{4/3}} = \frac{4}{9}\frac{(3x^2 + 3 - 2x^2)}{(x^2 + 1)^{4/3}}$$
$$= \frac{4}{9}\frac{(x^2 + 3)}{(x^2 + 1)^{4/3}}.$$

So the function is concave up everywhere, decreasing for x < 0, and increasing for $x > 0$. Also $x = 0$ is a local min.

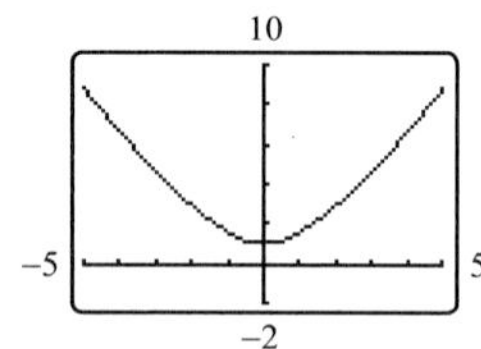

29.

$$y = x^5 - 2x^3 + 1$$
$$y' = 5x^4 - 6x^2 = x^2(5x^2 - 6)$$

Hence, critical numbers are $x = 0$ and $x = \pm\sqrt{6/5}$.

$$y' \begin{cases} > 0 \text{ on } (-\infty, -\sqrt{6/5}) \cup (\sqrt{6/5}, \infty) \\ < 0 \text{ on } (-\sqrt{6/5}, 0) \cup (0, \sqrt{6/5}) \end{cases}.$$

So the function is increasing on $(-\infty, -\sqrt{6/5})$ and on $(\sqrt{6/5}, \infty)$; decreasing on $(-\sqrt{6/5}, \sqrt{6/5})$. Thus $x = -\sqrt{6/5}$ is a local max and $x = \sqrt{6/5}$ is a local min.

$$y'' = 20x^3 - 12x = 4x(5x^2 - 3).$$

$$y'' \begin{cases} > 0 \text{ on } (-\sqrt{3/5}, 0) \cup (\sqrt{3/5}, \infty) \\ < 0 \text{ on } (-\infty, -\sqrt{3/5}) \cup (0, \sqrt{3/5}). \end{cases}$$

So the function is concave up on $(-\sqrt{3/5}, 0)$ and on $(\sqrt{3/5}, \infty)$; concave down on $(-\infty, -\sqrt{3/5})$ and on $(0, \sqrt{3/5})$. There are inflection points at $x = 0$ and at $x = \pm\sqrt{3/5}$.

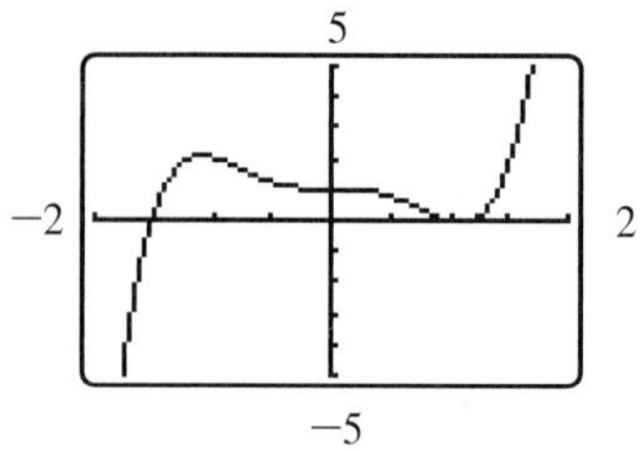

31.

$$y = \frac{x^2 + 1}{3x^2 - 1}$$

Note that $x = \pm\sqrt{1/3}$ are not in the domain of the function, but yield vertical asymptotes.

$$y' = \frac{2x(3x^2 - 1) - (x^2 + 1)(6x)}{(3x^2 - 1)^2}$$
$$= \frac{(6x^3 - 2x) - (6x^3 + 6x)}{(3x^2 - 1)^2} = \frac{-8x}{(3x^2 - 1)^2}.$$

So the only critical point is $x = 0$.

Now $y' \begin{cases} > 0 \text{ for } x < 0 \\ < 0 \text{ for } x > 0, \end{cases}$

so y is increasing on $(-\infty, -\sqrt{1/3})$ and on $(-\sqrt{1/3}, 0)$; decreasing on $(0, \sqrt{1/3})$ and on $(\sqrt{1/3}, \infty)$. Thus there is a local max at $x = 0$.

$$y'' = 8\frac{9x^2 + 1}{(3x^2 - 1)^3}.$$

$$y'' \begin{cases} > 0 \text{ on } (-\infty, -\sqrt{1/3}) \cup (\sqrt{1/3}, \infty) \\ < 0 \text{ on } (-\sqrt{1/3}, \sqrt{1/3}) \end{cases}.$$

Hence, y is concave up on $(-\infty, -\sqrt{1/3})$ and on $(\sqrt{1/3}, \infty)$; concave down on $(-\sqrt{1/3}, \sqrt{1/3})$. Finally, when $|x|$ is large, the function approached 1/3, so $y = 1/3$ is a horizontal asymptote.

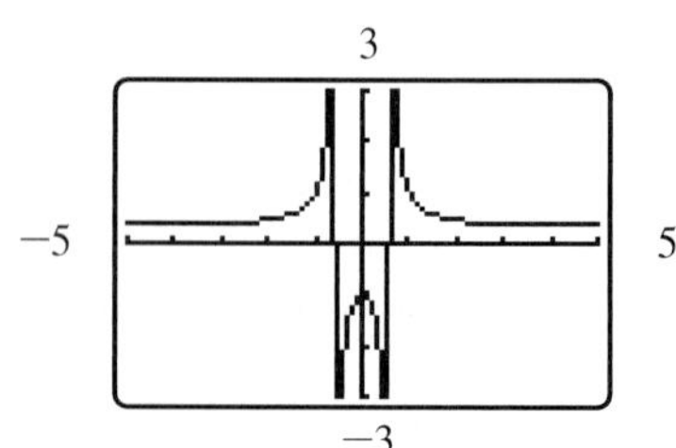

33. $$y = \frac{5x}{x^3 - x + 1}$$

Looking at the graph of $x^3 - x + 1$, we see that there is one real root, at approximately -1.325; so the domain of the function is all x except for this one point, and $x = -1.325$ will be a vertical asymptote.

$$y' = 5\frac{1 - 2x^3}{(x^3 - x - 1)^2}.$$

So the only critical point is $x = \sqrt[3]{1/2}$. By the first derivative test, this is a local max

$$y'' = 10\frac{3x^5 + x^3 - 6x^2 + 1}{(x^3 - x + 1)^3}.$$

The numerator of y'' has three real roots, which are approximately $x = -.39018$, $x = .43347$, and $x = 1.1077$.

$$y'' \begin{cases} > 0 \text{ on } (-\infty, -1.325) \\ \quad \cup(-.390, .433) \cup (1.108, \infty) \\ < 0 \text{ on } (-1.325, -.390) \cup (.433, 1.108) \end{cases}.$$

So y is concave up on

$$(-\infty, -1.325) \cup (-.390, .433) \cup (1.108, \infty)$$

and concave down on

$$(-1.325, -.390) \cup (.433, 1.108).$$

Hence, $x = -.39018$, $x = .43347$, and $x = 1.1077$ are inflection points.

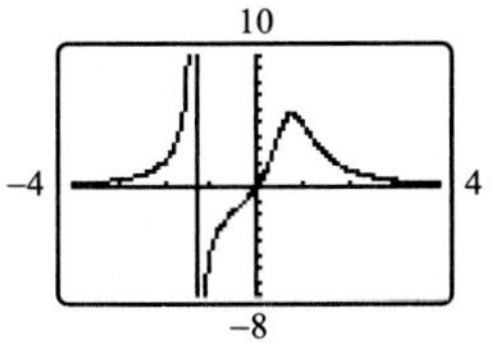

35. $f(x) = x^2\sqrt{x^2 - 9}$; f is undefined on $(-3, 3)$.

$$f'(x) = 2x\sqrt{x^2 - 9} + x^2\left(\frac{1}{2}(x^2 - 9)^{-1/2} \cdot 2x\right)$$
$$= 2x\sqrt{x^2 - 9} + \frac{x^3}{\sqrt{x^2 - 9}}$$
$$= \frac{2x(x^2 - 9) + x^3}{\sqrt{x^2 - 9}}$$
$$= \frac{3x^3 - 18x}{\sqrt{x^2 - 9}}$$
$$= \frac{3x(x^2 - 6)}{\sqrt{x^2 - 9}}$$
$$= \frac{3x(x + \sqrt{6})(x - \sqrt{6})}{\sqrt{x^2 - 9}}.$$

Critical points ± 3. (Note that f is undefined at $x = 0, \pm\sqrt{6}$.)

$f''(x)$

$$= \frac{(9x^2 - 18)\sqrt{x^2 - 9} - (3x^3 - 18x) \cdot \frac{1}{2}(x^2 - 9)^{-1/2} \cdot 2x}{x^2 - 9}$$
$$= \frac{(9x^2 - 18)(x^2 - 9) - x(3x^3 - 18x)}{(x^2 - 9)^{3/2}}$$
$$= \frac{(6x^4 - 81x^2 + 162)}{(x^2 - 9)^{3/2}};$$

$$f''(x) = 0 \text{ when } x^2 = \frac{81 \pm \sqrt{81^2 - 4(6)(162)}}{2(6)}$$
$$= \frac{81 \pm \sqrt{2673}}{12} = \frac{1}{4}(27 \pm \sqrt{297}).$$

So $x \approx \pm 3.325$ or $x \approx \pm 1.562$, but these latter values are not in the same domain. So only ± 3.325 are potential inflection points.

$$f'(x) \begin{cases} > 0 \text{ on } (3, \infty) \\ < 0 \text{ on } (-\infty, -3) \end{cases}.$$

$$f''(x) \begin{cases} > 0(-\infty, -3.3) \cup (3.3, \infty) \\ < 0(-3.3, 3) \cup (3, 3.3) \end{cases}.$$

f is increasing on $(3, \infty)$, decreasing on $(-\infty, -3)$, concave up on $(-\infty, -3.3) \cup (3.3, \infty)$, concave down on $(-3.3, -3) \cup (3, 3.3)$. $x = \pm 3.3$ are inflection points.

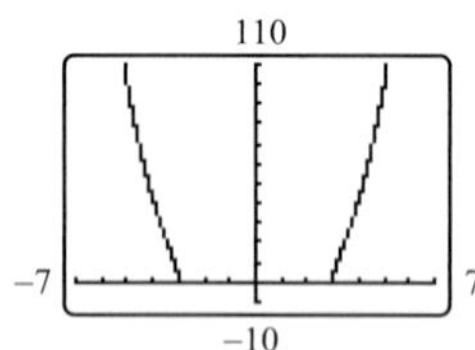

37.
$$f(x) = e^{-2x} \sin x$$
$$f'(x) = e^{-2x}(\cos x - 2 \sin x)$$
$$f''(x) = e^{-2x}(3 \sin x - 4 \cos x)$$

$f'(x) = 0$ when $\cos x = 2 \sin x$; that is, when $\tan x = \frac{1}{2}$; that is, when $x = k\pi + \tan^{-1}\frac{1}{2}$, where k is any integer. $f'(x) < 0$, and f is decreasing, on intervals of the form

$$\left(2k\pi + \tan^{-1}\frac{1}{2}, (2k+1)\pi + \tan^{-1}\frac{1}{2}\right);$$

$f'(x) > 0$ and f is increasing, on intervals of the form

$$\left((2k-1)\pi + \tan^{-1}\frac{1}{2}, 2k\pi + \tan^{-1}\frac{1}{2}\right).$$

Hence, f has a local max at

$$x = 2k\pi + \tan^{-1}\frac{1}{2}$$

and a local min at

$$x = (2k+1)\pi + \tan^{-1}\frac{1}{2}.$$

$f''(x) = 0$ when $3 \sin x = 4 \cos x$; that is, when $\tan x = \frac{4}{3}$; that is, when

$$x = k\pi + \tan^{-1}\frac{4}{3}.$$

The sign of f'' changes at each of these points, so all of them are inflection points.

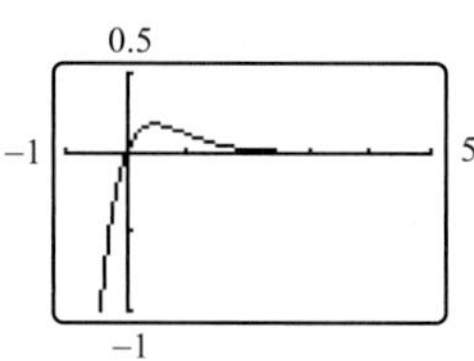

39.
$$f(x) = x^4 - 16x^3 + 42x^2 - 39.6x + 14$$
$$f'(x) = 4x^3 - 48x^2 + 84x - 39.6$$
$$f''(x) = 12x^2 - 96x + 84$$
$$= 12(x^2 - 8x + 7)$$
$$= 12(x-7)(x-1)$$

$$f'(x) \begin{cases} > 0 \text{ on } (.8952, 1.106) \cup (9.9987, \infty) \\ < 0 \text{ on } (-\infty, .8952) \cup (1.106, 9.9987) \end{cases}$$

$$f''(x) \begin{cases} > 0 \text{ on } (-\infty, 1) \cup (7, \infty) \\ < 0 \text{ on } (1, 7) \end{cases}$$

f is increasing on $(.8952, 1.106)$ and on $(9.9987, \infty)$, decreasing on $(-\infty, .8952)$ and on $(1.106, 9.9987)$, concave up on $(-\infty, 1) \cup (7, \infty)$, concave down on $(1, 7)$, $x = .8952, 9.9987$ are local min, $x = 1.106$ is local max, $x = 1, 7$ are inflection points.

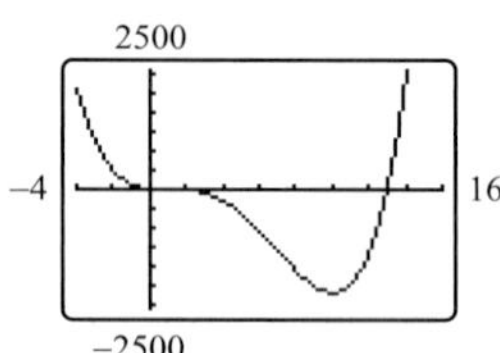

41. $f(x) = x\sqrt{x^2 - 4}$; f undefined on $(-2, 2)$.

$$f'(x) = \sqrt{x^2 - 4} + x\left(\frac{1}{2}\right)(x^2 - 4)^{-1/2}(2x)$$
$$= \sqrt{x^2 - 4} + \frac{x^2}{\sqrt{x^2 - 4}}$$
$$= \frac{2x^2 - 4}{\sqrt{x^2 - 4}};$$

$$f''(x) = \frac{4x\sqrt{x^2-4} - (2x^2-4)\frac{1}{2}(x^2-4)^{-1/2}(2x)}{x^2-4}$$

$$= \frac{4x(x^2-4)x - (2x^2-4)}{(x^2-4)^{3/2}} = \frac{2x^3-12x}{(x^2-4)^{3/2}}$$

$$= \frac{2x(x^2-6)}{(x^2-4)^{3/2}}.$$

$f'(x) > 0(-\infty, -2) \cup (2, \infty)$;

$$f''(x) \begin{cases} > 0 \text{ on } \left(-\sqrt{6}, 2\right) \cup \left(\sqrt{6}, \infty\right) \\ < 0 \text{ on } \left(-\infty, -\sqrt{6}\right) \cup \left(2, \sqrt{6}\right). \end{cases}$$

f is increasing on $(-\infty, -2)$ and on $(2, \infty)$;
concave up on $\left(-\sqrt{6}, -2\right) \cup \left(\sqrt{6}, \infty\right)$;
concave down on $\left(-\infty, -\sqrt{6}\right) \cup \left(2, \sqrt{6}\right)$,
$x = \pm\sqrt{6}$ are inflection points;

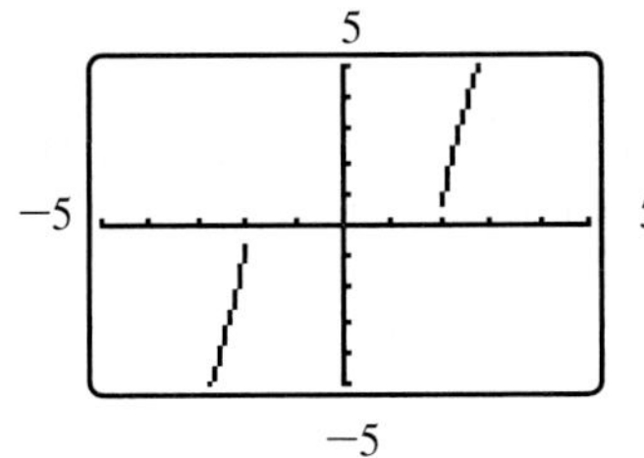

43.

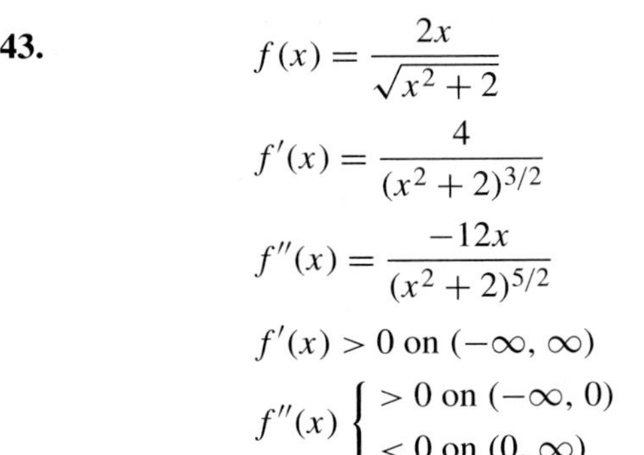

$$f(x) = \frac{2x}{\sqrt{x^2+2}}$$

$$f'(x) = \frac{4}{(x^2+2)^{3/2}}$$

$$f''(x) = \frac{-12x}{(x^2+2)^{5/2}}$$

$f'(x) > 0$ on $(-\infty, \infty)$

$$f''(x) \begin{cases} > 0 \text{ on } (-\infty, 0) \\ < 0 \text{ on } (0, \infty) \end{cases}$$

f increasing on $(-\infty, \infty)$, concave up on $(-\infty, 0)$, concave down on $(0, \infty)$; $x = 0$ inflection point.

$$\lim_{x\to\infty} \frac{2x}{\sqrt{x^2+2}} = \lim_{x\to\infty} \frac{x \cdot 2}{\sqrt{x^2} \cdot \sqrt{1+\frac{2}{x^2}}}$$

$$= \lim_{x\to\infty} \frac{x \cdot 2}{x \cdot \sqrt{1+\frac{2}{x^2}}}$$

$$= \lim_{x\to\infty} \frac{2}{\sqrt{1+\frac{2}{x^2}}}$$

$$= 2;$$

$$\lim_{x\to-\infty} \frac{2x}{\sqrt{x^2+2}} = \lim_{x\to-\infty} \frac{x \cdot 2}{\sqrt{x^2} \cdot \sqrt{1+\frac{2}{x^2}}}$$

$$= \lim_{x\to-\infty} \frac{x \cdot 2}{(-x) \cdot \sqrt{1+\frac{2}{x^2}}}$$

$$= \lim_{x\to-\infty} \frac{-2}{\sqrt{1+\frac{2}{x^2}}}$$

$$= -2.$$

So f has horizontal asymptotes at $y = 2$ and $y = -2$.

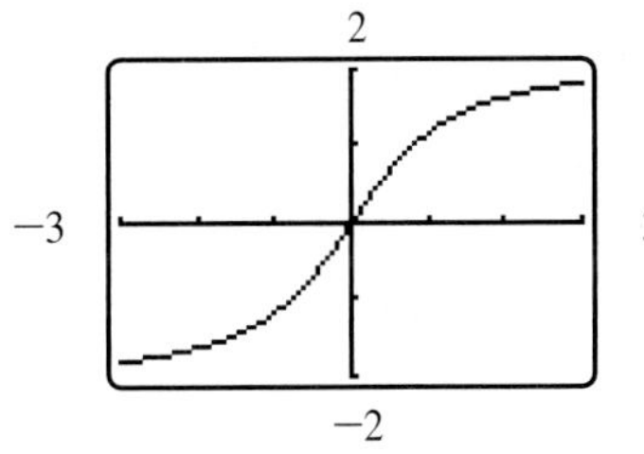

45.

$$y = \frac{25 - 50\sqrt{x^2+0.25}}{x}$$

$$= 25\left(\frac{1 - 2\sqrt{x^2+0.25}}{x}\right)$$

$$= 25\left(\frac{1 - \sqrt{4x^2+1}}{x}\right)$$

Note that $x = 0$ is not in the domain of the function.

$$y' = 25\left(\frac{1 - \sqrt{4x^2+1}}{x^2\sqrt{4x^2+1}}\right),$$

and we see that there are no critical points. Indeed, $y' < 0$ wherever y is defined. One can verify that

$$y'' \begin{cases} > 0 \text{ on } (0, \infty) \\ < 0 \text{ on } (-\infty, 0). \end{cases}$$

Hence, the function is concave up on $(0, \infty)$ and concave down on $(-\infty, 0)$.

$$\begin{aligned}
&\lim_{x\to\infty} \frac{25 - 50\sqrt{x^2 + 0.25}}{x} \\
&= \lim_{x\to\infty} \frac{25}{x} - \frac{50\sqrt{x^2 + 0.25}}{x} \\
&= \lim_{x\to\infty} 0 - 50\frac{x\sqrt{1 + \frac{0.25}{x^2}}}{x} \\
&= \lim_{x\to\infty} -50\sqrt{1 + \frac{0.25}{x^2}} \\
&= -50;
\end{aligned}$$

$$\begin{aligned}
&\lim_{x\to-\infty} \frac{25 - 50\sqrt{x^2 + 0.25}}{x} \\
&= \lim_{x\to\infty} \frac{25}{x} - \frac{50\sqrt{x^2 + 0.25}}{x} \\
&= \lim_{x\to-\infty} 0 - 50\frac{(-x)\sqrt{1 + \frac{0.25}{x^2}}}{x} \\
&= \lim_{x\to\infty} 50\sqrt{1 + \frac{0.25}{x^2}} \\
&= 50.
\end{aligned}$$

So f has horizontal asymptotes at $y = 50$ and $y = -50$.

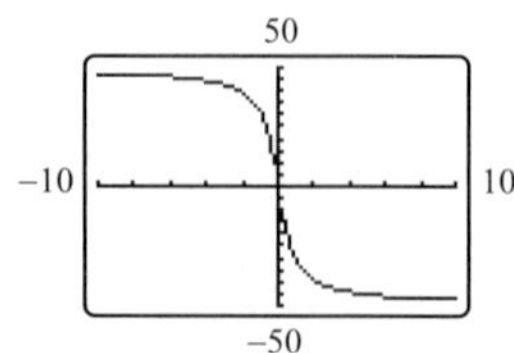

47.

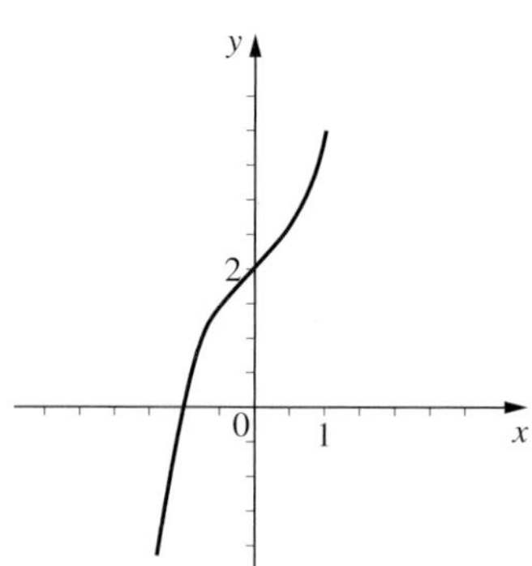

49.

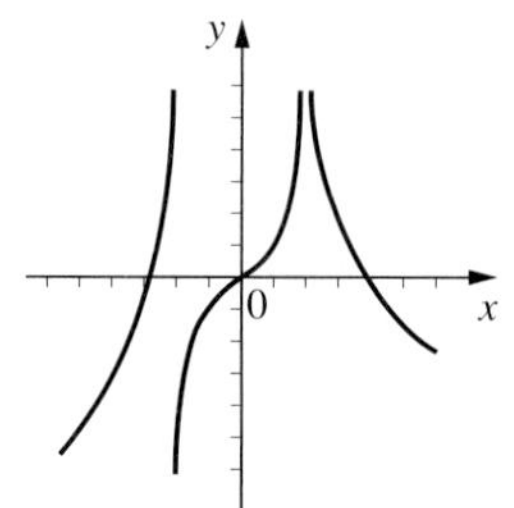

51.

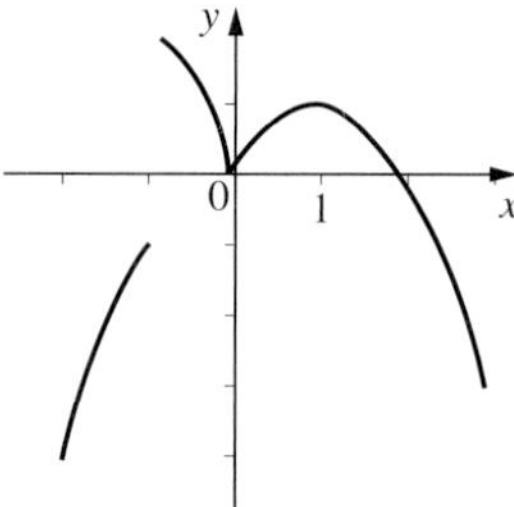

53. f is increasing on $(-\infty, 1/2)$ and on $(3/2, \infty)$; decreasing on $(1/2, 3/2)$; concave up on $(1, \infty)$, concave down on $(-\infty, 1)$; local max at $x = 1/2$, local min at $x = 3/2$; inflection point at $x = 1$.

55. For the graph in Exercise 53, f is increasing where the graph is positive, that is, on $(0, 1)$ and on $(2, \infty)$; decreasing where the graph is negative, that is, on $(-\infty, 0)$ and on $(1, 2)$; concave up where the graph is increasing, that is, on $(-\infty, 1/2)$ and on $(3/2, \infty)$; concave down where the graph is decreasing, that is on $(1/2, 3/2)$. There are local min where the graph goes from negative to positive, that is, at $x = 0$ and $x = 2$; local max where the graph goes from positive to negative, that is, at $x = 1$; and inflection points where the graph changes from increasing to decreasing or from decreasing to increasing, that is, at $x = 1/2$ and $x = 3/2$.

57.

$$f(x) = x^4 + cx^2$$
$$f'(x) = 4x^3 + 2cx$$
$$f''(x) = 12x^2 + 2c$$

$c = 0$: 1 extremum, 0 inflection points
$c < 0$: 3 extrema, 2 inflection points
$c > 0$: 1 extremum, 0 inflection points
$c \to -\infty$: the graph widens and lowers
$c \to +\infty$: the graph narrows

59.

$$f(x) = \frac{x^2}{x^2 + c^2}$$
$$f'(x) = \frac{2c^2x}{(x^2 + c^2)^2}$$
$$f''(x) = \frac{2c^4 - 6c^2x^2}{(x^2 + c^2)^3}$$

If $c = 0$: $f(x) = 1$, except that f is undefined at $x = 0$.
$c < 0$, $c > 0$: horizontal asymptote at $y = 1$, local min at $x = 0$, since the derivative changes sign from negative to positive at $x = 0$; also there are inflection points at

$$x = \pm\frac{c}{\sqrt{3}}.$$

As $c \to -\infty$, $c \to +\infty$: the graph widens.

61.

$$f(x) = \frac{x^n}{x^n + c^n}$$

If $c = 0$, then $f(x) = 1$, except that f is undefined at $x = 0$.
If $c > 0$,

$$f'(x) = \frac{nx^{n-1}c^n}{(x^n + c^n)^2}$$
$$f''(x) = nc^n x^{n-2}\left(\frac{nc^n - c^n - x^n n - x^n}{(x^n + c^n)^3}\right)$$

We need to consider separately the cases n even and n odd.

If n is even, the situation is similar to Exercise 59. Again there is a local min at $x = 0$. To find the inflection points, we solve $f''(x) = 0$ and get three solutions,

$$x = 0, \pm c\left(\frac{n-1}{n+1}\right)^{1/n}.$$

There is no inflection point at $x = 0$, since the second derivative does not change sign at $x = 0$, but there are inflection points at the other two values. Also, $y = 1$ is a horizontal asymptote, and as $c \to \infty$ the graph widens.

If n is odd, then $f(x)$ is not defined at $x = -c$, and in fact $x = -c$ is a vertical asymptote. Now there is no local extremum at $x = 0$, since $f'(x)$ is positive on both sides of 0; however, $f''(0)$ does change sign at 0 and so there is an inflection point at 0. Also, there is an inflection point at

$$x = c\left(\frac{n-1}{n+1}\right)^{1/n}.$$

Finally, $y = 1$ is a horizontal asymptote.

63.

$$f(x) = \frac{3x^2 - 1}{x} = 3x - \frac{1}{x}$$

$y = 3x$ is a slant asymptote.

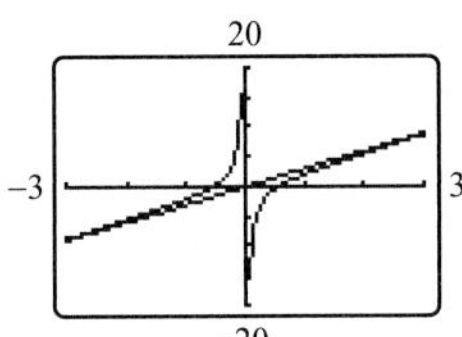

65.

$$f(x) = \frac{x^3 - 2x^2 + 1}{x^2} = x - 2 + \frac{1}{x^2}$$

$y = x - 2$ is a slant asymptote.

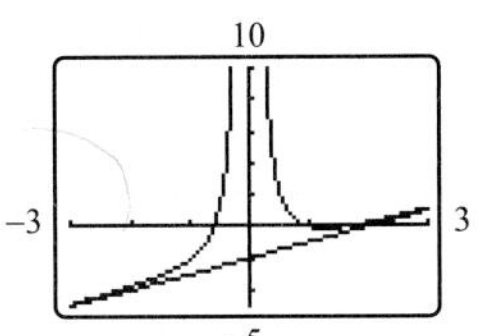

67. $$f(x) = \frac{x^4}{x^3+1} = x - \frac{x}{x^3+1}$$

$y = x$ is a slant asymptote.

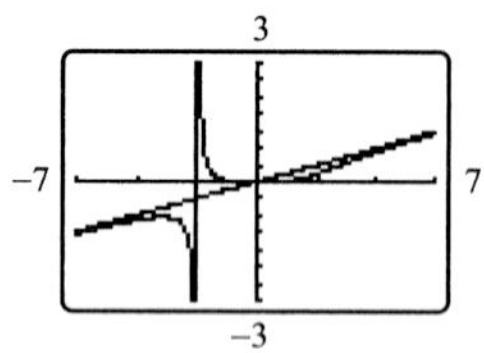

69. asymptotes $x = 1$, $x = 2$, $y = 3$

$$f(x) = \frac{3x^2}{(x-1)(x-2)}$$

71. asymptotes $x = -1$, $x = 1$, $y = -2$, $y = 2$

$$f(x) = \frac{2x}{\sqrt{(x-1)(x+1)}}$$

73. We need to know $w'(0)$ to know if the depth is increasing.

75.
$$s(x) = -3x^3 + 270x^2 - 3600x + 18000$$
$$s'(x) = -9x^2 + 540x - 3600$$
$$s''(x) = -18x + 540 = 0$$

$x = 30$. This is a max because the graph of $s'(x)$ is a parabola opening down. So spend \$30,000 on advertising to maximize the rate of change of sales.

77.
$$C(x) = .01x^2 + 40x + 3600$$
$$\overline{C}(x) = \frac{C(x)}{x} = .01x + 40 + 3600x^{-1}$$
$$\overline{C}'(x) = .01 - 3600x^{-2} = 0$$

$x = 600$. This is a min because $\bar{C}''(x) = 7200x^{-3} > 0$ for $x > 0$, so the graph is concave up. So manufacture 600 units to minimize average cost.

79.
$$f(x) = x^4 + cx^3$$
$$f'(x) = 4x^3 + 3cx^2 = x^2(4x + 3c)$$
$$f''(x) = 12x^2 + 6cx = 6x(2x + c)$$

So (0, 0) is inflection point since the sign of the second derivative changes at $x = 0$. There is a local min at $x = -\frac{3}{4}c$ and an inflection point when $x = -\frac{1}{2}c$.

81. Let $f(x) = -1 - x^2$. Then
$$f'(x) = -2x$$
$$f''(x) = -2,$$
so f is concave down for all x.
But $-1 - x^2 = 0$ has no solution.

83. Since the tangent line points above the sun, the sun appears higher in the sky than it really is.

85.
$$f(x) = xe^{-bx}$$
$$f(0) = 0$$
$$f(x) > 0 \text{ for } x > 0$$
$$\lim_{x\to\infty} xe^{-bx} = \lim_{x\to\infty} \frac{x}{e^{bx}} = \lim_{x\to\infty} \frac{1}{be^{bx}}$$
(by L'Hôpital's Rule)
$$= 0.$$

$f'(x) = e^{-bx}(1 - bx)$, so there is a unique critical point at $x = 1/b$, which must be the maximum. The bigger b is, the closer the max is to the origin. For time since conception, $\frac{1}{b}$ represents the most common gestation time. For survival time, $\frac{1}{b}$ represents the most common life span.

87. Both functions are increasing for $x > 0$ and have the same asymptote, $y = 1$, so that is no help. However, if
$$f(x) = \frac{x}{27 + x},$$
then
$$f'(x) = \frac{27}{(27 + x)^2}.$$
Hence, $f'(x)$ is decreasing for $x > 0$, and so f has no inflection points on this interval. On the other hand, if
$$f(x) = \frac{x^3}{c^3 + x^3},$$

then

$$f'(x) = \frac{3cx^2}{(c^3 + x^3)^2}$$

and

$$f''(x) = \frac{6c^3x(c^3 - 2x^3)}{(c^3 + x^3)^3},$$

and so there is an inflection point at $x = \frac{c}{\sqrt[3]{2}}$. When $c = 27$, $\frac{27}{\sqrt[3]{2}} \approx 21.4$, in excellent agreement with the given graph.

3.6 Optimization

1. $f(x) = x^2 + 1$ has a minimum at $x = 0$, while $\sin(x^2 + 1)$ has minima where

$$x^2 + 1 = \frac{3\pi}{2} + 2np\pi.$$

3.

$$A = xy = 1800$$

$$y = \frac{1800}{x}$$

$$P = 2x + y = 2x + \frac{1800}{x}$$

$$P' = 2 - \frac{1800}{x^2} = 0$$

$$2x^2 = 1800$$

$$x = 30$$

$$P'(x) \begin{cases} > 0 \text{ for } x > 30 \\ < 0 \text{ for } 0 < x < 30 \end{cases}$$

So $x = 30$ is min.

$$y = \frac{1800}{x} = \frac{1800}{30} = 60$$

So the dimensions are $30' \times 60'$ and the minimum perimeter is 120 ft.

5.

$$P = 2x + 3y = 120$$

$$3y = 120 - 2x$$

$$y = 40 - \frac{2}{3}x$$

$$A = x \cdot y$$

$$A(x) = x\left(40 - \frac{2}{3}x\right)$$

$$A'(x) = 1\left(40 - \frac{2}{3}x\right) + x\left(-\frac{2}{3}\right)$$

$$= 40 - \frac{4}{3}x = 0$$

$$40 = \frac{4}{3}x$$

$$x = 30$$

$$A'(x) > 0 \text{ for } 0 < x < 30$$

$$A'(x) < 0 \text{ for } x > 30$$

So $x = 30$ is max, $y = 40 - \frac{2}{3} \cdot 30 = 20$.
So the dimensions are $20' \times 30'$.

7.

$$A = xy$$

$$P = 2x + 2y$$

$$2y = P - 2x$$

$$y = \frac{P}{2} - x$$

$$A(x) = x\left(\frac{P}{2} - x\right)$$

$$A'(x) = 1 \cdot \left(\frac{P}{2} - x\right) + x(-1) = \frac{P}{2} - 2x = 0$$

$$P = 4x$$

$$x = \frac{P}{4}$$

$$A'(x) \begin{cases} > 0 \text{ for } 0 < x < \frac{P}{4} \\ < 0 \text{ for } x > \frac{P}{4} \end{cases}$$

So $x = \frac{P}{4}$ is max,

$$y = \frac{P}{2} - x = \frac{P}{2} - \frac{P}{4} = \frac{P}{4}.$$

So the dimensions are $\frac{P}{4} \times \frac{P}{4}$. Thus we have a square.

9.

$$d = \sqrt{(x-0)^2 + (y-1)^2}$$
$$y = x^2$$
$$d = \sqrt{x^2 + (x^2-1)^2}$$
$$f(x) = [d(x)]^2 = x^2 + (x^2-1)^2, \ -1 \le x \le 1$$
$$f'(x) = 2x + 2(x^2-1)\cdot 2x$$
$$= 2x(1 + 2x^2 - 2)$$
$$= 2x(2x^2-1) = 0$$
$$x = 0, \pm\sqrt{\frac{1}{2}}$$
$$f(0) = 1$$
$$f(1) = 1 \qquad f(-1) = 1$$
$$f\left(\sqrt{\frac{1}{2}}\right) = \frac{3}{4} \qquad f\left(-\sqrt{\frac{1}{2}}\right) = \frac{3}{4}$$

Thus $x = \pm\sqrt{\frac{1}{2}}$ are min, and the points on $y = x^2$ closest to (0, 1) are

$$\left(\sqrt{\frac{1}{2}}, \frac{1}{2}\right) \quad \text{and} \quad \left(-\sqrt{\frac{1}{2}}, \frac{1}{2}\right).$$

11.

$$d = \sqrt{(x-0)^2 + (y-0)^2}$$
$$y = \cos x$$
$$d = \sqrt{x^2 + \cos^2 x}$$
$$f(x) = [d(x)]^2 = x^2 + \cos^2 x, \ 0 \le x \le \frac{\pi}{2}$$
$$f'(x) = 2x + 2\cos x(-\sin x)$$
$$= 2x - 2\cos x \sin x = 0$$
$$x = \cos x \sin x$$
$$x = 0$$
$$f(0) = 0^2 + \cos^2 0 = 1$$
$$f\left(\frac{\pi}{2}\right) = \left(\frac{\pi}{2}\right)^2 + \cos^2\frac{\pi}{2} = \frac{\pi^2}{4}$$

So $x = 0$ is min and the point on $y = \cos x$ closest to (0, 0) is (0, 1).

13. For (0, 1), $\left(\sqrt{\frac{1}{2}}, \frac{1}{2}\right)$ on $y = x^2$, we have

$$y' = 2x, \ y'\left(\sqrt{\frac{1}{2}}\right) = 2\cdot\sqrt{\frac{1}{2}} = \sqrt{2}$$

and $m = \dfrac{\frac{1}{2} - 1}{\sqrt{\frac{1}{2}} - 0} = -\dfrac{1}{\sqrt{2}}$

For (0, 1), $\left(-\sqrt{\frac{1}{2}}, \frac{1}{2}\right)$ on $y = x^2$, we have

$$y'\left(-\sqrt{\frac{1}{2}}\right) = 2\left(-\sqrt{\frac{1}{2}}\right) = -\sqrt{2}$$

and $m = \dfrac{\frac{1}{2} - 1}{-\sqrt{\frac{1}{2}} - 0} = \dfrac{1}{\sqrt{2}}.$

For (3, 4), (2.06, 4.2436) on $y = x^2$, we have

$$y'(2.06) = 2(2.06) = 4.12$$

and

$$m = \frac{4.2436 - 4}{2.06 - 3} = -0.2591 \approx -\frac{1}{4.12}.$$

15.

$$V = l\cdot w\cdot h$$
$$V(x) = (10-2x)(6-2x)\cdot x, \ 0 \le x \le 3$$
$$V'(x) = -2(6-2x)\cdot x + (10-2x)(-2)\cdot x + (10-2x)(6-2x)$$
$$= 60 - 64x + 12x^2 = 4(3x^2 - 16x + 15)$$
$$= 0$$
$$x = \frac{16 \pm \sqrt{(-16)^2 - 4\cdot 3\cdot 15}}{6} = \frac{8}{3} \pm \frac{\sqrt{19}}{3}$$
$$x = \frac{8}{3} + \frac{\sqrt{19}}{3} > 3$$
$$V'(x) > 0 \text{ for } x < \frac{8}{3} - \frac{\sqrt{19}}{3}$$
$$V'(x) < 0 \text{ for } x > \frac{8}{3} - \frac{\sqrt{19}}{3}$$

So $x = \frac{8}{3} - \frac{\sqrt{19}}{3}$ is a max.

$$x = \frac{8}{3} - \frac{\sqrt{19}}{3} \approx 1.2137.$$

17. $f(x) = \sqrt{3^2 + (5-x)^2} + \sqrt{4^2 + x^2},\ 0 \le x \le 5$

$$f'(x) = \frac{1}{2}(9 + (5-x)^2)^{-1/2}(2)(5-x)(-1)$$
$$+ \frac{1}{2}(16 + x^2)^{-1/2}(2x)$$
$$= \frac{x-5}{\sqrt{9+(5-x)^2}} + \frac{x}{\sqrt{16+x^2}}$$
$$= 0$$

$x = \frac{20}{7} \approx 2.857$

$f(0) = 4 + \sqrt{34} \approx 9.831$

$f\left(\frac{20}{7}\right) = \sqrt{74} \approx 8.602$

$f(5) = 3 + \sqrt{41} \approx 9.403$

So $x = \frac{20}{7}$ is min.

The water line should be $\frac{20}{7}$ miles west of the second development, $5 - \frac{20}{7} = \frac{15}{7}$ miles east of the first development.

19. $C(x) = 5\sqrt{16 + x^2} + 2\sqrt{36 + (8-x)^2},\ 0 \le x \le 8$

$C(x) = 5\sqrt{16 + x^2} + 2\sqrt{100 - 16x + x^2}$

$$C'(x) = 5\left(\frac{1}{2}\right)(16 + x^2)^{-1/2} \cdot 2x +$$
$$2\left(\frac{1}{2}\right)(100 - 16x + x^2)^{-1/2}(2x - 16)$$
$$= \frac{5x}{\sqrt{16+x^2}} + \frac{2x-16}{\sqrt{100-16x+x^2}}$$
$$= 0$$

$x \approx 1.2529$

$C(0) = 40$

$C(1.2529) \approx 39.0162$

$C(8) \approx 56.7214$

The highway should emerge from the marsh 1.2529 miles east of the bridge. If we build a straight line to the interchange, we have $x = (3.2)$.
Since $C(3.2) - C(1.2529) \approx 1.963$, we save \$1.963 million.

21. $C(x) = 5\sqrt{16 + x^2} + 3\sqrt{36 + (8-x)^2},\ 0 \le x \le 8$

$$C'(x) = \frac{5x}{\sqrt{16+x^2}} + \frac{3x-24}{\sqrt{100-16x+x^2}} = 0$$

$x \approx 1.8941$

$C(0) = 50$

$C(1.8941) \approx 47.8104$

$C(8) \approx 62.7214$

The highway should emerge from the marsh 1.8941 miles east of the bridge. So if we must use the path from Exercise 21, the extra cost is

$$C(1.2529) - C(1.8941) = 48.0452 - 47.8104$$
$$= 0.2348,$$

or about \$234.8 thousand.

23. $T(x) = \frac{\sqrt{1+x^2}}{v_1} + \frac{\sqrt{1+(2-x)^2}}{v_2}$

$$T'(x) = \frac{1}{v_1} \cdot \frac{1}{2}(1 + x^2)^{-1/2}$$
$$\cdot 2x + \frac{1}{v_2}(1 + (2-x)^2)^{-1/2} \cdot (2-x)(-1)$$
$$= \frac{x}{v_1\sqrt{1+x^2}} + \frac{x-2}{v_2\sqrt{1+(2-x)^2}}$$

Note that

$$T'(x) = \frac{1}{v_1} \cdot \frac{x}{\sqrt{1+x^2}} - \frac{1}{v_2} \cdot \frac{(2-x)}{\sqrt{1+(2-x)^2}}$$
$$= \frac{1}{v_1}\sin\theta_1 - \frac{1}{v_2}\sin\theta_2.$$

When $T'(x) = 0$, we have

$$\frac{1}{v_1}\sin\theta_1 = \frac{1}{v_2}\sin\theta_2,$$
$$\frac{\sin\theta_1}{\sin\theta_2} = \frac{v_1}{v_2}.$$

25. Cost $= 2(2\pi r^2) + 2\pi r h$.

$$12 \text{ fl oz} = 12 \text{ fl oz} \cdot \frac{1.80469 \text{ in}^3}{\text{fl oz}} = 21.65628 \text{ in}^3.$$

$$\text{Vol} = \pi r^2 h,\ h = \frac{\text{Vol}}{\pi r^2} = \frac{21.65628}{\pi r^2}, \text{ so}$$

$$\text{Cost} = 4\pi r^2 + 2\pi r \left(\frac{21.65628}{\pi r^2}\right) \text{ or}$$

$$C(r) = 4\pi r^2 + 43.31256 r^{-1},$$

so

$$C'(r) = 8\pi r - 43.31256 r^{-2} = \frac{8\pi r^3 - 43.31256}{r^2}.$$

$$r = \sqrt[3]{\frac{43.31256}{8\pi}} = 1.1989'' \text{ when } C'(r) = 0.$$

$$C'(r) \begin{cases} < 0 \text{ on } (0, 1.1989) \\ > 0 \text{ on } (1.1989, \infty). \end{cases}$$

Thus $r = 1.1989$ minimizes the cost.

$$h = \frac{21.65628}{\pi (1.1989)^2} = 4.7957''.$$

27.

$$\begin{aligned} V(r) &= cr^2(r_0 - r) \\ V'(r) &= 2cr(r_0 - r) + cr^2(-1) \\ &= 2crr_0 - 3cr^2 \\ &= cr(2r_0 - 3r) \\ &= 0 \end{aligned}$$

$$r = \frac{2r_0}{3} \text{ when } V'(r) = 0$$

$$V'(r) \begin{cases} > 0 \text{ on } \left(0, \frac{2r_0}{3}\right) \\ < 0 \text{ on } \left(\frac{2r_0}{3}, \infty\right) \end{cases}$$

Thus $r = \dfrac{2r_0}{3}$ maximizes the velocity.

$r = \dfrac{2r_0}{3} < r_0$, so the windpipe contracts.

29.

$$\begin{aligned} p(x) &= \frac{V^2 x}{(R+x)^2} \\ p'(x) &= \frac{V^2(R+x)^2 - V^2 x \cdot 2(R+x)}{(R+x)^4} \\ &= \frac{V^2 R^2 - V^2 x^2}{(R+x)^4} \\ &= 0 \end{aligned}$$

$x = R$ when $p'(x) = 0$

$$p'(x) \begin{cases} > 0 \text{ on } (0, R) \\ < 0 \text{ on } (R, \infty) \end{cases}$$

Thus $x = R$ maximizes the power absorbed.

31.

$$\pi r + 4r + 2w = 8 + \pi$$

$$w = \frac{8 + \pi - r(\pi + 4)}{2}$$

$$\begin{aligned} A(r) &= \frac{\pi r^2}{2} + 2rw \\ &= \frac{\pi r^2}{2} + r(8 + \pi - r(\pi + 4)) \\ &= r^2\left(-4 - \frac{\pi}{2}\right) + r(8 + \pi) \end{aligned}$$

$$A'(r) = -2r\left(4 + \frac{\pi}{2}\right) + (8 + \pi) = 0$$

$r = 1,\ 2r = 2$ when $A'(r) = 0$

$$A'(r) \begin{cases} > 0 \text{ on } (0, 1) \\ < 0 \text{ on } (1, \infty) \end{cases}$$

Thus $r = 1$ maximizes the area.
So $w = \dfrac{8 + \pi - (\pi + 4)}{2} = 2.$
Thus the dimensions of the rectangle are 2×2.

33.

$$l \times w = 92,\ w = \frac{92}{l}$$

$$\begin{aligned} A(l) &= (l + 4)(w + 2) \\ &= (l + 4)\left(\frac{92}{l} + 2\right) \\ &= 92 + \frac{368}{l} + 2l + 8 \\ &= 100 + 368 l^{-1} + 2l \end{aligned}$$

$$A'(l) = -368l^{-2} + 2$$
$$= \frac{2l^2 - 368}{l^2}$$
$$l = \sqrt{184} = 2\sqrt{46} \text{ when } A'(l) = 0$$
$$A'(l) \begin{cases} < 0 \text{ on } (0, 2\sqrt{46}) \\ > 0 \text{ on } (2\sqrt{46}, \infty) \end{cases}$$

Therefore, $l = 2\sqrt{46}$ minimizes the total area.
When $l = 2\sqrt{46}$, $w = \dfrac{92}{2\sqrt{46}} = \sqrt{46}$.
For the minimum total area, the printed area has width $\sqrt{46}$ in. and length $2\sqrt{46}$ in., and the advertisement has overall width $\sqrt{46} + 2$ in. and overall length $2\sqrt{46} + 4$ in.

35. $$I(A) = 2A^3 - 33A^2 + 108A - 310$$
$$I'(A) = 6A^2 - 66A + 108 = 6(A^2 - 11A + 18)$$
$$= 6(A - 9)(A - 2)$$

The critical numbers are 2 and 9; the endpoints are 0 and 10.

$$I(0) = -310$$
$$I(2) = -210$$
$$I(9) = -553$$
$$I(10) = -530$$

The farmer should plant 2 acres.

37. $$R(x) = \frac{35x - x^2}{x^2 + 35}$$
$$R'(x) = -35\frac{x^2 + 2x - 35}{(x^2 + 35)^2}$$
$$= -35\frac{(x - 5)(x + 7)}{(x^2 + 35)^2}$$

Hence, the only critical number for $x \geq 0$ is $x = 5$ (i.e., 5000 items). This must correspond to the absolute maximum, since $R(0) = 0$ and $R(x)$ is negative for large x. So maximum revenue is $R(5) = 2.5$ (i.e., $2500).

39. $Q'(t)$ is efficiency because it represents the number of additional items produced in unit time.

$$Q(t) = -t^3 + 12t^2 + 60t$$
$$Q'(t) = -3t^2 + 24t + 60$$
$$= 3(-t^2 + 8t + 20).$$

This is the quantity we want to maximize. $Q''(t) = 3(-2t + 8)$, so the only critical number is $t = 4$ hours. This must be the maximum since the function $Q'(t)$ is a parabola opening down.

41. $$C(t) = (\text{price per ticket})(\text{number of tickets})$$
$$= (40 - (t - 20))(t)$$
$$= (60 - t)(t) = 60t - t^2 \text{ for } \underline{20} < t < \underline{50}.$$

Then $C'(t) = 60 - 2t$, so $t = 30$ is the only critical number. This must correspond to the maximum since $C(t)$ is a parabola opening down.

43. $$R = \frac{2v^2 \cos^2\theta}{g}(\tan\theta - \tan\beta)$$
$$R'(\theta) = \frac{2v^2}{g}\left[2\cos\theta(-\sin\theta)(\tan\theta - \tan\beta) + \cos^2\theta \cdot \sec^2\theta\right]$$
$$= \frac{2v^2}{g}\left[-2\cos\theta\sin\theta \cdot \frac{\sin\theta}{\cos\theta} + 2\cos\theta\sin\theta\tan\beta + \cos^2\theta \cdot \frac{1}{\cos^2\theta}\right]$$
$$= \frac{2v^2}{g}\left[-2\sin^2\theta + \sin(2\theta)\tan\beta + 1\right]$$
$$= \frac{2v^2}{g}\left[-2\sin^2\theta + \sin(2\theta)\tan\beta + (\sin^2\theta + \cos^2\theta)\right]$$
$$= \frac{2v^2}{g}\left[\sin(2\theta)\tan\beta + (\cos^2\theta - \sin^2\theta)\right]$$
$$= \frac{2v^2}{g}\left[\sin(2\theta)\tan\beta + \cos(2\theta)\right]$$
$$R'(\theta) = 0 \text{ when } \tan\beta = \frac{-\cos(2\theta)}{\sin(2\theta)} = -\cot(2\theta)$$
$$= -\tan\left(\frac{\pi}{2} - 2\theta\right) = \tan\left(2\theta - \frac{\pi}{2}\right)$$

Hence, $\beta = 2\theta - \frac{\pi}{2}$, so

$$\theta = \frac{1}{2}\left(\beta + \frac{\pi}{2}\right) = \frac{\beta}{2} + \frac{\pi}{4} = \frac{\beta^\circ}{2} + 45^\circ.$$

(a) $\beta = 10^\circ, \theta = 50^\circ$;
(b) $\beta = 0^\circ, \theta = 45^\circ$;
(c) $\beta = -10^\circ, \theta = 40^\circ$.

45. $T = \frac{-1}{c} \ln\left(1 - c \cdot \frac{b-a}{v_0}\right)$

$b = 300,\ a = 0,\ v_0 = 125,\ c = 0.1$

$T = \frac{-1}{0.1} \ln\left(1 - 0.1 \cdot \frac{300-0}{125}\right) = 2.744 \text{ sec}$

$$\begin{aligned} T(x) &= -10 \ln(1 - 0.0008(300 - x)) \\ &\quad - 10 \ln(1 - 0.0008x) + 0.1 \end{aligned}$$

$T'(x) = -10\left(\frac{0.0008}{0.76 + 0.0008x} - \frac{0.0008}{1 - 0.0008x}\right) = 0$

$0.0008(1 - 0.0008x) = 0.0008(0.76 + 0.0008x)$

$x = \frac{1 - 0.76}{0.0016} = 150$ ft when $T'(x) = 0$.

$T'(x) \begin{cases} < 0 \text{ on } (0, 150) \\ > 0 \text{ on } (150, 300) \end{cases}$

Hence, $x = 150$ minimizes the total time.

$$\begin{aligned} T(150) &= -10 \ln(1 - 0.0008(300 - 150)) \\ &\quad - 10 \ln(1 - 0.0008(150)) + 0.1 \\ &= 2.656 \text{ sec.} \end{aligned}$$

So the relay is faster.

If the delay is 0.2 sec, the relay takes longer.

47.
$$\begin{aligned} T(x) &= -10 \ln\left(1 - 0.1\frac{300-x}{125}\right)\Bigg) \\ &\quad - 10 \ln\left(1 - 0.1\frac{x}{100}\right) + 0.1 \\ &= -10(\ln(1 - 0.0008(300 - x)) \\ &\quad - 10 \ln(1 - .001x) + 0.01 \end{aligned}$$

$T'(x) = -10\left(\frac{0.0008}{0.76 + 0.0008x} - \frac{0.001}{1 - 0.001x}\right) = 0$

$0.0008(1 - 0.001x) = 0.001(0.76 + 0.0008x)$

$x = 25$ ft when $T'(x) = 0$.

$T'(x) \begin{cases} < 0 \text{ on } (0, 25) \\ > 0 \text{ on } (25, 300) \end{cases}$

Hence, $x = 25$ minimizes the total time.

$$\begin{aligned} T(25) &= -10 \ln(1 - 0.0008(300 - 25)) \\ &\quad - 10 \ln(1 - 0.001(25)) + 0.1 \\ &= 2.838 \text{ sec.} \end{aligned}$$

So the relay takes longer. Without the delay, the relay would take 2.738 sec, so a delay of $2.744 - 2.738 = .006$ sec makes the two times equal.

3.7 Rates of Change in Economics and the Sciences

1. $C(x) = x^3 + 20x^2 + 90x + 15$

$C'(x) = 3x^2 + 40x + 90$

$C'(50) = 3(50)^2 + 40 \cdot 50 + 90 = 9590$

$C(50) = 50^3 + 20(50)^2 + 90 \cdot 50 + 15 = 179515$

$C(49) = 49^3 + 20(49)^2 + 90 \cdot 49 + 15 = 170094$

$C(50) - C(49) = 9421$

3. $C(x) = x^3 + 21x^2 + 110x + 20$

$C'(x) = 3x^2 + 42x + 110$

$C'(100) = 3(100)^2 + 42 \cdot 100 + 110 = 34310$

$$\begin{aligned} C(100) &= 100^3 + 21(100)^2 + 110 \cdot 100 + 20 \\ &= 1221020 \end{aligned}$$

$C(99) = 99^3 + 21(99)^2 + 110 \cdot 99 + 20 = 1187030$

$C(100) - C(99) = 33990$

5.
$$\begin{aligned} C(x) &= x^3 - 30x^2 + 300x + 100 \\ C'(x) &= 3x^2 - 60x + 300 \\ C''(x) &= 6x - 60 = 0 \end{aligned}$$

$x = 10$ is the inflection point because $C''(x)$ changes from negative to positive at this value. After this point, cost rises more sharply.

7.
$$\overline{C}(x) = C(x)/x = 0.1x + 3 + \frac{2000}{x}$$
$$\overline{C}'(x) = 0.1 - \frac{2000}{x^2}$$

Critical number is $x = 100\sqrt{2} \approx 141.4$.
$\overline{C}'(x)$ is negative to the left of the critical number and positive to the right, so this must be the minimum.

9.
$$\overline{C}(x) = C(x)/x = 10\frac{e^{0.02x}}{x}$$
$$\overline{C}'(x) = 10e^{.02x}\left(\frac{.02x - 1}{x^2}\right)$$

Critical number is $x = 50$.
$\overline{C}'(x)$ is negative to the left of the critical number and positive to the right, so this must be the minimum.

11.
$$C(x) = 0.01x^2 + 40x + 3600$$
$$C'(x) = 0.02x + 40$$
$$\overline{C}(x) = \frac{C(x)}{x} = 0.01x + 40 + \frac{3600}{x}$$
$$C'(100) = 42$$
$$\overline{C}(100) = 77$$
$$\text{so } C'(100) < \overline{C}(100)$$
$$\overline{C}(101) = 76.65 < \overline{C}(100)$$

13.
$$\overline{C}'(x) = 0.01 - \frac{3600}{x^2} = 0$$
so $x = 600$ is min and
$$\bar{C}'(600) = 52$$
$$\overline{C}(600) = 52$$

15.
$$P(x) = R(x) - C(x)$$
$$P'(x) = R'(x) - C'(x) = 0$$
$$R'(x) = C'(x)$$

17. $f(p) = 200(30 - p), \quad p < 30$

$$E = \frac{p}{f(p)}f'(p) = \frac{p}{200(30 - p)}(-200) = \frac{p}{p - 30}$$

To solve $\frac{p}{p - 30} < -1$, multiply both sides by the negative quantity $p - 30$ to get $p > (-1)(p - 30)$ or $p > 30 - p$, so $2p > 30$, so $15 < p < 30$.

19. $f(p) = 100p(20 - p) = 100(20p - p^2), \quad p < 20$

$$E = \frac{p}{f(p)}f'(p) = \frac{p}{100p(20 - p)}(100)(20 - 2p)$$
$$= \frac{20 - 2p}{20 - p}$$

To solve $\frac{20 - 2p}{20 - p} < -1$, multiply both sides by the positive quantity $20 - p$ to get $20 - 2p < (-1)(20 - p)$, or $20 - 2p < p - 20$, so $40 < 3p$, so $40/3 < p < 20$.

21. Elasticity of demand at price $p = 15$ is, by definition, the relative change in demand divided by the relative change in price as price increases from 15 to an amount slightly larger than 15. So if (rel change in demand)/(rel change in price) is less than (-1), then rel change in demand is less than (-1)(rel change in price). This means that demand goes down more than price goes up, so revenue should decrease. (See Exercise 23.)

23.
$$[pf(p)]' < 0$$
$$\text{if and only if } p'f(p) + pf'(p) < 0$$
$$\text{if and only if } f(p) + pf'(p) < 0$$
$$\text{if and only if } pf'(p) < -f(p)$$
$$\text{if and only if } \frac{pf'(p)}{f(p)} < -1.$$

25. $x'(t) = 2x(t)[4 - x(t)] = 0$
$$f(x) = 2x(4x)$$
$$f'(x) = 2(4 - x) + 2x(-1) = 8 - 4x = 4(2 - x) = 0$$

$x = 2$, so $x(t) = 2$.
This is a maximum since $f(x)$ is a downward opening parabola.

27. $2x'(t) = 2x(t)[4 - x(t)] = 0$
$x(t) = 0,\ x(t) = 4$ are critical numbers.

$$x'(t) \begin{cases} > 0 \text{ for } 0 < x(t) < 4 \\ < 0 \text{ for } x(t) > 4. \end{cases}$$

So $x(t) = 4$ is the maximum concentration.

$$x'(t) = 0.5x(t)[5 - x(t)]$$

$x(t) = 0,\ x(t) = 5$ are critical numbers.

$$x'(t) \begin{cases} > 0 \text{ for } 0 < x(t) < 5 \\ < 0 \text{ for } x(t) > 5. \end{cases}$$

So $x(t) = 5$ is the maximum concentration.

29.
$$\begin{aligned} y'(t) &= c \cdot y(t)[K - y(t)] \\ y(t) &= Kx(t) \\ y'(t) &= Kx'(t) \\ Kx'(t) &= c \cdot Kx(t)[K - Kx(t)] \\ x'(t) &= c \cdot Kx(t)[1 - x(t)] \\ &= rx(t)[1 - x(t)] \\ r &= cK \end{aligned}$$

31.
$$\begin{aligned} x'(t) &= [a - x(t)][b - x(t)] \\ &\text{for } x(t) = a, \\ x'(t) &= [a - a][b - a] = 0. \end{aligned}$$

So the concentration of product is staying the same. If $a < b$ and $x(0) = 0$, then

$$x'(t) \begin{cases} > 0 \text{ for } 0 < x < a \text{ and } x > b \\ < 0 \text{ for } a < x < b. \end{cases}$$

Thus $x(t) = a$ is a maximum.

33.
$$x(t) = \frac{a[1 - e^{-(b-a)t}]}{1 - \left(\frac{a}{b}\right) e^{-(b-a)t}},\quad a < b$$

$$x(0) = \frac{a[1 - e^{-(b-a)\cdot 0}]}{1 - \left(\frac{a}{b}\right) e^{-(b-a)\cdot 0}} = \frac{a[1 - 1]}{1 - \left(\frac{a}{b}\right)} = 0$$

$$\lim_{t\to\infty} x(t) = \frac{a[1 - 0]}{1 - 0} = a$$

35.
$$p(f) = c + \ln \frac{f}{1 - f},\quad 0 < f < 1$$

$$\begin{aligned} p'(f) &= \frac{1}{\frac{f}{1-f}} \left(\frac{1 - f - f(-1)}{(1 - f)^2} \right) = \frac{1}{f(1 - f)} \\ &= (f - f^2)^{-1} \end{aligned}$$

$$p''(f) = -(f - f^2)^{-2}(1 - 2f) = \frac{2f - 1}{(f - f^2)^2} = 0$$

$$f = \frac{1}{2} \cdot p''(f) \begin{cases} > 0 \text{ when } \frac{1}{2} < f < 1 \\ < 0 \text{ when } 0 < f < \frac{1}{2} \end{cases}$$

so $f = \dfrac{1}{2}$ is a min.

As $f \to 1$, $p(f) = c + \ln \dfrac{f}{1 - f} \to \infty$.

37.
$$R(x) = \frac{rx}{k + x},\ x \geq 0$$

$$R'(x) = \frac{rk}{(k + x)^2},$$

so there are no critical numbers. Any possible maximum would have to be at the endpoint $x = 0$, but in fact R is increasing on $[0, \infty)$, so there is no maximum (although as x goes to infinity, R approaches r).

39.
$$\begin{aligned} &PV^{7/5} = c \\ &\frac{d}{dP}\left(PV^{7/5}\right) = \frac{d}{dP}(c) = 0 \\ &\left(\frac{d}{dP}(P)\right) V^{7/5} + P\frac{7}{5}V^{2/5}\frac{dV}{dP} = 0 \\ &V^{7/5} + \frac{7}{5}PV^{2/5}\frac{dV}{dP} = 0 \\ &V + \frac{7}{5}P\frac{dV}{dP} = 0 \\ &\frac{dV}{dP} = \frac{-5}{7}\frac{V}{P} \end{aligned}$$

But $V^{7/5} = \dfrac{c}{P}$, so $V = \left(\dfrac{c}{P}\right)^{5/7}$. Hence,

$$\frac{dV}{dP} = \frac{-5}{7}\frac{V}{P} = \frac{-5}{7}\frac{(c/P)^{5/7}}{P} = \frac{-5c^{5/7}}{7P^{12/7}}.$$

As pressure increases, volume decreases.

41. $m(x) = 4x - \sin x$ for $0 \leq x \leq 6$
$m'(x) = 4 - \cos x$
So the rod is less dense at the ends.

43. $m(x) = 4x$ for $0 \leq x \leq 2$
$m'(x) = 4$
So the rod is homogeneous.

45.
$$\begin{aligned} Q(t) &= e^{-2t}(\cos 3t - 2\sin 3t) \text{ Coulombs} \\ Q'(t) &= e^{-2t} \cdot (-2)(\cos 3t - 2\sin 3t) \\ &\quad + e^{-2t}((-\sin 3t \cdot 3) - 2\cos 3t \cdot 3) \\ &= e^{-2t}(-8\cos 3t + \sin 3t) \text{ amps} \end{aligned}$$

47. $Q(t) = e^{-3t}\cos 2t + 4\sin 3t$. As $t \to \infty$, $Q(t) \to 4\sin 3t$, so $e^{-3t}\cos 2t$ is called the transient term and $4\sin 3t$ is called the steady-state value.

$$\begin{aligned} Q'(t) &= e^{-3t} \cdot (-3)\cos 2t \\ &\quad + e^{-3t}(-\sin 2t \cdot 2) + 4\cos 3t \cdot 3 \\ &= e^{-3t}(-3\cos 2t - 2\sin 2t) + 12\cos 3t. \end{aligned}$$

The transient term is

$$e^{-3t}(-3\cos 2t - 2\sin 2t).$$

The steady-state value is $12\cos 3t$.

49. The rate of population growth is given by

$$\begin{aligned} f(p) &= 4p(5 - p) = 4(5p - p^2) \\ f'(p) &= 4(5 - 2p), \end{aligned}$$

so the only critical number is $p = 2.5$. Since the graph of f is a parabola opening down, this must be a max.

51.
$$\begin{aligned} p(t) &= \frac{B}{1 + Ae^{-kt}} \\ p'(t) &= \frac{-B(1 + Ae^{-kt})'}{(1 + Ae^{-kt})^2} \\ &= \frac{-B(-kAe^{-kt})}{(1 + Ae^{-kt})^2} = \frac{kABe^{-kt}}{(1 + Ae^{-kt})^2} \\ &= \frac{kABe^{-kt}}{1 + 2Ae^{-kt} + A^2e^{-2kt}} \\ &= \frac{kAB}{e^{kt} + 2A + A^2e^{-kt}} \end{aligned}$$

As t goes to infinity, the exponential term goes to 0, and so the limiting population is

$$\frac{B}{1 + A(0)} = B.$$

53.
$$\begin{aligned} f(t) &= \frac{a}{1 + 3e^{-bt}} \text{ for } a = 70,\ b = 0.2 \\ f(t) &= \frac{70}{1 + 3e^{-0.2t}} = 70(1 + 3e^{-0.2t})^{-1} \\ f(2) &= \frac{70}{1 + 3e^{-0.2 \cdot 2}} \approx 23 \\ f'(t) &= 70(-1)(1 + 3e^{-0.2t})^{-2}(3e^{-0.2t})(-0.2) \\ &= \frac{42e^{-0.2t}}{(1 + 3e^{-0.2t})^2} \\ f'(2) &= \frac{42e^{0.2 \cdot 2}}{(1 + 3e^{-0.2 \cdot 2})^2} \approx 3.105 \end{aligned}$$

So about 3% of the population will hear the rumor in the 3rd hour.

$$\lim_{t \to \infty} f(t) = \frac{70}{1 + 0} = 70,$$

so 70% of the population will eventually hear the rumor.

55.
$$\begin{aligned} f(x) &= \frac{160x^{-0.4} + 90}{4x^{-0.4} + 15} \\ f'(x) &= \frac{\left(\begin{array}{c} -64x^{-1.4}(4x^{-0.4} + 15) \\ -(160x^{-0.4} + 90)(-1.6x^{-1.4}) \end{array} \right)}{(4x^{-0.4} + 15)^2} \\ &= \frac{-816x^{-1.4}}{(4x^{-0.4} + 15)^2} < 0 \end{aligned}$$

So $f(x)$ is decreasing. This shows that pupils shrink as light increases.

57. If for some x marginal revenue equals marginal cost, then

$$P'(x) = R'(x) - C'(x) = 0,$$

so x is a critical number, but it may not be a maximum.

59. If v is not greater than c, the fish will never make any headway. (You know this feeling from studying calculus.)

$$E(v) = \frac{v^2}{v-c}, \quad v > c$$

$$E'(v) = \frac{v(v-2c)}{(v-c)^2},$$

so the only critical number is $v = 2c$. When v is large, $E(v)$ is large, and when v is just a little bigger than c, $E(v)$ is large, so we must have a minimum.

3.8 Related Rates and Parametric Equations

1. $V(t) = \text{(depth)(area)} = \frac{\pi}{48}[r(t)]^2$
(units in cubic feet per min)
$V'(t) = \frac{\pi}{48}2r(t)r'(t) = \frac{\pi}{24}r(t)r'(t)$
We are given $V'(t) = \frac{120}{7.5} = 16$.
Hence $16 = \frac{\pi}{24}r(t)r'(t)$,
so $r'(t) = \frac{(16)(24)}{\pi r(t)}$.

When $r = 100$,

$$r'(t) = \frac{(16)(24)}{100\pi} = \frac{96}{25\pi} \approx 1.2223 \text{ ft/min;}$$

When $r = 200$,

$$r'(t) = \frac{(16)(24)}{200\pi} = \frac{48}{25\pi} \approx 0.61115 \text{ ft/min.}$$

3. Again $V'(t) = \frac{\pi}{48}2r(t)r'(t) = \frac{\pi}{24}r(t)r'(t)$,
so $\frac{g}{7.5} = \frac{\pi}{24}(100)(.6) = 2.5\pi$,
so $g = (7.5)(2.5)\pi = 18.75\pi \approx 58.905$ gal/min.

5. t = hours elapsed since injury
r = radius of the infected area
A = area of the infection
$A = \pi r^2$

$$\frac{d}{dt}(A) = \frac{d}{dt}(\pi r^2)$$
$$A' = 2\pi r \cdot r'$$

When $r = 3$ mm, $r' = 1$ mm/hr, so

$$A' = 2\pi(3)(1) = 6\pi \text{ mm}^2\text{/hr.}$$

7.

$$V(t) = \frac{4}{3}\pi[r(t)]^3$$
$$V'(t) = 4\pi[r(t)]^2 r'(t) = Ar'(t)$$

If $V'(t) = kA(t)$, then

$$r'(t) = \frac{V'(t)}{A(t)} = \frac{kA(t)}{A(t)} = k.$$

9. We know $[x(t)]^2 + 4^2 = [s(t)]^2$. Hence, $2x(t)x'(t) = 2s(t)s'(t)$, so

$$x'(t) = \frac{s(t)s'(t)}{x(t)} = \frac{-240s(t)}{x(t)}.$$

When $x = 40$, $s = \sqrt{40^2 + 4^2} = 4\sqrt{101}$, so at that moment

$$x'(t) = \frac{(-240)(4\sqrt{101})}{40} = -24\sqrt{101}.$$

So speed is $24\sqrt{101} \approx 241.2$mph.

11. If the police car is not moving, then $x'(t) = 0$, but all the other data are unchanged. So

$$d'(t) = \frac{x(t)x'(t) + y(t)y'(t)}{\sqrt{[x(t)]^2 + [y(t)]^2}} = \frac{-(1/2)(50)}{\sqrt{1/4 + 1/16}} = \frac{-100}{\sqrt{5}} \approx -44.721.$$

This is more accurate.

13.

$$d'(t) = \frac{x(t)x'(t) + y(t)y'(t)}{\sqrt{[x(t)]^2 + [y(t)]^2}} = \frac{-(1/2)(\sqrt{2}-1)(50) - (1/2)(50)}{\sqrt{1/4 + 1/4}} = -50.$$

15.
$$\overline{C}(x) = 100 + \frac{100}{x}$$
$$\overline{C}'(x) = \frac{-100}{x^2}$$

$\overline{C}(10) = -1$ dollars per item, but production is increasing at the rate of 2 items per year, so average cost is decreasing at the rate of \$2 per year.

17. From the table, we see that the recent trend is for advertising to increase by \$2000 per year. A good estimate is then $x'(2) \approx 2$ (in units of thousands). Starting with the sales equation

$$s(t) = 60 - 40e^{-0.05x(t)},$$

we use the chain rule to obtain

$$\begin{aligned} s'(t) &= -40e^{-0.05x(t)}[-0.05x'(t)] \\ &= 2x'(t)e^{-0.05x(t)}. \end{aligned}$$

Using our estimate that $x'(2) \approx 2$, and since $x(2) = 20$, we get $s'(2) \approx 2(2)e^{-1} \approx 1.471$. Thus, sales are increasing at the rate of approximately \$1471 per year.

19. We have $\tan\theta = \frac{x}{2}$, so

$$\frac{d}{dt}(\tan\theta) = \frac{d}{dt}\left(\frac{x}{2}\right)$$
$$\sec^2\theta \cdot \theta' = \frac{1}{2}x'$$
$$\theta' = \frac{1}{2\sec^2\theta} \cdot x' = \frac{x'\cos^2\theta}{2}.$$

At $x = 0$ we have $\tan\theta = \frac{x}{2} = \frac{0}{2}$, so $\theta = 0$; and we have $x' = -130$ft/s, so

$$\theta' = \frac{(-130)\cdot\cos^2 0}{2} = -65 \text{ rad/s}.$$

21. t = number of seconds since launch
x = height of rocket in miles after t seconds
θ = camera angle in radians after t seconds

$$\tan\theta = \frac{x}{2}$$
$$\frac{d}{dx}(\tan\theta) = \frac{d}{dx}\left(\frac{x}{2}\right)$$
$$\sec^2\theta \cdot \theta' = \frac{1}{2}x'$$
$$\theta' = \frac{\cos^2\theta \cdot x'}{2}$$

When $x = 3$, $\tan\theta = \frac{3}{2}$, so $\cos\theta = \frac{2}{\sqrt{13}}$,

$$\theta' = \frac{\left(\frac{2}{\sqrt{13}}\right)^2(.2)}{2} \approx .03 \text{ rad/s}.$$

23. Let θ be the angle between the end of the shadow and the top of the lamppost. Then

$$\tan\theta = \frac{6}{s} \quad\text{and}\quad \tan\theta = \frac{18}{s+x}.$$

So
$$\frac{x+s}{18} = \frac{s}{6}$$
$$\frac{d}{dx}\left(\frac{x+s}{18}\right) = \frac{d}{dx}\left(\frac{s}{6}\right)$$
$$\frac{x'+s'}{18} = \frac{s'}{6}$$
$$x' + s' = 3s'$$
$$s' = \frac{x'}{2}$$
$$x' = 2.$$

So $s' = \frac{2}{2} = 1$ ft/s.

25. $x' = -2\sin t$, $y' = 2\cos t$

(a) At $t = 0$, $x' = -2\sin 0 = 0$, $y' = 2\cos 0 = 2$.
Speed $= \sqrt{0^2 + 2^2} = 2$.
Motion is up.

(b) At $t = \frac{\pi}{2}$, $x' = -2\sin\frac{\pi}{2} = -2$, $y' = 2\cos\frac{\pi}{2} = 0$.
Speed $= \sqrt{(-2)^2 + 0^2} = 2$.
Motion is to the left.

27. $x' = 20$, $y' = -2 - 32t$

(a) At $t = 0$, $x' = 20$, $y' = -2$.
Speed $= \sqrt{20^2 + (-2)^2} = 2\sqrt{101}$.
Motion is to the right and slightly down.

(b) At $t = 2$, $x' = 20$, $y' = -2 - 64 = -66$.
Speed $= \sqrt{20^2 + (-66)^2} = 2\sqrt{1189}$.
Motion is to the right and down.

29. $x' = -4 \sin 2t + 5 \cos 5t$, $y' = 4 \cos 2t - 5 \sin 5t$

(a) At $t = 0$, $x' = -4 \sin 0 + 5 \cos 0 = 5$,
$y' = 4 \cos 0 - 5 \sin 0 = 4$.
Speed $= \sqrt{5^2 + 4^2} = \sqrt{41}$.
Motion is to the right and up.

(b) At $t = \dfrac{\pi}{2}$, $x' = -4 \sin \pi + 5 \cos \dfrac{5\pi}{2} = 0$,
$y' = 4 \cos \pi - 5 \sin \dfrac{5\pi}{2} = -9$.
Speed $= \sqrt{0^2 + (-9)^2} = 9$.
Motion is down.

31. Crossing x-axis $\Rightarrow y = 0 \Rightarrow t = n\pi$.

$x' = -4 \cos t \sin t - 2 \sin t$,

$y' = 2 \sin^2 t + 2(1 - \cos t) \cos t$.

At $t = n\pi$, n even, $(x, y) = (3, 0)$ and

$x' = 0$, $y' = 0$, speed $= 0$.

At $t = n\pi$, n odd, $(x, y) = (-1, 0)$ and

$x' = 0$, $y' = -4$, speed $= 4$.

33. The $3t$ and $5t$ indicate a ratio of 5-to-3.

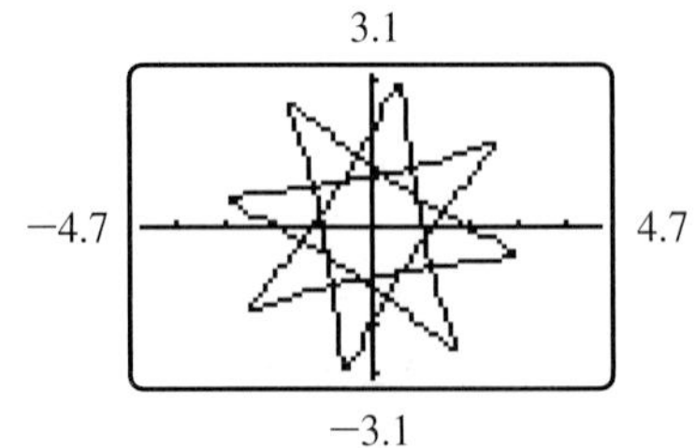

35. Use $x = 2 \cos t + \sin 3t$, $y = 2 \sin t + \cos 3t$.

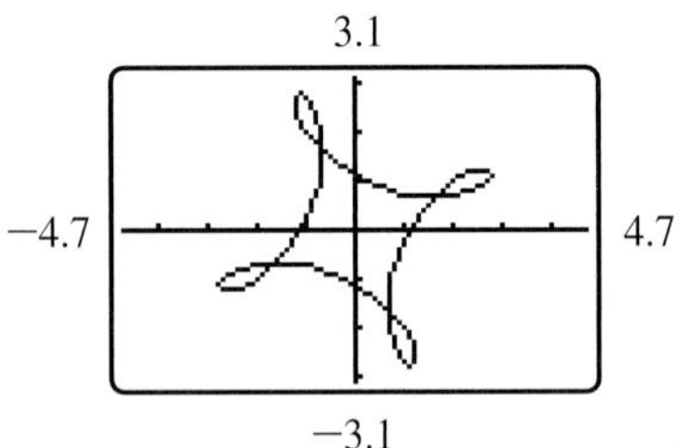

$$x' = -2 \sin t + 3 \cos 3t,\ y' = 2 \cos t - 3 \sin 3t$$

$$\begin{aligned} s(t) &= \sqrt{(-2 \sin t + 3 \cos 3t)^2 + (2 \cos t - 3 \sin 3t)^2} \\ &= \sqrt{\begin{pmatrix} 4 \sin^2 t - 12 \sin t \cos 3t + 9 \cos^2 3t \\ +4 \cos^2 t - 12 \cos t \sin 3t + 9 \sin^2 3t \end{pmatrix}} \\ &= \sqrt{4 + 9 - 12(\sin t \cos 3t + \cos t \sin 3t)} \\ &= \sqrt{13 - 12 \sin 4t}. \end{aligned}$$

Minimum speed $= \sqrt{13 - 12} = 1$; maximum speed $= \sqrt{13 + 12} = 5$.

37. $x' = 4 \cos 4t$, $y' = 4 \sin 4t$, and

$$\begin{aligned} s(t) &= \sqrt{(4 \cos 4t)^2 + (4 \sin 4t)^2} \\ &= \sqrt{16(\sin^2 4t + \cos^2 4t)} = 4. \end{aligned}$$

The slope of the tangent line is $\dfrac{y'}{x'} = \tan 4t$, and the slope of the origin-to-object line is $\dfrac{y}{x} = -\cot 4t$, and the product of the two slopes is -1.

39.

$$P(t) \cdot V'(t) + P'(t)V(t) = 0$$

$$\frac{P'(t)}{V'(t)} = -\frac{P(t)}{V(t)} = -\frac{c}{V(t)^2}$$

41. Let $r(t)$ be the length of the rope at time t, and let $x(t)$ be the distance (along the water) between the boat and the dock.

$$r(t)^2 = 36 + x(t)^2$$
$$2r(t)r'(t) = 2x(t)x'(t)$$
$$x'(t) = \frac{r(t)r'(t)}{x(t)} = \frac{-2r(t)}{x(t)} = \frac{-2\sqrt{36+x^2}}{x}.$$

When $x = 20$, $x' = -2.088$; when $x = 10$, $x' = -2.332$.

43. First let's ask how rapidly the wheel is rotating. The circumference of the wheel is $5\pi/3$ feet, and the wheel is moving along the ground at the rate of 4 feet per second, so the wheel makes one complete revolution in $5\pi/12$ seconds. So if the wheel were rotating at this rate, but the center were held fixed at the origin, the equation of the dot would be

$$x = \frac{5}{6}\sin(4.8t),\ y = \frac{5}{6}\cos(4.8t).$$

But the center of the wheel is not fixed at the origin. The y-coordinate of the center is 5/6, and the x-coordinate of the center is $4t$. Hence, the equation of the dot is

$$x = 4t + \frac{5}{6}\sin(4.8t),\ y = \frac{5}{6} + \frac{5}{6}\cos(4.8t).$$

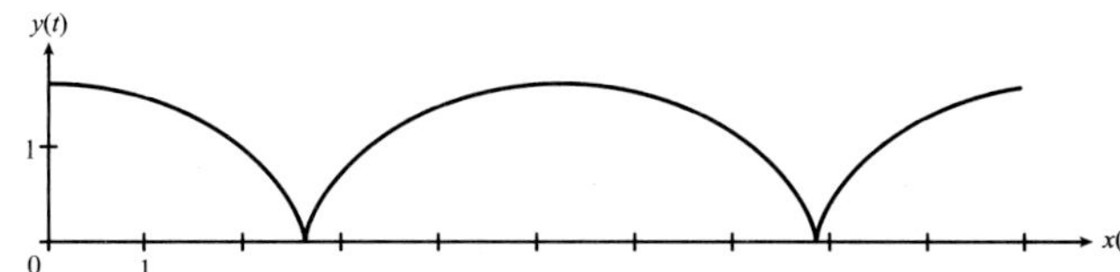

45. When we calculate $\left[\frac{y(t)}{x(t)}\right]'$, the denominator will be $[x(t)]^2$, which is never negative. Hence, the sign of $\left[\frac{y(t)}{x(t)}\right]'$ is the same as the sign of the numerator of this expression. A long but routine calculation shows that the numerator is

$$8\pi(4 + 5\cos 16\pi t\cos 4\pi t + 5\sin 16\pi t\sin 4\pi t).$$

Using the trig identity

$$\cos(x - y) = \cos x\cos y + \sin x\sin y,$$

the numerator can be converted to $8\pi(4 + 5\cos 12\pi t)$. This quantity will be positive, and the motion will be counter-clockwise, when $\cos 12\pi t > -\frac{4}{5}$. Now $\cos^{-1}(-4/5) \approx 2.4981$, so $\cos\theta > 0$ on the intervals $(2\pi n - 2.498, 2\pi n + 2.498)$, where n is any integer. Hence, $\cos 12\pi t$ is positive, and the motion is counter-clockwise, on the intervals

$$\left(\frac{2\pi n - 2.498}{12\pi}, \frac{2\pi n + 2.498}{12\pi}\right)$$
$$= \left(\frac{n}{6} - .066, \frac{n}{6} + .066\right).$$

47.

$$V = \frac{4}{3}\pi r^3$$
$$\frac{dV}{dt} = \frac{4}{3}\pi(3r^2)\frac{dr}{dt} = 4\pi r^2\frac{dr}{dt}$$
$$1 = 4\pi r^2\frac{dr}{dt}$$
$$\frac{dr}{dt} = \frac{1}{4\pi r^2}$$

when $r = .01$, $\frac{dr}{dt} = \frac{2500}{\pi}$

when $r = .1$, $\frac{dr}{dt} = \frac{25}{\pi}$.

At first, the radius expands rapidly; later it expands more slowly.

Chapter 3 Review

1. See Example 1.1.

$$f(x) = e^{3x},\ x_0 = 0$$
$$f'(x) = 3e^{3x}$$
$$\begin{aligned}L(x) &= f(x_0) + f'(x_0)(x - x_0)\\ &= f(0) + f'(0)(x - 0)\\ &= e^{3\cdot 0} + 3e^{3\cdot 0}x\\ &= 1 + 3x\end{aligned}$$

3. See Example 1.3.

$$f(x) = \sqrt[3]{x} = x^{1/3}, \ x_0 = 8$$
$$f'(x) = \frac{1}{3}x^{-2/3}$$
$$\begin{aligned} L(x) &= f(x_0) + f'(x_0)(x - x_0) \\ &= f(8) + f'(8)(x - 8) \\ &= \sqrt[3]{8} + \frac{1}{3}(8)^{-2/3}(x - 8) \\ &= 2 + \frac{1}{12}(x - 8) \end{aligned}$$
$$L(7.96) = 2 + \frac{1}{12}(7.96 - 8) \approx 1.99666$$

5. See Example 1.5.

$$f(x) = x^3 + 5x - 1 = 0$$
$$f'(x) = 3x^2 + 5$$
$$x_0 = 0$$
$$x_{n+1} = x_n - \frac{f(x_n)}{f'(x_n)} \text{ for } n = 0, 1, 2, 3, \ldots$$

n	x_n
1	.2
2	.198437
3	.198437

$x \approx .198437$

7. See Example 1.8.

$$f(x) = x^3 + 2 = 0, \ x_0 = 1$$
$$f'(x) = 3x^2 - 3x$$
$$x_1 = x_0 - \frac{f(x_0)}{f'(x_0)} = 1 - \frac{1 - 3 + 2}{3 - 3} = 1 - \frac{0}{0}$$

Newton's Method fails because $f'(1) = 0$.

For Exercises 9–17, see the Examples in Sections 3.3, 3.4, and 3.5.

9.
$$\lim_{x\to 1} \frac{x^3 - 1}{x^2 - 1} = \lim_{x\to 1} \frac{3x^2}{2x} = \frac{3}{2}$$

11.
$$\begin{aligned} \lim_{x\to 0} \frac{e^{2x}}{x^4 + 2} &= \lim_{x\to\infty} \frac{2e^{2x}}{4x^3} \\ &= \lim_{x\to\infty} \frac{4e^{2x}}{12x^2} = \lim_{x\to\infty} \frac{8e^{2x}}{24x} \\ &= \lim_{x\to\infty} \frac{16e^{2x}}{24} = \infty \end{aligned}$$

13.
$$y = \left(1 + \frac{3}{x}\right)^x$$
$$\ln y = x \ln\left(1 + \frac{3}{x}\right)$$
$$\begin{aligned} \lim_{x\to\infty} x \ln\left(1 + \frac{3}{x}\right) &= \lim_{x\to\infty} \frac{\ln\left(1 + \frac{3}{x}\right)}{1/x} \\ &= \lim_{x\to\infty} \frac{\left(1 + \frac{3}{x}\right)^{-1}(3)(-x^{-2})}{-x^2} \\ &= \lim_{x\to\infty} \frac{\left(1 + \frac{3}{x}\right)^{-1}(3)}{1} = (1)(3) = 3 \end{aligned}$$

Thus as $x \to \infty$, $\ln y \to 3$.
Hence, $\lim_{x\to\infty} y = \lim_{x\to\infty} e^{\ln y} = e^3$.

15.
$$\lim_{x\to 0} \frac{\tan x}{x} = \lim_{x\to 0} \frac{\sin x}{x \cos x} =$$
$$\lim_{x\to 0} \frac{\sin x}{x} \frac{1}{\cos x} = (1)(1) = 1$$

17.
$$f(x) = x^3 + 3x^2 - 9x$$
$$\begin{aligned} f'(x) &= 3x^2 + 6x - 9 = 3(x^2 + 2x - 3) \\ &= 3(x + 3)(x - 1) \end{aligned}$$

So the critical numbers are $x = 1, x = -3$.

$$f'(x) \begin{cases} > 0 \text{ on } (-\infty, -3) \cup (1, \infty) \\ < 0 \text{ on } (-3, 1). \end{cases}$$

Hence, f is increasing on $(-\infty, -3)$ and on $(1, \infty)$, and f is decreasing on $(-3, 1)$. Hence there is a local max at $x = -3$ and a local min at $x = 1$.

Now $f''(x) = 3(2x + 2) = 6(x + 1)$,

$$f''(x) \begin{cases} > 0 \text{ on } (-1, \infty) \\ < 0 \text{ on } (-\infty, -1). \end{cases}$$

Hence, f is concave up on $(-1, \infty)$ and concave down on $(-\infty, -1)$, and there is an inflection point at $x = -1$.

19. $$f(x) = x^4 - 4x^3 + 2$$
$$f'(x) = 4x^3 - 12x^2 = 4x^2(x-3)$$
$$f''(x) = 12x^2 - 24x = 12x(x-2)$$
$$f'(x) \begin{cases} > 0 \text{ on } (3, \infty) \\ < 0 \text{ on } (-\infty, 0) \cup (0, 3) \end{cases}$$
$$f''(x) \begin{cases} > 0 \text{ on } (-\infty, 0) \cup (2, \infty) \\ < 0 \text{ on } (0, 2) \end{cases}$$
$$f''(3) > 0$$

(a) $x = 0, 3$ are critical numbers;

(b) f increasing on $(3, \infty)$, decreasing on $(-\infty, 3)$;

(c) $x = 3$ is a local min;

(d) f is concave up on $(-\infty, 0) \cup (2, \infty)$, concave down on $(0, 2)$;

(e) $x = 0, 2$ are inflection points.

21. $$f(x) = xe^{-4x}$$
$$f'(x) = 1 \cdot e^{-4x} + xe^{-4x}(-4) = e^{-4x}(1-4x)$$
$$f''(x) = e^{-4x}(-4)(1-4x) + e^{-4x}(-4)$$
$$= -4e^{-4x}(2-4x)$$
$$f'(x) \begin{cases} > 0 \text{ on } \left(-\infty, \frac{1}{4}\right) \\ < 0 \text{ on } \left(\frac{1}{4}, \infty\right) \end{cases}$$
$$f''(x) \begin{cases} > 0 \text{ on } \left(\frac{1}{2}, \infty\right) \\ < 0 \text{ on } \left(-\infty, \frac{1}{2}\right) \end{cases}$$
$$f''\left(\frac{1}{4}\right) < 0$$

(a) $x = \frac{1}{4}$ is a critical number;

(b) f increasing on $\left(-\infty, \frac{1}{4}\right)$, decreasing on $\left(-\frac{1}{4}, \infty\right)$;

(c) $x = \frac{1}{4}$ is a local max;

(d) f is concave up on $\left(\frac{1}{2}, \infty\right)$, concave down on $\left(-\infty, \frac{1}{2}\right)$;

(e) $x = \frac{1}{2}$ is inflection point.

23. $$f(x) = x\sqrt{x^2-4};\ f \text{ undefined on } (-2, 2)$$
$$f'(x) = \sqrt{x^2-4} + x \cdot \frac{1}{2}(x^2-4)^{-1/2} \cdot 2x$$
$$= \frac{2x^2-4}{\sqrt{x^2-4}}$$

$f'(x) = 0$ when $x = \pm\sqrt{2}$, but $\pm\sqrt{2}$ are not in the domain.

$$f''(x) = \frac{4x\sqrt{x^2-4} - (2x^2-4)\frac{1}{2}(x^2-4)^{-1/2} \cdot 2x}{x^2-4}$$
$$= \frac{2x^3 - 12x}{(x^2-4)^{3/2}}.$$

$f'(x) > 0$ for all x in the domain.

$$f''(x) \begin{cases} > 0 \text{ on } (-\sqrt{6}, -2) \cup (\sqrt{6}, \infty) \\ < 0 \text{ on } (-\infty, -\sqrt{6}) \cup (2, \sqrt{6}). \end{cases}$$

(a) none;

(b) f increasing on $(-\infty, -2)$ and on $(2, \infty)$;

(c) no local min/max;

(d) f is concave up on $\left(-\sqrt{6}, -2\right) \cup \left(\sqrt{6}, \infty\right)$, concave down on $\left(-\infty, -\sqrt{6}\right) \cup \left(2, \sqrt{6}\right)$;

(e) $x = \pm\sqrt{6}$ are inflection points.

25. $$f(x) = \frac{x}{x^2+4}$$
$$f'(x) = \frac{x^2+4-x(2x)}{(x^2+4)^2} = \frac{4-x^2}{(x^2+4)^2}$$
$$f''(x) = \frac{-2x(x^2+4)^2 - (4-x^2)[2(x^2+4) \cdot 2x]}{(x^2+4)^4}$$
$$= \frac{2x^3 - 24x}{(x^2+4)^3}$$

$$f'(x)\begin{cases} > 0 \text{ on } (-2, 2) \\ < 0 \text{ on } (-\infty, -2) \cup (2, \infty) \end{cases}$$

$$f''(x)\begin{cases} > 0 \text{ on } \left(-\sqrt{12}, 0\right) \cup \left(\sqrt{12}, \infty\right) \\ < 0 \text{ on } \left(-\infty, -\sqrt{12}\right) \cup \left(0, \sqrt{12}\right) \end{cases}$$

(a) $x = \pm 2$ are critical numbers;

(b) f increasing on $(-2, 2)$, decreasing on $(-\infty, -2)$ and on $(2, \infty)$;

(c) local min at $x = -2$, local max at $x = 2$;

(d) f is concave up on $\left(-\sqrt{12}, 0\right) \cup \left(\sqrt{12}, \infty\right)$, concave down on $\left(-\infty, -\sqrt{12}\right) \cup \left(0, \sqrt{12}\right)$;

(e) $x = \pm\sqrt{12}, 0$ are inflection points.

27. See Example 3.11.

$$f(x) = x^3 + 3x^2 - 9x \text{ on } [0, 4],$$
$$f'(x) = 3x^2 + 6x - 9$$
$$= 3(x^2 + 2x - 3)$$
$$= 3(x + 3)(x - 1).$$

$x = -3$, $x = 1$ are critical numbers.
$x = -3 \in [0, 4]$.

$$f(0) = 0^3 + 3 \cdot 0^2 - 9 \cdot 0 = 0,$$
$$f(4) = 4^3 + 3 \cdot 4^2 - 9 \cdot 4 = 76,$$
$$f(1) = 1^3 + 3 \cdot 1^2 - 9 \cdot 1 = -5.$$

So $f(4) = 76$ is absolute max on $[0, 4]$, $f(1) = -5$ is absolute min.

29. See Example 3.12.

$$f(x) = x^{4/5} \text{ on } [-2, 3]$$
$$f'(x) = \frac{4}{5}x^{-1/5}$$

$x = 0$ is critical number.

$$f(-2) = (-2)^{4/5} \approx 1.74,$$
$$f(3) = (3)^{4/5} \approx 2.41,$$
$$f(0) = (0)^{4/5} = 0.$$

$f(0) = 0$ is absolute min, $f(3) = 3^{4/5}$ is absolute max.

31. See Example 3.6.

$$f(x) = x^3 + 4x^2 + 2x,$$
$$f'(x) = 3x^2 + 8x + 2 = 0,$$
$$x = \frac{-8 \pm \sqrt{64 - 24}}{6} = -\frac{4}{3} \pm \frac{\sqrt{10}}{3}.$$

$x = -\frac{4}{3} - \frac{\sqrt{10}}{3}$ is local max, $x = -\frac{4}{3} + \frac{\sqrt{10}}{3}$ is local min.

33. See Example 3.6.

$f(x) = x^5 - 2x^2 + x$
$f'(x) = 5x^4 - 4x + 1 = 0$
$x \approx 0.2553,\ 0.8227$
local min at $x \approx 0.8227$
local max at $x \approx 0.2553$

35. See Theorem 4.1.

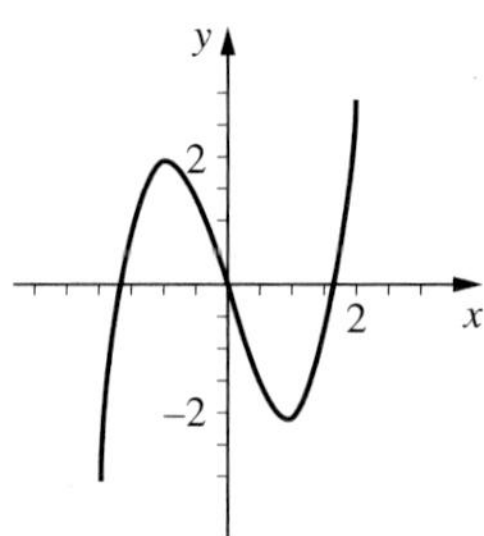

37. See Example 5.3.

$$f(x) = x^4 + 4x^3$$
$$f'(x) = 4x^3 + 12x^2 = 4x^2(4x + 3)$$
$$f''(x) = 12x^2 + 24x = 12x(x + 2)$$

$$f'(x)\begin{cases} > 0 \text{ on } (-3, 0) \cup (0, \infty) \\ < 0 \text{ on } (-\infty, -3) \end{cases}$$

$$f''(x)\begin{cases} > 0 \text{ on } (-\infty, -2) \cup (0, \infty) \\ < 0 \text{ on } (-2, 0) \end{cases}$$

f increasing on $(-3, \infty)$, decreasing on $(-\infty, -3)$, concave up on $(-\infty, -2) \cup (0, \infty)$, concave down on $(-2, 0)$, local min at $x = -3$, inflection points at $x = -2, 0$.

$$x = -\frac{1}{2} - \frac{\sqrt{3}}{2}.$$

39. See Example 5.3.

$$f(x) = x^4 + 4x,$$

$$f'(x) = 4x^3 + 4 = 4(x^3 + 1),$$

$$f''(x) = 12x^2.$$

$$f'(x) \begin{cases} > 0 \text{ on } (-1, \infty) \\ < 0 \text{ on } (-\infty, -1), \end{cases}$$

$$f''(x) > 0 \text{ on } (-\infty, 0) \cup (0, \infty).$$

f increasing on $(-1, \infty)$, decreasing on $(-\infty, -1)$, concave up on $(-\infty, \infty)$, local min at $x = -1$.

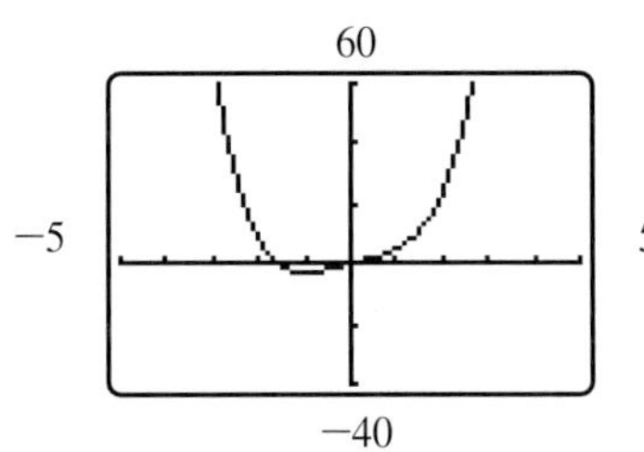

41. See Example 5.4.

$$f(x) = \frac{x}{x^2+1},$$

$$f'(x) = \frac{x^2 + 1 - x(2x)}{(x^2+1)^2} = \frac{1 - x^2}{(x^2+1)^2},$$

$$f''(x) = \frac{-2x(x^2+1)^2 - (1 - x^2)2(x^2+1)2x}{(x^2+1)^4}$$

$$= \frac{2x(x^2 - 3)}{(x^2+1)^4}.$$

$$f'(x) \begin{cases} > 0 \text{ on } (-1, 1) \\ < 0 \text{ on } (-\infty, -1) \cup (1, \infty), \end{cases}$$

$$f''(x) \begin{cases} > 0 \text{ on } \left(-\sqrt{3}, 0\right) \cup \left(\sqrt{3}, \infty\right) \\ < 0 \text{ on } \left(-\infty, -\sqrt{3}\right) \cup \left(0, \sqrt{3}\right). \end{cases}$$

f increasing on $(-1, 1)$, decreasing on $(-\infty, -1)$ and on $(1, \infty)$, concave up on

$$\left(-\sqrt{3}, 0\right) \cup \left(\sqrt{3}, \infty\right),$$

concave down on

$$\left(-\infty, -\sqrt{3}\right) \cup \left(0, \sqrt{3}\right),$$

local min at $x = -1$, local max at $x = 1$, inflection points at $0, \pm\sqrt{3}$.

$$\lim_{x\to\infty} \frac{x}{x^2+1} = \lim_{x\to-\infty} \frac{x}{x^2+1} = 0.$$

So f has a horizontal asymptote at $y = 0$.

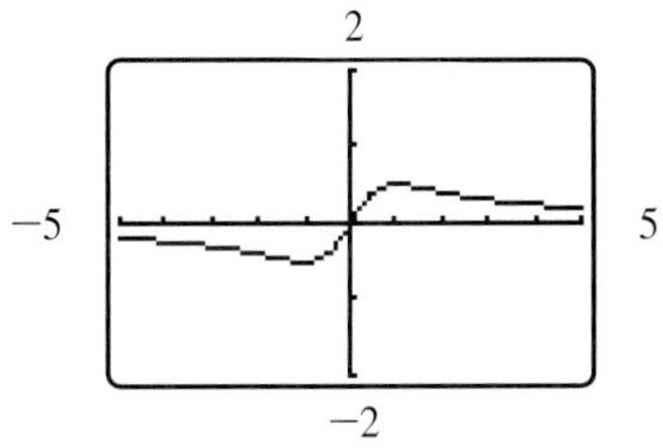

43. See Example 5.4.

$$f(x) = \frac{x^2}{x^2+1},$$

$$f'(x) = \frac{(2x)(x^2+1) - x^2(2x)}{(x^2+1)^2} = \frac{2x}{(x^2-1)^2},$$

$$f''(x) = \frac{2(x^2+1)^2 - 2x \cdot 2(x^2+1)2x}{(x^2+1)^4}$$

$$= \frac{2 - 6x^2}{(x^2+1)^3}.$$

$$f'(x)\begin{cases} > 0 \text{ on } (0, \infty) \\ < 0 \text{ on } (-\infty, 0), \end{cases}$$

$$f''(x)\begin{cases} > 0 \text{ on } \left(-\sqrt{\frac{1}{3}}, \sqrt{\frac{1}{3}}\right) \\ < 0 \text{ on } \left(-\infty, -\sqrt{\frac{1}{3}}\right) \cup \left(\sqrt{\frac{1}{3}}, \infty\right). \end{cases}$$

f increasing on $(0, \infty)$ decreasing on $(-\infty, 0)$, concave up on

$$\left(-\sqrt{\frac{1}{3}}, \sqrt{\frac{1}{3}}\right),$$

concave down on

$$\left(-\infty, -\sqrt{\frac{1}{3}}\right) \cup \left(\sqrt{\frac{1}{3}}, \infty\right),$$

local min at $x = 0$, inflection points at $x = \pm\sqrt{\frac{1}{3}}$.

$$\lim_{x\to\infty} \frac{x^2}{x^2+1} = \lim_{x\to-\infty} \frac{x^2}{x^2+1} = 1.$$

So f has a horizontal asymptote at $y = 1$.

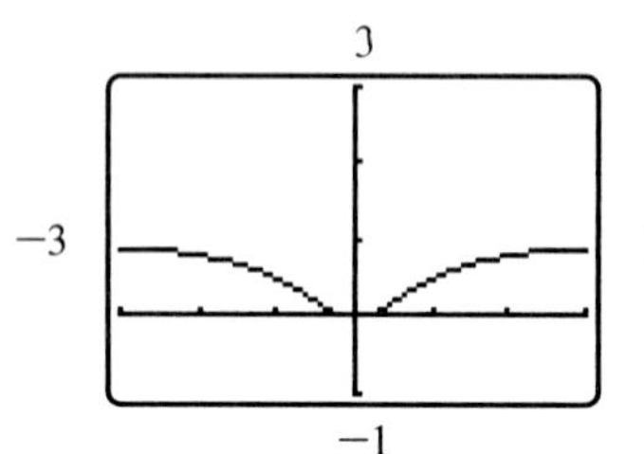

45. See Example 5.4.

$$f(x) = \frac{x^3}{x^2-1},$$

$$f'(x) = \frac{3x^2(x^2-1) - x^3(2x)}{(x^2-1)^2} = \frac{x^4-3x^2}{(x^2-1)^2},$$

$$f''(x) = \frac{(4x^3-6x)(x^2-1)^2 - (x^4-3x^2)2(x^2+1)2x}{(x^2-1)^4}$$

$$= \frac{2x^3+6x}{(x^2-1)^4}.$$

$$f'(x)\begin{cases} > 0 \text{ on } \left(-\infty, -\sqrt{3}\right) \cup \left(\sqrt{3}, \infty\right) \\ < 0 \text{ on } \left(-\sqrt{3}, -1\right) \cup (-1, 0) \cup (0, 1) \cup \left(1, \sqrt{3}\right), \end{cases}$$

$$f''(x)\begin{cases} > 0 \text{ on } (-1, 0) \cup (1, \infty) \\ < 0 \text{ on } (-\infty, -1)(0, 1). \end{cases}$$

f increasing on $(-\infty, -\sqrt{3})$ and on $(\sqrt{3}, \infty)$; decreasing on $(-\sqrt{3}, -1)$, on $(-1, 1)$ and on $(1, \sqrt{3})$; concave up on $(-1, 0) \cup (1, \infty)$, concave down on $(-\infty, -1) \cup (0, 1)$; $x = -\sqrt{3}$ local max; $x = \sqrt{3}$ local min; $x = 0$ inflection point. f is undefined at $x = -1$ and $x = 1$.

$$\lim_{x\to1^+} \frac{x^3}{x^2-1} = \infty, \quad \text{and} \quad \lim_{x\to1^-} \frac{x^3}{x^2-1} = -\infty.$$

So f has vertical asymptotes at $x = 1$ and $x = -1$.

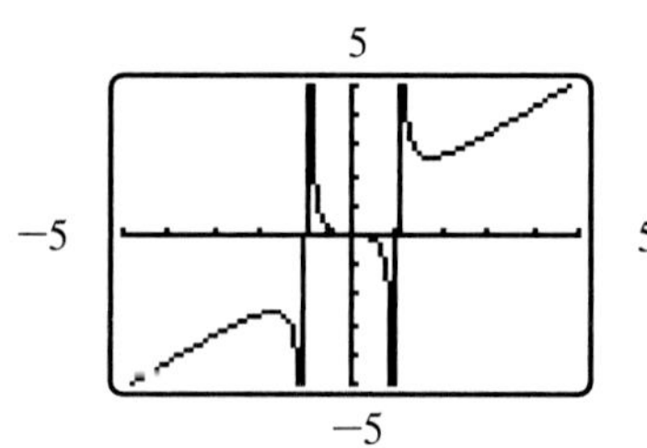

47. See Example 6.3.

$y = 2x^2$, $(2, 1)$

$d = \sqrt{(x-2)^2 + (y-1)^2} = \sqrt{(x-2)^2 + (2x^2-1)^2}$

$f(x) = (x-2)^2 + (2x^2-1)^2$

$f'(x) = 2(x-2) + 2(2x^2-1)4x$

$\quad = 16x^3 - 6x - 4$

$x \approx 0.8237$ when $f'(x) = 0$

$$f'(x)\begin{cases} < 0 \text{ on } (-\infty, 0.8237) \\ > 0 \text{ on } (0.8237, \infty) \end{cases}$$

So $x \approx 0.8237$ corresponds to the closest point.

$$y = 2x^2 = 2(0.8237)^2 = 1.3570.$$

$(0.8237, 1.3570)$ is closest to $(2, 1)$.

49. See Example 6.5.

$$C(x) = 6\sqrt{4^2 + (4 - x)^2} + 2\sqrt{2^2 + x^2}$$

$$C'(x) = 6 \cdot \frac{1}{2}[16 + (4 - x)^2]^{-1/2} \cdot 2(4 - x)(-1)$$

$$+ 2\frac{1}{2}(4 + x^2)^{-1/2} \cdot 2x$$

$$= \frac{6(x - 4)}{\sqrt{16 + (4 - x)^2}} + \frac{2x}{\sqrt{4 + x^2}}$$

$x \approx 2.864$ when $C'(x) = 0$

$$C'(x) \begin{cases} < 0 \text{ on } (0, 2.864) \\ > 0 \text{ on } (2.864, 4) \end{cases}$$

So $x \approx 2.864$ gives the minimum cost. Locate highway corner $4 - 2.864 = 1.136$ miles east of point A.

51. See Example 6.4.

$$\text{Area } = 2\pi r^2 + 2\pi rh$$

$$16 \text{ fl oz} = 16 \text{ fl oz} \cdot 1.80469 \frac{\text{in}^3}{\text{fl oz}}$$

$$= 28.87504 \text{ in}^3$$

$$\text{Vol} = \pi r^2 h$$

$$h = \frac{\text{Vol}}{\pi r^2} = \frac{28.87504}{\pi r^2}$$

$$A(r) = 2\pi \left(r^2 + \frac{28.87504}{\pi r}\right)$$

$$A'(r) = 2\pi \left(2r - \frac{28.87504}{\pi r^2}\right)$$

$$2\pi r^3 = 28.87504$$

$$r = \sqrt[3]{\frac{28.87504}{2\pi}} \approx 1.663$$

$$A'(r) \begin{cases} < 0 \text{ on } (0, 1.663) \\ > 0 \text{ on } (1.663, \infty) \end{cases}$$

So $r \approx 1.663$ gives the minimum surface area.

$$h = \frac{28.87504}{\pi(1.663)^2} \approx 3.325$$

53. See Example 7.7.

$$Q(t) = e^{-3t} \sin 2t$$

$$Q'(t) = -3e^{-3t} \sin 2t + e^{-3t} \cos 2t \cdot 2$$

$$= e^{-3t}(2 \cos 2t - 3 \sin 2t) \text{ amps}$$

55. See Example 7.6.

$$m(x) = 20 + x^2,\ 0 \le x \le 4$$

$$\rho(x) = m'(x) = 2x$$

As you move along the rod to the right, its density increases.

57. See Example 7.1.

$$C(x) = 0.02x^2 + 20x + 1800$$

$$C'(x) = 0.04x + 20$$

$$C'(20) = 0.04(20) + 20 = 20.8$$

$$C(20) - C(19) = 0.02(20)^2 + 20(20) + 1800$$

$$- [0.02(19)^2 + 20(19) + 1800]$$

$$= 20.78$$

Chapter 4

Integration

4.1 Area Under a Curve

1. $\sum_{i=1}^{50} i^2 = \frac{50(50+1)(2\cdot50+1)}{6} = 42{,}925$

3.
$$\sum_{i=1}^{10} \sqrt{i} = \sqrt{1} + \sqrt{2} + \sqrt{3} + \sqrt{4} + \sqrt{5}$$
$$+ \sqrt{6} + \sqrt{7} + \sqrt{8} + \sqrt{9} + \sqrt{10} \approx 22.47$$

5.
$$\sum_{i=1}^{6} 3i^2 = 3\cdot1^2 + 3\cdot2^2 + 3\cdot3^2$$
$$+ 3\cdot4^2 + 3\cdot5^2 + 3\cdot6^2$$
$$= 3 + 12 + 27 + 48 + 75 + 108$$
$$= 273$$

7.
$$\sum_{i=o}^{4} (4i - \sin i\pi)$$
$$= (4\cdot0 - \sin 0\pi) + (4\cdot1 - \sin\pi)$$
$$+ (4\cdot2 - \sin 2\pi) + (4\cdot3 - \sin 3\pi)$$
$$+ (4\cdot4 - \sin 4\pi)$$
$$= (4)(0+1+2+3+4) - (0+0+0+0+0)$$
$$= 40$$

9.
$$\sum_{i=1}^{70} (3i - 1) = 3\sum_{i=1}^{70} i - \sum_{i=1}^{70} 1$$
$$= 3\cdot\frac{70(71)}{2} - 70\cdot1$$
$$= 7385$$

11.
$$\sum_{i=1}^{40} (4 - i^2) = \sum_{i=1}^{40} 4 - \sum_{i=1}^{40} i^2$$
$$= 4\cdot40 - \frac{40(41)(81)}{6}$$
$$= -21{,}980$$

13.
$$\sum_{i=1}^{100} (i^2 - 3i + 2)$$
$$= \sum_{i=1}^{100} i^2 - 3\sum_{i=1}^{100} i + \sum_{i=1}^{100} 2$$
$$= \frac{100(101)(201)}{6} - \frac{3\cdot100(101)}{2} + 2\cdot100$$
$$= 323{,}400$$

15.
$$\sum_{i=1}^{n} \frac{1}{n}\left[\left(\frac{i}{n}\right)^2 + 2\left(\frac{i}{n}\right)\right]$$
$$= \frac{1}{n^3}\sum_{i=1}^{n} i^2 + \frac{2}{n^2}\sum_{i=1}^{n} i$$
$$= \frac{1}{n^3}\cdot\frac{n(n+1)(2n+1)}{6} + \frac{2}{n^2}\cdot\frac{n(n+1)}{2}$$
$$= \frac{(n+1)(2n+1)}{6n^2} + \frac{n+1}{n} \to \frac{2}{6} + 1$$
$$= \frac{4}{3} \text{ as } n \to \infty$$

17.
$$\sum_{i=1}^{n} \frac{1}{n}\left[4\left(\frac{2i}{n}\right)^2 - \left(\frac{2i}{n}\right)\right]$$
$$= \frac{16}{n^3}\sum_{i=1}^{n} i^2 - \frac{2}{n^2}\sum_{i=1}^{n} i$$
$$= \frac{16}{n^3}\cdot\frac{n(n+1)(2n+1)}{6} - \frac{2}{n^2}\cdot\frac{n(n+1)}{2}$$
$$= \frac{8}{3}\frac{(n+1)(2n+1)}{n^2} - \frac{(n+1)}{n} \to \frac{16}{3} - 1$$
$$= \frac{13}{3} \text{ as } n \to \infty$$

19.
$$\begin{aligned} d &= 50(2) + 60(1) + 70(.5) + 60(3) \\ &= 100 + 60 + 35 + 180 \\ &= 375 \text{ miles} \end{aligned}$$

21.
$$\begin{aligned} d &= 15\left(\frac{1}{3}\right) + 18\left(\frac{1}{2}\right) + 16\left(\frac{1}{6}\right) + 12\left(\frac{2}{3}\right) \\ &= 5 + 9 + \frac{8}{3} + 8 \\ &= \frac{74}{3} \text{ miles} \end{aligned}$$

23. $f(x) = x^2 + 1$, $[0, 1]$, $n = 4$.
The evaluation points are .125, .375, .625, .875.

$$A \approx A_4 = \sum_{i=1}^{4} f(x_i)\Delta x =$$
$$(.25)(f(.125) + f(.375) + f(.625) + f(.875))$$
$$= 1.328125.$$

25. $f(x) = x^3 - 1$, $[1, 2]$, $n = 4$.
The evaluation points are 1.125, 1.375, 1.625, 1.875.

$$A \approx A_4 = \sum_{i=1}^{4} f(x_i)\Delta x =$$
$$(.25)[f(1.125) + f(1.375) + f(1.625) + f(1.875)]$$
$$= 2.7265625.$$

27. $f(x) = \sin x$, $[0, \pi]$, $n = 4$.
The evaluation points are $\frac{\pi}{8}, \frac{3\pi}{8}, \frac{5\pi}{8}, \frac{7\pi}{8}$.

$$\begin{aligned} A \approx A_4 &= \sum_{i=1}^{4} f(x_i)\Delta x \\ &= \left(\frac{\pi}{4}\right)\left(f\left(\frac{\pi}{8}\right) + f\left(\frac{3\pi}{8}\right)\right. \\ &\quad \left. + f\left(\frac{5\pi}{8}\right) + f\left(\frac{7\pi}{8}\right)\right) \\ &= 2.05234. \end{aligned}$$

29. $y = x^2$ on $[-1, 1]$, $n = 8$, left-endpoint evaluation.

$$\begin{aligned} \Delta x &= \frac{1 - (-1)}{8} = .25 \\ x_i &= -1 + .25i,\ i = 0, 1, \ldots, 7 \\ A \approx A_8 &\approx \sum_{i=0}^{7} f(x_i)(.25) \\ &= .25 \sum_{i=0}^{7} x_i^2 \\ &= .25((-1)^2 + (-.75)^2 + (-.5)^2 \\ &\quad + (-.25)^2 + 0^2 + .25^2 + .5^2 + .75^2) \\ &= .6875 \end{aligned}$$

31. $y = \sqrt{x + 2}$ on $[1, 4]$, $n = 8$, midpoint evaluation.

$$\begin{aligned} \Delta x &= \frac{4 - 1}{8} = .375 \\ x_i &= 1.1875 + .375i,\ i = 0, 1, \ldots, 7 \\ A \approx A_8 &\approx \sum_{i=0}^{7} f(x_i)(.375) = .375 \sum_{i=0}^{7} \sqrt{x_i + 2} \\ &= .375(\sqrt{3.1875} + \sqrt{3.5625} + \sqrt{3.9375} \\ &\quad + \sqrt{4.3125} + \sqrt{4.6875} + \sqrt{5.0625} \\ &\quad + \sqrt{5.4375} + \sqrt{5.8125}) \approx 6.3343 \end{aligned}$$

33. $y = e^{2x}$ on $[-1, 1]$, $n = 16$, midpoint evaluation.

$$\begin{aligned} \Delta x &= \frac{1 - (-1)}{16} = .125 \\ x_i &= (-1 + 0.0625) + .125i,\ i = 0, 1, \ldots, 15 \\ A \approx A_{16} &\approx \sum_{i=0}^{15} f(x_i)(.125) = .125 \sum_{i=0}^{15} e^{-2x_i} \\ &= .125\left(e^{-2(-.9375)} + e^{-2(-.8125)} + e^{-2(-.6875)}\right. \\ &\quad + e^{-2(-.5625)} + e^{-2(-.4375)} + e^{-2(-.3125)} \\ &\quad + e^{-2(-.1875)} + e^{-2(-.0625)} + e^{-2(.0625)} \\ &\quad + e^{-2(.1875)} + e^{-2(.3125)} + e^{-2(.4375)} \\ &\quad \left. + e^{-2(.5625)} + e^{-2(.6875)} + e^{-2(.8125)} + e^{-2(.9375)}\right) \\ &\approx 3.6174. \end{aligned}$$

35. $y = 3x - 2$ on $[1, 4]$, $n = 4$, midpoint evaluation.

$$\Delta x = \frac{4-1}{4} = .75$$

$$x_i = 1.375 + .75i,\ i = 0, 1, 2, 3$$

$$A \approx A_4 \approx \sum_{i=0}^{3} f(x_i)(.75) = .75 \sum_{i=0}^{3} (3x_i - 2)$$

$$= 2.25 \sum_{i=0}^{3} x_i + .75 \sum_{i=0}^{3} -2$$

$$= 2.25(1.375 + 2.125 + 2.875 + 3.625)$$

$$- 2(.75)(4)$$

$$= 16.5.$$

37.

n	left endpoint	midpoint	right endpoint
10	10.56	10.72	10.56
50	10.662	10.669	10.662
100	10.6656	10.6672	10.6656
500	10.6666	10.6667	10.6666
1000	10.6667	10.6667	10.6667

39. $f(x) = x^2 + 2$; $a = 0$, $b = 1$.
The Riemann sum is

$$\frac{b-a}{n} \sum_{k=0}^{n-1} f\left(\frac{b-a}{n} \cdot k + a\right)$$

$$= \frac{1}{n} \sum_{k=0}^{n-1} \left(\left(\frac{k}{n}\right)^2 + 2\right).$$

$$\lim_{n\to\infty} \frac{1}{n} \sum_{k=0}^{n-1} \left(\frac{k^2}{n^2} + 2\right)$$

$$= \lim_{n\to\infty} \frac{2n^2 - 3n + 1}{6n^2} + 2$$

$$= \lim_{n\to\infty} \frac{14n^2 - 3n + 1}{6n^2}$$

$$= \frac{7}{3}.$$

41. $f(x) = 2x^2 + 1$; $a = 1$, $b = 3$.
The Riemann sum is

$$\frac{b-a}{n} \sum_{k=0}^{n-1} f\left(\frac{b-a}{n} \cdot k + a\right)$$

$$= \frac{2}{n} \sum_{k=0}^{n-1} \left(2\left(\frac{2k}{n} + 1\right)^2 + 1\right).$$

$$\lim_{n\to\infty} \frac{2}{n} \sum_{k=0}^{n-1} \left(\frac{8k^2}{n^2} + \frac{8k}{n} + 3\right)$$

$$= \lim_{n} \frac{2}{n} \left(\frac{8(2n^2 - 3n + 1)}{6n} + \frac{8(n-1)}{2} + 3n\right)$$

$$= \lim_{n\to\infty} \frac{2(29n^2 - 24n + 4)}{3n^2}$$

$$= \frac{58}{3}.$$

43. left endpoint evaluation

$$A \approx .1(2.0 + 2.4 + 2.6 + 2.7 +$$
$$2.6 + 2.4 + 2.0 + 1.4) = 1.81$$

right endpoint evaluation

$$A \approx .1(2.4 + 2.6 + 2.7 + 2.6$$
$$+ 2.4 + 2.0 + 1.4 + 0.6) = 1.67$$

45. left endpoint evaluation

$$A \approx .1(1.8 + 1.4 + 1.1 + .7$$
$$+ 1.2 + 1.4 + 1.8 + 2.4) = 1.18$$

right endpoint evaluation

$$A \approx .1(1.4 + 1.1 + .7 + 1.2$$
$$+ 1.4 + 1.8 + 2.4 + 2.6) = 1.26$$

47. Using left endpoints gives an approximate area of 0.07133, which is less than the true area since the function is increasing. Using right endpoints gives an approximate area of 0.1139, which is more than the true area. Assuming the graph of the function consists of line segments connecting the given values, and using the midpoint evaluation, we get an approximate area of $(0.07133 + 0.1139)/2 = 0.0926$.

49. Since the function is increasing, the value of the function at the left endpoint will be smaller than the value at other points in the interval. Hence, the rectangle constructed using the left endpoint will be under the curve. Thus the left-endpoint Riemann sum will be less than the area under the curve. Similarly, since the function is increasing, the Riemann sum using right endpoints will give a value greater than the area under the curve.

It is harder to analyze the situation with the midpoint rule. To the left of the midpoint, the graph of the function will be below the edge of the rectangle, while to the right of the midpoint the graph will be above the edge of the rectangle. However, since the function is concave up, the amount that the function is above the rectangle will be greater than the amount the function is below the rectangle. Hence, the midpoint rule gives an estimate which is less than the true area.

51. Since the function is decreasing, the situation is the reverse of Exercise 49; now the left-endpoint estimate is too large and the right endpoint estimate is too small. Since the function is concave up, the amount the graph is above the midpoint line is more then the amount the graph is below the midpoint line, so the midpoint estimate is too small.

53. For example, use $x = \dfrac{1}{\sqrt{6}}$ on [0, 0.5] and use $x = \dfrac{1}{\sqrt{2}}$ on [0.5, 1].

55. $\Delta x = \dfrac{(b-a)}{n}$, $[a, b]$ interval for right endpoint evaluation, the evaluation points are

$$\begin{aligned}
c_1 &= a + \Delta x \\
c_2 &= c_1 + \Delta x = (a + \Delta x) + \Delta x = a + 2\Delta x \\
c_3 &= c_2 + \Delta x = (a + 2\Delta x) + \Delta x = a + 3\Delta x \\
&\vdots \\
c_i &= a + i \cdot \Delta x,\ i = 1, \ldots, n.
\end{aligned}$$

57. $\Delta x = \dfrac{(b-a)}{n}$, $[a, b]$ interval for midpoint evaluation, the evaluation points are

$$\begin{aligned}
c_1 &= a + \frac{\Delta x}{2} \\
c_2 &= c_1 + \Delta x = \left(a + \frac{\Delta x}{2}\right) + \Delta x = a + \frac{3}{2}\Delta x \\
c_3 &= c_2 + \Delta x = \left(a + \frac{3}{2}\Delta x\right) + \Delta x = a + \frac{5}{2}\Delta x \\
&\vdots \\
c_i &= a + \left(i - \frac{1}{2}\right)\Delta x,\ i = 1, \ldots, n.
\end{aligned}$$

4.2 The Definite Integral

1.

$$\begin{aligned}
\int_0^3 \left(x^3 + x\right) dx &\approx \sum_{i=1}^{n} \left(c_i^3 + c_i\right) \Delta x \\
&= \sum_{i=1}^{n} \left(c_i^3 + c_i\right) \left(\frac{3-0}{n}\right) \\
&= \sum_{i=1}^{n} \frac{3}{n} \left(c_i^3 + c_i\right),
\end{aligned}$$

$$c_i = \frac{x_i + x_{i-1}}{2}$$

$n \geq 50 \Rightarrow$ Riemann sum ≈ 24.75

3.

$$\begin{aligned}
\int_2^4 \frac{1}{x^2}\, dx &\approx \sum_{i=1}^{n} \left(\frac{1}{c_i^2}\right) \Delta x \\
&= \sum_{i=1}^{n} \frac{1}{c_i^2} \left(\frac{4-2}{n}\right) \\
&= \sum_{i=1}^{n} \frac{2}{n} \frac{1}{c_i^2},
\end{aligned}$$

$$c_i = \frac{x_i + x_{i-1}}{2}$$

$n \geq 10 \Rightarrow$ Riemann sum $\approx .25$

5.

$$\begin{aligned}
\int_0^{\pi} \sin x^2\, dx &= \sum_{i=1}^{n} \sin c_i^2 \Delta x \\
&= \sum_{i=1}^{n} \sin c_i^2 \left(\frac{\pi - 0}{n}\right) \\
&= \sum_{i=1}^{n} \frac{\pi}{n} \sin c_i^2,
\end{aligned}$$

$$c_i = \frac{x_i + x_{i-1}}{2}$$

$n \geq 40 \Rightarrow$ Riemann sum $\approx .77$

7. $$\int_0^1 2x\,dx$$

$$\Delta x = \frac{1-0}{n} = \frac{1}{n}$$

$$x_0 = 0,\ x_1 = x_0 + \Delta x = \frac{1}{n},$$

$$x_2 = x_1 + \Delta x = \frac{1}{n} + \frac{1}{n} = \frac{2}{n},\ x_i = \frac{i}{n}$$

$$R_n = \sum_{i=1}^{n} f(x_i)\,\Delta x$$

$$= \sum_{i=1}^{n} (2x_i)\Delta x$$

$$= \sum_{i=1}^{n} (2x_i)\frac{1}{n}$$

$$= \sum_{i=1}^{n} \frac{2i}{n} \cdot \frac{1}{n}$$

$$= \sum_{i=1}^{n} \frac{2i}{n^2}$$

$$= \frac{2}{n^2} \sum_{i=1}^{n} i$$

$$= \frac{2}{n^2} \frac{(n)(n+1)}{2}$$

$$= \frac{n^2+n}{n^2}$$

$$\int_0^1 2x\,dx = \lim_{n\to\infty} \left(\frac{n^2+n}{n^2}\right) = 1$$

9. $$\int_0^2 x^2\,dx$$

$$\Delta x = \frac{2-0}{n} = \frac{2}{n}$$

$$x_0 = 0,\ x_1 = \frac{2}{n},\ x_2 = \frac{4}{n},\ x_i = \frac{2i}{n}$$

$$R_n = \sum_{i=1}^{n} f(x_i)\,\Delta x$$

$$= \sum_{i=1}^{n} x_i^2 \Delta x$$

$$= \sum_{i=1}^{n} \left(\frac{2i}{n}\right)^2 \left(\frac{2}{n}\right)$$

$$= \frac{8}{n^3} \sum_{i=1}^{n} i^2$$

$$= \frac{8}{n^3} \frac{n(n+1)(2n+1)}{6}$$

$$\int_0^2 x^2\,dx = \lim_{n\to\infty} \left(\frac{8}{6n^2}(n+1)(2n+1)\right)$$

$$= \frac{8}{3}$$

11. $$\int_1^3 (2x-1)\,dx$$

$$\Delta x = \frac{3-1}{n} = \frac{2}{n}$$

$$x_0 = 1,\ x_1 = 1 + \frac{2}{n},\ x_2 = 1 + \frac{4}{n},$$

$$x_i = 1 + \frac{2i}{n} = \frac{n+2i}{n}$$

$$R_n = \sum_{i=1}^{n} f(x_i)\,\Delta x$$

$$= \sum_{i=1}^{n} (2x_i - 1)\Delta x$$

$$= \sum_{i=1}^{n} \left[2\left(\frac{n+2i}{n}\right) - 1\right] \cdot \frac{2}{n}$$

$$= \frac{2}{n} \sum_{i=1}^{n} 1 + \frac{8}{n^2} \sum_{i=1}^{n} i$$

$$= \frac{2}{n} \cdot n + \frac{8}{n^2} \cdot \frac{n(n+1)}{2}$$

$$= 2 + 4 + \frac{4}{n}$$

$$= 6 + \frac{4}{n}$$

$$\int_1^3 (2x-1)\,dx = \lim_{n\to\infty} \left(6 + \frac{4}{n}\right) = 6$$

13. $\int_{-2}^{2}\left(4-x^2\right)dx$

15. $-\int_{-2}^{2}\left(x^2-4\right)dx$

17. $\int_{0}^{2}x^2\,dx$

19. $\int_{0}^{\pi}\sin x\ dx$

21. $\int_{0}^{1}\left(x^3-3x^2+2x\right)dx-\int_{1}^{2}\left(x^3-3x^2+2x\right)dx$

23. $v(t)=60-16t$, $[0,2]$
distance traveled over $[0,2]$

$$\begin{aligned}&=\int_0^2 (60-16t)\,dt\\ &\approx\sum_{i=1}^{n}(60-16c_i)\Delta t\\ &=\sum_{i=1}^{n}(60-16c_i)\frac{2-0}{n}\\ &=\sum_{i=1}^{n}\frac{2}{n}(60-16c_i),\end{aligned}$$

$$c_i=\frac{t_i+t_{i-1}}{2}$$

Any value of n gives a Riemann sum = 88. Since the starting position is 2, the final position is $2+88=90$.

25. $v(t)=40\left(1-e^{2t}\right)$, $[0,4]$
distance traveled over $[0,4]$

$$\begin{aligned}&=\int_0^4 40\left(1-e^{-2t}\right)dt\\ &\approx 40\sum_{i=1}^{n}\left(1-e^{-2c_i}\right)\Delta t\\ &=40\sum_{i=1}^{n}\left(1-e^{-2c_i}\right)\frac{4}{n},\ c_i=\frac{t_i+t_{i-1}}{2}\end{aligned}$$

$n\geq 40\Rightarrow$ Riemann sum ≈ 140.0. Since the starting position is 0, the final position is 140.0.

27. $\int_0^2 f(x)\,dx+\int_2^3 f(x)\,dx=\int_0^3 f(x)\,dx$

29. $\int_0^2 f(x)\,dx+\int_2^1 f(x)\,dx=\int_0^1 f(x)\,dx$

31.

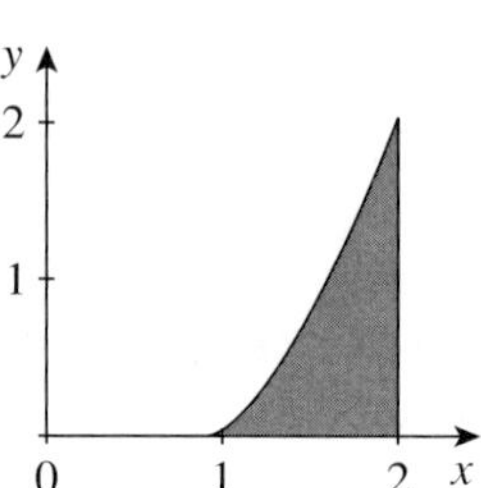

33.

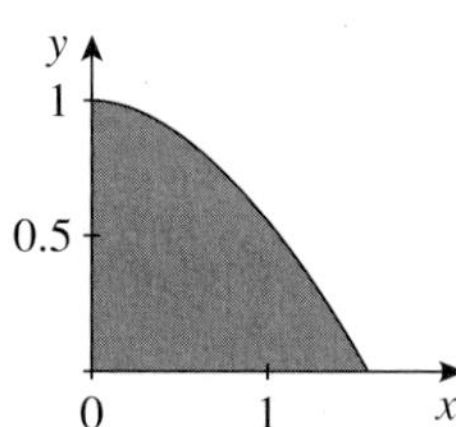

35. Max of the integrand is ≈ 1.36981;
min of the integrand is ≈ -2.343636.
The interval of integration has length $\frac{\pi}{2}-\frac{\pi}{3}=\frac{\pi}{6}$.

$$-2.343636\cdot\frac{\pi}{6}\leq I\leq 1.36981\cdot\frac{\pi}{6}$$
$$-1.227\leq I\leq 0.718$$

37. Max of the integrand is 3;
min of the integrand is 1.
The interval of integration has length 2.

$$1\cdot 2\leq I\leq 3\cdot 2$$
$$2\leq I\leq 6$$

39. We need $3x^2=4$, so $x=\frac{2}{\sqrt{3}}$.

41. Substituting g' for f in the Integral Mean Value Theorem, the theorem says that for g' continuous on $[a,b]$, there is a c in (a,b) such that

$$g'(c)=\frac{1}{b-a}\int_a^b g'(x)\,dx.$$

The Mean Value Theorem says that if g is continuous on $[a, b]$ and differentiable on (a, b), there is some c in (a, b) (perhaps a different c) such that

$$g'(c) = \frac{g(b) - g(a)}{b - a}.$$

This suggests that there might be some relationship between $\int_a^b g'(x)\,dx$ and $g(b) - g(a)$. Note that since g' satisfies the hypotheses of the Integral Mean Value theorem, g' is continuous on $[a, b]$, so certainly g will be continuous on $[a, b]$, and thus the conditions of the Mean Value Theorem are fulfilled for the function g.

43. Since f is integrable on $[a, b]$, we have

$$\int_a^b f(x)\,dx = \lim_n \sum_{i=1}^{n} f(c_i)\,\Delta x,$$

where

$$c_i \in [a + (i-1)\Delta x, a + i\Delta x]$$

and

$$\Delta x = \frac{b-a}{n}.$$

So for n sufficiently large, we have

$$\int_a^b f(x)\,dx \approx \sum_{i=1}^{n} f(c_i)\Delta x,$$

and we might as well take n to be even. When n is even, then

$$c = \frac{a+b}{2} = a + \frac{n}{2}\Delta x,$$

so we can rewrite the Riemann sum

$$\sum_{i=1}^{n} f(c_i)\Delta x = \sum_{i=1}^{n/2} f(c_i)\Delta x + \sum_{i=\frac{n}{2}+1}^{n} f(c_i)\Delta x.$$

Note:

$$\Delta x = \frac{b-a}{n} = \frac{(2c-a)-a}{n} = \frac{2(c-a)}{n} = \frac{c-a}{\left(\frac{n}{2}\right)},$$

and

$$\Delta x = \frac{b-a}{n} = \frac{b-(2c-b)}{n} = \frac{2(b-a)}{n} = \frac{b-c}{\left(\frac{n}{2}\right)}.$$

Thus,

$$\sum_{i=1}^{n/2} f(c_i)\Delta x = \sum_{i=1}^{n/2} f(c_i)\cdot\frac{c-a}{\left(\frac{n}{2}\right)} \approx \int_a^c f(x)\,dx,$$

and

$$\sum_{i=\frac{n}{2}+1}^{n} f(c_i)\Delta x = \sum_{i=\frac{n}{2}+1}^{n} f(c_i)\cdot\frac{b-c}{\left(\frac{n}{2}\right)} \approx \int_c^b f(x)\,dx.$$

So as $n \to \infty$, we have

$$\int_a^b f(x)\,dx = \int_a^c f(x)\,dx + \int_c^b f(x)\,dx.$$

45. positive

47. negative

49. If n is even and we use midpoint evaluation, we never have an evaluation at $x = 1$, which is the only x for which the functions have different values. Since the approximate Riemann sums with midpoint evaluation are equal for all even values of n, and since we are assuming the integrals exist,

$$\lim_n \sum_{i=1}^{n} f(c_i)\,\Delta x = \lim_n \sum_{i=1}^{n} g(c_i)\,\Delta x,$$

so the integrals are equal.

51. $\int_0^4 f(x)\,dx = \int_0^1 2x\,dx + \int_1^4 4\,dx;$

$\int_0^1 2x\,dx = 1$ (by calculating area of triangle formed by $y = 2x$, $x = 1$, and the x-axis);

$\int_1^4 4\,dx = 12$ (by calculating the area of the rectangle). So $\int_0^4 f(x)\,dx = 1 + 12 = 13.$

53. $\int_0^4 f(x)\,dx = \int_0^3 4\,dx + \int_3^4 (x+1)\,dx;$

$\int_0^3 4\,dx = 12$, (by calculating the area of the rectangle); $\int_3^4 (x+1)\,dx = 4.5$, since it equals the area of a trapezoid with base 1 and sides of height 4 and 5; and the area of the trapezoid is $\frac{1}{2}(1)(4+5) = 4.5$.

So $\int_0^4 f(x)\,dx = 12 + 4.5 = 16.5$.

55. $\int_0^{12} [b(t) - a(t)]\,dt$ is the births minus deaths in the first year, which is the net change in population.

$$\begin{aligned} b(t) &= 410 - .3t \\ a(t) &= 390 + .2t \\ b(t) &> a(t) \quad \Leftrightarrow \\ 410 - .3t &> 390 + .2t \quad \Leftrightarrow \\ 20 &> .5t \quad \Leftrightarrow \\ 40 &> t. \end{aligned}$$

The population is increasing for $t < 40$, decreasing for $t > 40$, max at $t = 40$.

57. $PV = 10$, $P = \frac{10}{v}$

$$\int_2^4 P(V)dV = \int_2^4 \frac{10}{V} dV \approx 6.93$$

(using midpoint evaluation with $n = 10$)

59.
$$f(x) = 2x + 1, \ [0, 4]$$
$$f_{ave} = \frac{1}{4-0} \int_0^4 (2x+1)\,dx$$
$$= \frac{1}{4} \int_0^4 (2x+1)\,dx = 5$$

(by calculating area of trapezoid formed by $y = 2x + 1$, $x = 0$, $x = 4$ and x-axis)

61.
$$f(x) = x^2 - 1, \ [1, 3]$$
$$f_{ave} = \frac{1}{3-1} \int_1^3 (x^2 - 1)\,dx$$
$$= \frac{1}{2} \int_1^3 (x^2 - 1)\,dx \approx 3.33$$

(using midpoint evaluation with $n = 10$)

63. The integral of the function over the interval $[0, 2]$ is 10; the integral over the interval $[2, 6]$ is 44. Hence, the integral over the interval $[0, 6]$ is 54, and the average value over that interval is $54/6 = 9$.

65. Since $F(t) = 9 - 10^8(t - 0.0003)^2$ is a quadratic function, with a graph that is a parabola opening downward, $F(t)$ has a maximum value and this occurs at its vertex, that is, when $t = 0.0003$. So $F(0.0003) = 9$ thousand pounds is the maximum force on the ball.

$$m\Delta v = \int_a^b F(t)\,dt$$
$$m = 0.01, \ a = 0, \ b = 0.0006$$
$$v(0.0006) - v(0)$$
$$= \frac{1}{0.01} \int_0^{0.0006} (9 - 10^8(t - 0.0003)^2)\,dt$$
$$\approx 0.36 \text{ thousand ft/s}$$
$$= 360 \text{ ft/s}$$

(using midpoint evaluation with $n = 20$)

67. If f is continuous on $[a, b]$, there is a $c \in [a, b]$ such that

$$f(c) = \frac{1}{b-a} \int_a^b f(x)\,dx.$$

So let $n = 1$, and $x_i = c$, $\Delta x = b - a$. Then

$$\sum_{i=1}^{n} f(x_i)\Delta x = f(c)(b-a) = \int_a^b f(x)\,dx.$$

69. In each interval $[x_{i-1}, x_i]$, we can choose an evaluation point c_i that is a rational number. Then $f(c_i) = 2$ for every i and so the Riemann sum = 2 and hence the limit as $n \to \infty$ of the Riemann sums = 2. On the other hand, in each interval $[x_{i-1}, x_i]$, we can choose an evaluation point c_i that is an irrational number. Then $f(c_i) = 4$ for every i and so the Riemann sum = 4 and hence the limit as $n \to \infty$ of the Riemann sums = 4. Thus the limit is not the same for these two choices of evaluation points, and so the integral does not exist.

71. This integral corresponds to the area of a triangle with base 2 and height 6, so the value of the integral is 6.

73. This integral corresponds to the area of a quarter circle of radius 2, so the value of the integral is

$$\frac{1}{4}\pi r^2 = \frac{1}{4}\pi(2)^2 = \pi.$$

75. This integral corresponds to the area of a triangle located below the x-axis with base 2 and height 2, so the value of the integral is -2.

4.3 Antiderivatives

1. $\int 3x^4\,dx = \frac{3x^5}{5} + c$

3. $$\int \left(3x^4 - 3x\right) dx = \int 3x^4\,dx - \int 3x\,dx = \frac{3x^5}{5} - \frac{3x^2}{2} + c$$

5. $\int 3\sqrt{x}\,dx = 2x^{3/2} + c$

7. $$\int \left(3 - \frac{1}{x^4}\right) dx = \int 3\,dx - \int x^{-4}\,dx = 3x + \frac{x^{-3}}{3} + c$$

9. $$\int \frac{x^{1/3} - 3}{x^{2/3}}\,dx = \int x^{-1/3}\,dx - \int 3x^{-2/3}\,dx = \frac{3}{2}x^{2/3} - 9x^{1/3} + c$$

11. $$\int 2\sin x + \cos x\,dx = \int 2\sin x\,dx + \int \cos x\,dx = -2\cos x + \sin x + c$$

13. $\int 2\sec x \tan x\,dx = 2\sec x + c$

15. $\int 5\sec^2 x\ dx = 5\tan x + c$

17. $$\int (3e^x - 2)\,dx = 3\int e^x\,dx - \int 2\,dx = 3e^x - 2x + c$$

19. $$\int \left(3\cos x - \frac{1}{x}\right) dx = \int 3\cos x\,dx - \int \frac{1}{x}\,dx = 3\sin x \ln|x| + c$$

21. $$\int \frac{4x}{x^2+4}\,dx = 2\ln\left|x^2+4\right| + c = 2\ln(x^2+4) + c$$

23. $$\int \left(5x - \frac{3}{e^x}\right) dx = \int 5x\,dx - \int 3e^{-x}\,dx = \frac{5x^2}{2} + 3e^{-x} + c$$

25. $$\int \left(e^{3x} - x\right) dx = \frac{e^{3x}}{3} - \frac{x^2}{2} + c$$

27. $$\int \frac{e^x}{e^x+3}\,dx = \ln\left|e^x+3\right| + c = \ln(e^x+3) + c$$

29. $$\int \frac{e^x+3}{e^x}\,dx = \int dx + \int 3e^{-x}\,dx = x - 3e^{-x} + c$$

31. $$\int \frac{4}{\sqrt{1-x^2}}\,dx = 4\int \frac{1}{\sqrt{1-x^2}}\,dx = 4\sin^{-1}x + c$$

33. $$\int 5\sin 2x\,dx = -\frac{5}{2}\cos 2x + c$$

35. $$\int 3\sec 2x \tan 2x\,dx = \frac{3}{2}\sec 2x + c$$

37. $\int \sqrt{x^3+4}\,dx$: N/A

39. $$\int \frac{3x^2-4}{x^2}\,dx = \int 3\,dx - \int 4x^{-2}\,dx = 3x + 4x^{-1} + c$$

41. $\int 2\sec x\,dx$: N/A

43. $$\int 2\sin 4x\,dx = -\frac{1}{2}\cos 4x + c$$

45. $\int e^{x^2}\,dx$: N/A

47. $$\int \left(\frac{1}{x^2} - 1\right) dx = \int x^{-2}\,dx - \int dx = -x^{-1} - x + c$$

49. for part (b), $\ln|\sec x + \tan x| + c$

for part (f), $\frac{1}{4}\sin 2x - \frac{1}{2}x\cos 2x + c$

(The CAS does not show the absolute value bars or the constant c.)

51.
$$f(x) = \int \left(4x^2 - 1\right)\, dx = \frac{4}{3}x^3 - x + c$$
$$f(0) = \frac{4}{3}\cdot 0^3 - 0 + c = 2$$
$$c = 2$$
$$f(x) = \frac{4}{3}x^3 - x + 2$$

53.
$$f(x) = \int (3e^x + x)\, dx = 3e^x + \frac{x^2}{2} + c$$
$$f(0) = 3e^0 + \frac{0^2}{2} + c = 4$$
$$c = 1$$
$$f(x) = 3e^x + \frac{x^2}{2} + 1$$

55.
$$f''(x) = 12,\ f'(0) = 2,\ f(0) = 3$$
$$f'(x) = \int 12\, dx = 12x + c_1$$
$$f'(0) = 12\cdot 0 + c_1 = 2,\ \text{ so } c_1 = 2$$
and $f'(x) = 12x + 2$
$$f(x) = \int (12x + 2)\, dx = 6x^2 + 2x + c_2$$
$$f(0) = 6\cdot 0^2 + 2\cdot 0 + c_2 = 3,\ \text{ so } c_2 = 3$$
$$f(x) = 6x^2 + 2x + 3$$

57.
$$f''(x) = 3\sin x + 4x^2$$
$$f'(x) = \int \left(3\sin x + 4x^2\right)\, dx$$
$$= -3\cos x + \frac{4}{3}x^3 + c_1$$
$$f(x) = \int \left(-3\cos x + \frac{4}{3}x^3 + c_1\right)\, dx$$
$$= -3\sin x + \frac{1}{3}x^4 + c_1 x + c_2$$

59.
$$f'''(x) = 4 - \frac{2}{x^3} = 4 - 2x^{-3}$$
$$f''(x) = \int \left(4 - 2x^{-3}\right)\, dx = 4x + x^{-2} + c_1$$
$$f'(x) = \int \left(4x + x^{-2} + c_1\right)\, dx$$
$$= 2x^2 - x^{-1} + c_1 x + c_2$$
$$f(x) = \int \left(2x^2 - \frac{1}{x} + c_1 x + c_2\right)\, dx$$
$$= \frac{2}{3}x^3 - \ln|x| + \frac{c_1}{2}x^2 + c_2 x + c_3$$

61. $v(t) = 3 - 12t,\ s(0) = 3$
$$s(t) = \int v(t)\, dt = \int (3 - 12t)\, dt = 3t - 6t^2 + c$$
$$s(0) = 3\cdot 0 - 6\cdot 0^2 + c = 3$$
$$c = 3$$
$$s(t) = 3t - 6t^2 + 3$$

63.
$$a(t) = 3\sin t + 1,\ v(0) = 0,\ s(0) = 4$$
$$v(t) = \int a(t)\, dt$$
$$= \int (3\sin t + 1)\, dt$$
$$= -3\cos t + t + c_1$$
$$v(0) = -3\cos 0 + 0 + c = 0$$
$$c_1 = 3$$
$$v(t) = -3\cos t + t + 3$$
$$s(t) = \int v(t)\, dt$$
$$= \int (-3\cos t + t + 3)\, dt$$
$$= -3\sin t + \frac{1}{2}t^2 + 3t + c_2$$
$$s(0) = -3\sin 0 + \frac{1}{2}0^2 + 3\cdot 0 + c_2 = 4$$
$$c_2 = 4$$
$$s(t) = -3\sin t + \frac{1}{2}t^2 + 3t + 4$$

65. $v(0) = 30 \text{ mph } = 30 \cdot \dfrac{1 \text{ hr}}{3600 \text{ sec}} = \dfrac{1}{120}\text{miles/sec}$

$v(4) = 50\text{mph} = 50 \cdot \dfrac{1 \text{ hr}}{3600 \text{ sec}} = \dfrac{1}{72}\text{miles/sec}$

$a = \dfrac{v(4) - v(0)}{4 - 0} = \dfrac{\frac{1}{72} - \frac{1}{120}}{4} = \dfrac{1}{720}\text{miles/sec}^2$

$v(t) = \int a(t)\,dt = \int \dfrac{1}{720}\,dt = \dfrac{t}{720} + c_1$

$v(0) = \dfrac{1}{120} = \dfrac{0}{720} + c_1$

$c_1 = \dfrac{1}{120}$

$v(t) = \dfrac{t}{720} + \dfrac{1}{120}$

$s(t) = \int v(t)\,dt$

$= \int \left(\dfrac{t}{720} + \dfrac{1}{120}\right) dt$

$= \dfrac{t^2}{1440} + \dfrac{t}{120} + c_2$

$s(0) = 0, \text{ so } c_2 = 0$

$s(t) = \dfrac{t^2}{1440} + \dfrac{t}{120}$

$s(4) = \dfrac{4^2}{1440} + \dfrac{4}{120} = \dfrac{1}{90} + \dfrac{4}{120} = \dfrac{4+12}{360}$

$= \dfrac{2}{45}$ miles

67.

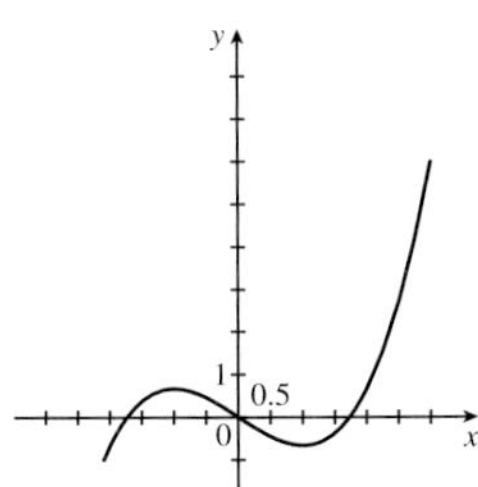

69. All functions that have the derivative shown in Exercise 67 are vertical translations of the graph given as the answer for Exercise 67.

71. Use derivative formulas that you know.

73. Distance = (velocity) × (time). To estimate the distance fallen, we first must estimate the velocity in each time interval, which we do by taking the average of the velocity at the start and end of each interval. For example, in the first time interval, we estimate the velocity as the average of -4.0 and -19.8, which is -11.9. However, since the time interval is half a second, the distance traveled is $(11.9)/2 = 5.95$ feet. Similar calculations give these distances in the other time intervals: 12.925, 17.4, 19.3 ,19.875, 20.

Acceleration = (change in velocity)/time. To estimate the acceleration in each time interval, take the velocity at the end of the interval and subtract the velocity at the beginning of the interval. For example, in the first interval, we estimate the acceleration as $-19.8 - (-4.0) = -15.8$. Since the time interval is half a second, the acceleration is $(-15.8)(2) = -31.6$. Similar calculations give these accelerations in the other time intervals: -24.2, -11.6, -3.6, -1, 0.

75. We are given that the speed at time $t = 0$ is 70. To estimate the speed at time $t = 0.5$, we use the formula

change in speed = (acceleration) × (time).

We estimate the acceleration as the average of the acceleration at the beginning and the acceleration at the end of each interval. So we estimate the acceleration in the first time interval as the average of -4.2 and 2.4, which is -0.9. Hence, we estimate the change in speed in the first half-second as $(-0.9)/2 = -0.45$. Thus we estimate the speed at time $t = 0.5$ as $70 - 0.45 = 69.55$.

To estimate the distance traveled at time $t = 0.5$, we use the formula

distance = velocity × time.

We estimate the velocity in the time interval as the average of the beginning velocity and the end velocity; that is, we estimate the velocity during this time interval as the average of 70 and 69.55, which is 69.775. Thus the displacement at the end of this half-second is (69.775)/2, or about 34.89.

Similar calculations apply to the other intervals.

77. $f''(x) = x - 1$

$f'(x) = \int x - 1\,dx = \frac{x^2}{2} - x + c_1$

$f'(1) = \frac{1^2}{2} - 1 + c_1 = 3,$

so $c_1 = \frac{7}{2}$

$f'(x) = \frac{x^2}{2} - x + \frac{7}{2}$

$f(x) = \int \frac{x^2}{2} - x + \frac{7}{2}\,dx = \frac{x^3}{6} - \frac{x^2}{2} + \frac{7}{2}x + c_2$

$f(1) = \frac{1^3}{6} - \frac{1^2}{2} + \frac{7}{2} \cdot 1 + c_2 = 2,$

so $c_2 = -\frac{7}{6}$

$f(x) = \frac{x^3}{6} - \frac{x^2}{2} + \frac{7}{2}x - \frac{7}{6}$

4.4 The Fundamental Theorem

1.
$$\int_0^2 (2x-3)\,dx = \left(x^2 - 3x\right)\Big|_0^2 = (4-6) - (0) = -2$$

3.
$$\int_1^1 \left(x^3 + 2x\right) dx = \left(\frac{x^4}{4} + x^2\right)\Big|_{-1}^1 = \left(\frac{1}{4} + 1\right) - \left(\frac{1}{4} + 1\right) = 0$$

5.
$$\int_0^4 (\sqrt{x} + 3x)\,dx = \left(\frac{2}{3}x^{3/2} + \frac{3}{2}x^2\right)\Big|_0^4 = \left(\frac{16}{3} + 24\right) - (0) = \frac{88}{3}$$

7.
$$\int_0^1 \left(x\sqrt{x} + x^{-1/2}\right) dx = \left(\frac{2}{5}x^{5/2} + 2x^{1/2}\right)\Big|_0^1 = \left(\frac{2}{5} + 2\right) - (0) = \frac{12}{5}$$

9.
$$\int_0^{\pi/2} 2\sin x\,dx = -2\cos x\big|_0^{\pi/2} = -2(0) - (-2)(1) = 2$$

11.
$$\int_0^{\pi/4} \sec x \tan x\,dx = \sec x\big|_0^{\pi/4} = \sqrt{2} - 1$$

13.
$$\int_{\pi/2}^{\pi} (2\sin x - \cos x)\,dx = (-2\cos x - \sin x)\big|_{\pi/2}^{\pi} = (2-0) - (0-1) = 3$$

15.
$$\int_0^1 (e^x - e^{-x})\,dx = (e^x + e^{-x})\big|_0^1 = (e^1 + e^{-1}) - (e^0 + e^0) = e + \frac{1}{e} - 2$$

17.
$$\int_0^1 \frac{e^{2x} - 1}{e^x}\,dx = \int_0^1 e^x - e^{-x}\,dx$$

(which is the same as Exercise 15)

19.
$$\int_0^{1/2} \frac{3}{\sqrt{1-x^2}}\,dx = 3\int_0^{1/2} \frac{1}{\sqrt{1-x^2}}\,dx = 3\sin^{-1} x\Big|_0^{1/2} = 3(\sin^{-1}\frac{1}{2} - \sin^{-1} 0) = 3\left(\frac{\pi}{6} - 0\right) = \frac{\pi}{2}$$

21.
$$\int_1^4 \frac{x-3}{x}\,dx = \int_1^4 (1 - 3x^{-1})\,dx = (x - 3\ln|x|)\Big|_1^4 = (4 - 3\ln 4) - (1 - 3\ln 1) = 3 - 3\ln 4$$

23.
$$\int_0^4 x(x-2)\,dx = \int_0^4 \left(x^2-2x\right)\,dx\Big|_0^4$$
$$= \left(\frac{x^3}{3}-x^2\right)$$
$$= \left(\frac{64}{3}-16\right)-(0)$$
$$= \frac{16}{3}$$

25.
$$\int_0^3 (x^3-\sin x)\,dx = \left(\frac{x^4}{4}+\cos x\right)\Big|_0^3$$
$$= \left(\frac{81}{4}+\cos 3\right)-(0+1)$$
$$= \frac{77}{4}+\cos 3$$

27. $\displaystyle\int_0^2 \sqrt{x^2+1}\,dx \approx \sum_{i=1}^n \sqrt{c_i^2+1}\left(\frac{2}{n}\right),\ c_i = \frac{x_i+x_{i-1}}{2}$

$n \geq 10 \Rightarrow$ Riemann sum ≈ 2.96

29.
$$\int_0^1 (e^x+1)^2\,dx = \int_0^1 (e^{2x}+2e^x+1)\,dx$$
$$= \left(\frac{e^{2x}}{2}+2e^x+x\right)\Big|_0^1$$
$$= \left(\frac{e^2}{2}+2e+1\right)-\left(\frac{1}{2}+2+0\right)$$
$$= \frac{e^2}{2}+2e-\frac{3}{2}$$

31. $\displaystyle\int_1^4 \frac{x^2}{x^2+4}\,dx \approx \sum_{i=1}^n \frac{c_i^2}{c_i^2+4c_i}\left(\frac{3}{n}\right),\ c_i = \frac{x_i+x_{i-1}}{2}$

$n \geq 10 \Rightarrow$ Riemann sum ≈ 1.12

33. $\displaystyle\int_0^\pi \sin x^2\,dx \approx \sum_{i=1}^n \sin c_i^2\left(\frac{\pi}{n}\right),\ c_i = \frac{x_i+x_{i-1}}{2}$

$n \geq 40 \Rightarrow$ Riemann sum $\approx .77$

35.
$$\int_0^{\pi/4} \frac{\sin x}{\cos}{}^2 x\,dx = \int_0^{\pi/4} \sec x \tan x\ dx$$
$$= \sec x|_0^{\pi/4}$$
$$= \sqrt{2}-1$$

37.
$$f(x) = \int_0^x \left(t^2-3t+2\right)dt$$
$$f'(x) = x^2-3x+2$$

39.
$$f(x) = \int_0^{x^2} \left(e^{-t^2}+1\right)dt$$

Let $u = x^2$, then

$$f(x) = \int_0^u \left(e^{-t^2}+1\right)dt,$$
$$f'(x) = \left(e^{-u^2}+1\right)\frac{du}{dx} = \left(e^{-x^4}+1\right)2x.$$

41. $\displaystyle f(x) = \int_x^{-1} \ln\left(t^2+1\right)\ dt = -\int_{-1}^x \ln\left(t^2+1\right)\ dt$

$f'(x) = -\ln\left(x^2+1\right)$

43.
$$\int_{-2}^2 4-x^2\,dx = \frac{32}{3}$$

45.
$$-\int_{-2}^2 x^2-4\,dx = \frac{32}{3}$$

47.
$$\int_0^2 x^2\,dx = \frac{8}{3}$$

49.
$$\int_0^\pi \sin x\,dx = 2$$

51.
$$y = \int_0^x \sin\sqrt{t^2+\pi^2}\,dt,\ x=0$$
$$y'(x) = \sin\sqrt{x^2+\pi^2}$$
$$y'(0) = \sin\sqrt{0+\pi^2} = 0$$
$$y(0) = \int_0^0 \sin\sqrt{t^2+\pi^2}\,dt = 0$$

The tangent line is $y = 0$.

53.
$$y = \int_2^x \cos\left(\pi t^3\right)\ dt,\ x=2$$
$$y'(x) = \cos(\pi x^3)$$
$$y'(2) = \cos 8\pi = 1$$
$$y(2) = \int_2^2 \cos(\pi t^3)\,dt = 0$$

The tangent line is $y = x-2$.

55. $$f(x) = \int_0^x \left(t^2 - 3t + 2\right)\,dt$$
$$f'(x) = x^2 - 3x + 2 = (x-2)(x-1)$$
$x = 2, 1$ are critical numbers.
$$f''(x) = 2x - 3$$
$$f''(2) = 1$$
$$f''(1) = -1$$
so $x = 2$ is local min, $x = 1$ is local max.

57. **(a)** $\frac{1}{x^2} > 0$, so the area can't be negative.

(b) The integral is incorrect because $\frac{1}{x^2}$ is not defined at $x = 0$.

59. $$v(t) = 40 - \sin t,\ s(0) = 2$$
$$s(t) = \int v(t)\,dt = \int (40 - \sin t)\,dt$$
$$= 40t + \cos t + c$$
$$s(0) = 40 \cdot 0 + \cos 0 + c = 2$$
$$c = 1$$
$$s(t) = 40t + \cos t + 1$$

61. $$v(t) = 25(1 - e^{-2t}),\ s(0) = 0$$
$$s(t) = \int v(t)\,dt$$
$$= \int 25(1 - e^{-2t})\,dt$$
$$= 25t + \frac{25}{2}e^{-2t} + c$$
$$s(0) = 0 + \frac{25}{2} + c$$
$$c = -\frac{25}{2}$$
$$s(t) = 25t + \frac{25}{2}e^{-2t} - \frac{25}{2}$$

63. $$a(t) = 4 - t,\ v(0) = 8,\ s(0) = 0$$
$$v(t) = \int a(t)\,dt = \int (4 - t)\,dt = 4t - \frac{t^2}{2} + c_1$$
$$v(0) = 0 - 0 + c_1 = 8,\ c_1 = 8$$
$$v(t) = 4t - \frac{t^2}{2} + 8$$
$$s(t) = \int v(t)\,dt$$
$$= \int \left(4t - \frac{t^2}{2} + 8\right)dt$$
$$= 2t^2 - \frac{t^3}{6} + 8t + c_2$$
$$s(0) = 0 - 0 + 0 + c_2 = 0,\ c_2 = 0$$
$$s(t) = 2t^2 - \frac{t^3}{6} + 8t$$

65. $$a(t) = 24 + e^{-t},\ v(0) = 0,\ s(0) = 0$$
$$v(t) = \int a(t)\,dt$$
$$= \int (24 + e^{-t})\,dt$$
$$= 24t - e^{-t} + c_1$$
$$v(0) = 0 - 1 + c_1 = 0,\ c_1 = 1$$
$$v(t) = 24t - e^{-t} + 1$$
$$s(t) = \int v(t)\,dt$$
$$= \int (24t - e^{-t} + 1)\,dt$$
$$= 12t^2 + e^{-t} + t + c_2$$
$$s(0) = 0 + 1 + 0 + c_2 = 0,\ c_2 = -1$$
$$s(t) = 12t^2 + e^{-t} + t - 1$$

67. $$\alpha(t) = 10 \text{ rad/s}^2$$
$$\omega(t) = \int a(t)\,dt = \int 10\,dt = 10t + c_1$$
$$\omega(0) = 10 \cdot 0 + c_1 = 0,\ \omega(t) = 10t$$
$$\omega(.8) = 10(.8) = 8 \text{ rad/s}$$
$$v = 3\omega = 3 \cdot 8 = 24 \text{ ft/s}$$
$$\theta(t) = \int \omega(t)\,dt = \int 10t\,dt = 5t^2 + c_2$$
$$\theta(0) = 5 \cdot 0^2 + c_2 = 0$$
$$\theta(t) = 5t^2$$
$$\theta(.8) = 5(.8)^2 = 3.2 \text{ rad}$$

69. $t = 4$ corresponds to 1974.

$$16.1e^{.07(4)} \approx 21.3024$$
$$21.3e^{.04(4-4)} \approx 21.3$$

$$\int_4^{10} 16.1e^{.07t}\, dt = \frac{16.1}{.07}e^{.07t}\Big|_4^{10} \approx 158.84$$

$$\int_4^{10} 21.3e^{.04(t-4)}\, dt = \frac{21.3}{.04}e^{.04(t-4)}\Big|_4^{10} \approx 144.44$$

$158.84 - 144.44 = 14.4$ million barrels saved.

71.
$$F(x) = 1000 \text{ lb}, \ b'(t) = 130 \text{ ft/s}$$
$$F\,(b(t))\,b'(t) = 1000\text{lb} \cdot 130\text{ft/s}$$
$$= 130{,}000 \text{ ft-lb/s}$$
$$= \frac{130{,}000}{550} \text{ hp}$$
$$\approx 236.36 \text{ hp}$$

73.
$$f(x) = x^2 - 1, \ [1, 3]$$
$$f_{ave} = \frac{1}{3-1}\int_1^3 (x^2 - 1)\, dx$$
$$= \frac{1}{2}\left(\frac{x^3}{3} - x\right)\Big|_1^3$$
$$= \frac{1}{2}\left[(9-3) - \left(\frac{1}{3} - 1\right)\right]$$
$$= \frac{10}{3}$$

75.
$$f(x) = 2x - 2x^2, \ [0, 1]$$
$$f_{ave} = \frac{1}{1-0}\int_0^1 \left(2x - 2x^2\right) dx$$
$$= \left(x^2 - \frac{2}{3}x^3\right)\Big|_0^1$$
$$= \left(1 - \frac{2}{3}\right) - (0)$$
$$= \frac{1}{3}$$

77.
$$f(x) = \cos x, \ \left[0, \frac{\pi}{2}\right]$$
$$f_{ave} = \frac{1}{\frac{\pi}{2} - 0}\int_0^{\pi/2} \cos x\, dx$$
$$= \frac{2}{\pi}(\sin x)\Big|_0^{\pi/2}$$
$$= \frac{2}{\pi}(1 - 0)$$
$$= \frac{2}{\pi}$$

79. Smallest to largest:

$$\int_0^3 f(x)\, dx, \ \int_0^2 f(x)\, dx, \ \int_0^1 f(x)\, dx.$$

81. Let $f(x) = \int_0^x e^{-t^2}\, dt$. Then, by the Fundamental Theorem, $f'(x) = e^{-x^2}$.

83.
$$\text{CS} = \int_0^Q D(q)dq - PQ$$
$$= \int_0^Q (150 - 2q - 3q^2)dq - PQ$$
$$= (150q - q^2 - q^3)\Big|_0^Q - PQ$$
$$= 150Q - Q^2 - Q^3 - PQ$$
$$= 150Q - Q^2 - Q^3 - (150 - 2Q - 3Q^2)Q$$
$$= Q^2 + 2Q^3.$$

When $Q = 4$, CS $= 144$; when $Q = 6$, CS $= 468$.

85. Let F be an antiderivative of f. Then, by the Fundamental Theorem,

$$\int_{a(x)}^{b(x)} f(t)\, dt = F(b(x))F(a(x)).$$

Now, by the chain rule,

$$\frac{d}{dx}\int_{a(x)}^{b(x)} f(t)\, dt$$
$$= \frac{d}{dx}(F(b(x)) - F(a(x)))$$
$$= f(b(x))b'(x) - f(a(x))a'(x).$$

4.5 Integration by Substitution

1. $u = x^3 + 2,\ du = 3x^2\,dx$

$$\int x^2\sqrt{x^3+2}\,dx = \frac{1}{3}\int \sqrt{u}\,du$$
$$= \frac{1}{3}\cdot\frac{2}{3}u^{3/2} + c$$
$$= \frac{2}{9}\left(x^3+2\right)^{3/2} + c$$

3. $u = \sqrt{x} + 2,\ du = \dfrac{1}{2\sqrt{x}}\,dx$

$$\int \frac{\left(\sqrt{x}+2\right)^3}{\sqrt{x}}\,dx = 2\int u^3\,du$$
$$= \frac{1}{2}u^4 + c$$
$$= \frac{1}{2}\left(\sqrt{x}+2\right)^4 + c$$

5. $u = x^3,\ du = 3x^2\,dx$

$$\int x^2\cos x^3\,dx = \frac{1}{3}\int \cos u\,du$$
$$= \frac{1}{3}\sin u + c$$
$$= \frac{1}{3}\sin x^3 + c$$

7. $u = \sqrt{x},\ du = \dfrac{1}{2\sqrt{x}}\,dx$

$$\int \frac{e^{\sqrt{x}}}{\sqrt{x}}\,dx = 2\int e^u\,du = 2e^u + c = 2e^{\sqrt{x}} + c$$

9. $u = x^2 + 4,\ du = 2x\,dx$

$$\int \frac{x}{x^2+4}\,dx = \frac{1}{2}\int \frac{du}{u}$$
$$= \frac{1}{2}\ln|u| + c = \frac{1}{2}\ln|x^2+4| + c$$
$$= \frac{1}{2}\ln(x^2+4) + c$$

11. $u = \sin^{-1}x,\ du = \dfrac{1}{\sqrt{1-x^2}}\,dx$

$$\int \frac{(\sin^{-1}x)^3}{\sqrt{1-x^2}}\,dx = \int u^3\,du$$
$$= \frac{u^4}{4} + c = \frac{(\sin^{-1}x)^4}{4} + c$$

13. $u = x - 5,\ du = dx$

$$\int \frac{x^2+2}{\sqrt{x-5}}\,dx = \int \frac{(u+5)^2+2}{\sqrt{u}}\,du$$
$$= \int \frac{u^2+10u+27}{\sqrt{u}}\,du$$
$$= \int u^{3/2} + 10u^{1/2} + 27u^{-1/2}\,du$$
$$= \frac{2}{5}u^{5/2} + \frac{20}{3}u^{3/2} + 54u^{1/2} + c$$
$$= \frac{2}{5}(x-5)^{5/2} + \frac{20}{3}(x-5)^{3/2} + 54(x-5)^{1/2} + c$$

15. $u = \cos x + 3,\ du = -\sin x\,dx$

$$\int \sin x(\cos x + 3)^{3/4}\,dx = -\int u^{3/4}\,du$$
$$= -\frac{4}{7}u^{7/4} + c$$
$$= -\frac{4}{7}(\cos x + 3)^{7/4} + c$$

17. $u = \cos x,\ du = -\sin x\,dx$

$$\int \frac{\sin x}{\sqrt{\cos x}}\,dx = -\int \frac{du}{\sqrt{u}}$$
$$= -2\sqrt{u} + c$$
$$= -2\sqrt{\cos x} + c$$

19. $u = \sqrt{x} + 1,\ du = \dfrac{1}{2\sqrt{x}}\,dx$

$$\int \frac{1}{\sqrt{x}\left(\sqrt{x}+1\right)^2}\,dx = 2\int u^{-2}\,du$$
$$= -2u^{-1} + c$$
$$= -2\left(\sqrt{x}+1\right)^{-1} + c$$
$$\text{or } -2\frac{\sqrt{x}-1}{x-1} + c$$

21. $u = \ln x,\ du = \frac{1}{x}\,dx$

$$\int \frac{\sqrt{\ln x}}{x}\,dx = \int \sqrt{u}\,du = \frac{2}{3}u^{3/2} + c = \frac{2}{3}(\ln x)^{3/2} + c$$

23. $u = x + 7,\ du = dx,\ x = u - 7$

$$\int \frac{2x+3}{x+7}\,dx = \int \frac{2(u-7)+3}{u}\,du = \int \left(2 - \frac{11}{u}\right) du$$
$$= 2u - 11\ln|u| + c$$
$$= 2(x+7) - 11\ln|x+7| + c$$
$$\text{or } 2x - 11\ln|x+7| + c$$

25. $u = 3 - 2x,\ du = -2\,dx$

$$\int e^{3-2x}\,dx = -\frac{1}{2}\int e^u\,du = -\frac{1}{2}e^u + c = -\frac{1}{2}e^{3-2x} + c$$

27. $u = \ln x + 2,\ du = \frac{1}{x}\,dx$

$$\int \frac{(\ln x + 2)^2}{x}\,dx = \int u^2\,du = \frac{1}{3}u^3 + c = \frac{1}{3}(\ln x + 2)^3 + c$$

29. $u = e^x + e^{-x},\ du = \left(e^x - e^{-x}\right)\,dx$

$$\int \frac{e^x - e^{-x}}{e^x + e^{-x}}\,dx = \int \frac{1}{u}\,du = \ln|u| + c = \ln\left(e^x + e^{-x}\right) + c$$

31. $u = x^3,\ du = 3x^2\,dx$

$$\int \frac{x^2}{1+x^6}\,dx = \frac{1}{3}\int \frac{du}{1+u^2} = \frac{1}{3}\tan^{-1}u + c = \frac{1}{3}\tan^{-1}(x^3) + c$$

33. $u = x^2,\ du = 2x\,dx$

$$\int \frac{x}{\sqrt{1-x^4}}\,dx = \frac{1}{2}\int \frac{du}{\sqrt{1-u^2}} = \frac{1}{2}\sin^{-1}u + c = \frac{1}{2}\sin^{-1}(x^2) + c$$

35.

$$\int (x^2+4)^2\,dx = \int x^4 + 8x^2 + 16\,dx = \frac{x^5}{5} + \frac{8}{3}x^3 + 16x + c$$

37. $u = x^2 + 1,\ du = 2x\,dx,\ u(0) = 1,\ u(2) = 5$

$$\int_0^2 x\sqrt{x^2+1}\,dx = \frac{1}{2}\int_1^5 \sqrt{u}\,du = \frac{1}{2}\cdot\frac{2}{3}u^{3/2}\Big|_1^5 = \frac{1}{3}\left(\sqrt{125} - 1\right) = \frac{5}{3}\sqrt{5} - \frac{1}{3}$$

39. $u = x^3,\ du = 3x^2\,dx$

$$\int_0^2 x^2 e^{x^3}\,dx = \frac{1}{3}\int_0^8 e^u\,du = \frac{1}{3}e^u\Big|_0^8 = \frac{1}{3}(e^8 - 1)$$

41. $u = \ln x,\ du = \frac{1}{x}\,dx$

$$\int_1^e \frac{\ln x}{x}\,dx = \int_0^1 u\,du = \frac{u^2}{2}\Big|_0^1 = \frac{1}{2}$$

43.

$$\int_1^4 \frac{x-1}{\sqrt{x}}\,dx = \int_1^4 \left(x^{1/2} - x^{-1/2}\right)dx = \left(\frac{2}{3}x^{3/2} - 2x^{1/2}\right)\Big|_1^4 = \left(\frac{16}{3} - 4\right) - \left(\frac{2}{3} - 2\right) = \frac{8}{3}$$

45. $u = e^x,\ du = e^x\,dx$

$$\int_0^2 \frac{e^x}{1+e^{2x}}\,dx = \int_1^{e^2} \frac{du}{1+u^2}$$
$$= \tan^{-1} u\Big|_1^{e^2} = \tan^{-1}(e^2) - \frac{\pi}{4}$$

47. $u = -x^2,\ du = -2x\,dx$

$$\int_{-2}^0 xe^{-x^2}\,dx = -\frac{1}{2}\int_{-4}^0 e^u\,du$$
$$= -\frac{1}{2}e^u\Big|_{-4}^0 = -\frac{1}{2}(1-e^{-4})$$
$$= \frac{1}{2}(e^{-4}-1)$$

49.
$$\int_0^\pi \sin x^2\,dx \approx .77$$

using midpoint evaluation with $n > 40$

51. $u = x^2,\ du = 2x\,dx,\ u(-1) = 1,\ u(1) = 1$

$$\int_{-1}^1 xe^{-x^2}\,dx = \frac{1}{2}\int_1^1 d^{-u}\,du = 0$$

53.
$$\int_0^2 \frac{4}{(x^2+1)^2}\,dx \approx 3.01$$

using midpoint evaluation with $n \geq 10$

55.
$$\int_0^{\pi/4} \sec x\,dx \approx .88$$

using midpoint evaluation with $n \geq 10$

57. $u = 1+x^2,\ du = 2x\,dx$

$$\int x^3\sqrt{1+x^2\,dx} = \frac{1}{2}\int x^2\sqrt{1+x^2}\,(2x)\,dx$$
$$= \frac{1}{2}\int (u-1)\sqrt{u}\,du = \frac{1}{2}\int u^{3/2} - u^{1/2}\,du$$
$$= \frac{1}{2}\left(\frac{2}{5}u^{5/2} - \frac{2}{3}u^{3/2}\right) + c$$
$$= \frac{1}{5}\left(1+x^2\right)^{5/2} - \frac{1}{3}\left(1+x^2\right)^{3/2} + c$$
$$\text{or } \frac{1}{15}\left(1+x^2\right)^{3/2}(3x^2-2) + c$$

59. $u = \frac{1}{2}x - 1, \quad du = \frac{dx}{2}$

$$\int \frac{2}{\sqrt{4x-x^2}}\,dx = \int \frac{2}{\sqrt{4-(x-2)^2}}\,dx$$
$$= \int \frac{4}{\sqrt{4-(2u)^2}}\,du = 2\int \frac{du}{\sqrt{1-u^2}} = 2\sin^{-1} u + c$$
$$= 2\sin^{-1}\left(\frac{1}{2}x - 1\right) + c$$

61. $u = \sqrt{x-1},\ du = \frac{(x-1)^{-1/2}}{2}\,dx$

$$\int \frac{1}{x\sqrt{x-1}}\,dx = \int \frac{2}{1+u^2}\,du$$
$$= 2\tan^{-1} u + c = 2\tan^{-1}(\sqrt{x-1}) + c$$

63. $u = \sqrt{x}+2,\ du = \frac{x^{-1/2}}{2}\,dx$

$$\int \frac{1}{2\sqrt{x}+x}\,dx = \int \frac{1}{\sqrt{x}(\sqrt{x}+2)}\,dx$$
$$= \int \frac{2}{u}\,du = 2\ln|u| + c = 2\ln(\sqrt{x}+2) + c$$

65. $u = 3 + \sqrt{x},\ du = \dfrac{1}{2\sqrt{x}}\,dx$

$$\int \frac{1}{(3+\sqrt{x})^3}\,dx = 2\int \frac{u-3}{u^3}\,du$$
$$= 2\int u^{-2} - 3u^{-3}\,du = 2\left(-u^{-1} - 3\frac{u^{-2}}{-2}\right) + c$$
$$= 3u^{-2} - 2u^{-1} + c$$
$$= 3\left(\frac{1}{3+\sqrt{x}}\right)^2 - \frac{2}{3+\sqrt{x}} + c$$

67. $$\frac{1}{2}\int_0^4 f(u)\,du$$

69. $$\int_0^1 f(u)\,du$$

71. $$\int \frac{5}{3+x^2}\,dx = 5\int \frac{1}{3+x^2}\,dx$$
$$= \frac{5}{3}\int \frac{1}{1+\left(x/\sqrt{3}\right)^2}\,dx$$

Let $u = x/\sqrt{3},\ du = dx/\sqrt{3}$.
Then the integral becomes

$$\frac{5}{3}\int \frac{1}{1+u^2}\sqrt{3}\,du = \frac{5}{\sqrt{3}}\tan^{-1} u + c$$
$$= \frac{5}{\sqrt{3}}\tan^{-1}\left(\frac{x}{\sqrt{3}}\right) + c.$$

73. $u = \ln x,\ du = \dfrac{1}{x}\,dx$

$$\int \frac{\ln x}{2x}\,dx = \int \frac{u}{2}\,du = \frac{u^2}{4} + c$$
$$= \frac{(\ln x)^2}{4} + c$$

75. $u = \ln\sqrt{x},\ du = \dfrac{1}{2x}\,dx$

$$\int \frac{1}{x\ln\sqrt{x}}\,dx = 2\int \frac{du}{u} = 2\ln|u| + c$$
$$= 2\ln|\ln\sqrt{x}| + c$$

Or,

$$\int \frac{1}{x\ln\sqrt{x}}\,dx = \int \frac{1}{x(\frac{1}{2}\ln x)}\,dx = 2\int \frac{1}{x\ln x}\,dx$$

$u = \ln x,\ du = \dfrac{1}{x}\,dx$

$$2\int \frac{1}{x\ln x}\,dx = 2\int \frac{du}{u} = 2\ln|u| + c$$
$$= 2\ln|\ln x| + c$$

Two answers are equivalent if they differ by a constant. Note that

$$2\ln|\ln\sqrt{x}| = 2\ln|\tfrac{1}{2}(\ln x)|$$
$$= 2\left[\ln\left(\frac{1}{2}\right) + \ln|\ln x|\right]$$
$$= 2\ln|\ln x| + 2\ln\left(\frac{1}{2}\right).$$

77. When x is negative, $x = -u^{1/4}$, not $u^{1/4}$, so in the long line with four equalities, the first equality is incorrect. The second equality is correct. The third equality would be correct if 32/5 were replaced by 128/5. The easiest way to do this problem is to use

$$\int_{-2}^{1} 4x^4\,dx = \frac{4}{5}x^5\Big|_{-2}^{1} = \frac{4}{5}(1-(-32)) = \frac{132}{5}.$$

If one uses the given substitution, then

$$\int_{-2}^{1} x(4x^3)\,dx$$
$$= \int_{-2}^{0} x(4x^3)\,dx + \int_0^1 x(4x^3)\,dx$$
$$= -\int_{16}^{0} u^{1/4}\,du + \int_0^1 u^{1/4}\,du$$
$$= -\frac{4}{5}u^{5/4}\Big|_{16}^{0} + \frac{4}{5}u^{5/4}\Big|_0^1$$
$$= -\frac{4}{5}(0-32) + \frac{4}{5}(1-0)$$
$$= \frac{4}{5} + \frac{128}{5} = \frac{132}{5}$$

79.
$$V(t) = V_p \sin(2\pi f t)$$
$$V^2(t) = V_p^2 \sin^2(2\pi f t)$$
$$= V_p^2 \left(\frac{1}{2} - \frac{1}{2}\cos(4\pi f t)\right)$$
$$= \frac{V_p^2}{2}(1 - \cos(4\pi f t))$$
$$rms = \sqrt{f \int_0^{1/f} V^2(t)\, dt}$$
$$= \sqrt{f \int_0^{1/f} \frac{V_p^2}{2}(1 - \cos(4\pi f t))\, dt}$$
$$= \frac{V_p\sqrt{f}}{\sqrt{2}}\sqrt{\int_0^{1/f} (1 - \cos(4\pi f t))\, dt}$$
$$= \frac{V_p\sqrt{f}}{\sqrt{2}}\sqrt{\left(t - \frac{\sin(4\pi f t)}{4\pi f}\right)\Bigg|_0^{1/f}}$$
$$= \frac{V_p\sqrt{f}}{\sqrt{2}}\sqrt{\frac{1}{f}} = \frac{V_p}{\sqrt{2}}$$

81.
$$\int_{-2}^{2} f(x)\, dx = \int_{-2}^{0} xe^{x^2}\, dx + \int_0^2 x^2 e^{x^3}\, dx$$
$$= \frac{1}{2}e^{x^2}\Big|_{-2}^{0} + \frac{1}{3}e^{x^3}\Big|_0^2$$
$$= \frac{1}{2}\left(1 - e^4\right) + \frac{1}{3}\left(e^8 - 1\right)$$
$$= \frac{1}{6} - \frac{e^4}{2} + \frac{e^8}{3}$$

83.
$$\int \frac{1}{|x|\sqrt{x^2-1}}\, dx = \int \frac{1}{x^2\sqrt{1-1/x^2}}\, dx$$
$$= -\int \frac{1}{\sqrt{1-u^2}}\, du = -\sin^{-1} u + c$$
$$= -\sin^{-1}(1/x) + c$$

Thus $-\sin^{-1}(1/x)$ and $\sec^{-1} x$ have the same derivative; hence $\sec^{-1} x = -\sin^{-1}(1/x) + c$ (or, equivalently, $\sec^{-1} x + \sin^{-1}(1/x) = c$) on each interval over which they are defined, namely $(-\infty, -1)$ and $(1, \infty)$. In fact, $c = \pi/2$ on both intervals.

4.6 Integration by Parts

1.
$$\int x \cos\, dx \qquad u = x \qquad dv = \cos x\, dx$$
$$du = dx \qquad v = \sin x$$
$$= x \sin x - \int \sin x\, dx$$
$$= x \sin x + \cos x + c$$

3.
$$\int xe^{2x}\, dx \qquad u = x \qquad dv = e^{2x}\, dx$$
$$du = dx \qquad v = \frac{1}{2}e^{2x}$$
$$= \frac{1}{2}xe^{2x} - \int \frac{1}{2}e^{2x}\, dx$$
$$= \frac{1}{2}xe^{2x} - \frac{1}{4}e^{2x} + c$$

5.
$$\int x^2 \ln x\, dx \qquad u = \ln x \qquad dv = x^2\, dx$$
$$du = \frac{1}{x}\, dx \qquad v = \frac{1}{3}x^3$$
$$= \frac{1}{3}x^3 \ln x - \int \frac{1}{3}x^3 \cdot \frac{1}{x}\, dx$$
$$= \frac{1}{3}x^3 \ln x - \frac{1}{3}\int x^2\, dx$$
$$= \frac{1}{3}x^3 \ln x - \frac{1}{9}x^3 + c$$

7.
$$\int x^2 e^{-3x}\, dx \qquad u = x^2 \qquad dv = e^{-3x}\, dx$$
$$du = 2x\, dx \qquad v = -\frac{1}{3}e^{-3x}$$
$$= -\frac{1}{3}x^2 e^{-3x} - \int \left(-\frac{1}{3}e^{-3x}\right) \cdot 2x\, dx$$
$$= -\frac{1}{3}x^2 e^{-3x} + \frac{2}{3}\int xe^{-3x}\, dx$$

$$u = x \qquad dv = e^{-3x}\,dx$$
$$du = dx \qquad v = -\frac{1}{3}e^{-3x}$$

$$= -\frac{1}{3}x^2e^{-3x} + \frac{2}{3}\left[-\frac{1}{3}xe^{-3x} - \int\left(-\frac{1}{3}e^{-3x}\right)dx\right]$$
$$= -\frac{1}{3}x^2e^{-3x} - \frac{2}{9}xe^{-3x} + \frac{2}{9}\int e^{-3x}\,dx$$
$$= -\frac{1}{3}x^2e^{-3x} - \frac{2}{9}xe^{-3x} - \frac{2}{27}e^{-3x} + c$$

9. $\int x^2 \cos 2x\,dx \qquad u = x^2 \qquad dv = \cos 2x\,dx$

$$du = 2x\,dx \qquad v = \frac{1}{2}\sin 2x$$

$$= \frac{1}{2}x^2\sin 2x - \int \frac{1}{2}(\sin 2x)2x\,dx$$
$$= \frac{1}{2}x^2\sin 2x - \int x\sin 2x\,dx$$

$$u = x \qquad dv = \sin 2x\,dx$$
$$du = dx \qquad v = -\frac{1}{2}\cos 2x$$

$$= \frac{1}{2}x^2\sin 2x$$
$$- \left[-\frac{1}{2}x\cos 2x - \int\left(-\frac{1}{2}\cos 2x\right)dx\right]$$
$$= \frac{1}{2}x^2\sin 2x + \frac{1}{2}x\cos 2x - \frac{1}{4}\sin 2x + c$$

11. $\int e^x \sin 4x\,dx$

$$u = e^x \qquad dv = \sin 4x\,dx$$
$$du = e^x\,dx \qquad v = -\frac{1}{4}\cos 4x$$

$$= -\frac{1}{4}e^x\cos 4x - \int\left(-\frac{1}{4}\cos 4x\right)e^x\,dx$$
$$= -\frac{1}{4}e^x\cos 4x + \frac{1}{4}\int e^x\cos 4x\,dx$$

$$u = e^x \qquad dv = \cos 4x\,dx$$
$$du = e^x\,dx \qquad v = \frac{1}{4}\sin 4x$$

$$= -\frac{1}{4}e^x\cos 4x$$
$$+ \frac{1}{4}\left(\frac{1}{4}e^x\sin 4x - \int\frac{1}{4}(\sin 4x)e^x\,dx\right)$$
$$\int e^x\sin 4x\,dx = -\frac{1}{4}e^x\cos 4x$$
$$+ \frac{1}{16}e^x\sin 4x - \frac{1}{16}\int e^x\sin 4x\,dx, \text{ so}$$
$$\frac{17}{16}\int e^x\sin 4x\,dx = -\frac{1}{4}e^x\cos 4x$$
$$+ \frac{1}{16}e^x\sin 4x + K, \text{ so}$$
$$\int e^x\sin 4x\,dx$$
$$= -\frac{4}{17}e^x\cos 4x + \frac{1}{17}e^x\sin 4x + c$$

13. $\int \cos x\cos 2x\,dx$

$$u = \cos x \qquad dv = \cos 2x\,dx$$
$$du = \sin x\,dx \qquad v = \frac{1}{2}\sin 2x$$

$$= \frac{1}{2}\cos x\sin 2x - \int\frac{1}{2}\sin 2x(-\sin x)\,dx$$
$$= \frac{1}{2}\cos x\sin 2x + \frac{1}{2}\int\sin x\sin 2x\,dx$$

$$u = \sin x \qquad dv = \sin 2x\,dx$$
$$du = \cos x\,dx \qquad v = -\frac{1}{2}\cos 2x$$

$$= \frac{1}{2}\cos x\sin 2x$$
$$+ \frac{1}{2}\left[-\frac{1}{2}\cos 2x\sin x - \int\left(-\frac{1}{2}\cos 2x\right)\cos x\,dx\right]$$

$$\int \cos x \cos 2x\, dx = \frac{1}{2} \cos x \sin 2x$$
$$-\frac{1}{4} \cos 2x \sin x + \frac{1}{4} \int \cos x \cos 2x\, dx, \text{ so}$$
$$\frac{3}{4} \int \cos x \cos 2x\, dx = \frac{1}{2} \cos x \sin 2x$$
$$-\frac{1}{4} \cos 2x \sin x + K, \text{ so}$$
$$\int \cos x \cos 2x\, dx$$
$$= \frac{2}{3} \cos x \sin 2x - \frac{1}{3} \cos 2x \sin x + c$$

15. $\int x \sec^2 x\, dx \qquad u = x \qquad dv = \sec^2 x\, dx$

$$du = dx \qquad v = \tan x$$
$$= x \tan x - \int \tan x\, dx$$
$$= x \tan x - \int \frac{\sin x}{\cos x}\, dx$$

Let $u = \cos x,\ du = -\sin x\, dx$

$$= x \tan x + \int \frac{1}{u}\, du$$
$$= x \tan x + \ln |u| + c$$
$$= x \tan x + \ln |\cos x| + c.$$

17. $\int (\ln x)^2\, dx \qquad u = (\ln x)^2 \qquad dv = dx$

$$du = 2\frac{\ln x}{x}\, dx \qquad v = x$$
$$= x(\ln x)^2 - \int x \cdot 2\frac{\ln x}{x}\, dx$$
$$= x(\ln x)^2 - 2 \int \ln x\, dx \qquad u = \ln x \qquad dv = dx$$
$$du = \frac{1}{x}\, dx \qquad v = x$$
$$= x(\ln x)^2 - 2\left[x \ln x - \int x \cdot \frac{1}{x}\, dx\right]$$
$$= x(\ln x)^2 - 2x \ln x + 2 \int dx$$
$$= x(\ln x)^2 - 2x \ln x + 2x + c$$

19. $\int \cos x \ln(\sin x)\, dx$

$$u = \ln(\sin x) \qquad dv = \cos x\, dx$$
$$du = \frac{1}{\sin x} \cdot \cos x\, dx \qquad v = \sin x$$
$$= \sin x \ln(\sin x) - \int \sin x \cdot \frac{1}{\sin x} \cdot \cos x\, dx$$
$$= \sin x \ln(\sin x) - \int \cos x\, dx$$
$$= \sin x \ln(\sin x) - \sin x + c$$

21. It will be easier if we begin with the substitution

$$p = \sqrt{x},\ dp = \frac{1}{2\sqrt{x}}\, dx.$$

Then

$$\int \sqrt{x} \sin \sqrt{x}\, dx = 2 \int p^2 \sin p\,dp.$$

From Example 6.4, we know that

$$2 \int p^2 \sin p\,dp$$
$$= 2(-p^2 \cos p + 2p \sin p + 2 \cos p) + c.$$

Hence,

$$\int \sin \sqrt{x}\, dx$$
$$= 2(-x \cos \sqrt{x} + 2\sqrt{x} \sin \sqrt{x} + 2 \cos \sqrt{x}) + c.$$

23. $\int_0^1 x \sin 2x\, dx \qquad u = x \qquad dv = \sin 2x\, dx$

$$du = dx \qquad v = -\frac{1}{2} \cos 2x$$

$$= -\frac{1}{2}x\cos 2x\Big|_0^1 - \int_0^1 \left(-\frac{1}{2}\cos 2x\right)dx$$

$$= -\frac{1}{2}(1\cos 2 - 0\cos 0) + \frac{1}{2}\int_0^1 \cos 2x\,dx$$

$$= -\frac{1}{2}\cos 2 + \frac{1}{2}\left(\frac{1}{2}\sin 2x\right)\Big|_0^1$$

$$= -\frac{1}{2}\cos 2 + \frac{1}{4}(\sin 2 - \sin 0)$$

$$= -\frac{1}{2}\cos 2 + \frac{1}{4}\sin 2$$

25. $\int_0^1 x\cos \pi x\,dx \qquad u = x \qquad dv = \cos\pi x\,dx$

$$du = dx \qquad v = \frac{1}{\pi}\sin \pi x$$

$$= \frac{1}{\pi}x\sin\pi x\Big|_0^1 - \int_0^1 \frac{1}{\pi}\sin\pi x\,dx$$

$$= \frac{1}{\pi}\sin\pi - 0 + \frac{1}{\pi^2}\cos\pi x\Big|_0^1$$

$$= 0 + \frac{1}{\pi^2}(\cos\pi - \cos 0)$$

$$= \frac{1}{\pi^2}(-1 - 1)$$

$$= -\frac{2}{\pi^2}$$

27. $\int_0^1 x\sin\pi x\,dx \qquad u = x \qquad dv = \sin\pi x\,dx$

$$du = dx \qquad v = -\frac{1}{\pi}\cos\pi x$$

$$= -\frac{1}{\pi}x\cos\pi x\Big|_0^1 - \int_0^1\left(-\frac{1}{\pi}\cos\pi x\right)dx$$

$$= -\frac{1}{\pi}\cos\pi - 0 + \frac{1}{\pi^2}\sin\pi x\Big|_0^1$$

$$= -\frac{1}{\pi}(-1) + \frac{1}{\pi^2}(\sin\pi - \sin 0)$$

$$= \frac{1}{\pi}$$

29. $\int_1^{10} \ln x\,dx \qquad u = \ln x \qquad dv = dx$

$$du = \frac{1}{x}dx \qquad v = x$$

$$= x\ln x|_1^{10} - \int_1^{10} x\cdot\frac{1}{x}dx$$

$$= 10\ln 10 - 1\ln 1 - \int_1^{10} dx$$

$$= 10\ln 10 - 0 - x|_1^{10}$$

$$= 10\ln 10 - (10 - 1)$$

$$= 10\ln 10 - 9$$

31. $\int \cos^{-1}x\,dx \qquad u = \cos^{-1}x \qquad dv = dx$

$$du = -\frac{1}{\sqrt{1-x^2}}dx \qquad v = x$$

$$= x\cos^{-1}x - \int x\left(-\frac{1}{\sqrt{1-x^2}}\right)dx$$

$$= x\cos^{-1}x + \int\frac{x}{\sqrt{1-x^2}}dx$$

Let $u = 1 - x^2,\ du = -2x\,dx$

$$= x\cos^{-1}x + \int\frac{1}{\sqrt{u}}\left(-\frac{1}{2}du\right)$$

$$= x\cos^{1}x - \frac{1}{2}\int u^{-1/2}du$$

$$= x\cos^{-1}x - \frac{1}{2}\cdot 2u^{1/2} + c$$

$$= x\cos^{-1}x - \sqrt{1-x^2} + c.$$

33. It will be easier if we begin with the substitution

$$p = \sqrt{x},\ dp = \frac{1}{2\sqrt{x}}dx.$$

Then

$$\int \sin\sqrt{x}\,dx = 2\int p\sin p dp$$

From Example 6.4, we know that

$$2\int p\sin p dp = 2(-p\cos p + \sin p) + c.$$

Hence,

$$\int \sin \sqrt{x}\, dx = 2(-\sqrt{x} \cos \sqrt{x} + \sin \sqrt{x}) + c.$$

35. n times; each integration reduces the power of x by 1

37. $\int \cos^n x\, dx$

$$u = \cos^{n-1} x \qquad dv = \cos x\, dx$$
$$du = (n-1)(\cos^{n-2} x)(-\sin x)\, dx \qquad v = \sin x$$

$$= \sin x \cos^{n-1} x - \int (\sin x)(n-1)(\cos^{n-2} x)(-\sin x)\, dx$$

$$= \sin x \cos^{n-1} x + \int (n-1)(\cos^{n-2} x)(\sin^2 x)\, dx$$

$$= \sin x \cos^{n-1} x + \int (n-1)(\cos^{n-2} x)(1 - \cos^2 x)\, dx$$

$$= \sin x \cos^{n-1} x + \int (n-1)(\cos^{n-2} x - \cos^n x)\, dx.$$

Thus $\int \cos^n x\, dx$

$$= \sin x \cos^{n-1} x + \int (n-1) \cos^{n-2} x\, dx - (n-1) \int \cos^n x\, dx.$$ So

$$n \int \cos^n x\, dx = \sin x \cos^{n-1} x + (n-1) \int \cos^{n-2} x\, dx$$

$$\int \cos^n x\, dx = \frac{1}{n} \sin x \cos^{n-1} x + \frac{n-1}{n} \int \cos^{n-2} x\, dx.$$

To evaluate $\int \sin^2 x\, dx$, note that

$$\cos 2x = \cos^2 x - \sin^2 x = (1 - \sin^2 x) - \sin^2 x$$
$$= 1 - 2\sin^2 x, \text{ so } \sin^2 x = \frac{1}{2}(1 - \cos 2x).$$

Then $\int \sin^2 x = \frac{1}{2} \int (1 \cos 2x)\, dx$

$$= \frac{1}{2}\left(x - \frac{\sin 2x}{1}\right) + c.$$

39. $\int x^3 e^x\, dx$

Three applications of the formula in Equation (6.4) produce, after factoring out e^x,

$$e^x(x^3 - 3x^2 + 6x - 6) + c.$$

41. $$\int \cos^3 x\, dx = \frac{1}{3} \cos^2 x \sin x + \frac{2}{3} \int \cos x\, dx$$
$$= \frac{1}{3} \cos^2 x \sin x + \frac{2}{3} \sin x + c$$

43. Four applications of the formula in Equation (6.4) produce, after factoring out e^x,

$$e^x(x^4 - 4x^3 + 12x^2 - 24x + 24) + c;$$
$$e^x(x^4 - 4x^3 + 12x^2 - 24x + 24)\Big|_0^1 = 9e - 24.$$

45. $$\int_0^{\pi/2} \sin^5 x\, dx$$

$$= -\frac{1}{5} \sin^4 x \cos x \Big|_0^{\pi/2} + \frac{4}{5} \int_0^{\pi/2} \sin^3 x\, dx$$

$$= -\frac{1}{5} \sin^4 x \cos x \Big|_0^{\pi/2} + \frac{4}{5}\left(-\frac{1}{3} \sin^2 x \cos x - \frac{2}{3} \cos x\right)\Big|_0^{\pi/2}$$

(Using Exercise 38)

$$= -\frac{1}{5}\left(\sin^4\left(\frac{\pi}{2}\right)\cos\frac{\pi}{2} - \sin^4 0 \cos 0\right)$$

$$+\frac{4}{5}\left(-\frac{1}{3}\sin^2\left(\frac{\pi}{2}\right)\cos\frac{\pi}{2} - \frac{2}{3}\cos\frac{\pi}{2}\right)$$

$$-\frac{4}{5}\left(-\frac{1}{3}\sin^2 0\cos 0 - \frac{2}{3}\cos 0\right)$$

$$= -\frac{1}{5}(1\cdot 0 - 0\cdot 1)$$

$$+\frac{4}{5}\left[\left(-\frac{1}{3}\cdot 1\cdot 0 - \frac{2}{3}\cdot 0\right)\right.$$

$$\left. - \left(-\frac{1}{3}\cdot 0\cdot 1 - \frac{2}{3}\cdot 1\right)\right]$$

$$= \frac{4}{5}\left(\frac{2}{3}\right) = \frac{8}{15}$$

47. m even : $\dfrac{(m-1)(m-3)\ldots 1}{m(m-2)\ldots 2}\cdot\dfrac{\pi}{2}$;

m odd: $\dfrac{(m-1)(m-3)\ldots 2}{m(m-2)\ldots 3}$

49. First column: derivatives; second column: antiderivatives

51.

	$\cos x$	
x^4	$\sin x$	+
$4x^3$	$-\cos x$	−
$12x^2$	$-\sin x$	+
$24x$	$\cos x$	−
24	$\sin x$	+

$$\int x^4\cos x\,dx = x^4\sin x + 4x^3\cos x$$
$$-12x^2\sin x - 24x\cos x + 24\sin x + c$$

53.

	e^{2x}	
x^4	$\frac{1}{2}e^{2x}$	+
$4x^3$	$\frac{1}{4}e^{2x}$	−
$12x^2$	$\frac{1}{8}e^{2x}$	+
$24x$	$\frac{1}{16}e^{2x}$	−
24	$\frac{1}{32}e^{2x}$	+

$$\int x^4e^{2x}\,dx = \left(\frac{1}{2}x^4 - x^3 + \frac{3}{2}x^2 - \frac{3}{2}x + \frac{3}{4}\right)e^{2x} + c$$

55.

	e^{-3x}	
x^3	$-\frac{1}{3}e^{-3x}$	+
$3x^2$	$\frac{1}{9}e^{-3x}$	−
$6x$	$-\frac{1}{27}e^{-3x}$	+
6	$\frac{1}{81}e^{-3x}$	−

$$\int x^3e^{-3x}\,dx = \left(-\frac{1}{3}x^3 - \frac{1}{3}x^2 - \frac{2}{9}x - \frac{2}{27}\right)e^{-3x} + c$$

57. $\displaystyle\int_{-\pi}^{\pi}\cos(mx)\cos(nx)\,dx$

$u = \cos(mx) \qquad dv = \cos(nx)\,dx$

$du = -m\sin(mx) \qquad v = \frac{1}{n}\sin(nx)$

$$= \frac{1}{n}\cos(mx)\sin(nx)\Big|_{-\pi}^{\pi}$$

$$+\frac{m}{n}\int_{-\pi}^{\pi}\sin(mx)\sin(nx)\,dx$$

$u = \sin(mx) \qquad dv = \sin(nx)\,dx$

$du = m\cos(mx)\,dx \qquad v = -\frac{1}{n}\cos(nx)$

$$= \frac{1}{n}\left[\cos(m\pi)\sin(m\pi) - \cos(-m\pi)\sin(-m\pi)\right]$$

$$+ \frac{m}{n}\left[-\frac{1}{n}\cos(nx)\sin(mx)\bigg|_{-\pi}^{\pi}\right.$$

$$\left. + \frac{m}{n}\int_{-\pi}^{\pi}\cos(mx)\cos(nx)\,dx\right]$$

$$\left(1 - \frac{m^2}{n^2}\right)\int_{-\pi}^{\pi}\cos(mx)\cos(nx)\,dx$$

$$= 0 - \frac{m}{n^2}[\cos(n\pi)\sin(m\pi)$$

$$- \cos(-n\pi)\sin(-m\pi)]$$

$= 0(\sin(n\pi) = 0$ (when n is an integer).

Since $m^2 \neq n^2$, the integral equals 0.

$$\left(1 - \frac{m^2}{n^2}\right)\int_{-\pi}^{\pi}\cos(mx)\sin(nx)\,dx$$

$$= -\frac{1}{n}[\cos(m\pi)\cos(n\pi) - \cos(-m\pi)\cos(-n\pi)]$$

$$- \frac{m}{n^2}[\sin(m\pi)\sin(n\pi) - \sin(-m\pi)\sin(-n\pi)]$$

$= 0(\cos x = \cos(-x)$ and $\sin(n\pi) = 0$ (when n is an integer).

If the positive integers m and n are not equal, this shows the integral equals 0. On the other hand, if $m = n$, then one integration by parts gives

$$\int_{-\pi}^{\pi}\cos mx \sin mx\,dx = -\int_{-\pi}^{\pi}\cos mx \sin mx\,dx,$$

so again the integral equals 0.

59. $\displaystyle\int_{-\pi}^{\pi}\cos(mx)\sin(nx)\,dx$

$$u = \cos(mx) \qquad dv = \sin(nx)\,dx$$

$$du = -m\sin(mx)\,dx \qquad v = -\frac{1}{n}\cos(nx)$$

$$= -\frac{1}{n}\cos(mx)\cos(nx)\bigg|_{-\pi}^{\pi}$$

$$- \frac{m}{n}\int_{-\pi}^{\pi}\sin(mx)\cos(nx)\,dx$$

$$u = \sin(mx) \qquad dv = \cos(nx)\,dx$$

$$du = m\cos(mx)\,dx \qquad v = \frac{1}{n}\sin(nx)$$

$$= -\frac{1}{n}\cos(mx)\cos(nx)\bigg|_{-\pi}^{\pi}$$

$$- \frac{m}{n}\left[\frac{1}{n}\sin(mx)\sin(nx)\bigg|_{-\pi}^{\pi}\right.$$

$$\left. - \frac{m}{n}\int_{-\pi}^{\pi}\cos(mx)\sin(nx)\,dx\right]$$

61. If $c(t) = R$, then

$$\frac{RT}{\int_0^T c(t)\,dt} = \frac{RT}{\int_0^T R\,dt} = \frac{RT}{RT} = 1.$$

If $c(t) = 3te^{2Tt}$, then

$$\int_0^T 3te^{2Tt}\,dt \qquad u = 3t \qquad dv = e^{2Tt}dv$$

$$du = 3\,dt \qquad v = \frac{e^{2Tt}}{2T}$$

$$= 3t\frac{e^{2Tt}}{2T}\bigg|_0^T - \int_0^T 3\frac{e^{2Tt}}{2T}\,dt$$

$$= 3T\frac{e^{2T^2}}{2T} - \frac{3e^{2Tt}}{4T^2}\bigg|_0^T = \frac{3e^{2T^2}}{2} - \frac{3e^{2T^2}}{4T^2} + \frac{3}{4T^2}$$

$$= \frac{6T^2e^{2T^2} - 3e^{2T^2} + 3}{4T^2}.$$

Hence,

$$\frac{RT}{\displaystyle\int_0^T c(t)\,dt} = (RT)\left(\frac{4T^2}{6T^2e^{2T^2} - 3e^{2T^2} + 3}\right).$$

4.7 Other Techniques of Integration

1. $\int \frac{x-5}{x^2-1} = \frac{x-5}{(x+1)(x-1)} = \frac{A}{x+1} + \frac{B}{x-1}$

$x - 5 = A(x-1) + B(x+1)$

$x = -1 : -6 = -2A; A = 3$

$x = 1 : -4 = 2B; B = -2$

$\frac{x-5}{x^2-1} = \frac{3}{x+1} - \frac{2}{x-1}$

$\int \frac{x-5}{x^2-1} dx = \int \left(\frac{3}{x+1} \frac{2}{x-1} \right) dx$

$= 3 \ln |x+1| - 2 \ln |x-1| + c$

3. $\frac{6x}{x^2-x-2} = \frac{6x}{(x-2)(x+1)} = \frac{A}{x-2} + \frac{B}{x+1}$

$6x = A(x+1) + B(x-2)$

$x = 2 : 12 = 3A; A = 4$

$x = -1 : -6 = -3B; B = 2$

$\frac{6x}{x^2-x-2} = \frac{4}{x-2} + \frac{2}{x+1}$

$\int \frac{6x}{x^2-x-2} dx = \int \left(\frac{4}{x-2} + \frac{2}{x+1} \right) dx$

$= 4 \ln |x-2| + 2 \ln |x+1| + c$

5. $\frac{x+1}{x^2-x-6} = \frac{x+1}{(x-3)(x+2)} = \frac{A}{x-3} + \frac{B}{x+2}$

$x + 1 = A(x+2) + B(x-3)$

$x = 3 : 4 = 5A; A = \frac{4}{5}$

$x = -2 : -1 = -5B; B = \frac{1}{5}$

$\frac{x+1}{x^2-x-6} = \frac{\frac{4}{5}}{x-3} + \frac{\frac{1}{5}}{x+2}$

$\int \frac{x+1}{x^2-x-6} dx = \int \left(\frac{\frac{4}{5}}{x-3} + \frac{\frac{1}{5}}{x+2} \right) dx$

$= \frac{4}{5} \ln |x-3| + \frac{1}{5} \ln |x+2| + c$

7. $\frac{-x+5}{x^3-x^2-2x} = \frac{-x+5}{x(x^2-x-2)} = \frac{-x+5}{x(x-2)(x+1)}$

$= \frac{A}{x} + \frac{B}{x-2} + \frac{C}{x+1}$

$-x + 5 = A(x-2)(x+1)$

$+ Bx(x+1) + Cx(x-2)$

$x = 0 : 5 = -2A : A = -\frac{5}{2}$

$x = 2 : 3 = 6B : B = \frac{1}{2}$

$x = -1 : 6 = 3C : C = 2$

$\frac{-x+5}{x^3-x^2-2x} = -\frac{\frac{5}{2}}{x} + \frac{\frac{1}{2}}{x-2} + \frac{2}{x+1}$

$\int \frac{-x+5}{x^3-x^2-2x} dx$

$= \int \left(-\frac{\frac{5}{2}}{x} + \frac{\frac{1}{2}}{x-2} + \frac{2}{x+1} \right) dx$

$= -\frac{5}{2} \ln |x| + \frac{1}{2} \ln |x-2| + 2 \ln |x+1| + c$

9.

$$\begin{array}{r} x - 2 \\ x^2 + 2x - 8 \overline{\smash{\big)}\, x^3 \qquad\qquad + x + 2} \\ -\,(x^3 + 2x^2 - 8x) \\ \hline -2x^2 + 9x + 2 \\ -\,(2x^2 - 4x + 16) \\ \hline 13x - 14 \end{array}$$

$\frac{x^3+x+2}{x^2+2x-8} = x - 2 + \frac{13x-14}{x^2+2x-8}$

$= x - 2 + \frac{13x-14}{(x+4)(x-2)}$

$= x - 2 + \frac{A}{x+4} + \frac{B}{x-2}$

$13x - 14 = A(x-2) + B(x+4)$

$x = -4 : -66 = -6A; A = 11$

$x = 2 : 12 = 6B; B = 2$

$$\frac{x^3+x+2}{x^2+2x-8} = x-2+\frac{11}{x+4}+\frac{2}{x-2}$$

$$\int \frac{x^3+x+2}{x^2+2x-8}\,dx = \int \left(x-2+\frac{11}{x+4}+\frac{2}{x-2}\right)dx$$

$$= \frac{1}{2}x^2 - 2x + 11\ln|x+4| + 2\ln|x-2| + c$$

11. $$\frac{-3x-1}{x^3-x} = \frac{-3x-1}{x(x^2-1)} = \frac{-3x-1}{x(x+1)(x-1)}$$

$$= \frac{A}{x}+\frac{B}{x+1}+\frac{C}{x-1}$$

$$-3x-1 = A(x+1)(x-1) + Bx(x-1) + Cx(x+1)$$

$$x=0: -1=-A; A=1$$

$$x=-1: 2=2B; B=1$$

$$x=1: -4=2C; C=-2$$

$$\frac{-3x-1}{x^3-x} = \frac{1}{x}+\frac{1}{x+1}-\frac{2}{x-1}$$

$$\int \frac{-3x-1}{x^3-x}\,dx = \int\left(\frac{1}{x}+\frac{1}{x+1}-\frac{2}{x-1}\right)dx$$

$$= \ln|x| + \ln|x+1| - 2\ln|x-1| + c$$

13. $$\frac{2x+3}{(x+2)^2} = \frac{A}{x+2}+\frac{B}{(x+2)^2}$$

$$2x+3 = A(x+2)+B = Ax+2A+B$$

$$A=2; B=-1$$

$$\frac{2x+3}{(x+2)^2} = \frac{2}{x+2}-\frac{1}{(x+2)^2}$$

$$\int \frac{2x+3}{(x+2)^2}\,dx = \int\left(\frac{2}{x+2}-\frac{1}{(x+2)^2}\right)dx$$

$$= 2\ln|x+2| + \frac{1}{x+2} + c$$

15. $$\frac{x-1}{x^3+4x^2+4x} = \frac{x-1}{x(x^2+4x+4)} = \frac{x-1}{x(x+2)^2}$$

$$= \frac{A}{x}+\frac{B}{x+2}+\frac{C}{(x+2)^2}$$

$$x-1 = A(x+2)^2 + Bx(x+2) + Cx$$

$$x=0: -1=4A; A=-\frac{1}{4}$$

$$x=-2: -3=-2C; C=\frac{3}{2}$$

$$x=1: 0=9A+3B+C = -\frac{9}{4}+3B+\frac{3}{2}; B=\frac{1}{4}$$

$$\frac{x-1}{x^3+4x^2+4x} = -\frac{\frac{1}{4}}{x}+\frac{\frac{1}{4}}{x+2}+\frac{\frac{3}{2}}{(x+2)^2}$$

$$\int \frac{x-1}{x^3+4x^2+4x}\,dx$$

$$= \int\left(-\frac{\frac{1}{4}}{x}+\frac{\frac{1}{4}}{x+2}+\frac{\frac{3}{2}}{(x+2)^2}\right)dx$$

$$= -\frac{1}{4}\ln|x| + \frac{1}{4}\ln|x+2| - \frac{3}{2(x+2)} + c$$

17. $$\frac{x+4}{x^3+3x^2+2x} = \frac{x+4}{x(x^2+3x+2)} = \frac{x+4}{x(x+2)(x+1)}$$

$$= \frac{A}{x}+\frac{B}{x+2}+\frac{C}{x+1}$$

$$x+4 = A(x+2)(x+1) + Bx(x+1) + Cx(x+2)$$

$$x=0: 4=2A; A=2$$

$$x=-2: 2=2B; B=1$$

$$x=-1: 3=-C; C=-3$$

$$\frac{x+4}{x^3+3x^2+2x} = \frac{2}{x}+\frac{1}{x+2}-\frac{3}{x+1}$$

$$\int \frac{x+4}{x^3+3x^2+2x}\,dx = \int\left(\frac{2}{x}+\frac{1}{x+2}-\frac{3}{x+1}\right)dx$$

$$= 2\ln|x| + \ln|x+2| - 3\ln|x+1| + c$$

19. $$\frac{x+2}{x^3+x} = \frac{x+2}{x(x^2+1)} = \frac{A}{x}+\frac{Bx+C}{x^2+1}$$

$$x+2 = A(x^2+1) + (Bx+C)x$$

$$= Ax^2 + A + Bx^2 + Cx = (A+B)x^2 + Cx + A$$

$$A=2; C=1; B=-2$$

$$\frac{x+2}{x^3+x} = \frac{2}{x}+\frac{-2x+1}{x^2+1}$$

$$\int \frac{x+2}{x^3+x}\,dx = \int \left(\frac{2}{x} + \frac{-2x+1}{x^2+1}\right) dx$$
$$= \int \left(\frac{2}{x} - \frac{2x}{x^2+1} + \frac{1}{x^2+1}\right) dx$$
$$= 2\ln|x| - \ln(x^2+1) + \tan^{-1} x + c$$

21. $$\frac{4x-2}{x^4-1} = \frac{4x-2}{(x^2+1)(x^2-1)} = \frac{4x-2}{(x^2+1)(x+1)(x-1)}$$
$$= \frac{Ax+B}{x^2+1} + \frac{C}{x+1} + \frac{D}{x-1}$$
$$4x-2 = (Ax+B)(x+1)(x-1) + C(x^2+1)(x-1) + D(x^2+1)(x+1)$$
$$4x-2 = Ax^3 - Ax + Bx^2 - B + Cx^3 - Cx^2 + Cx - C + Dx^3 + Dx^2 + Dx + D$$
$$x = 1: 2 = 4D;\ D = \frac{1}{2}$$
$$x = -1: -6 = -4C;\ C = \frac{3}{2}$$
$$A + C + D = 0: A = -2$$
$$B - C + D = 0: B = 1$$
$$\frac{4x-2}{x^4-1} = \frac{-2x+1}{x^2+1} + \frac{\frac{3}{2}}{x+1} + \frac{\frac{1}{2}}{x-1}$$
$$\int \frac{4x-2}{x^4-1}\,dx = \int \left(\frac{-2x+1}{x^2+1} + \frac{\frac{3}{2}}{x+1} + \frac{\frac{1}{2}}{x-1}\right) dx$$
$$= \int \left(-\frac{2x}{x^2+1} + \frac{1}{x^2+1} + \frac{\frac{3}{2}}{x+1} + \frac{\frac{1}{2}}{x-1}\right) dx$$
$$= -\ln(x^2+1) + \tan^{-1} x + \frac{3}{2}\ln|x+1| + \frac{1}{2}\ln|x-1| + c$$

23. $$\frac{3x^2-6}{x^2-x-2}$$

$$\begin{array}{r} 3 \\ x^2-x-2 \overline{\smash{)}\,3x^2 \qquad -6} \\ -\,(3x^3-3x-6) \\ \hline 3x \end{array}$$

$$\frac{3x^2-6}{x^2-x-2} = 3 + \frac{3x}{x^2-x-2}$$
$$= 3 + \frac{3x}{(x-2)(x+1)}$$
$$= 3 + \frac{A}{x-2} + \frac{B}{x+1}$$
$$3x = A(x+1) + B(x-2)$$
$$x = -1: -3 = -3B;\ B = 1$$
$$x = 2: 6 = 3A;\ A = 2$$
$$\frac{3x^2-6}{x^2-x-2} = 3 + \frac{2}{x-2} + \frac{1}{x+1}$$
$$\int \frac{3x^2-6}{x^2-x-2}\,dx = \int \left(3 + \frac{2}{x-2} + \frac{1}{x+1}\right) dx$$
$$= 3x + 2\ln|x-2| + \ln|x+1| + c$$

25. $$\frac{2x+3}{x^2+2x+1} = \frac{2x+3}{(x+1)^2} = \frac{A}{x+1} + \frac{B}{(x+1)^2}$$
$$2x+3 = A(x+1) + B$$
$$x = -1: B = 1;\ A = 2$$
$$\frac{2x+3}{x^2+2x+1} = \frac{2}{x+1} + \frac{1}{(x+1)^2}$$
$$\int \frac{2x+3}{x^2+2x+1}\,dx = \int \left(\frac{2}{x+1} + \frac{1}{(x+1)^2}\right) dx$$
$$= 2\ln|x+1| - \frac{1}{x+1} + c$$

27. $$\frac{x^2+2x+1}{x^3+x} = \frac{x^2+2x+1}{x(x^2+1)} = \frac{A}{x} + \frac{Bx+C}{x^2+1}$$
$$x^2+2x+1 = A(x^2+1) + (Bx+C)x$$
$$x^2+2x+1 = Ax^2 + A + Bx^2 + Cx$$
$$A = 1;\ C = 2;\ B = 0$$
$$\int \frac{x^2+2x+1}{x^3+x} = \frac{1}{x} + \frac{2}{x^2+1}$$
$$\int \frac{x^2+2x+1}{x^3+x}\,dx = \int \left(\frac{1}{x} + \frac{2}{x^2+1}\right) dx$$
$$= \ln|x| + 2\tan^{-1} x + c$$

29. $\dfrac{4x^2+3}{x^3+x^2+x} = \dfrac{4x^2+3}{x(x^2+x+1)} = \dfrac{A}{x} + \dfrac{Bx+C}{x^2+x+1}$

$$4x^2+3 = A(x^2+x+1)+(Bx+C)x$$
$$= Ax^2+Ax+A+Bx^2+Cx$$
$$A=3;\ C=-3;\ B=1$$

$$\frac{4x^2+3}{x^3+x^2+x} = \frac{3}{x} + \frac{x-3}{x^2+x+1}$$

$$\int \frac{4x^2+3}{x^3+x^2+x}\,dx$$
$$= \int \left(\frac{3}{x} + \frac{x-3}{x^2+x+1}\right) dx$$
$$= \int \left(\frac{3}{x} + \frac{x+\frac{1}{2}}{x^2+x+1} - \frac{\frac{7}{2}}{x^2+x+1}\right) dx$$
$$= 3\ln|x| + \frac{1}{2}\ln\left|x^2+x+1\right|$$
$$-\frac{7}{\sqrt{3}}\tan^{-1}\left(\frac{2x+1}{\sqrt{3}}\right)+c$$

31.

$$\begin{array}{r} 3 \\ x^3-x^2+x-1\overline{)\,3x^3 \qquad\qquad +1} \\ -(3x^3-3x^2+3x-3) \\ \hline 3x^2-3x+4 \end{array}$$

$$\frac{3x^3+1}{x^3-x^2+x-1} = 3+\frac{3x^2-3x+4}{x^3-x^2+x-1}$$
$$= 3+\frac{3x^2-3x+4}{x^2(x-1)+x-1}$$
$$= 3+\frac{3x^2-3x+4}{(x^2+1)(x-1)}$$
$$= 3+\frac{Ax+B}{x^2+1}+\frac{C}{x-1}$$

$$3x^2-3x+4 = (Ax+B)(x-1)+C(x^2+1)$$
$$= Ax^2-Ax+Bx-B+Cx^2+C$$
$$x=1: 4=2C;\ C=2$$
$$A+C=3: A=1$$
$$-A+B=-3: B=-2$$

$$\frac{3x^3+1}{x^3-x^2+x-1} = 3+\frac{x-2}{x^2+1}+\frac{2}{x-1}$$
$$\int \frac{3x^3+1}{x^3-x^2+x-1}\,dx$$
$$= \int\left(3+\frac{x-2}{x^2+1}+\frac{2}{x-1}\right)dx$$
$$= \int\left(3+\frac{x}{x^2+1}-\frac{2}{x^2+1}+\frac{2}{x-1}\right)dx$$
$$= 3x+\frac{1}{2}\ln(x^2+1)-2\tan^{-1}x+2\ln|x-1|+c$$

33.
$$A+C=0: C=0$$
$$B+D=2: D=-2$$
$$\frac{4x^2+2}{(x^2+1)^2} = \frac{4}{x^2+1}-\frac{2}{(x^2+1)^2}$$

35. $\dfrac{2x^2+4}{(x^2+4)^2} = \dfrac{Ax+B}{x^2+4}+\dfrac{Cx+D}{(x^2+4)^2}$

$$2x^2+4 = (Ax+B)(x^2+4)+Cx+D$$
$$= Ax^3+4Ax+Bx^2+4B+Cx+D$$
$$A=0;\ B=2$$
$$4A+C=0: C=0$$
$$4B+D=4: D=-4$$
$$\frac{2x^2+4}{(x^2+4)^2} = \frac{2}{x^2+4}-\frac{4}{(x^2+4)^2}$$

37. $\dfrac{4x^2+3}{(x^2+x+1)^2} = \dfrac{Ax+B}{x^2+x+1}+\dfrac{Cx+D}{(x^2+x+1)^2}$

$$4x^2+3 = (Ax+B)(x^2+x+1)+Cx+D$$
$$= Ax^3+Ax^2+Ax+Bx^2+Bx+B+Cx+D$$
$$A=0$$
$$A+B=4: B=4$$
$$A+B+C=0: C=-4$$
$$B+D=3: D=-1$$
$$\frac{4x^2+3}{(x^2+x+1)^2} = \frac{4}{x^2+x+1}-\frac{4x+1}{(x^2+x+1)^2}$$

39. $u=x^3+1,\ du=3x^2\,dx$

$$\int\frac{3}{x^4+x}\,dx = \int\frac{3}{x(x^3+1)}\,dx$$
$$= \int\frac{1}{(u-1)u}\,du = \int\frac{1}{u-1}-\frac{1}{u}\,du$$

$$= \ln|u-1| - \ln|u| + c$$
$$= \ln\left|\frac{u-1}{u}\right| + c = \ln\left|\frac{x^3}{x^3+1}\right| + c$$
$$u = \frac{1}{x}, \quad du = -\frac{1}{x^2}\,dx$$
$$\int \frac{3}{x^4+x}\,dx = -\int \frac{3u^2}{1+u^3}\,du$$
$$= -\ln|1+u^3| + c = -\ln|1+(1/x)^3| + c$$

To see that the two answers are equivalent, note that

$$\ln\left|\frac{x^3}{x^3+1}\right| = -\ln\left|\frac{x^3+1}{x^3}\right| = -\ln|1+(1/x)^3|.$$

41. $$\frac{1}{y(1-y)} = \frac{A}{y} + \frac{B}{1-y}$$
$$1 = A(1-y) + By$$
$$A = 1; \; B = 1$$
$$\frac{1}{y(1-y)} = \frac{1}{y} + \frac{1}{1-y}$$
$$\int \frac{1}{y(1-y)}dy = \int \left(\frac{1}{y} + \frac{1}{1-y}\right) dy$$
$$= \ln|y| - \ln|1-y| + c$$
$$= \ln\frac{y}{1-y} + c$$
$$at - c = \ln\frac{y}{1-y}$$
$$e^{at-c} = \frac{y}{1-y}$$
$$e^{c-at} = \frac{1-y}{y} = \frac{1}{y} - 1$$
$$1 + e^{c-at} = \frac{1}{y}$$
$$\frac{1}{1+e^{c-at}} = y$$
$$y = \frac{ce^{at}}{1+ce^{at}}$$

43. $\int \cos x \sin^4 x\,dx$
Let $u = \sin x, \; du = \cos x\,dx$.

$$\int u^4\,du = \frac{1}{5}u^5 + c$$
$$= \frac{1}{5}\sin^5 x + c$$

45. $\int_0^{\pi/4} \cos x \sin^3 x\,dx$
Let $u = \sin x, \; du = \cos x\,dx$.
The new limits are $\sin 0 = 0$ and $\sin(\pi/4) = 1/\sqrt{2}$.

$$\int_0^{1/\sqrt{2}} u^3\,du = \frac{1}{4}u^4\Big|_0^{1/\sqrt{2}}$$
$$= \frac{1}{4\cdot(\sqrt{2})^4}$$
$$= \frac{1}{16}$$

47. $$\int \cos^2 x\,dx = \frac{1}{2}\int (1+\cos 2x)\,dx$$
$$= \frac{1}{2}x + \frac{1}{4}\sin 2x + c$$

49. $\int \tan x \sec^3 x\,dx = \int \tan x \sec x \sec^2 x\,dx$
Let $u = \sec x, \; du = \sec x \tan x\,dx$.

$$\int u^2\,du = \frac{1}{3}u^3 + c$$
$$= \frac{1}{3}\sec^3 x + c$$

51. $$\int_0^{\pi/4} \tan^4 x \sec^4 x\,dx$$
$$= \int_0^{\pi/4} \tan^4 x \sec^2 x \sec^2 x\,dx$$
$$= \int_0^{\pi/4} \tan^4 x(1+\tan^2 x)\sec^2 x\,dx$$

Let $u = \tan x, \; du = \sec^2 x\,dx$.
The new limits are $\tan 0 = 0$ and $\tan(\pi/4) = 1$.

$$\int_0^1 u^4(1+u^2)\,du = \int_0^1 u^4 + u^6\,du$$
$$\frac{u^5}{5} + \frac{u^7}{7}\Big|_0^1 = \frac{1}{5} + \frac{1}{7} = \frac{12}{35}$$

53. $\int \cos^2 x \sin x\, dx$

Let $u = \cos x$, $du = -\sin x\, dx$.

$$\int u^2(-du) = -\frac{1}{3}u^3 + c$$
$$= -\frac{1}{3}\cos^3 x + c$$

55. $$\int \cos^2 x \sin^2 x\, dx$$
$$= \int \frac{1}{2}(1+\cos 2x)\cdot\frac{1}{2}(1-\cos 2x)\, dx$$
$$= \frac{1}{4}\int (1-\cos^2 2x)\, dx$$
$$= \frac{1}{4}\int \left[1 - \frac{1}{2}(1+\cos 4x)\right] dx$$
$$= \frac{1}{4}\left(\frac{1}{2}x - \frac{1}{8}\sin 4x\right) + c$$
$$= \frac{1}{8}x - \frac{1}{32}\sin 4x + c$$

57. $$\int_{-\pi/3}^{0} \sqrt{\cos x}\sin^3 x\, dx$$
$$\int_{-\pi/3}^{0} \sqrt{\cos x}(1-\cos^2 x)\sin x\, dx$$

Let $u = \cos x$, $du = -\sin x\, dx$.
The new limits are $\cos(-\pi/3) = 1/2$ and $\cos 0 = 1$.

$$\int_{1/2}^{1} \sqrt{u}(1-u^2)(-du) = \int_{1/2}^{1}(u^{5/2} - u^{1/2})du$$
$$= \frac{2}{7}u^{7/2} - \frac{2}{3}u^{3/2}\Big|_{1/2}^{1}$$
$$= \frac{25}{168}\sqrt{2} - \frac{8}{21}$$

$$\int \cot^4 x \csc^4 x\, dx = \int \cot^4 x \csc^2 x \csc^2 x\, dx$$
$$\int \cot^4 x(1+\cot^2 x)\csc^2 x\, dx$$

$u = \cot x$, $du = -\csc^2 x\, dx$

$$-\int u^4(1+u^2)\, du = -\int u^4 + u^6\, du$$
$$= -\frac{u^5}{5} - \frac{u^7}{7} + c = -\frac{\cot^5 x}{5} - \frac{\cot^7 x}{7} + c$$

61. $\int \frac{1}{x^2\sqrt{16-x^2}}\, dx$

Let $x = 4\sin\theta$, $-\frac{\pi}{2} < \theta < \frac{\pi}{2}$;
$dx = 4\cos\theta d\theta$.

$$\int \frac{4\cos\theta}{16\sin^2\theta\sqrt{16-16\sin^2\theta}}d\theta$$
$$= \int \frac{\cos\theta}{16\sin^2\theta\sqrt{1-\sin^2\theta}}d\theta$$
$$= \int \frac{\cos\theta}{16\sin^2\theta\cos\theta}d\theta$$
$$= \frac{1}{16}\int \csc^2\theta d\theta$$
$$= -\frac{1}{16}\cot\theta + c$$

$$\cot\theta = \frac{\cos\theta}{\sin\theta} = \frac{\sqrt{16-x^2}}{x}$$

$$\int \frac{1}{x^2\sqrt{9-x^2}}\, dx = -\frac{1}{16}\cot\theta + c$$
$$= -\frac{1}{16}\frac{\sqrt{16-x^2}}{x} + c$$

63. $u = 4 - x^2$, $du = -2x\, dx$

$$\int_0^1 \frac{x}{\sqrt{4-x^2}}\, dx = -\int_4^3 \frac{du}{2\sqrt{u}}$$
$$= -u^{1/2}\Big|_4^3 = -(\sqrt{3}-2) = 2-\sqrt{3}$$

65. $u = x^2 - 1$, $du = 2x\, dx$

$$\int x^3\sqrt{x^2-1}\, dx = \frac{1}{2}\int x^2\sqrt{x^2-1}(2x)\, dx$$
$$= \frac{1}{2}\int (u+1)\sqrt{u}\, du = \frac{1}{2}\int u^{3/2} + u^{1/2}\, du$$

$$= \frac{1}{2}\left(\frac{2u^{5/2}}{5} + \frac{2u^{3/2}}{3}\right) + c$$
$$= \frac{1}{5}(x^2 - 1)^{5/2} + \frac{1}{3}(x^2 - 1)^{3/2} + c$$

67. $\int \sqrt{x^2 + 16}\, dx$

Let $x = 4 \tan \theta$, $-\frac{\pi}{2} < \theta < \frac{\pi}{2}$;
$dx = 4 \sec^2 \theta d\theta$.

$$\int \sqrt{16 \tan^2 \theta + 16} \cdot 4 \sec^2 \theta d\theta$$
$$= 16 \int \sqrt{\tan^2 \theta + 1} \sec^2 \theta d\theta$$
$$= 16 \int \sec^3 \theta d\theta$$
$$= 16\left(\frac{1}{2} \sec \theta \tan \theta + \frac{1}{2} \int \sec \theta d\theta\right)$$
$$= 8 \sec \theta \tan \theta + 8 \int \sec \theta d\theta$$
$$= 8 \sec \theta \tan \theta + 8 \ln |\sec \theta + \tan \theta| + c$$
$$\tan \theta = \frac{x}{4},\ \sec \theta = \sqrt{1 + \left(\frac{x}{4}\right)^2} = \frac{1}{4}\sqrt{16 + x^2}$$
$$\int \sqrt{x^2 + 16}\, dx$$
$$= 8 \sec \theta \tan \theta + 8 \ln |\sec \theta + \tan \theta| + c$$
$$= \frac{1}{2} x \sqrt{16 + x^2} + 8 \ln \left(\frac{1}{4}\sqrt{16 + x^2} + \frac{x}{4}\right) + c$$

69. $\int \tan x \sec^4 x\, dx$

Let $u = \tan x,\ du = \sec^2 x\, dx$.

$$\int \tan x (1 + \tan^2 x) \sec^2 x\, dx$$
$$= \int u(1 + u^2)\, du$$
$$= \int (u + u^3)\, du$$
$$= \frac{1}{2}u^2 + \frac{1}{4}u^4 + c$$
$$= \frac{1}{2} \tan^2 x + \frac{1}{4} \tan^4 x + c$$

Let $u = \sec x,\ du = \sec x \tan x\, dx$.

$$\int \tan x \sec x \sec^3 x\, dx = \int u^3\, du$$
$$= \frac{1}{4}u^4 + c = \frac{1}{4} \sec^4 x + c$$

71. $\frac{1}{\frac{2\pi}{\omega}} \int_0^{2\pi/\omega} RI^2 \cos^2(\omega t)\, dt$

$$= \frac{\omega RI^2}{2\pi} \int_0^{2\pi/\omega} \frac{1}{2} [1 + \cos(2\omega t)]\, dt$$
$$= \frac{\omega RI^2}{4\pi} \left[t + \frac{1}{2\omega} \sin(2\omega t)\right]\Bigg|_0^{2\pi/\omega}$$
$$= \frac{\omega RI^2}{4\pi} \left[\frac{2\pi}{\omega} + \frac{1}{2\omega} \sin\left(\frac{4\omega\pi}{\omega}\right) - 0\right]$$
$$= \frac{1}{2}RI^2$$

4.8 Integration Tables and Computer Algebra Systems

1. $\int \frac{x}{(2 + 4x)^2}\, dx$

$$\int \frac{u\, du}{(a + bu)^2} = \frac{a}{b^2(a + bu)} + \frac{1}{b^2} \ln |a + bu| + c$$
$$\int \frac{x}{(2 + 4x)^2}\, dx = \frac{2}{16(2 + 4x)} + \frac{1}{16} \ln |2 + 4x| + c$$
$$= \frac{1}{8(2 + 4x)} + \frac{1}{16} \ln |2 + 4x| + c$$

3. $\int e^{2x} \sqrt{1 + e^x}\, dx$

Let $u = 1 + e^x,\ du = e^x\, dx,\ e^x = u - 1$.

$$\int (u - 1)\sqrt{u}\, du = \int (u^{3/2} - u^{1/2})\, du$$
$$= \frac{2}{5}u^{5/2} - \frac{2}{3}u^{3/2} + c$$
$$= \frac{2}{5}(1 + e^x)^{5/2} - \frac{2}{3}(1 + e^x)^{3/2} + c$$

5. $\displaystyle\int \frac{x^2}{\sqrt{1+4x^2}}\,dx$

Let $u = 2x,\ du = 2\,dx,\ x = \frac{u}{2}$.

$$\int \frac{\left(\frac{u}{2}\right)^2}{\sqrt{1+u^2}} \cdot \frac{1}{2}\,du = \frac{1}{8}\int \frac{u^2}{\sqrt{1+u^2}}\,du$$

From the table,

$$\int \frac{u^2\,du}{\sqrt{a^2+u^2}}$$
$$= \frac{u}{2}\sqrt{a^2+u^2} - \frac{a^2}{2}\ln\left(u+\sqrt{a^2+u^2}\right) + c;$$
$$\frac{1}{8}\int \frac{u^2}{\sqrt{1+u^2}}\,du$$
$$= \frac{1}{8}\left[\frac{u}{2} - \sqrt{1+u^2} - \frac{1}{2}\ln(u+\sqrt{1+u^2})\right] + c$$
$$= \frac{1}{8}x\sqrt{1+4x^2} - \frac{1}{16}\ln(2x+\sqrt{1+4x^2}) + c.$$

7. $\displaystyle\int x^8\sqrt{4-x^6}\,dx$

Let $u = x^3,\ du = 3x^2\,dx$

$$\int u^2\sqrt{4-u^2}\cdot\frac{1}{3}\,du = \frac{1}{3}\int u^2\sqrt{4-u^2}\,du$$

From the table,

$$\int u^2\sqrt{a^2-u^2}\,du$$
$$= \frac{u}{8}(2u^2-a^2)\sqrt{a^2-u^2} + \frac{a^4}{8}\sin^{-1}\frac{u}{a} + c;$$
$$\frac{1}{3}\int u^2\sqrt{4-u^2}\,du$$
$$= \frac{1}{3}\left[\frac{u}{8}(2u^2-4)\sqrt{4-u^2} + \frac{16}{8}\sin^{-1}\frac{u}{2}\right] + c$$
$$= \frac{1}{24}x^3(2x^6-4)\sqrt{4-x^6} + \frac{2}{3}\sin^{-1}\frac{x^3}{2} + c.$$

Using a CAS to evaluate the definite integral, we get

$$\int_0^1 x^8\sqrt{4-x^6}\,dx = \frac{\pi}{9} - \frac{\sqrt{3}}{12}.$$

9. $\displaystyle\int \frac{e^x}{\sqrt{e^{2x}+4}}\,dx$

Let $u = e^x,\ du = e^x\,dx$.

$$\int \frac{1}{\sqrt{u^2+4}}\,du$$

From the table,

$$\int \frac{du}{\sqrt{a^2+u^2}} = \ln(u+\sqrt{a^2+u^2}) + c;$$
$$\int \frac{1}{\sqrt{u^2+4}}\,du = \ln(u+\sqrt{4+u^2}) + c$$
$$= \ln(e^x+\sqrt{4+e^{2x}}) + c.$$

Using a CAS to evaluate the definite integral, we get

$$\int_0^{\ln 2} \frac{e^x}{\sqrt{e^{2x}+4}}\,dx = \ln\left(\frac{2\sqrt{2}+2}{1+\sqrt{5}}\right).$$

11. $$\int \frac{\sqrt{6x-x^2}}{(x-3)^2}\,dx = \int \frac{\sqrt{x(6-x)}}{(x-3)^2}\,dx$$

Let $u = x-3,\ du = dx,\ x = u+3$.

$$\int \frac{\sqrt{(u+3)(6-(u+3))}}{u^2}\,du$$
$$= \int \frac{\sqrt{(u+3)(3-u)}}{u^2}\,du$$
$$= \int \frac{\sqrt{9-u^2}}{u^2}\,du$$

From the table,

$$\int \frac{\sqrt{a^2-u^2}}{u^2}\,du$$
$$= -\frac{1}{u}\sqrt{a^2-u^2} - \sin^{-1}\frac{u}{a} + c;$$
$$\int \frac{\sqrt{9-u^2}}{u^2}\,du = -\frac{1}{u}\sqrt{9-u^2} - \sin^{-1}\frac{u}{3} + c$$
$$= -\frac{1}{x-3}\sqrt{9-(x-3)^2} - \sin^{-1}\left(\frac{x-3}{3}\right) + c.$$

13. $\int \tan^6 x\,dx$

$$\int \tan^n u\,du = \frac{1}{n-1}\tan^{n-1} u - \int \tan^{n-2} u\,du$$

$$\int \tan^6 x\,dx = \frac{1}{5}\tan^5 x - \int \tan^4 x\,dx$$

$$= \frac{1}{5}\tan^5 x - \left[\frac{1}{3}\tan^3 x - \int \tan^2 x\,dx\right]$$

$$= \frac{1}{5}\tan^5 x - \frac{1}{3}\tan^3 x + \tan x - x + c$$

(Using the formula for $\int \tan^2 x\,dx$.)

15. $\int \frac{\cos x}{\sin x\sqrt{4+\sin x}}\,dx$

Let $u = \sin x,\ du = \cos x\,dx$.

$$\int \frac{1}{u\sqrt{4+u}}\,du$$

From the table,

$$\int \frac{du}{u\sqrt{a+bu}} = \frac{1}{\sqrt{a}}\ln\left|\frac{\sqrt{a+bu}-\sqrt{a}}{\sqrt{a+bu}+\sqrt{a}}\right| + c;$$

$$\int \frac{1}{u\sqrt{4+u}}\,du = \frac{1}{\sqrt{4}}\ln\left|\frac{\sqrt{4+u}-2}{\sqrt{4+u}+2}\right| + c$$

$$= \frac{1}{2}\ln\left|\frac{\sqrt{4+\sin x}-2}{\sqrt{4+\sin x}+2}\right| + c.$$

17. $\int x^3\cos x^2\,dx$

Let $u = x^2,\ du = 2x\,dx$.

$$\int u\cos u\cdot\frac{1}{2}\,du = \frac{1}{2}\int u\cos u\,du$$

From the table,

$$\int u\cos u\,du = \cos u + u\sin u + c;$$

$$\frac{1}{2}\int u\cos u\,du = \frac{1}{2}(\cos u + u\sin u) + c$$

$$= \frac{1}{2}\cos x^2 + \frac{1}{2}x^2\sin x^2 + c.$$

19. $\int \frac{\sin x\cos x}{\sqrt{1+\cos x}}\,dx$

Let $u = \cos x,\ du = -\sin x\,dx$.

$$-\int \frac{u}{\sqrt{1+u}}\,du$$

From the table,

$$\int \frac{u\,du}{\sqrt{a+bu}} = \frac{2}{3b^2}(bu-2a)\sqrt{a+bu} + c;$$

$$-\int \frac{u}{\sqrt{1+u}}\,du = -\frac{2}{3}(u-2)\sqrt{1+u} + c$$

$$= -\frac{2}{3}(\cos x - 2)\sqrt{1+\cos x} + c.$$

21. $\int \frac{\sin^2 x\cos x}{\sqrt{\sin^2 x + 4}}\,dx$

Let $u = \sin x,\ du = \cos x\,dx$.

$$\int \frac{u^2}{\sqrt{u^2+4}}\,du$$

From the table,

$$\int \frac{u^2\,du}{\sqrt{a^2+u^2}}$$

$$= \frac{u}{2}\sqrt{a^2+u^2} - \frac{a^2}{2}\ln(u + \sqrt{a^2+u^2}) + c;$$

$$\int \frac{u^2}{\sqrt{u^2+4}}\,du$$

$$= \frac{u}{2}\sqrt{4+u^2} - \frac{4}{2}\ln(u + \sqrt{4+u^2}) + c$$

$$= \frac{1}{2}\sin x\sqrt{4+\sin^2 x}$$

$$- 2\ln(\sin x + \sqrt{4+\sin^2 x}) + c.$$

23. $\int \frac{e^{-2/x^2}}{x^3}\,dx$

Let $u = -\frac{2}{x^2},\ du = \frac{4}{x^3}\,dx$.

$$\frac{1}{4}\int e^u\,du = \frac{1}{4}e^u + c$$

$$= \frac{1}{4}e^{-2/x^2} + c$$

25. $\int \frac{x}{\sqrt{4x-x^2}}\,dx$

From the table,

$$\int \frac{u\,du}{\sqrt{2au-u^2}}$$
$$= -\sqrt{2au-u^2} + a\cos^{-1}\left(\frac{a-u}{a}\right) + c;$$
$$\int \frac{x}{\sqrt{4x-x^2}}\,dx = -\sqrt{4x-x^2} + 2\cos^{-1}\left(\frac{2-x}{2}\right) + c.$$

27. $\int e^x \tan^{-1}(e^x)\,dx$

Let $u = e^x$, $du = e^x\,dx$.

$$\int \tan^{-1} u\,du = u\tan^{-1} u - \frac{1}{2}\ln(1+u^2) + c$$
$$= e^x \tan^{-1} e^x - \frac{1}{2}\ln(1+e^{2x}) + c$$

29. Answer depends on CAS used.

31. Any answer is wrong because the integrand is undefined for all $x \neq 1$.

33. Answer depends on CAS used.

35. Answer depends on CAS used.

4.9 Numerical Integration

1. $\int_0^1 \left(x^2+1\right)\,dx$

Midpoint:

$$\frac{1}{4}\left[f\left(\frac{1}{8}\right)+f\left(\frac{3}{8}\right)+f\left(\frac{5}{8}\right)+f\left(\frac{7}{8}\right)\right]$$
$$= \frac{85}{64}.$$

Trapezoidal:

$$T_4(f)$$
$$= \frac{1-0}{2(4)}\left[f(0)+2f\left(\frac{1}{4}\right)+2f\left(\frac{1}{2}\right)\right.$$
$$\left.+2f\left(\frac{3}{4}\right)+f(1)\right]$$
$$= \frac{43}{32}.$$

Simpson:

$$S_4(f)$$
$$= \frac{1-0}{3(4)}\left[f(0)+4f\left(\frac{1}{4}\right)+2f\left(\frac{1}{2}\right)\right.$$
$$\left.+4f\left(\frac{3}{4}\right)+f(1)\right]$$
$$= \frac{4}{3}.$$

3. $\int_1^3 \frac{1}{x}\,dx$

Midpoint:

$$\frac{3-1}{4}\left[f\left(\frac{5}{4}\right)+f\left(\frac{7}{4}\right)+f\left(\frac{9}{4}\right)+f\left(\frac{11}{4}\right)\right]$$
$$= \frac{1}{2}\left(\frac{4}{5}+\frac{4}{7}+\frac{4}{9}+\frac{4}{11}\right) = \frac{3776}{3465}.$$

Trapezoidal:

$$T_4(f)$$
$$= \frac{3-1}{2(4)}\left[f(1)+2f\left(\frac{3}{2}\right)+2f(2)\right.$$
$$\left.+2f\left(\frac{5}{2}\right)+f(3)\right]$$
$$= \frac{1}{4}\left(1+\frac{4}{3}+1+\frac{4}{5}+\frac{1}{3}\right) = \frac{67}{60}.$$

Simpson:

$$S_4(f) = \frac{3-1}{3(4)}\left[f(1) + 4f\left(\frac{3}{2}\right) + 2f(2) + 4f\left(\frac{5}{2}\right) + f(3)\right] = \frac{1}{6}\left(1 + \frac{8}{3} + 1 + \frac{8}{5} + \frac{1}{3}\right) = \frac{11}{10}.$$

5. $\int_0^2 f(x)\,dx$

(a) $\frac{2-0}{4}\left[f(0) + f(.5) + f(1) + f(1.5)\right] = \frac{1}{2}(1 + .25 + 0 + .25) = .75$

(b) $\frac{2-0}{4}\left[f(.25) + f(.75) + f(1.25) + f(1.75)\right] = \frac{1}{2}(.65 + .15 + .15 + .65) = .8$

(c) $\frac{2-0}{2(4)}\left[f(0) + 2f(.5) + 2f(1) + 2f(1.5) + f(2)\right] = \frac{1}{4}(1 + .5 + 0 + .5 + 1) = .75$

7. $\int_0^{\pi} \cos x^2\,dx \approx .565694$

N	Midpoint	Trapezoidal	Simpson
10	.5538	.5889	.5660
20	.5629	.5713	.5655
50	.5652	.5666	.5657

9. $\int_0^2 e^{-x^2}\,dx \approx .882081$

N	Midpoint	Trapezoidal	Simpson
10	.88220	.88184	.88207
20	.88211	.88202	.88208
50	.88209	.88207	.88208

11. $\int_0^{\pi} e^{\cos x}\,dx \approx 3.97746$

N	Midpoint	Trapezoidal	Simpson
10	3.9775	3.9775	3.9775
20	3.9775	3.9775	3.9775
50	3.9775	3.9775	3.9775

13. $\int_0^1 5x^4\,dx = x^5\Big|_0^1 = 1 - 0 = 1$

N	Midpoint error	Trapezoidal error	Simpson error
10	.00832	−.01665	−.00007
20	.00208	−.00417	-4.2×10^{-6}
40	.00052	−.00104	-2.6×10^{-7}
80	.00013	−.00026	-1.6×10^{-8}

15. $\int_0^{\pi} \cos x\,dx = \sin x|_0^{\pi} = 0$

N	Midpoint error	Trapezoidal error	Simpson error
10	0	0	0
20	0	0	0
40	0	0	0
80	0	0	0

The graph of $\cos x$ on the interval $[\pi/2, \pi]$ looks like the graph on the interval $[0, \pi/2]$, but flipped about the x-axis. Hence, any positive contribution to the estimate will be cancelled by a corresponding negative contribution.

17. 4, 4 and 16

19. **(a)** 1.366162
(b) 1.428091
(c) 1.391621

21. **(a)** 0.843666
(b) 0.837084
(c) 0.841489

23. **(a)** 0.282
(b) 0.141
(c) 0.127

25. **(a)** 0.0052

(b) 0.0026

(c) 0.000022

27. Using Theorems 9.1 and 9.2, and the calculation in Example 9.10, we find the following lower bounds for the number of steps needed to guarantee accuracy of 10^{-7} in Exercise 19:

Midpoint: $\sqrt{\dfrac{2\cdot 3^3}{24\cdot 10^{-7}}} \approx 4744;$

Trapezoidal: $\sqrt{\dfrac{2\cdot 3^3}{14\cdot 10^{-7}}} \approx 6709;$

Simpson's: $\sqrt[4]{\dfrac{24\cdot 3^5}{180\cdot 10^{-7}}} \approx 135.$

29. $f(x) = \cos x$, $f''(x) = -\cos x$, $f^{(4)}(x) = \cos x$, so we can pick bounds $K = L = 1$.

Midpoint: $|EM_n| K\dfrac{(b-a)^3}{24n^2} = \dfrac{1}{24n^2}.$

We want $\dfrac{1}{24n^2} \le 10^7$

$$24n^2 \ge 10^7$$

$$n^2 \ge \frac{10^7}{24}$$

$$n\sqrt{\ge}\frac{10^7}{24} \approx 645.5,$$

so we need $n \ge 646$.

Trapezoid: $|ET_n| K\dfrac{(b-a)^3}{12n^2} = \dfrac{1}{12n^2}.$

We want $n^2 \ge \dfrac{10^7}{12}$

$$n \ge \sqrt{\frac{10^7}{12}} \approx 912.87,$$

so we need $n \ge 913$.

Simpson: $|ES_n| L\dfrac{(b-a)^5}{180n^4} = \dfrac{1}{180n^4}.$

We want $\dfrac{1}{180n^4} \le 10^{-7}$

$$180n^4 \ge 10^7$$

$$n^4 \ge \frac{10^7}{180}$$

$$n \ge \sqrt[4]{\frac{10^7}{180}} \approx 15.4,$$

so we need $n \ge 16$.

31. Error bounds and actual errors in Exercise 13:

	Midpoint	
Number of steps	Error bound	Error
10	0.025	0.008319
20	0.00625	0.008319
40	0.001563	0.002082
80	0.000391	0.00013

	Trapezoidal	
Number of steps	Error bound	Error
10	0.05	0.01665
20	0.0125	0.00416
40	0.003125	0.001042
80	0.000781	0.00026

	Simpson's	
Number of steps	Error bound	Error
10	6.6667×10^{-5}	6.6667×10^{-5}
20	4.1667×10^{-6}	4.1667×10^{-6}
40	2.6042×10^{-7}	2.6042×10^{-7}
80	1.6276×10^{-8}	1.6276×10^{-8}

33. $\int_0^2 f(x)\,dx$

(a) Trapezoidal:

$$T_8(f) = \frac{2-0}{2(8)}\big[f(0) + 2f(.25) + 2f(.5) + 2f(.75) + 2f(1) + 2f(1.25) + 2f(1.5) + 2f(1.75) + f(2)\big]$$
$$= \frac{1}{8}[4.0 + 9.2 + 10.4 + 9.6 + 10 + 9.2 + 8.8 + 7.6 + 4.0]$$
$$= 9.1$$

(b) Simpson:

$$S_8(f) = \frac{2-0}{3(8)}\big[f(0) + 4f(.25) + 2f(.5) + 4f(.75) + 2f(1) + 4f(1.25) + 2f(1.5) + 4f(1.75) + f(2)\big]$$
$$= \frac{1}{12}(4.0 + 18.4 + 10.4 + 19.2 + 10 + 18.4 + 8.8 + 15.2 + 4.0)$$
$$\approx 9.033$$

35. $\int_0^2 f(x)\,dx$

(a) Trapezoidal:

$$T_{10}(f) = \frac{2-0}{2(10)}\big[f(0) + 2f(.2) + 2f(.4) + 2f(.6) + 2f(.8) + 2f(1) + 2f(1.2) + 2f(1.4) + 2f(1.6) + 2f(1.8) + f(2)\big]$$
$$= \frac{1}{10}(2.4 + 5.2 + 5.8 + 6.4 + 6.8 + 7.2 + 7.6 + 7.8 + 8 + 8.2 + 4.2)$$
$$= 6.96$$

(b) Simpson:

$$S_{10}(f) = \frac{2-0}{3(10)}\big[f(0) + 4f(.2) + 2f(.4) + 4f(.6) + 2f(.8) + 4f(1) + 2f(1.2) + 4f(1.4) + 2f(1.6) + 4f(1.8) + f(2)\big]$$
$$= \frac{1}{15}(2.4 + 10.4 + 5.8 + 12.8 + 6.8 + 14.4 + 7.6 + 15.6 + 8 + 16.4 + 4.2)$$
$$= 6.96$$

37. $$S_{12}(f) = \frac{120-0}{3(12)}\big[f(0) + 4f(10) + 2f(20) + 4f(30) + 2f(40) + 4f(50) + 2f(60) + 4f(70) + 2f(80) + 4f(90) + 2f(100) + 4f(110) + f(120)\big]$$
$$= \frac{10}{3}(56 + 216 + 116 + 248 + 116 + 232 + 124 + 224 + 104 + 192 + 80 + 128 + 22)$$
$$\approx 6193 \text{ ft}^2$$

39. $$S_{12}(f) = \frac{12-0}{3(12)}\big[f(0) + 4f(1) + 2f(2) + 4f(3) + 2f(4) + 4f(5) + 2f(6) + 4f(7) + 2f(8) + 4f(9) + 2f(10) + 4f(11) + f(12)\big]$$
$$= \frac{1}{3}(40 + 168 + 80 + 176 + 96 + 200 + 92 + 184 + 84 + 176 + 80 + 168 + 42)$$
$$\approx 529 \text{ ft}$$

41. $$S_{12}(f) = \frac{2.4-0}{3(12)}\big[f(0) + 4f(.2) + 2f(.4) + 4f(.6) + 2f(.8) + 4f(1) + 2f(1.2) + 4f(1.4) + 2f(1.6) + 4f(1.8) + 2f(2) + 4f(2.2) + f(2.4)\big]$$
$$= \frac{1}{15}(0 + .8 + .8 + 4 + 3.2 + 8 + 4.4 + 8 + 3.2 + 4.8 + 1.2 + .8 + 0)$$
$$\approx 2.6 \text{ liters}$$

43. $f''(x) > 0,\ f'(x) > 0$
So f is increasing and concave up.

(a) Midpoint rule would underestimate.
(b) Trapezoidal rule would overestimate.
(c) Simpson's rule—not enough information.

45. $f''(x) < 0,\ f'(x) > 0$
So f is increasing and concave down.

(a) Midpoint rule would overestimate.
(b) Trapezoidal rule would underestimate.
(c) Simpson's rule—not enough information.

47. $f''(x) = 4,\ f'(x) > 0$
So f is quadratic, concave up, and increasing.

(a) Midpoint rule would underestimate.
(b) Trapezoidal rule would overestimate.
(c) Simpson's rule would be exact.

49. Let's assume h_1 is greater than h_2. Divide the trapezoid into a rectangle with base L and height h_2 and a triangle with base L and height $(h_1 - h_2)$. Then the area of the trapezoid is

$$\begin{aligned} A &= Lh_2 + \frac{L(h_1 - h_2)}{2} \\ &= Lh_2 + \frac{Lh_1}{2} - \frac{Lh_2}{2} \\ &= \frac{Lh_1 + Lh_2}{2} \\ &= \frac{L(h_1 + h_2)}{2}. \end{aligned}$$

51.

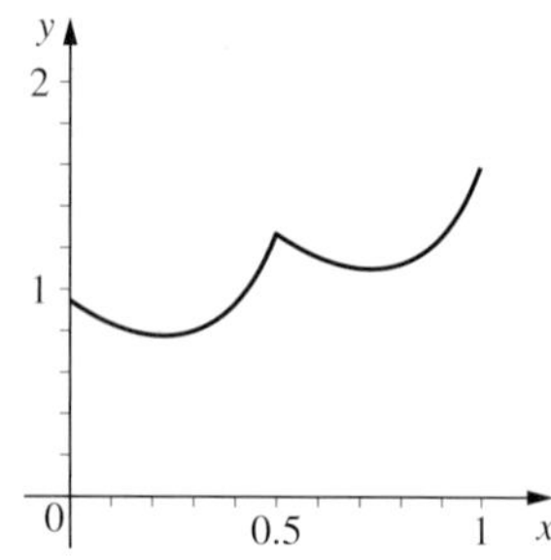

The Simpson's Rule approximation is smaller because parabolas curve.

53. Let I be the exact integral. Then we have

$$\begin{aligned} T_n - I &\approx -2(M_n - I) \\ T_n - I &= 2I - 2M_n \\ T_n + 2M_n &\approx 3I \\ \frac{T_n}{3} + \frac{2}{3}M_n &= I. \end{aligned}$$

55. $\int_0^1 \sqrt{1 - x^2}\, dx$ represents the area of a quarter circle, and so equals $\pi/4$. (You can also evaluate this integral with the substitution $x = \sin\theta$.)

$$\begin{aligned} &\int_0^1 \frac{1}{1 + x^2}\, dx = \tan^{-1} x \Big|_0^1 \\ &= \tan^{-1} 1 - \tan^{-1} 0 = \frac{\pi}{4} - 0 = \frac{\pi}{4}. \end{aligned}$$

Integrating $\frac{1}{1 + x^2}$ provides better accuracy for a given n. This is to be expected. The error is bounded by the fourth derivative. If $f(x) = \frac{1}{1 + x^2}$, then the fourth derivative on the interval $[0, 1]$ is bounded by 24. If $f(x) = \sqrt{1 - x^2}$, then the fourth derivative is unbounded on $[0, 1]$.

57. For x and x^3, the integral is zero, because the area of the region above the x-axis cancels with the area of the region below the x-axis. Similarly,

$$f\left(\frac{-1}{\sqrt{3}}\right) = -f\left(\frac{1}{\sqrt{3}}\right),$$

so the approximation is also zero. For x^2, the integral equals 2/3, while

$$f\left(\frac{1}{\sqrt{3}}\right) + f\left(\frac{-1}{\sqrt{3}}\right) = \frac{1}{3} + \frac{1}{3} = \frac{2}{3}.$$

4.10 Improper Integrals

1. improper, function not defined at $x = 0$

3. not improper, function continuous on entire interval

5. improper, function not defined at 0

7. improper, interval is infinite

9.
$$\begin{aligned}\int_0^1 x^{-1/3}\,dx &= \lim_{R\to 0^+}\int_R^1 x^{-1/3}\,dx\\ &= \lim_{R\to 0^+}\frac{3}{2}x^{2/3}\Big|_R^1\\ &= \lim_{R\to 0^+}\frac{3}{2}(1-R^{2/3})\\ &= \frac{3}{2}\end{aligned}$$

11.
$$\begin{aligned}\int_0^1 x^{-4/3}\,dx &= \lim_{R\to 0^+}\int_R^1 x^{-4/3}\,dx\\ &= \lim_{R\to 0^+}(-3x^{-1/3})\Big|_R^1\\ &= \lim_{R\to 0^+}(-3)(1-R^{-1/3})\\ &= \infty\end{aligned}$$

The integral diverges.

13.
$$\begin{aligned}\int_1^\infty x^{-1/3}\,dx &= \lim_{R\to\infty}\int_1^R x^{-1/3}\,dx\\ &= \lim_{R\to\infty}\frac{3}{2}x^{2/3}\Big|_1^R\\ &= \lim_{R\to\infty}\frac{3}{2}(R^{2/3}-1)\\ &= \infty\end{aligned}$$

The integral diverges.

15.
$$\begin{aligned}\int_1^\infty x^{-4/3}\,dx &= \lim_{R\to\infty}\int_1^R x^{-4/3}\,dx\\ &= \lim_{R\to\infty}(-3x^{-1/3})\Big|_1^R\\ &= \lim_{R\to\infty}(-3)(R^{-1/3}-1)\\ &= 3\end{aligned}$$

17.
$$\begin{aligned}\int_1^3 \frac{1}{\sqrt{x-1}}\,dx &= \lim_{R\to 1^+}\int_R^3 \frac{1}{\sqrt{x-1}}\,dx\\ &= \lim_{R\to 1^+} 2\sqrt{x-1}\Big|_R^3\\ &= \lim_{R\to 1^+} 2(\sqrt{2}-\sqrt{R-1})\\ &= 2\sqrt{2}\end{aligned}$$

19.
$$\int_0^1 \ln x\,dx = \lim_{R\to 0^+}\int_R^1 \ln x\,dx$$

$$u=\ln x \qquad dv = dv$$
$$du = \frac{1}{x}\,dx \qquad v = x$$

$$\begin{aligned}&= \lim_{R\to 0^+}\left[x\ln x|_R^1 - \int_R^1 dx\right]\\ &= \lim_{R\to 0^+}(x\ln x - x)|_R^1\\ &= \lim_{R\to 0^+}[(0-1)-(R\ln R - R)]\\ &= -1 - \lim_{R\to 0^+}(R\ln R - R)\\ &= -1 - \lim_{R\to 0^+}\frac{\ln R}{\frac{1}{R}} + \lim_{R\to 0} R\\ &= -1 - \lim_{R\to 0^+}\frac{\frac{1}{R}}{-\frac{1}{R^2}}\\ &= -1 - \lim_{R\to 0^+}(-R)\\ &= -1\end{aligned}$$

21.
$$\begin{aligned}&\int_0^3 \frac{1}{x^2-1}\,dx\\ &= \int_0^3 \frac{1}{(x-1)(x+1)}\,dx\\ &= \lim_{R\to 1^-}\int_0^R \frac{1}{(x-1)}(x+1)\,dx\\ &\quad + \lim_{R\to 1^+}\int_R^3 \frac{1}{(x-1)(x+1)}\,dx\end{aligned}$$

Both of these integrals behave like

$$\begin{aligned}\lim_{R\to 0^+}\int_R^1 \frac{1}{x}\,dx &= \lim_{R\to 0^+}(\ln 1 - \ln R)\\ &= \lim_{R\to 0^+}\ln\left(\frac{1}{R}\right) = \infty,\end{aligned}$$

so the original integral diverges.

23.
$$\int_0^\infty xe^x\,dx = \lim_{R\to\infty}\int_0^R xe^x\,dx$$
$$u = x \qquad dv = e^x\,dx$$
$$du = dx \qquad v = e^x$$
$$= \lim_{R\to\infty}\left(xe^x\Big|_0^R - \int_0^R e^x\,dx\right)$$
$$= \lim_{R\to\infty}\left(\mathrm{Re}^R - 0 - e^x\Big|_0^R\right)$$
$$= \lim_{R\to\infty}[(\mathrm{Re}^R - (e^R - 1)]$$
$$= \lim_{R\to\infty}[e^R(R-1)+1]$$
$$= \infty$$

The integral diverges.

25.
$$\int_{-\infty}^0 xe^x\,dx = \lim_{R\to-\infty}\int_R^0 xe^x\,dx$$
$$u = x \qquad dv = e^x\,dx$$
$$du = dx \qquad v = e^x$$
$$= \lim_{R\to-\infty}\left(xe^x\Big|_R^0 - \int_R^0 e^x\,dx\right)$$
$$= \lim_{R\to-\infty}(xe^x - e^x)\Big|_R^0$$
$$= \lim_{R\to-\infty}[(0-1) - (\mathrm{Re}^R - e^R)]$$
$$= \lim_{R\to-\infty}[-1 - e^R(R-1)]$$
$$= -1 - \lim_{R\to-\infty}\frac{R-1}{e^{-R}}$$
$$= -1 - \lim_{R\to-\infty}\frac{1}{-e^{-R}}$$
$$= -1$$

27.
$$\int_{-\infty}^\infty \frac{x}{\sqrt{x^2+1}}\,dx$$
$$= \int_{-\infty}^0 \frac{x}{\sqrt{x^2+1}}\,dx + \int_0^\infty \frac{x}{\sqrt{x^2+1}}\,dx$$
$$= \lim_{R\to-\infty}\int_R^0 \frac{x}{\sqrt{x^2+1}}\,dx + \lim_{R\to\infty}\int_0^R \frac{x}{\sqrt{x^2+1}}\,dx$$
$$= \lim_{R\to-\infty}\sqrt{x^2+1}\Big|_R^0 + \lim_{R\to\infty}\sqrt{x^2+1}\Big|_0^R$$

Both of these limits are infinite (one positive, the other negative), so the integral diverges.

29.
$$\int_{-\infty}^\infty e^{-2x}\,dx = \int_{-\infty}^0 e^{-2x}\,dx + \int_0^\infty e^{-2x}\,dx$$
$$= \lim_{R\to-\infty}\int_R^0 e^{-2x}\,dx + \lim_{R\to-\infty}\int_0^R e^{-2x}\,dx$$
$$= \lim_{R\to-\infty}\left(-\frac{1}{2}e^{-2x}\right)\Big|_R^0 + \lim_{R\to\infty}\left(-\frac{1}{2}e^{-2x}\right)\Big|_0^R$$
$$= \lim_{R\to-\infty}\left[-\frac{1}{2}(1-e^{-2R})\right] + \lim_{R\to\infty}\left[-\frac{1}{2}(e^{-2R}-1)\right]$$
$$= \infty + \lim_{R\to\infty}\left[-\frac{1}{2}(e^{-2R}-1)\right]$$
$$= \infty$$

The integral diverges.

31.
$$\int_{-\infty}^\infty \frac{1}{1+x^2}\,dx = \int_{-\infty}^0 \frac{1}{1+x^2}\,dx + \int_0^\infty \frac{1}{1+x^2}\,dx$$
$$= \lim_{R\to-\infty}\int_R^0 \frac{1}{1+x^2}\,dx + \lim_{R\to\infty}\int_0^R \frac{1}{1+x^2}\,dx$$
$$= \lim_{R\to\infty}\tan^{-1}x\Big|_R^0 + \lim_{R\to\infty}\tan^{-1}x\Big|_0^R$$
$$= \lim_{R\to-\infty}(\tan^{-1}0 - \tan^{-1}R)$$
$$+ \lim_{R\to\infty}(\tan^{-1}R - \tan^{-1}0)$$
$$= 0 - \left(-\frac{\pi}{2}\right) + \frac{\pi}{2} - 0$$
$$= \pi$$

33.
$$\int_0^\pi \tan x\,dx = \int_0^{\pi/2}\tan x\,dx + \int_{\pi/2}^\pi \tan x\,dx$$
$$= \lim_{R\to\frac{\pi}{2}^-}\int_0^R \tan x\,dx + \lim_{R\to\frac{\pi}{2}^+}\int_R^\pi \tan x\,dx$$
$$= \lim_{R\to\frac{\pi}{2}^-}\ln|\sec x|\big|_0^R + \lim_{R\to\frac{\pi}{2}^+}\ln|\sec x|\big|_R^\pi$$
$$= \lim_{R\to\frac{\pi}{2}^-}(\ln|\sec R| - \ln|\sec 0|)$$
$$+ \lim_{R\to\frac{\pi}{2}^+}(\ln|\sec\pi| - \ln|\sec R|)$$

Both of these limits are infinite (one positive, the other negative), so the integral diverges.

35. $\displaystyle\int_0^2 \frac{x}{x^2-1}\,dx = \int_0^1 \frac{x}{x^2-1}\,dx + \int_1^2 \frac{x}{x^2-1}\,dx$

$$= \lim_{R\to 1^-}\int_0^R \frac{x}{x^2-1}\,dx + \lim_{R\to 1^+}\int_R^2 \frac{x}{x^2-1}\,dx$$

$$= \lim_{R\to 1^-}\frac{1}{2}\ln\left|x^2-1\right|\Big|_0^R + \lim_{R\to 1^+}\frac{1}{2}\ln\left|x^2-1\right|\Big|_R^2$$

$$\lim_{R\to 1^-}\frac{1}{2}\ln\left|x^2-1\right|\Big|_0^R$$

$$= \lim_{R\to 1^-}\left(\frac{1}{2}\ln\left|R^2-1\right| - \frac{1}{2}\ln|-1|\right) = -\infty$$

The integral diverges.

37. $\displaystyle\int_0^1 \frac{2}{\sqrt{1-x^2}}\,dx = \lim_{R\to 1^-}\int_0^R \frac{2}{\sqrt{1-x^2}}\,dx$

$$= \lim_{R\to 1^-} 2\sin^{-1}x\Big|_0^R$$

$$= \lim_{R\to 1^-} 2(\sin^{-1}R - \sin^{-1}0)$$

$$= 2\left(\frac{\pi}{2}-0\right)$$

$$= \pi$$

39. $\displaystyle\int_1^\infty \left(\frac{1}{x-1}-\frac{1}{x+1}\right)dx =$

$$\int_1^2 \left(\frac{1}{x-1}-\frac{1}{x+1}\right)dx + \int_2^\infty \left(\frac{1}{x-1}-\frac{1}{x+1}\right)dx$$

$$= \lim_{R\to 1^+}\int_R^2 \left(\frac{1}{x-1}-\frac{1}{x+1}\right)dx$$

$$+ \lim_{R\to\infty}\int_2^R \left(\frac{1}{x-1}-\frac{1}{x+1}\right)dx$$

The first integral equals

$$\lim_{R\to 1^+}\int_R^2 \left(\frac{1}{x-1}-\frac{1}{x+1}\right)dx$$

$$= \lim_{R\to 1^+}(\ln(x-1)-\ln(x+1))|_R^2$$

$$\lim_{R\to 1^+}(\ln 1 - \ln 3 - \ln(R-1) + \ln(R+1))$$

$$= \ln 2 - \ln 3 - \lim_{R\to 1^+}\ln(R-1)$$

which diverges to ∞. Hence, the original integral diverges.

41. $\displaystyle\int_0^\infty \frac{1}{\sqrt{x}e^{\sqrt{x}}}\,dx$

$$= \lim_{R\to 0^+}\int_R^1 \frac{1}{\sqrt{x}e^{\sqrt{x}}}\,dx + \lim_{R\to\infty}\int_1^R \frac{1}{\sqrt{x}e^{\sqrt{x}}}\,dx$$

Let $u=\sqrt{x},\ du = \dfrac{1}{2\sqrt{x}}\,dx$.

$$\int \frac{1}{\sqrt{x}e^{\sqrt{x}}}\,dx = 2\int e^{-u}\,du = -2e^{-u}$$

So the original integral is equal to

$$\lim_{R\to 0^+}\left(\frac{-2}{e^u}\Big|_R^1\right) + \lim_{R\to\infty}\left(\frac{-2}{e^u}\Big|_1^R\right)$$

$$= \lim_{R\to 0^+}\left(\frac{-2}{e}+\frac{2}{e^R}\right) + \lim_{R\to\infty}\left(\frac{-2}{e}+\frac{2}{e^R}\right)$$

$$= 1+1 = 2.$$

43. $\displaystyle\int_0^\infty \cos x\,dx = \lim_{R\to\infty}\int_0^R \cos x\,dx$

$$= \lim_{R\to\infty}\sin x|_0^R$$

$$= \lim_{R\to\infty}(\sin R - \sin 0)$$

The integral diverges.

45. $r=1$

47. $\displaystyle\int_0^\infty xe^{cx}\,dx$ converges for $c<0$

$\displaystyle\int_{-\infty}^0 xe^{cx}\,dx$ converges for $c>0$

49.

$$\int_1^\infty \frac{x}{1+x^3}\,dx$$

$$0 < \frac{x}{1+x^3} < \frac{x}{x^3} = \frac{1}{x^2}$$

$$\int_1^\infty \frac{1}{x^2}\,dx = \lim_{R\to\infty} \int_1^R \frac{1}{x^2}\,dx = \lim_{R\to\infty} \left(-\frac{1}{x}\right)\Big|_1^R = \lim_{R\to\infty} \left(-\frac{1}{R}+1\right) = 1$$

So $\int_1^\infty \frac{x}{1+x^3}\,dx$ converges.

51.

$$\int_2^\infty \frac{x}{x^{3/2}-1}\,dx$$

$$\frac{x}{x^{3/2}-1} > \frac{x}{x^{3/2}} = \frac{1}{x^{1/2}} > 0$$

$$\int_2^\infty x^{-1/2}\,dx = \lim_{R\to\infty} \int_2^R x^{-1/2}\,dx = \lim_{R\to\infty} 2\sqrt{x}\Big|_2^R = \lim_{R\to\infty} (2\sqrt{R} - 2\sqrt{2}) = \infty$$

So $\int_2^\infty \frac{x}{x^{3/2}-1}\,dx$ diverges.

53.

$$\int_0^\infty \frac{3}{x+e^x}\,dx$$

$$0 < \frac{3}{x+e^x} < \frac{3}{e^x}$$

$$\int_0^\infty \frac{3}{e^x}\,dx = \lim_{R\to\infty} \int_0^R \frac{3}{e^x}\,dx = \lim_{R\to\infty} \left(-\frac{3}{e^x}\right)\Big|_0^R = \lim_{R\to\infty} \left(-\frac{3}{e^R}+3\right) = 3$$

So $\int_0^\infty \frac{3}{x+e^x}\,dx$ converges.

55.

$$\int_0^\infty \frac{\sin^2 x}{1+e^x}\,dx$$

$$\frac{\sin^2 x}{1+e^x} \le \frac{1}{1+e^x} < \frac{1}{e^x}$$

$$\int_0^\infty \frac{1}{e^x}\,dx = \lim_{R\to\infty} \int_0^R \frac{1}{e^x}\,dx = \lim_{R\to\infty} \int (-e^{-x})\Big|_0^R = \lim_{R\to\infty} (-e^{-R}+1) = 1$$

So $\int_0^\infty \frac{\sin^2 x}{1+e^x}\,dx$ converges.

57.

$$\int_2^\infty \frac{x^2 e^x}{\ln x}\,dx$$

$$\frac{x^2 e^x}{\ln x} > e^x$$

$$\int_2^\infty e^x\,dx = \lim_{R\to\infty} \int_2^R e^x\,dx = \lim_{R\to\infty} e^x\Big|_2^R = \lim_{R\to\infty} (e^R - e^2) = \infty$$

So $\int_2^{\infty} \frac{x^2 e^x}{\ln x}\, dx$ diverges.

59. $$\int_0^1 x \ln 4x\, dx = \lim_{R\to 0^+} \int_R^1 x \ln 4x\, dx$$

$$u = \ln 4x \qquad dv = x\, dx$$

$$du = \frac{4}{4x}\, dx \qquad v = \frac{x^2}{2}$$

$$= \lim_{R\to 0^+} \left[\frac{1}{2}x^2 \ln 4x \Big|_R^1 = \int_R^1 \frac{1}{2} x\, dx \right]$$

$$= \lim_{R\to 0^+} \left[\left(\frac{1}{2}\ln 4 - \frac{1}{2}R^2 \ln 4R \right) - \frac{1}{4}x^2 \Big|_R^1 \right]$$

$$= \lim_{R\to 0^+} \left[\frac{1}{2}\ln 4 - \frac{\ln 4R}{\frac{2}{R^2}} - \left(\frac{1}{4} - \frac{R^2}{4} \right) \right]$$

$$= \frac{1}{2}\ln 4 - \frac{1}{4} + \lim_{R\to 0^+} \frac{\frac{4}{4R}}{-\frac{4}{R^3}}$$

$$= \frac{1}{2}\ln 4 - \frac{1}{4} + \lim_{R\to 0^+} \left(-\frac{R^2}{4} \right)$$

$$= \frac{1}{2}\ln 4 - \frac{1}{4}$$

61. $\int_{-R}^{R} x^3\, dx = 0$, so the limit = 0. On the other hand, $\int_0^{\infty} x^3\, dx$ diverges, so $\int_{-\infty}^{\infty} x^3\, dx$ diverges.

63. False; see Exercise 61. The improper integral may diverge even though the limit exists.

65. False. See Example 10.6 and the discussion preceding it.

67. False; see Exercise 61 for a counterexample.

Chapter 4 Review

1. See Example 4.3.

$$\int_1^3 4x^2 - 3\, dx = 4\frac{x^3}{3} - 3x \Big|_1^3$$

$$= (36 - 9) - \left(\frac{4}{3} - 3 \right) = \frac{86}{3}$$

3. See page 363.

$$\int \frac{4}{x}\, dx = 4 \ln |x| + c$$

5. See Example 3.9.

$$\int_0^{\pi/8} 2 \sin 4x\, dx = -\frac{1}{2}\cos 4x \Big|_0^{\pi/8}$$

$$= -\frac{1}{2}\left[\cos\frac{\pi}{2} - \cos 0 \right] = -\frac{1}{2}(0 - 1) = \frac{1}{2}$$

7. See Example 3.9.

$$\int \frac{1}{\sqrt{3x+1}}\, dx = \frac{2}{3}(3x+1)^{1/2} + c$$

9. See Section 4.3.

$$\int \frac{x^2+4}{x}\, dx = \int \left(x + \frac{4}{x} \right) dx$$

$$= \frac{x^2}{2} + 4 \ln |x| + c$$

11. See Section 4.3.

$$\int e^x \left(1 - e^{-x} \right) dx = \int (e^x - 1)\, dx = e^x - x + c$$

13. See Example 5.2.
$u = x^2 + 4,\ du = 2x\, dx$

$$\int_0^2 x\sqrt{x^2+4} = \frac{1}{2}\int_4^8 \sqrt{u}\, du$$

$$= \frac{1}{3}u^{3/2}\Big|_4^8 = \frac{1}{3}(8^{3/2} - 8)$$

$$= \frac{1}{3}(16\sqrt{2} - 8)$$

15. See Example 5.3.
$u = x^3,\ du = 3x^2\,dx$

$$\int 6x^2 \cos x^3\,dx = 2\int \cos u\,du$$
$$= 2\sin u + c$$
$$= 2\sin x^3 + c$$

17. See Example 5.2.
$u = \frac{1}{x},\ du = -\frac{1}{x^2}\,dx$

$$\int \frac{e^{1/x}}{x^2}\,dx = -\int e^u\,du$$
$$= -e^u + c$$
$$= -e^{1/x} + c$$

19. See Example 5.6.
$u = \cos x,\ du = -\sin x\,dx$

$$\int \tan x\,dx = \int \frac{\sin x}{\cos x}\,dx$$
$$= -\int \frac{1}{u}\,du$$
$$= -\ln|u| + c$$
$$= -\ln|\cos x| + c$$

21. See Example 4.3.

$$\int_0^{10} (1 - e^{-t/4})\,dt = (t + 4e^{-t/4})\Big|_0^{10}$$
$$= (10 + 4e^{-10/4}) - (0 + 4)$$
$$= 6 + 4e^{-5/2}$$

23. See Section 4.5.
$u = e^x,\ du = e^x\,dx$

$$\int \frac{e^x}{e^{2x} + 4}\,dx = \int \frac{du}{u^2 + 4}$$
$$= \frac{1}{2}\tan^{-1}\left(\frac{u}{2}\right) + c = \frac{1}{2}\tan^{-1}\left(\frac{e^x}{2}\right) + c$$

25. See Section 4.7.

$$\int \sqrt{\sin x}\cos^3 x\,dx$$
$$= \sqrt{\sin x}(1 - \sin^2 x)\cos x\,dx$$

Let $u = \sin x,\ du = \cos x\,dx$.

$$\int \sqrt{u}(1 - u^2)\,du$$
$$= \int (^{1/2} - u^{5/2})\,du$$
$$= \frac{2}{3}u^{3/2} - \frac{2}{7}u^{7/2} + c$$
$$= \frac{2}{3}(\sin x)^{3/2} - \frac{2}{7}(\sin x)^{7/2} + c$$

27. See Section 4.6.

$$\int_0^1 x\sin 3x\,dx \qquad u = x \qquad dv = \sin 3x\,dx$$
$$du = dx \qquad v = -\frac{1}{3}\cos 3x$$
$$= -\frac{1}{3}x\cos 3x\Big|_0^1 + \frac{1}{3}\int_0^1 \cos 3x\,dx$$
$$= -\frac{1}{3}\cos 3 + \frac{1}{3}\cdot\frac{1}{3}\sin 3x\Big|_0^1$$
$$= -\frac{1}{3}\cos 3 + \frac{1}{9}\sin 3$$

29. See Section 4.5.
$u = \ln x, \quad du = \frac{1}{x}\,dx$

$$\int \frac{\sin(\ln x)}{x}\,dx = \int \sin u\,du$$
$$= -\cos u + c = -\cos(\ln x) + c$$

31. See Example 5.7.
$u = \tan^{-1} x, \quad du = \frac{1}{x^2 + 1}\,dx$

$$\int \frac{3}{(x^2 + 1)\tan^{-1} x}\,dx = \int \frac{3}{u}\,du$$
$$= 3\ln|u| + c = 3\ln|(\tan^{-1} x)| + c$$

33.
$$\int 3^{4x}\,dx = \int e^{(\ln 3)(4x)}\,dx = \frac{e^{(\ln 3)(4x)}}{4\ln 3} + c$$
$$= \frac{3^{4x}}{4\ln 3} + c$$

35.
$$\int_0^1 \frac{3}{x^2+9}\,dx = \tan^{-1}\left(\frac{x}{3}\right)\Big|_0^1$$
$$\tan^{-1}\left(\frac{1}{3}\right) - \tan^{-1}(0) = \tan^{-1}\left(\frac{1}{3}\right)$$

37. See Section 4.5.
$u = x^3, \quad dx = 3x^2\,dx$
$$\int \frac{x^2}{\sqrt{1-x^6}}\,dx = \frac{1}{3}\int \frac{1}{\sqrt{1-u^2}}\,du$$
$$= \frac{1}{3}\sin^{-1} u + c = \frac{1}{3}\sin^{-1}(x^3) + c$$

39. See Section 4.5.
$u = \sqrt{x}, \quad du = \dfrac{1}{2\sqrt{x}}\,dx$
$$\int \frac{e^{\sqrt{x}}}{\sqrt{x}}\,dx = 2\int e^u\,du = e^u + c = 2e^{\sqrt{x}} + c$$

41. See Section 4.6.
$$\int_{-1}^1 x\sin\pi x\,dx \qquad u = x \qquad dv = \sin\pi x\,dx$$
$$du = dx \qquad v = -\frac{1}{\pi}\cos\pi x$$
$$= -\frac{1}{\pi}x\cos\pi x\Big|_{-1}^1 + \frac{1}{\pi}\int_{-1}^1 \cos\pi x\,dx$$
$$= -\frac{1}{\pi}\cos\pi + \frac{1}{\pi}(-1)\cos(-\pi) + \frac{1}{\pi^2}\sin\pi x\Big|_{-1}^1$$
$$= \frac{1}{\pi} + \frac{1}{\pi} + \frac{1}{\pi^2}(\sin\pi - \sin(-\pi))$$
$$= \frac{2}{\pi}$$

43. See Section 4.7.
$$\int \cos x\sin^2 x\,dx$$
Let $u = \sin x,\ du = \cos x\,dx.$
$$\int u^2\,du = \frac{1}{3}u^3 + c$$
$$= \frac{1}{3}\sin^3 x + c$$

45. See Section 4.10.
$$\int_1^\infty \frac{3}{x^2+1}\,dx = 3\lim_{R\to\infty}\int_1^R \frac{1}{x^2+1}\,dx$$
$$= 3\lim_{R\to\infty}(\tan^{-1} R - \tan^{-1} 1)$$
$$= 3\left(\frac{\pi}{2} - \frac{\pi}{4}\right) = \frac{3\pi}{4}$$

47. See Section 4.7.
$$\int \frac{x^2}{\sqrt{1-x^2}}\,dx$$
Let $x = \sin\theta,\ dx = \cos\theta\,d\theta.$
$$\int \frac{\sin^2\theta}{\sqrt{1-\sin^2\theta}}\cos\theta\,d\theta = \int \sin^2\theta\,d\theta$$
$$= \frac{1}{2}\int (1-\cos 2\theta)\,d\theta$$
$$= \frac{1}{2}\left(\theta - \frac{1}{2}\sin 2\theta\right) + c$$
$$\sin 2\theta = 2\sin\theta\cos\theta = 2x\sqrt{1-x^2}$$
$$\int \frac{x^2}{\sqrt{1-x^2}}\,dx = \frac{1}{2}\theta - \frac{1}{4}\sin 2\theta + c$$
$$= \frac{1}{2}\sin^{-1} x - \frac{1}{2}x\sqrt{1-x^2} + c$$

49. See Example 7.1.
$$\int \frac{x+4}{x^2+3x+2}\,dx = \int \frac{x+4}{(x+2)(x+1)}\,dx$$
$$\frac{x+4}{(x+2)(x+1)} = \frac{A}{x+2} + \frac{B}{x+1}$$
$$x+4 = A(x+1) + B(x+2)$$
$$x = -1 : 3 = B$$
$$x = -2 : 2 = -A;\ A = -2$$
$$\frac{x+4}{(x+2)(x+1)} = \frac{-2}{x+2} + \frac{3}{x+1}$$
$$\int \frac{x+4}{x^2+3x+2}\,dx = \int \left(-\frac{2}{x+2} + \frac{3}{x+1}\right)dx$$
$$= -2\ln|x+2| + 3\ln|x+1| + c$$

51. See Section 4.6.

$$\int x^2e^{-3x}\,dx \qquad u=x^2 \qquad dv=e^{-3x}\,dx$$

$$du=2x\,dx \qquad v=-\frac{1}{3}e^{-3x}$$

$$=-\frac{1}{3}x^2e^{-3x}+\frac{2}{3}\int xe^{-3x}\,dx$$

$$u=x \qquad dv=e^{-3x}\,dx$$

$$du=dx \qquad v=-\frac{1}{3}e^{-3x}$$

$$=-\frac{1}{3}x^2e^{-3x}+\frac{2}{3}\left(-\frac{1}{3}xe^{-3x}+\frac{1}{3}\int e^{-3x}\,dx\right)$$

$$=-\frac{1}{3}x^2e^{-3x}-\frac{2}{9}xe^{-3x}-\frac{2}{27}e^{-3x}+c$$

53. See Section 4.10.

$$\int_0^1 \frac{x}{x^2-1}\,dx=\lim_{R\to 1^-}\int_0^R \frac{x}{x^2-1}\,dx$$

$$=\lim_{R\to 1^-}\frac{1}{2}\ln\left|x^2-1\right|\Big|_0^R$$

$$\lim_{R\to 1^-}\frac{1}{2}\ln\left|x^2-1\right|\Big|_0^R$$

$$=\lim_{R\to 1^-}\left(\frac{1}{2}\ln\left|R^2-1\right|-\frac{1}{2}\ln|-1|\right)=-\infty$$

The integral diverges.

55. See Section 7.1.

$$\int \frac{x}{1+x^4}\,dx$$

Let $u=x^2,\ du=2x\,dx$.

$$\int \frac{1}{1+u^2}\cdot\frac{1}{2}\,du=\frac{1}{2}\tan^{-1}u+c$$

$$=\frac{1}{2}\tan^{-1}x^2+c$$

57. $\int e^{2\ln x}\,dx=\int x^2\,dx=\frac{x^3}{3}+c$

59. See Section 4.6.

$$\int_1^2 x^3\ln x\,dx \qquad u=\ln x \qquad dv=x^3\,dx$$

$$du=\frac{1}{x}\,dx \qquad v=\frac{1}{4}x^4$$

$$=\frac{1}{4}x^4\ln x\Big|_1^2-\frac{1}{4}\int_1^2 x^3\,dx$$

$$=\frac{1}{4}\cdot 16\ln 2-\frac{1}{4}\ln 1-\frac{1}{16}x^4\Big|_1^2$$

$$=4\ln 2-\frac{1}{16}(16-1)$$

$$=4\ln 2-\frac{15}{16}$$

61. See Example 7.11.

$$\int \tan^2 x\sec^4 x\,dx$$

$$=\int \tan^2 x(1+\tan^2 x)\sec^2 x\,dx$$

Let $u=\tan x,\ du=\sec^2 x\,dx$.

$$\int u^2(1+u)^2\,du=\int (u^2+u^4)\,du$$

$$=\frac{1}{3}u^3+\frac{1}{5}u^5+c$$

$$=\frac{1}{3}\tan^3 x+\frac{1}{5}\tan^5 x+c$$

63. See Example 7.6.

$$\int \frac{2}{8+4x+x^2}=\int \frac{2}{x^2+4x+4+8-4}\,dx$$

$$=\int \frac{2}{(x+2)^2+4}\,dx$$

$$=\int \frac{2}{4\left[\left(\frac{x+2}{2}\right)+1\right]}\,dx$$

Let $u = \frac{x+2}{2}$, $du = \frac{1}{2}\,dx$.

$$\int \frac{1}{u^2+1}\,du = \tan^{-1} u + c$$
$$= \tan^{-1}\left(\frac{x+2}{2}\right) + c$$

65. See Example 7.2.

$$\int \frac{4x^2+6x-12}{x^3-4x}\,dx = \int \frac{4x^2+6x-12}{x(x-2)(x+2)}\,dx$$

$$\frac{4x^2+6x-12}{x(x-2)(x+2)} = \frac{A}{x} + \frac{B}{x-2} + \frac{C}{x+2}$$

$4x^2+6x-12 =$

$A(x-2)(x+2) + Bx(x+2) + Cx(x-2)$

$x = 0 : -12 = -4A : A = 3$

$x = 2 : 16 = 8B : B = 2$

$x = -2 : -8 = 8C : C = -1$

$$\int \frac{4x^2+6x+12}{x^3-4x}\,dx = \int \frac{3}{x} + \frac{2}{x-2} - \frac{1}{x+2}\,dx$$
$$= 3\ln|x| + 2\ln|x-2| - \ln|x+2| + c$$

67. See Section 4.5.

$u = x^2, \quad du = 2x\,dx$

$$\int 4x \sec x^2 \tan x^2\,dx = 2\int \sec u \tan u\,du$$
$$= 2\sec u + c = 2\sec(x^2) + c$$

69. See Example 10.7.

$$\int_0^\infty xe^{-x}\,dx = \lim_{R\to\infty}\int_0^R xe^{-x}\,dx$$

$$u = x \qquad dv = e^{-x}\,dx$$
$$du = dx \qquad v = -e^{-x}$$

$$= \lim_{R\to\infty}\left(-xe^{-x}\Big|_0^R - \int_0^R -e^{-x}\,dx\right)$$
$$= \lim_{R\to\infty}\left(-\mathrm{Re}^{-R} - 0 - e^{-x}\Big|_0^R\right)$$
$$= \lim_{R\to\infty}[(-\mathrm{Re}^{-R} - (e^{-R} - 1)]$$
$$= 1 - \lim_{R\to\infty}[-\mathrm{Re}^{-R}]$$

But

$$\lim_{R\to\infty} -\mathrm{Re}^{-R} = -\lim_{R\to\infty}\frac{R}{e^R} = 0$$

(using L'Hopital's Rule, for example), so the original integral equals 1.

71. See Example 3.13.

$$f'(x) = 3x^2 + 1$$
$$f(x) = \int (3x^2+1)\,dx = x^3 + x + c$$
$$f(0) = 0^3 + 0 + c = 2$$
$$f(x) = x^3 + x + 2$$

73. See Example 3.13.

$$v(t) = -32t + 10$$
$$\int (-32t + 10)\,dt = -16t^2 + 10t + c$$
$$s(0) = -16\cdot 0^2 + 10\cdot 0 + c = 2$$
$$s(t) = -16t^2 + 10t + 2$$

75. See Example 1.2.

$$\sum_{i=1}^{6}\left(i^2+3i\right) = \left(1^2+3(1)\right) + \left(2^2+3(2)\right)$$
$$+ \left(3^2+3(3)\right) + \left(4^2+3(4)\right)$$
$$+ \left(5^2+3(5)\right) + \left(6^2+3(6)\right)$$
$$= 4 + 10 + 18 + 28 + 40 + 54$$
$$= 154$$

77. See Example 1.3.

$$\sum_{i=1}^{100}\left(i^2-1\right) = \sum_{i=1}^{100} i^2 - \sum_{i=1}^{100} 1$$
$$= \frac{100(101)(201)}{6} - 100 = 338{,}250$$

79. See Example 2.3.

$$\frac{1}{n^3}\sum_{i=1}^{n}\left(i^2-i\right)=\frac{1}{n^3}\left[\sum_{i=1}^{n} i^2-\sum_{i=1}^{n} i\right]$$

$$=\frac{1}{n^3}\left[\frac{n(n+1)(2n+1)}{6}-\frac{n(n+1)}{2}\right]$$

$$\lim_{n\to\infty}\left(\frac{1}{n^3}\right)\left[\frac{n(n+1)(2n+1)}{6}-\frac{n(n+1)}{2}\right]$$

$$=\lim_{n\to\infty}\left(\frac{1}{n^3}\right)\left[\frac{(n+1)(2n^2+n-3n)}{6}\right]$$

$$=\lim_{n\to\infty}\frac{(n+1)(2n^2-2n)}{6n^3}=\frac{1}{3}$$

81. See Example 2.1.
$f(x)=x^2$, $[0, 2]$, $n=8$, midpoint evaluation

$$R_8=\frac{2-0}{8}\left[f\left(\frac{1}{8}\right)+f\left(\frac{3}{8}\right)+f\left(\frac{5}{8}\right)+f\left(\frac{7}{8}\right)\right.$$
$$\left.+f\left(\frac{9}{8}\right)+f\left(\frac{11}{8}\right)+f\left(\frac{13}{8}\right)+f\left(\frac{15}{8}\right)\right]$$
$$=\frac{1}{4}(.015625+.140625+.390625+.765625$$
$$+1.265625+1.890625+2.640625+3.515625)$$
$$=2.65625$$

83. See Example 2.1.

$$f(x)=\sqrt{x+1}\text{ on }[0, 3],$$

$n=8$, midpoint evaluation

$$R_8=\frac{3-0}{8}\left[f\left(\frac{3}{16}\right)+f\left(\frac{9}{16}\right)+f\left(\frac{15}{16}\right)\right.$$
$$+f\left(\frac{21}{16}\right)+f\left(\frac{27}{16}\right)+f\left(\frac{33}{16}\right)$$
$$\left.+f\left(\frac{39}{16}\right)+f\left(\frac{45}{16}\right)\right]$$
$$\approx\frac{3}{8}(12.4483)\approx 4.668$$

85. See Section 9.

(a)
$$A_8=\frac{1.6-0}{8}(f(0)+f(.2)+f(.4)+f(.6)$$
$$+f(.8)+f(1)+f(1.2)+f(1.4))$$
$$=\frac{1}{5}(1+1.4+1.6+2+2.2+2.4+2+1.6)$$
$$=2.84$$

(b)
$$A_8=\frac{1.6-0}{8}(f(.2)+f(.4)+f(.6)+f(.8)$$
$$+f(1)+f(1.2)+f(1.4)+f(1.6))$$
$$=\frac{1}{5}(1.4+1.6+2+2.2+2.4+2$$
$$+1.6+1.4)$$
$$=2.92$$

(c)
$$T_8=\frac{1.6-0}{2(8)}[f(0)+2f(.2)+2f(.4)$$
$$+2f(.6)+2f(.8)+2f(1)+2f(1.2)$$
$$+2f(1.4)+f(1.6)]$$
$$=2.88$$

(d)
$$S_8=\frac{1.6-0}{3(8)}[f(0)+4f(.2)+2f(.4)$$
$$+4f(.6)+2f(.8)+4f(1)+2f(1.2)$$
$$+4f(1.4)+f(1.6)]$$
$$\approx 2.907$$

87. See Example 9.10.
Simpson's Rule is expected to be most accurate.

89. See Example 2.3.

$$\int_0^1 2x^2\,dx$$

$$\Delta x=\frac{1-0}{n}=\frac{1}{n},\ x_i=\frac{i}{n}$$

$$R_n = \sum_{i=1}^{n} f(x_i)\,\Delta x$$
$$= \sum_{i=1}^{n} \left(2x_i^2\right)\Delta x$$
$$= \sum_{i=1}^{n} \left(2\left(\frac{i}{n}\right)^2\right)\frac{1}{n}$$
$$= \frac{2}{n^3}\sum_{i=1}^{n} i^2$$
$$= \frac{2}{n^3}\cdot\frac{n(n+1)(2n+1)}{6}$$
$$\lim_{n\to\infty}(R_n) = \lim_{n\to\infty}\left(\frac{1}{3}\cdot\frac{(n+1)(2n+1)}{n^2}\right) = \frac{2}{3}$$

91. See Example 4.5.

$$\int_0^3 \left(3x - x^2\right)dx = \left(\frac{3x^2}{2} - \frac{x^3}{3}\right)\Bigg|_0^3$$
$$= \left(\frac{27}{2} - \frac{27}{3}\right) - (0 - 0)$$
$$= \frac{27}{6}$$
$$= \frac{9}{2}$$

93. See Example 4.10.

$$v(t) = 40 - 10t,\ [1, 2]$$
$$d = \int_1^2 (40 - 10t)\,dt$$
$$= \left(40t - 5t^2\right)\Big|_1^2$$
$$= (80 - 20) - (40 - 5)$$
$$= 25$$

95. See Example 2.7.

$$f_{ave} = \frac{1}{2}\int_0^2 e^x\,dx$$
$$= \frac{1}{2}e^x\Big|_0^2$$
$$= \frac{1}{2}\left(e^2 - e^0\right)$$
$$= \frac{e^2}{2} - \frac{1}{2}$$

97. See Example 4.7.

$$f(x) = \int_2^x \left(\sin t^2 - 2\right)dt$$
$$f'(x) = \sin x^2 - 2$$

99. See Section 4.9.

$$\int_0^1 \sqrt{x^2+4}\,dx$$

(a) Midpoint:

$$\frac{1-0}{4}\left[f\left(\frac{1}{8}\right) + f\left(\frac{3}{8}\right) + f\left(\frac{5}{8}\right) + f\left(\frac{7}{8}\right)\right]$$
$$\approx 2.079$$

(b) Trapezoidal:

$$\frac{1-0}{2(4)}\left[f(0) + 2f\left(\frac{1}{4}\right) + 2f\left(\frac{1}{2}\right) + 2f\left(\frac{3}{4}\right) + f(1)\right]$$
$$\approx 2.083$$

(c) Simpson:

$$\frac{1-0}{3(4)}\left[f(0) + 4f\left(\frac{1}{4}\right) + 2f\left(\frac{1}{2}\right) + 4f\left(\frac{3}{4}\right) + f(1)\right]$$
$$\approx 2.080$$

101. See Section 4.9.

N	Midpoint	Trapezoidal	Simpson
20	2.08041	2.08055	2.08046
40	2.08045	2.08048	2.08046

103. See Example 7.1.

$$\frac{4}{x^2-3x-4} = \frac{4}{(x-4)(x+1)}$$
$$= \frac{A}{x-4} + \frac{B}{x+1}$$
$$4 = A(x+1) + B(x-4)$$
$$x = 4 : A = \frac{4}{5}$$
$$x = -1 : B = -\frac{4}{5}$$
$$\frac{4}{x^2-3x-4} = \frac{\frac{4}{5}}{x-4} - \frac{\frac{4}{5}}{x+1}$$

105. See Example 7.2.

$$\frac{-6}{x^3+x^2-2x} = \frac{-6}{x(x^2+x-2)}$$
$$= \frac{-6}{x(x+2)(x-1)}$$
$$= \frac{A}{x} + \frac{B}{x+2} + \frac{C}{x-1}$$
$$-6 = A(x+2)(x-1) + Bx(x-1) + Cx(x+2)$$
$$x = 0 : -6 = -2A; A = 3$$
$$x = -2 : -6 = 6B; B = -1$$
$$x = 1 : -6 = 3C; C = -2$$
$$\frac{-6}{x^3+x^2-2x} = \frac{3}{x} - \frac{1}{x+2} - \frac{2}{x-1}$$

107. See Example 7.4.

$$\frac{x-2}{x^2+4x+4} = \frac{x-2}{(x+2)^2} = \frac{A}{x+2} + \frac{B}{(x+2)^2}$$
$$x - 2 = A(x+2) + B = Ax + 2A + B$$
$$A = 1; B = -2 - 2A = -4$$
$$\frac{x-2}{x^2+4x+4} = \frac{1}{x+2} - \frac{4}{(x+2)^2}$$

109. See Section 4.8.

$$\int e^{3x}\sqrt{4+e^{2x}}\,dx = \int e^x e^{2x}\sqrt{4+e^{2x}}\,dx$$

Let $u = e^x$, $du = e^x\,dx$.

$$\int u^2\sqrt{4+u^2}\,du$$

From the table,

$$\int u^2\sqrt{a^2+u^2}\,du = \frac{u}{8}(a^2+2u^2)\sqrt{a^2+u^2} - \frac{a^4}{8}\ln\left(u+\sqrt{a^2+u^2}\right) + c;$$

$$\int u^2\sqrt{4+u^2}\,du = \frac{u}{8}(4+2u^2)\sqrt{4+u^2} - 2\ln\left(u+\sqrt{4+u^2}\right) + c$$
$$= \frac{1}{8}e^x(4+2e^{2x})\sqrt{4+e^{2x}} - 2\ln\left(e^x+\sqrt{4+e^{2x}}\right) + c.$$

111. See Section 4.8.

$$\int \sec^4 x\,dx$$

From the table,

$$\int \sec^n x\,dx = \frac{1}{n-1}\tan u\sec^{n-2} u + \frac{n-2}{n-1}\int \sec^{n-2} u\,du;$$

$$\int \sec^4 x\,dx = \frac{1}{3}\tan x\sec^2 x + \frac{2}{3}\int \sec^2 x\,dx$$
$$= \frac{1}{3}\tan x\sec^2 x + \frac{2}{3}\tan x + c.$$

113. See Section 4.8.

$$\int \frac{4}{x(3-x)^2}\,dx = 4\int \frac{1}{x(3-x)^2}\,dx$$

From the table,

$$\int \frac{du}{u(a+bu)^2} = \frac{1}{a(a+bu)} - \frac{1}{a^2}\ln\left|\frac{a+bu}{u}\right| + c;$$

$$4\int \frac{1}{x(3-x)^2}\,dx = 4\left[\frac{1}{3(3-x)} - \frac{1}{9}\ln\left|\frac{3-x}{x}\right|\right] + c$$

$$= \frac{4}{3(3-x)} - \frac{4}{9}\ln\left|\frac{3-x}{x}\right| + c.$$

115. See Section 4.8.

$$\int \frac{\sqrt{9+4x^2}}{x^2}\,dx = \int \frac{2\sqrt{\frac{9}{4}+x^2}}{x^2}\,dx$$

From the table,

$$\int \frac{\sqrt{a^2+u^2}}{u^2}\,du = -\frac{\sqrt{a^2+u^2}}{u} + \ln\left(u+\sqrt{a^2+u^2}\right) + c;$$

$$2\int \frac{2\sqrt{\frac{9}{4}+x^2}}{x^2}\,dx = 2\left[-\frac{\sqrt{\frac{9}{4}+x^2}}{x} + \ln\left(x+\sqrt{\frac{9}{4}+x^2}\right)\right] + c$$

$$= -\frac{\sqrt{9+4x^2}}{x} + 2\ln\left(x+\sqrt{\frac{9}{4}+x^2}\right) + c.$$

117. See Example 5.1.

$$\int \frac{\sqrt{4-x^2}}{x}\,dx$$

From the table,

$$\int \frac{\sqrt{a^2-u^2}}{u}\,du = \sqrt{a^2-u^2} - a\ln\left|\frac{a+\sqrt{a^2-u^2}}{u}\right| + c;$$

$$\int \frac{\sqrt{4-x^2}}{x}\,dx = \sqrt{4-x^2} - 2\ln\left|\frac{2+\sqrt{4-x^2}}{x}\right| + c.$$

Chapter 5

Applications of the Definite Integral

5.1 Area of a Plane Region

1. The area below $y = x^3$ and above $y = x^2 - 1$ on the interval [1, 3]:

$$\int_1^3 \left[x^3 - \left(x^2 - 1\right)\right] dx$$
$$= \left(\frac{x^4}{4} - \frac{x^3}{3} + x\right)\Big|_1^3$$
$$= \left(\frac{81}{4} - \frac{27}{3} + 3\right) - \left(\frac{1}{4} - \frac{1}{3} + 1\right)$$
$$= \frac{160}{12} = \frac{40}{3}.$$

3. The area below $y = e^x$ and above $y = x - 1$ on the interval $[-2, 0]$:

$$\int_{-2}^0 \left[e^x - (x - 1)\right] dx$$
$$= \left(e^x - \frac{x^2}{2} + x\right)\Big|_{-2}^0$$
$$= (1 - 0 + 0) - \left(e^{-2} - \frac{4}{2} + (-2)\right)$$
$$= 5 - e^{-2}.$$

5. The area between $y = x^2 - 1$ and $y = 1 - x$ on the interval [0, 2]. Because the curves cross at $x = 1$, there are two pieces:

$$\int_0^1 \left[1 - x - \left(x^2 - 1\right)\right] dx +$$
$$\int_1^2 \left[x^2 - 1 - (1 - x)\right] dx$$
$$= \int_0^1 \left(2 - x - x^2\right) dx +$$
$$\int_1^2 \left(x^2 + x - 2\right) dx$$
$$= \left(2x - \frac{x^2}{2} - \frac{x^3}{3}\right)\Big|_0^1 +$$
$$\left(\frac{x^3}{3} + \frac{x^2}{2} - 2x\right)\Big|_1^2$$
$$= \left(2 - \frac{1}{2} - \frac{1}{3}\right) - (0 - 0 - 0) +$$
$$\left(\frac{8}{3} + \frac{4}{2} - 4\right) - \left(\frac{1}{3} + \frac{1}{2} - 2\right) = 3.$$

7. The area between the curves $y = x3 - 1$ and $y = 1 - x$ on the interval $[-2, 2]$. Again, the curves cross (at $x = 1$) and so there are two pieces:

$$\int_{-2}^1 \left[1 - x - \left(x^3 - 1\right)\right] dx +$$
$$\int_1^2 \left[x^3 - 1 - (1 - x)\right] dx$$
$$= \left(2x - \frac{x^2}{2} - \frac{x^4}{4}\right)\Big|_{-2}^1 +$$
$$\left(\frac{x^4}{4} - 2x + \frac{x^2}{2}\right)\Big|_1^2$$
$$= \left(2 - \frac{1}{2} - \frac{1}{4}\right) - (-4 - \frac{4}{2} - \frac{16}{4}) +$$
$$\left(\frac{16}{4} - 4 + \frac{4}{2}\right) - \left(\frac{1}{4} - 2 + \frac{1}{2}\right)$$
$$= \frac{5}{4} - (-10) + 2 - \left(-\frac{5}{4}\right) = \frac{29}{2}.$$

9. The area enclosed by $y = x^2 - 1$ and $y = 7 - x^2$. These curves cross at $x = 2$ and at $x = -2$. The area enclosed is

$$\int_{-2}^{2} \left[7 - x^2 - \left(x^2 - 1\right)\right] dx$$
$$= \left(8x - \frac{2x^3}{3}\right)\Big|_{-2}^{2}$$
$$= \left(16 - \frac{16}{3}\right) - \left(-16 + \frac{16}{3}\right)$$
$$= \frac{64}{3}.$$

11. $y = x^2 + 1$, $y = 3x - 1$. The curves cross at $x = 1$ and at $x = 2$. The area enclosed is

$$\int_{1}^{2} \left[3x - 1 - \left(x^2 + 1\right)\right] dx$$
$$= \left(\frac{3x^2}{2} - 2x - \frac{x^3}{3}\right)\Big|_{1}^{2}$$
$$= \left(\frac{12}{2} - 4 - \frac{8}{3}\right) - \left(\frac{3}{2} - 2 - \frac{1}{3}\right)$$
$$= \frac{1}{6}.$$

13. $y = xe^x$, $y = 2x$. The curves cross at $x = 0$ and again at $e^x = 2$ or $x = \ln 2$. The area enclosed is

$$\int_{0}^{\ln 2} 2x - xe^x \, dx = x^2\Big|_{0}^{\ln 2} - \int_{0}^{\ln 2} xe^x \, dx.$$

The first is $\ln^2 2$. The integral, after integration by parts, is

$$(xe^x - e^x)\Big|_{0}^{\ln 2} = (2 \ln 2 - 2) - (-1) = 2 \ln 2 - 1.$$

Final answer, $\ln^2 2 + 1 - 2 \ln 2 = 0.09416$ (rounded).

15. $y = x^3$, $y = x^2$. The curves meet only at $x = 0$ and $x = 1$. The area enclosed is

$$\int_{0}^{1} \left(x^2 - x^3\right) dx = \left(\frac{x^3}{3} - \frac{x^4}{4}\right)\Big|_{0}^{1} =$$
$$\left(\frac{1}{3} - \frac{1}{4}\right) - (0 - 0) = \frac{1}{12}.$$

17. $y = e^x$, $y = 1 - x^2$. The curves cross at $x = 0$ and near $x = -.7145$. The area enclosed is about

$$\int_{-.7145}^{0} (1 - x^2) - e^x \, dx$$
$$= \left(-e^x + x - \frac{x^3}{3}\right)\Big|_{-.7145}^{0}$$
$$\approx (-1 + 0 - 0) - (-1.08235)$$
$$\approx .08235.$$

19. $y = \sin x$, $y = x^2$. The curves cross at $x = 0$ and near $x = .8767$. The area enclosed is about

$$\int_{0}^{.8767} \left(\sin x - x^2\right) dx = \left(-\cos x - \frac{x^3}{3}\right)\Big|_{0}^{.8767}$$
$$\approx .135697.$$

21. $y = x^4$, $y = 2 + x$. The curves cross at $x = -1$ and near $x = 1.3532$. The area enclosed is about

$$\int_{-1}^{1.3532} \left(2 + x - x^4\right) dx = \left(2x + \frac{x^2}{2} - \frac{x^5}{5}\right)\Big|_{-1}^{1.3532}$$
$$\approx 4.01449.$$

23. $y = x$, $y = 2 - x$, $y = 0$. The middle equation, which is the right-hand boundary, is also $x = 2 - y$. The figure is a triangle of base 2, height 1, hence area 1. As an integral in y, its area is given by

$$\int_{0}^{1} [(2 - y) - y] dy = \int_{0}^{1} [2 - 2y] dy$$
$$= \left(2y - y^2\right)\Big|_{0}^{1}$$
$$= 1 - 0 = 1.$$

25. $x = 3y$, $x = 2 + y^2$. The curves cross at $y = 1$ and $y = 2$. The area enclosed is given, as an integral in y, by

$$\int_1^2 \left[3y - \left(2 + y^2\right)\right] dy$$
$$= \left(\frac{3}{2}y^2 - 2y - \frac{y^3}{3}\right)\Bigg|_1^2$$
$$= \left(6 - 4 - \frac{8}{3}\right) - \left(\frac{3}{2} - 2 - \frac{1}{3}\right)$$
$$= \frac{1}{6}.$$

27. $x = y,\ x = -y,\ x = 1$

$$\int_0^1 [x - (-x)]\, dx = 2\int_0^1 x\, dx = x^2\Big|_0^1 = 1 - 0 = 1$$

29. $y = x,\ y = 2,\ y = 6x,\ y = 0$. The figure is a trapezoid of lower base 6, upper base 2, and height 2. Its area is 8. As an integral in y, the area is given by

$$\int_0^2 [(6 - y) - y]\, dy = \int_0^2 (6 - 2y) dy$$
$$= \left(6y - y^2\right)\Big|_0^2$$
$$= (12 - 4) - (0 - 0)$$
$$= 8.$$

31. $$A = \frac{1}{b - a}\int_a^b f(x)\, dx,\ f(x) = x^2 \text{ on } [0, 3]$$

$$A = \frac{1}{3 - 0}\int_0^3 x^2\, dx = \left(\frac{1}{3}\cdot\frac{x^3}{3}\right)\Bigg|_0^3 = \frac{27}{9} - 0 = 3$$

Relative to the interval [0, 3], the inequality $x^2 < 3$ holds only on the subinterval $[0, \sqrt{3})$. We find

$$\int_0^{\sqrt{3}} \left(3 - x^2\right) dx = \left(3x - \frac{x^3}{3}\right)\Bigg|_0^{\sqrt{3}}$$
$$= (3\sqrt{3} - \sqrt{3}) - (0 - 0)$$
$$= 2\sqrt{3},\ \text{whereas}$$

$$\int_{\sqrt{3}}^3 \left(x^2 - 3\right) dx = \left(\frac{x^3}{3} - 3x\right)\Bigg|_{\sqrt{3}}^3$$
$$= (9 - 9) - (\sqrt{3} - 3\sqrt{3})$$
$$= 2\sqrt{3},\ \text{which is the same.}$$

33. $$f(4) = 16.1e^{.07(4)} = 21.3$$
$$g(4) = 21.3e^{.04(4-4)} = 21.3$$

21.3 represents the consumption rate (million barrels per year) at time $t = 4$ (1/1/74).

$$\int_4^{10} \left(16.1e^{.07t} - 21.3e^{.04(t-4)}\right) dt$$
$$= \left(230e^{.07t} - 532.5e^{.04(t-4)}\right)\Big|_4^{10}$$
$$= 14.4 \text{ million barrels saved.}$$

35. For $t \geq 0$,

$$b(t) = 2e^{.04t} \geq 2e^{.02t} = d(t)$$
$$\int_0^{10} \left(2e^{.04t} - 2e^{.02t}\right) dt$$
$$= \left(50e^{.04t} - 100e^{.02t}\right)\Big|_0^{10}$$
$$= 2.45 \text{ million people.}$$

This number represents births minus deaths, hence population growth over the ten-year interval.

37. $$\int_0^{.4} f_c(x) \approx \frac{.4}{3(4)}\{f_c(0) + 4f_c(.1) + 2f_c(.2) + 4f_c(.3) + f_c(.4)\} = 291.67.$$

$$\int_0^{.4} f_e(x) \approx \frac{.4}{3(4)}\{f_e(0) + 4f_e(.1) + 2f_e(.2) + 4f_e(.3) + f_e(.4)\} = 102.33.$$

$$\frac{\int_0^{.4} f_c(x) - \int_0^{.4} f_e(x)}{\int_0^{.4} f_c(x)} \approx \frac{291.67 - 102.33}{291.67}$$
$$= .6491\ldots.$$

$1 - .6491 = .3508$, so the proportion of energy retained is about 35.08%.

39. $$\int_0^3 f_s(x) \approx \frac{3}{3(4)}\{f_s(0) + 4f_s(.75) + 2f_s(1.5) + 4f_s(2.25) + f_s(3)\} = 860.$$

$$\int_0^3 f_r(x) \approx \frac{3}{3(4)}\{f_r(0) + 4f_r(.75) + 2f_r(1.5) + 4f_r(2.25) + f_r(3)\} = 800.$$

$$1 - \left(\frac{860 - 800}{860}\right) = .9302.$$

Energy returned by the tendon is 93.02%.

41. $$f(t) = -40 - 32t;\ g(t) = -30 - 32t$$

$$\int_0^{10} [g(t) - f(t)]\,dt = \int_0^{10} 10\,dt = 100$$

The numerical "area" between the curves is 100, but since the vertical units (for f and g) are feet per second, while the horizontal units are seconds, the units of the integral are *feet* rather than *square feet.* The number represents the 100 feet that separate the objects after 10 seconds. This can also be viewed as the (constant) difference in their velocities (10 feet per second) multiplied by 10 elapsed seconds.

43. $f(t) = 40(1 - e^{-t});\ g(t) = 20t.$
$f(t) = g(t)$ at $t = 0$ and $t \approx 1.593624$.
Therefore the largest lead is

$$\int_0^{1.594} [f(t) - g(t)]\,dt \approx 6.476 \text{ miles.}$$

We find more generally that

$$\int_0^T [f(t) - g(t)]\,dt = 40(T + e^{-T} - 1) - 10T^2.$$

This number represents the (signed) difference in distance after time T. The B-car catches up when this number is zero, or after about 2.557 hours (using solver).

45. Without formulae or tables, only rough or qualitative estimates are possible.

time	1	2	3	4	5
amount	397	403	401	412	455

(answers will vary)

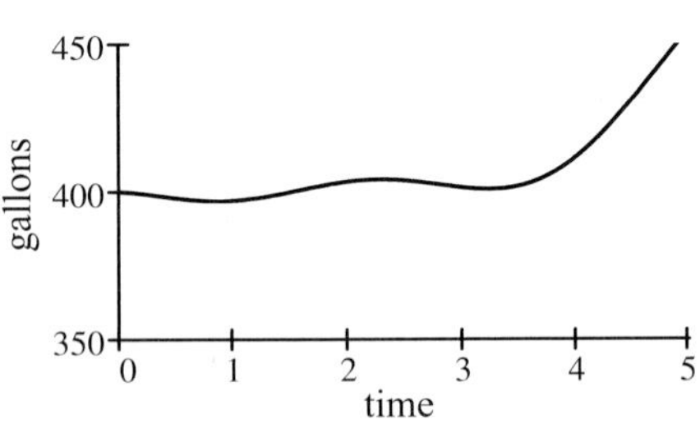

47. In this set-up, p is price and q is quantity. We find that $D(q) = S(q)$ only if $D(q) = S(q)$

$$10 - \frac{q}{40} = 2 + \frac{q}{120} + \frac{q^2}{1200}$$

$$12000 - 30q = 2400 + 10q + q^2$$

$$q^2 + 40q - 9600 = 0$$

$$(q - 80)(q + 120) = 0$$

within the range of the picture only at $q = 80$. Thus $q^* = 80$ and $p^* = D(q^*) = S(q^*) = 8$.

Consumer surplus, as an area, is that part of the picture below the D curve, above $p = p^*$, and to the left of $Q = q^*$.

Numerically the consumer surplus in this case is

$$\int_0^{q^*} \left[D(q) - p^*\right] dq$$

$$= \int_0^{80} \left(2 - \frac{q}{40}\right) dq$$

$$= 2q - \frac{q^2}{80}\Bigg|_0^{80} = 160 - 80 = 80.$$

The units are dollars (q counting items, p in dollars per item).

49. The curves, meeting as they do at 2 and 5, represent the derivatives C' and R'. The area (a) between the curves over the interval [0, 2] is the loss resulting from the production of the first 2000 items. The area (b) between the curves over the interval [2, 5] is the profit resulting from the production of the next 3000 items. The area (c), as the *sum* of the two previous (call it (a)+(b)), is *without meaning.* However, the *difference* (b)−(a) would be the total profit on the first 5000 items, or, if negative, would represent the loss. The area (d) between the curves over the interval [5, 6] represents the loss attributable to the (unprofitable) production of the next thousand items after the first 5000.

51. Let $y_1 = ax^2 + bx + c$, $y_2 = mx + n$, and $u = y_1 - y_2$. If we assume that $a < 0$, then $y_1 > y_2$ on (A, B) and the area between the curves is given by the integral

$$\int_A^B (y_1 - y_2)\,dx = \int_A^B u\,dx = ux|_A^B - \int_A^B xdu.$$

By assumption, u is zero ($y_1 = y_2$) at both A and B, so the first part of the last expression is zero. We must now show that

$$-\int_A^B x[2ax + (b - m)]\,dx (= -\int_A^B xdu)$$

is the same as

$$|a|(B - A)^3/6 =$$
$$|a|(B^3 - 3B^2A + 3BA^2 - A^3)/6.$$

But again because $u = 0$ at both A and B, we know that

$$aA^2 + bA + c = mA + n \text{ and}$$
$$aB^2 + bB + c = mB + n.$$

By subtraction of the first from second, factoring out (and cancelling) $B - A$, we learn $a(B + A) = m - b$, so that our target integral is also given by

$$-2a \int_A^B x(x - \frac{A + B}{2})\,dx =$$
$$|a|\{2(B^3 - A^3)/3 - (A + B)(B^2 - A^2)/2\},$$

and the student who cares enough can finish the details. The case in which $a > 0 (y_2 > y_1)$ is not essentially different.

53. Let the upper parabola be $y = y_1 = qx^2 + v + h$ and let the lower be $y = y_2 = px^2 + v$. They are to meet at $x = w/2$, so we must have $qw^2/4 + h = pw^2/4$, hence $h = (p - q)w^2/4$ or $(q - p)w^2 = -4h$. Using symmetry, the area between the curves is given by the integral

$$2\int_0^{w/2} (y_1 - y_2)\,dx$$
$$= 2\int_0^{w/2} [h + (q - p)x^2]\,dx$$
$$= 2[hw/2 + (q - p)w^3/24]$$
$$= w[h + (q - p)w^2/12]$$
$$= w[h - 4h/12] = (2/3)wh.$$

5.2 Volume

1. $A(x) = x + 2,\ -1 \le x \le 3$

$$V = \int_{-1}^3 A(x)\,dx = \int_{-1}^3 (x + 2)\,dx$$
$$= \left(\frac{x^2}{2} + 2x\right)\Big|_{-1}^3$$
$$= \left(\frac{9}{2} + 6\right) - \left(\frac{1}{2} - 2\right)$$
$$= 12$$

3. $A(x) = \pi(4 - x)^2,\ 0 \le x \le 2$

$$V = \pi \int_0^2 (4 - x)^2\,dx$$
$$= -\frac{\pi}{3}(4 - x)^3\Big|_0^2$$
$$= -\frac{\pi}{3}(8 - 64) = \frac{56\pi}{3}$$

5. At height y, the horizontal cross section is a square of side length proportional to $160 - y$. Hence, its area is proportional to $(160 - y)^2$ and actually equal to

$$\frac{300^2}{160^2}(160 - y)^2 = \left(\frac{300}{160}\right)^2 (160 - y)^2.$$

Integrating this from $y = 0$ to $y = 160$, we get

$$-\left(\frac{300}{160}\right)^2 \frac{(160 - y)^3}{3}\Big|_0^{160} = \frac{300^2 \cdot 160}{3}$$
$$= 4{,}800{,}000(\textit{cubic feet}).$$

This should be one eighth $((1/2) \times (1/2) \times (1/2))$ of the answer to Example 2.1 (38,400,000), and it is.

7. Replacing the 300 by 750 and the 160 by 500 in the previous solution will produce the answer $750^2 \cdot 500/3 = 93{,}750{,}500$ ft^3, which is one third the area of the base times the height.

9. The key observation in this problem is that by simple proportions, had the steeple continued to a point it would have had height 36, hence 6 extra feet. One can copy the integration method, integrating only to 30, or one can subtract the volume of the missing "point" from the full pyramid. Either way the answer is $3^2 36/3 - (1/2)^2 6/3 = 215/2$.

11.
$$V = \pi \int_0^{2\pi} \left(4 + \sin\frac{x}{2}\right)^2 dx$$
$$= \pi \int_0^{2\pi} \left(16 + 8\sin\frac{x}{2} + \sin^2\frac{x}{2}\right) dx$$
$$= \pi \left(16x - 16\cos\frac{x}{2} + \frac{1}{2}x - \frac{1}{2}\sin x\right)\Big|_0^{2\pi}$$
$$= 33\pi^2 + 32\pi \text{ in}^3$$

13.
$$V = \int_0^1 A(x)\,dx$$
$$\approx \frac{1}{3(10)} \begin{bmatrix} A(0) + 4A(.1) + 2A(.2) + \\ 4A(.3) + 2A(.4) + 4A(.5) + \\ 2A(.6) + 4A(.7) + 2A(.8) + \\ 4A(.9) + A(1.0) \end{bmatrix}$$
$$= (7.4)/30 \approx 0.247\text{cm}^3$$

15.
$$V = \int_0^2 A(x)\,dx \approx$$
$$\frac{2}{3(4)} \begin{bmatrix} A(0) + 4A(.5) + 2A(1) \\ +4A(1.5) + A(2) \end{bmatrix}$$
$$= 2.5 \text{ ft}^3$$

17. $y = 2 - x,\ y = 0,\ x = 0$

(a) about the x-axis

$$V = \pi \int_0^2 (2-x)^2\,dx$$
$$= -\pi \left(\frac{(2-x)^3}{3}\right)\Big|_0^2$$
$$= \frac{8\pi}{3}$$

(b) about $y = 3$

$$V = \pi \int_0^2 \left[3^2 - \{3 - (2-x)\}^2\right] dx$$
$$= \pi \int_0^2 \left[9 - \{1 + x\}^2\right] dx$$
$$= \pi \left[9x\big|_0^2 - \frac{(1+x)^3}{3}\Big|_0^2\right]$$
$$= \pi \left[18 - \frac{3^3 - 1^3}{3}\right] = \frac{28\pi}{3}$$

19. $y = \sqrt{x},\ y = 2,\ x = 0$

(a) about the y-axis

$$V = \pi \int_0^2 (y^2)^2 dy$$
$$= \pi \int_0^2 y^4 dy$$
$$= \pi \left(\frac{y^5}{5}\right)\Big|_0^2$$
$$= \frac{32\pi}{5}$$

(b) about $x = 4$

$$V = \pi \int_0^2 (4)^2 dy - \pi \int_0^2 (4 - y^2)^2 dy$$
$$= \pi \int_0^2 (-y^4 + 8y^2) dy$$
$$= \pi \left(-\frac{y^5}{5} + \frac{8y^3}{3}\right)\Big|_0^2$$
$$= \pi \left[\left(-\frac{32}{5} + \frac{64}{3}\right) - (0 + 0)\right]$$
$$= \frac{224\pi}{15}$$

21. $y = e^x,\ x = 0,\ x = 2,\ y = 0$

(a) about y-axis

$$V = 4\pi e^2 - \pi \int_1^{e^2} (\ln y)^2 dy$$

The integral is

$$[y(\ln y)^2 - 2y \ln y + 2y]\Big|_1^{e^2}$$
$$= 2e^2 - 2.$$

The final answer is $2\pi(e^2 + 1)$.

(b) same region about $y = -2$

$$V = \pi \int_0^2 (e^x + 2)^2 \, dx - \pi \int_0^2 (2)^2 \, dx$$
$$= \pi \int_0^2 \left(e^{2x} + 4e^x\right) dx$$
$$= \pi \left(\frac{e^{2x}}{2} + 4e^x\right)\Big|_0^2$$
$$= \pi \left[\left(\frac{e^4}{2} + 4e^2\right) - \left(\frac{1}{2} + 4\right)\right]$$
$$= \pi \left(\frac{e^4}{2} + 4e^2 - \frac{9}{2}\right)$$

23. $y = \sqrt{\dfrac{x}{x^2 + 2}}$, x-axis, $x = 0$, $x = 1$

(a) about the x-axis

$$V_a = \pi \int_0^1 \frac{x}{x^2 + 2} \, dx =$$
$$\frac{\pi}{2} \ln(x^2 + 2) \mid_0^1 =$$
$$\frac{\pi}{2} \ln(3/2) \approx 0.6369$$

(b) about $y = 3$

$$V_b = 9\pi - \pi \int_0^1 (3 - y)^2 \, dx =$$
$$\left(6\pi \int_0^1 y \, dx\right) - \pi \int_0^1 y^2 \, dx$$

The second integral is our friend V_a from part (a), whereas the software Scientific Workplace gives the value .4302 to the integral

$$\int_0^1 y \, dx = \int_0^1 \sqrt{\frac{x}{x^2 + 2}} \, dx,$$

and this leads to

$$V_b = 6\pi(.4302) - V_a \approx 7.472.$$

25. $y = 3 - x$, x-axis, y-axis

(a) about y-axis

$$V = \pi \int_0^3 (3 - y)^2 dy$$
$$= \pi \int_0^3 \left(9 - 6y + y^2\right) dy$$
$$= \pi \left(9y - 3y^2 + \frac{y^3}{3}\right)\Big|_0^3$$
$$= 9\pi$$

(b) about x-axis

$$V = \pi \int_0^3 (3 - x)^2 \, dx$$
$$= \pi \int_0^3 \left(9 - 6x + x^2\right) dx$$
$$= \pi \left(9x - 3x^2 + \frac{x^3}{3}\right)\Big|_0^3$$
$$= 9\pi$$

(c) about $y = 3$

$$V = \pi \int_0^3 [(3)^2 - (3 - (3 - x))^2] \, dx$$
$$= \pi \int_0^3 [9 - x^2] \, dx = \pi \left(9x - \frac{x^3}{3}\right)\Big|_0^3$$
$$= 18\pi$$

(d) about $y = -3$

$$V = \pi \int_0^3 [((3 - x) + 3)^2 - 3^2] \, dx$$
$$= \pi \int_0^3 [(6 - x)^2 - 9] \, dx$$
$$= \pi \int_0^3 \left(27 - 12x + x^2\right) dx$$
$$= \pi \left(27x - 6x^2 + \frac{x^3}{3}\right)\Big|_0^3$$
$$= 36\pi$$

(e) about $x = 3$

$$V = \pi \int_0^3 [(3)^2 - (3 - (3 - y))^2] dy$$
$$= \pi \int_0^3 \left(9 - y^2\right) dy$$
$$= \pi \left(9y - \frac{y^3}{3}\right)\Big|_0^3$$
$$= 18\pi$$

(f) about $x = -3$

$$V = \pi \int_0^3 [((3 - y) + 3)^2 - (3)^2] dy$$
$$= \pi \int_0^3 [(6 - y)^2 - 9] dy$$
$$= \pi \int_0^3 \left(27 - 12y + y^2\right) dy$$
$$= \pi \left(27y - 6y^2 + \frac{y^3}{3}\right)\Big|_0^3$$
$$= 36\pi$$

One should make geometric observations to confirm some of these results. The equalities of pairs (a,b), (c,e), (d,f) follow by symmetry. In (a) the figure is a cone, and the volume is indeed one third the area of the base times the height ($(1/3) \times (9\pi) \times 3 = 9\pi$). In (c) the figure is a cylinder with a cone of type (a) removed. Hence, its volume is $27\pi - 9\pi = 18\pi$.

27. $y = x^2$, $y = 0$, $x = 1$

(a) about y-axis

$$V = \int_0^1 \pi(1)^2 dy - \int_0^1 \pi \left(\sqrt{y}\right)^2 dy$$
$$= \pi \int_0^1 (1 - y) dy$$
$$= \pi \left(y - \frac{y^2}{2}\right)\Big|_0^1$$
$$= \frac{\pi}{2}$$

(b) about x-axis

$$V = \int_0^1 \pi \left(x^2\right)^2 dx = \pi \frac{x^5}{5}\Big|_0^1 = \frac{\pi}{5}$$

(c) about $x = 1$

$$V = \int_0^1 \pi \left(1 - \sqrt{y}\right)^2 dy$$
$$= \pi \int_0^1 \left(1 - 2y^{1/2} + y\right) dy$$
$$= \pi \left(y - \frac{4}{3}y^{3/2} + \frac{y^2}{2}\right)\Big|_0^1$$
$$= \frac{\pi}{6}$$

(d) about $y = 1$

$$V = \int_0^1 \pi(1)^2 dx - \int_0^1 \pi \left(1 - x^2\right)^2 dx$$
$$= \pi \int_0^1 \left(2x^2 - x^4\right) dx$$
$$= \pi \left(\frac{2}{3}x^3 - \frac{x^5}{5}\right)\Big|_0^1$$
$$= \frac{7\pi}{15}$$

(e) about $x = -1$

$$V = \int_0^1 \pi(2)^2 dy - \int \pi \left(1 + \sqrt{y}\right)^2 dy$$
$$= \pi \int_0^1 \left(3 - 2y^{1/2} - y\right) dy$$
$$= \pi \left(3y - \frac{4}{3}y^{3/2} - \frac{y^2}{2}\right)\Big|_0^1$$
$$= \frac{7\pi}{6}$$

(f) about $y = -1$

$$V = \int_0^1 \pi \left(x^2+1\right)^2 dx - \int_0^1 \pi(1)^2\,dx$$
$$= \pi \int_0^1 \left(x^4 + 2x^2\right) dx$$
$$= \pi \left(\frac{x^5}{5} + \frac{2}{3}x^3\right)\Big|_0^1$$
$$= \frac{13\pi}{15}$$

29. $y = ax^2$, $y = h$, y-axis, revolved about the y-axis

$$V = \pi \int_0^h \left(\sqrt{\frac{y}{a}}\right)^2 dy = \pi \int_0^h \frac{y}{a}dy$$
$$= \frac{\pi y^2}{2a}\Big|_0^h = \frac{\pi h^2}{2a}$$

For cylinder of height h, radius $\sqrt{\frac{h}{a}}$,

$$V_c = \pi\left(\sqrt{\frac{h}{a}}\right)^2 \cdot h = \frac{\pi h^2}{a}$$
$$\frac{1}{2}V_c = \frac{\pi h^2}{2a} = V.$$

31. $V = \pi \int_{-1}^{1} (1)^2 dy = \pi y\Big|_{-1}^{1} = 2\pi$

33.

$$V = \int_{-1}^1 \pi \left(\frac{1-y}{2}\right)^2 dy$$
$$= \pi \int_{-1}^1 \left(\frac{1}{4} - \frac{y}{2} + \frac{y^2}{4}\right)dy$$
$$= \pi\left(\frac{y}{4} - \frac{y^2}{4} + \frac{y^3}{12}\right)\Big|_{-1}^1$$
$$= \pi \left[\left(\frac{1}{4} - \frac{1}{4} + \frac{1}{12}\right) - \left(-\frac{1}{4} - \frac{1}{4} - \frac{1}{12}\right)\right] = \frac{2\pi}{3}$$

This is another instance of the volume-of-a-cone formula.

35.

$$V = \pi \int_{-r}^r \left(\sqrt{r^2 - y^2}\right)^2 dy$$
$$= \pi \int_{-r}^r (r^2 - y^2)dy$$
$$= \pi\ (r^2 y - \frac{y^3}{3})\Big|_{-r}^r$$
$$= \frac{4}{3}\pi r^3$$

37. If we compute the two volumes using disks parallel to the base, we have identical cross sections, so the volumes are the same.

39. The cross sections perpendicular to the x-axis intersect the base in the line segments

$$Ix = \left[-\sqrt{1-x^2}, \sqrt{1-x^2}\right] \quad (-1 \le x \le 1).$$

(a) If each of these line segments is the base of a square, then the cross-sectional area is evidently $A(x) = 4(1-x^2)$.

The volume would be $V_a =$

$$\int_{-1}^1 A(x)\,dx = 2\int_0^1 A(x)\,dx$$
$$= 8\left(x - \frac{x^3}{3}\right)\Big|_0^1 = 16/3.$$

(b) These segments I_x cannot be the literal "bases" of circles, because circles "sit" on a single point of tangency. They could however be *diameters*. Assuming so, the cross sectional area would be "pi times radius-squared," or $\pi(1-x^2)$. The resulting volume would be $\pi/4$ times the previous case, or $4\pi/3$. Since the figure is actually a sphere of radius 1, this well-known volume confirms that (a) is also correct.

41. Reasoning as in the solution to Exercise 39, the line segment I_x is $[x^2, 2-x^2]$, $(1 \le x \le 1)$. The length of this segment is $(2-x^2) - x^2 = 2(1-x^2)$, hence in case (a)

$$A(x) = 4(1-x^2)^2 = 4(1 - 2x^2 + x^4).$$

The volume would again be

$$2\int_0^1 A(x)\,dx = 8\left(x - \frac{2x^3}{3} + \frac{x^5}{5}\right)\Bigg|_0^1 =$$
$$8\left(1 - \frac{2}{3} + \frac{1}{5}\right) = \frac{64}{15}.$$

With the same provisos as in Exercise 39, the answer to (b) would be $\pi/4$ times the (a)-case, or $16\pi/15$. For (c), the volume would be $\sqrt{3}/4$ times the (a)-case, or $16\sqrt{3}/15$.

43. Continuing the theme of Exercises 39 and 41, this time the line segment I_x is $[0, e^{-2x}]$, $(0 \le x \le \ln 5)$. If (a), this is the base of a square, and the cross-sectional area is $A(x) = (e^{-2x})^2 = e^{-4x}$. The volume V_a would be the integral

$$\int_0^{\ln 5} A(x)\,dx = \int_0^{\ln 5} e^{-4x}\,dx = \frac{-e^{-4x}}{4}\Bigg|_0^{\ln 5} =$$
$$\frac{1-\left(\frac{1}{5}\right)^4}{4} = \frac{156}{625} = .2496.$$

In the (b)-case, the segment I_x is the base of a *semi*circle, so the cross-sectional area would be

$$\left(\frac{1}{2}\right)\pi\left(\frac{e^{-2x}}{2}\right)^2 = \left(\frac{\pi}{8}\right)e^{-4x}.$$

The resulting volume V_b would be $(\pi/8)V_a = \frac{39\pi}{1250} \approx .09802$.

45. We must estimate $\pi\int_0^3 [f(x)]^2\,dx$.

The given table can be extended to give these respective values for $f(x)2:4$, 1.44, .81, .16, 1.0, 1.96, and 2.56. Simpson's approximation to the integral would be

$$\frac{3}{(3)(6)}\left\{\begin{array}{l} 4+4(1.44)+2(.81) \\ +4(.16)+2(1.0) \\ +4(1.96)+2.56 \end{array}\right\}.$$

The sum in the braces is 24.42, and this must be multiplied by $\pi/6$, giving a final answer of 12.786 . . . .

47. In this problem, let $x = g(y)$ be the equation of the given curve describing the shape of the container. For each height y, let $V(y)$ be the volume of fluid in the container when the depth is y. Later we will estimate $V(y)$. For now, one knows that $V(y)$ is the integral of $\pi[g(y)]^2$, or by the fundamental theorem of calculus, that

$$\frac{dV}{dy} = \pi[g(y)]^2.$$

In actual practice, y and hence V are functions of t (time). Our primary interest is in y as a function of t, but we will obtain this information indirectly, first finding V as a function of y. It appears that $g(y)$ is about $2y$ for $0 < y < 1$, which leads to $[g(y)]^2 = 4y^2$, $V(y) = 4\pi y^3/3$ (on $0 < y < 1$), and $V(1) = 4\pi/3 = 4.2$. We'll keep the *formula* in mind for later, but for now will use the value at $y = 1$ and the crude trapezoidal estimate

$$V(y+1) = V(y) + \pi[g^2(y) + g^2(y+1)]/2$$

to compile the following table:

y	$g(y)$	$g^2(y)$	$V(y)$
1	2	4	4.2
2	3	9	24.6
3	3	9	52.9
4	3	9	81.2
5	4	16	120.4

The assumption of uniform flow rate amounts to $dV/dt =$ constant, and if we start the clock ($t = 0$) as we begin the flow, we get $V = kt$ for some k. The above table, supplemented by the formula when $y < 1$, can be read to give y (vertical) as a function of V (horizontal). But because $V = kt$, the graph looks exactly the same if the horizontal units are time. In the following picture, we have scaled it on the assumption of a flow rate of 120.4 cubic units per minute, a rate which requires one minute to fill the container. The previous formula, $4\pi y^3/3 = V(= kt = (120.4)t)$ (on $0 < y < 1$), becomes $y = (3.06)t^{1/3}$ for very small t and accounts for the (barely discernible) vertical tangent at $t = 0$.

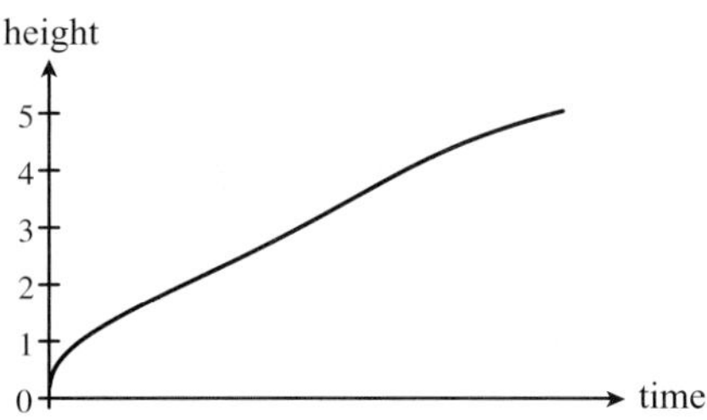

49. $y = x^2$, x-axis, $-1 \leq x \leq 1$, about $x = 2$, $r = 2 - x$, $h = x^2$

$$\begin{aligned} V &= \int_{-1}^{1} 2\pi(2-x)x^2\,dx \\ &= 2\pi\left(\frac{2x^3}{3} - \frac{x^4}{4}\right)\bigg|_{-1}^{1} \\ &= \frac{8\pi}{3} \end{aligned}$$

51. $y = x$, $y = -x$, $x = 1$, about y-axis, $r = x$, $h = 2x$

$$V = \int_0^1 2\pi x(2x)\,dx = \frac{4\pi}{3}x^3\bigg|_0^1 = \frac{4\pi}{3}$$

53. $y = x^2$, $y = 2 - x^2$, rotated

(a) about $x = -2$

$$\begin{aligned} V &= \int_{-1}^{1} 2\pi(x+2)\left((2-x^2) - x^2\right)\,dx \\ &= 2\pi\int_{-1}^{1}\left(4 + 2x - 4x^2 - 2x^3\right)dx \\ &= 2\pi\left(4x + x^2 - \frac{4x^3}{3} - \frac{x^4}{2}\right)\bigg|_{-1}^{1} \\ &= \frac{32\pi}{3} \end{aligned}$$

(b) about $x = 2$. The factor $(x + 2)$ above is replaced by $(2 - x)$. As it turns out, the "x" part in these factors contributes nothing to either definite integral, so the (b)-answer is the same as the previous answer, or $32\pi/3$.

55. $y = x$, $y = x^2 - 2$, rotated

(a) about $x = 2$

$$\begin{aligned} V &= \int_{-1}^{2} 2\pi(2-x)(x - (x^2-2))\,dx \\ &= 2\pi\int_{-1}^{2}\left(4 - 3x^2 + x^3\right)dx \\ &= 2\pi\left(4x - x^3 + \frac{x^4}{4}\right)\bigg|_{-1}^{2} \\ &= \frac{27\pi}{2} \end{aligned}$$

(b) about $x = 3$. Here, the factor of $(2 - x)$ is replaced by $(3 - x)$. The answer is $45\pi/2$.

57. $y = 4 - x$, $y = 4$, $y = x$ revolved around

(a) the x-axis

$$\begin{aligned} V &= \int_2^4 2\pi(y)\,(y - (4-y))\,dy \\ &= 2\pi\int_2^4\left(2y^2 - 4y\right)dy \\ &= 2\pi\left(\frac{2y^3}{3} - 2y^2\right)\bigg|_2^4 \\ &= \frac{80\pi}{3} \end{aligned}$$

(b) the y-axis

$$\begin{aligned} V &= \int_0^2 2\pi(x)\,(4 - (4-x))\,dx + \\ &\quad \int_2^4 2\pi(x)(4-x)\,dx \\ &= 2\pi\left(\frac{x^3}{3}\right)\bigg|_0^2 + 2\pi\left(2x^2 - \frac{x^3}{3}\right)\bigg|_2^4 \\ &= 2\pi\left(\frac{8}{3} + \frac{16}{3}\right) = 16\pi \end{aligned}$$

(c) $x = 4$

$$V = \int_2^4 \pi\,(4-(4-y))^2\,dy - \int_2^4 \pi(4-y)^2 dy$$
$$= \pi \int_2^4 y^2 dy - \pi \int_2^4 (16-8y+y^2)dy$$
$$= \pi \int_2^4 (-16+8y)dy$$
$$= \pi\,(-16y+4y^2)\Big|_2^4 = 16\pi$$

Clearly this should be the same as (b), and it is.

(d) $y = 4$

$$V = \int_2^4 2\pi(4-y)\,(y-(4-y))\,dy$$
$$= 2\pi \int_2^4 \left(-2y^2+12y-16\right)dy$$
$$= 2\pi \left(-\frac{2y^3}{3}+6y^2-16y\right)\Bigg|_2^4$$
$$= \frac{16\pi}{3}$$

59. $y = x^2$, $y = 2x$, $x = 0$ revolved about

(a) the x-axis

$$V = \int_0^1 \pi(2-x)^2\,dx - \int_0^1 \pi\left(x^2\right)^2\,dx$$
$$= \pi \int_0^1 (x^2-4x+4)\,dx - \pi \int_0^1 x^4\,dx$$
$$= \pi \int_0^1 (-x^4+x^2-4x+4)\,dx$$
$$= \pi \left(\frac{x^5}{5}+\frac{x^3}{3}-2x^2+4x\right)\Bigg|_0^1$$
$$= \frac{32\pi}{15}$$

(b) the y-axis

$$V = \int_0^1 2\pi x\left(2-x-x^2\right)\,dx$$
$$= 2\pi \int_0^1 \left(2x-x^2-x^3\right)\,dx$$
$$= 2\pi\left(x^2-\frac{x^3}{3}-\frac{x^4}{4}\right)\Bigg|_0^1$$
$$= \frac{5\pi}{6}$$

(c) $x = 1$

$$V = \int_0^1 2\pi(1-x)(2-x-x^2)\,dx$$
$$= 2\pi \int_0^1 \left(x^3-3x+2\right)dx$$
$$= 2\pi\left(\frac{x^4}{4}-\frac{3x^2}{2}+2x\right)\Bigg|_0^1$$
$$= \frac{3\pi}{2}$$

(d) $y = 2$

$$V = \int_0^1 \pi\left(2-2x^2\right)^2\,dx - \int_0^1 \pi\,(2-(2-x))^2\,dx$$
$$= \pi \int_0^1 (x^4-4x^2+4)\,dx - \pi \int_0^1 x^2\,dx$$
$$= \pi \int_0^1 (x^4-5x^2+4)\,dx$$
$$= \pi\left(\frac{x^5}{5}-\frac{5x^3}{3}+4x\right)\Bigg|_0^1 = \frac{38\pi}{15}$$

61. If the r-interval $[0, R]$ is partitioned by points ri, the circular band

$$\{r_i^2 \le x^2+y^2 \le r_{i+1}^2\}$$

has approximate area $c(r_i)\Delta r_i$ (length times thickness). The limit of the sum of these areas is

$$A = \lim \sum_{i=1}^{n} c(r_i)\Delta r_i = \int_0^R c(r)dr.$$

Because we know that $c(r) = 2\pi r$, we can evaluate the integral, getting

$$2\pi \frac{r^2}{2}\bigg|_0^R = \pi R^2.$$

63. To compile the volumes by the shell method, one needs the equations of the upper and lower boundaries:

$y = \pm b$ for the rectangle;

$y = \pm b\sqrt{1 - \frac{x^2}{a^2}}$ for the ellipse; and

$y = b\left(1 - \frac{x}{2a}\right)$, $y = -b$ for the triangle.

The total height is found by subtraction of the lower from the upper; the result is multiplied by $2\pi x$ and integrated in x from $x = 0$ to $x = a$. The resulting volumes are respectively $2\pi a^2 b$, $4\pi a^2 b/3$, and $2\pi a^2 b/3$. If these are all divided by the last one, the ratios are 3, 2, 1 respectively. (This is what it means to have relative volumes in the proportion 3 : 2 : 1.)

5.3 Arc Length and Surface Area

1. $y = 2x + 1$ $(0 \le x \le 1)$. This is a straight line segment from (0, 1) to (2, 5). As such, its length is

$$\sqrt{(5-1)^2 + (2-0)^2} = \sqrt{20} = 2\sqrt{5}.$$

The arc length integrand would be

$$\sqrt{1 + (y')^2} = \sqrt{1+4} = \sqrt{5}.$$

This constant, integrated from 0 to 2, gives the same result.

3. $y = 1 + 4x^{3/2}$ $(1 \le x \le 2)$. With $y' = 6x^{1/2}$, the arc length integrand is

$$\sqrt{1 + (y')^2} = \sqrt{1 + 36x}.$$

The change of variable $u = 1 + 36x$ converts the arc length integral from

$$\int_1^2 \sqrt{1+36x}\,dx$$

to the integral

$$\int_{37}^{73} \sqrt{u}\left(\frac{du}{36}\right) = \frac{2}{3(36)}\, u^{3/2}\bigg|_{37}^{73}$$
$$= \frac{1}{54}(73\sqrt{73} - 37\sqrt{37}) \approx 7.3824.$$

5.
$$y = \frac{x^2}{4} - \frac{\ln x}{2} \quad (1 \le x \le 2).$$

With

$$y' = \frac{2x}{4} - \frac{1}{2x} = \frac{1}{2}\left(x - \frac{1}{x}\right),$$

the arc length integrand (squared) is

$$1 + (y')^2 = 1 + \frac{1}{4}\left(x^2 - 2 + \frac{1}{x^2}\right) =$$
$$\frac{1}{4}\left(x^2 + 2 + \frac{1}{x^2}\right) = \left[\frac{1}{2}\left(x + \frac{1}{x}\right)\right]^2.$$

The arc length integral is

$$\frac{1}{2}\int_1^2 \left(x + \frac{1}{x}\right) dx = \frac{1}{2}\left(\frac{x^2}{2} + \ln x\right)\bigg|_1^2$$
$$= \frac{1}{2}\left(\frac{3}{2} + \ln 2\right) \approx 1.0965.$$

7.
$$y = \frac{x^4}{8} + \frac{1}{4x^2} \quad (-2 \le x \le -1).$$

With

$$y' = \frac{x^3}{2} - \frac{1}{2x^3} = \frac{1}{2}\left(x^3 - \frac{1}{x^3}\right),$$

one finds (much as in the solution to Exercise 5)

$$1 + (y')^2 = 1 + \frac{1}{4}\left(x^6 - 2 + \frac{1}{x^6}\right) =$$
$$\frac{1}{4}\left(x^6 + 2 + \frac{1}{x^6}\right) = \left[\frac{1}{2}(x^3 + \frac{1}{x^3})\right]^2$$

In this case, however, the quantity in the round bracket is negative, and in taking the square root, we get for the arc length

$$\int_{-2}^{-1} \sqrt{1+(y')^2}\,dx = -\frac{1}{2}\int_{-2}^{-1}\left(x^3+\frac{1}{x^3}\right)dx$$

$$= \frac{1}{2}\left(-\frac{x^4}{4}\Big|_{-2}^{-1} + \frac{1}{2x^2}\Big|_{-2}^{-1}\right) = \frac{1}{2}\left(\frac{15}{4}+\frac{3}{8}\right) = \frac{33}{16}.$$

9.

$$y = \frac{x^{3/2}}{3} - x^{1/2}\ (1 \le x \le 4).$$

With

$$y' = \frac{x^{1/2}}{2} - \frac{x^{-1/2}}{2} = \frac{1}{2}\left(\sqrt{x} - \frac{1}{\sqrt{x}}\right),$$

we find

$$1+(y')^2 = 1+\frac{1}{4}\left(x-2+\frac{1}{x}\right) =$$

$$\frac{1}{4}\left(x+2+\frac{1}{x}\right) = \left[\frac{1}{2}\left(\sqrt{x}+\frac{1}{\sqrt{x}}\right)\right]^2.$$

Hence, the arc length is

$$\int_1^4 \sqrt{1+(y')^2} = \frac{1}{2}\int_1^4\left(\sqrt{x}+\frac{1}{\sqrt{x}}\right)dx$$

$$= \frac{x^{3/2}}{3}\Big|_1^4 + \sqrt{x}\Big|_1^4 = \frac{7}{3}+1 = \frac{10}{3}.$$

11. $y = x^3$ on $(-1, 1)$

$$s = \int_{-1}^{1}\sqrt{1+\left(3x^2\right)^2}\,dx$$
$$= \int_{-1}^{1}\sqrt{1+9x^4}\,dx \approx 3.0957$$

13. $y = 2x - x^2$ on $(0, 2)$

$$s = \int_0^2 \sqrt{1+(2-2x)^2}\,dx \approx 2.9578$$

15. $y = x^3 + x$ on $(0, 3)$

$$s = \int_0^3 \sqrt{1+\left(3x^2+1\right)^2}\,dx \approx 30.3665$$

17. $y = \cos x$ on $(0, \pi)$

$$s = \int_0^{\pi}\sqrt{1+(-\sin x)^2}\,dx$$
$$= \int_0^{\pi}\sqrt{1+\sin^2 x}\,dx \approx 3.8201$$

19. $y = \int_0^x u \sin u du$ makes $y' = x \sin x$ and

$$s = \int_0^{\pi}\sqrt{1+(x\sin x)^2}\,dx = 4.6984.$$

21.

$$s = \int_{-10}^{10}\sqrt{1+\left(\frac{2}{5}\left(e^{x/10}-e^{-x/10}\right)\right)^2}\,dx$$
$$\approx 22.346 \text{ ft}$$

23. In Example 3.4,

$$y(x) = 5(e^{x/10}+e^{-x/10})$$
$$y(0) = 5(e^0+e^0) = 10$$
$$y(-10) = y(10) = 5(e^1+e^{-1}) = 15.43$$
$$\text{sag} = 15.43 - 10 = 5.43 \text{ ft.}$$

A lower estimate for the arc length given the sag would be

$$2\sqrt{(10)^2+(\text{sag})^2}$$
$$= 2\sqrt{100+29.4849} \approx 22.76.$$

This looks good against the calculated arc length of 23.504.

25. $y = \frac{1}{15}x(60-x)$. $y = 0$ when $x = 0$ and when $x = 60$, so the punt traveled 60 yards horizontally.

$$y'(x) = 4 - \frac{2}{15}x = \frac{2}{15}(30-x).$$

This is zero only when $x = 30$, at which point the punt was $(30)^2/15 = 60$ yards high.

$$s = \int_0^{60}\sqrt{1+\left(4-\frac{2}{15}x\right)^2}\,dx \approx 139.4 \text{ yards},$$

$$v = \frac{s}{4 \text{ sec}} = \frac{139.4 \text{ yards}}{4 \text{ sec}} \cdot \frac{3 \text{ feet}}{1 \text{ yard}} = 104.55 \text{ ft/s}.$$

27. All details are provided in the text.

29. Following a pattern of many of the earlier problems (e.g., Exercise 7 with $n = 3$), and with

$$y = \frac{1}{2}\left(\frac{x^{n+1}}{n+1} - \frac{x^{-n+1}}{n-1}\right),$$

one has

$$1 + (y')^2 = \left(\frac{x^n + x^{-n}}{2}\right)^2.$$

Assuming $0 < a < b$, the arc length over the interval $[a, b]$ would be

$$\int_a^b \frac{x^n + x^{-n}}{2}\,dx = \frac{1}{2}\left(\frac{x^{n+1}}{n+1} - \frac{x^{-n+1}}{n-1}\right)\Big|_a^b =$$

$$\frac{1}{2}\left[\frac{(b^{n+1} - a^{n+1})}{n+1} + \frac{(a^{-n+1} - b^{-n+1})}{n-1}\right].$$

31.

$$\frac{d}{dx}\sqrt{2}du \int_0^x \sqrt{1 - \frac{\sin^2 u}{3}}du$$

$$= \frac{1}{2}\sqrt{2} \cdot \sqrt{4 - 2\sin^2 x} = \sqrt{1 + \cos^2 x}$$

33. The antiderivatives returned by some CAS still include an integral, indicating that the CAS can't find an antiderivative in closed form (see Exercise 34). In this case, numerical integration is the method of choice.

35. $y = x^2$, $0 \le x \le 1$, revolved about x-axis

$$S = 2\pi \int_0^1 \text{yds} = 2\pi \int_0^1 x^2\sqrt{1 + (2x)^2}\,dx$$

$$\approx 3.8097$$

37. $y = 2x - x^2$, $0 \le x \le 2$, about x-axis. See Exercise 13.

$$S = 2\pi \int_2^2 \text{yds}$$

$$= 2\pi \int_0^2 (2x - x^2)\sqrt{1 + (2 - 2x)^2}\,dx$$

$$\approx 10.9654$$

39. $y = e^x$, $0 \le x \le 1$, about x-axis

$$S = 2\pi \int_0^1 \text{yds}$$

$$= 2\pi \int_0^1 e^x\sqrt{1 + e^{2x}}\,dx \approx 22.9430$$

41. $y = \cos x$, $0 \le x \le \frac{\pi}{2}$, about x-axis

$$S = 2\pi \int_0^{\pi/2} \text{yds}$$

$$= 2\pi \int_0^{\pi/2} \cos x\sqrt{1 + \sin^2 x}\,dx$$

$$\approx 7.2117$$

43. $$L_1 = \int_{-\pi/6}^{\pi/6} \sqrt{1 + \cos^2 x}\,dx \approx 1.44829$$

$$L_2 = \sqrt{\left(\sin\frac{\pi}{6} - \sin\left(-\frac{\pi}{6}\right)\right)^2 + \left(\frac{\pi}{6} - \left(-\frac{\pi}{6}\right)\right)^2}$$

$$\approx 1.44797$$

$$\frac{L_2}{L_1} = \frac{1.44797}{1.44829} \approx .9998$$

45. $$L_1 = \int_{\pi/6}^{\pi/2} \sqrt{1 + \cos^2 x}\,dx \approx 1.18595$$

$$L_2 = \sqrt{\left(\sin\frac{\pi}{2} - \sin\frac{\pi}{6}\right)^2 + \left(\frac{\pi}{2} - \frac{\pi}{6}\right)^2}$$

$$\approx 1.16044$$

$$\frac{L_2}{L_1} = \frac{1.16044}{1.18595} \approx .9785$$

47. 2π. See Exercise 48. It is not a coincidence. To do the problem the hard way, one might parameterize the position (x, y) of the midpoint by the angle Θ made by the ladder to the wall. Then one would have $x = 4\sin\Theta$, $y = 4\cos\Theta$, and $ds = 4d\Theta$.

49. $\int_0^1 \sqrt{1+(6x^5)^2}\,dx = \int_0^1 \sqrt{1+36x^{10}}\,dx \approx 1.672$

$\int_0^1 \sqrt{1+(8x^7)^2}\,dx = \int_0^1 \sqrt{1+64x^{14}}\,dx \approx 1.720$

$\int_0^1 \sqrt{1+(10x^9)^2}\,dx = \int_0^1 \sqrt{1+100x^{18}}\,dx \approx 1.75$

As $n \to \infty$, the length approaches 2, since one can see that the graph of $y = x^n$ on [0, 1] approaches a path consisting of the horizontal line segment from (0, 0) to (1, 0) followed by the vertical line segment from (1, 0) to (1, 1).

51.

$$y_1 = x^4,\ y_1' = 4x^3$$
$$y_2 = x^2,\ y_2' = 2x$$

Since both are increasing for positive x, y_1 is "steeper" (y_2 is "flatter") if and only if $y_1' > y_2'$, i.e.,

$$4x^3 > 2x,\ x^2 > \frac{1}{2},\ x > \sqrt{\frac{1}{2}}.$$

53.

$$2\pi \int_0^1 x^6\sqrt{1+(6x^5)^2}\,dx \approx 3.313$$
$$2\pi \int_0^1 x^{10}\sqrt{1+(10x^9)^2}\,dx \approx 3.2233$$
$$2\pi \int_0^1 x^{14}\sqrt{1+(14x^{13})^2}\,dx \approx 3.1902$$

As $n \to \infty$, the area approaches π. See the comment following the solution to Exercise 49. Here, the line segment from (0, 0) to (1, 0), even after rotation about the x-axis, has zero area. The other segment, after rotation, fills out a disk of radius 1 and area π.

55. Considering only the vertical segment $x = 1$, $(-1 < y < 1)$, the area after rotation, as an integral in y, would be

$$2\pi \int_{y=-1}^{y=1} x\,ds(y) = 2\pi \int_{-1}^{1} (1)\sqrt{1+0^2}dy$$
$$= 2\pi y|_{-1}^{1} = 4\pi$$

(height times circumference).

The full solid of revolution is a cylinder with radius 1, and its top and bottom each have area $\pi(1)^2 = \pi$. Hence, the total surface area is $4\pi + \pi + \pi = 6\pi$.

57. The equation for the right segment of the triangle is $x = (1-y)/2$. Hence, the resulting area is

$$2\pi \int_{y=-1}^{y=1} x\,ds(y)$$
$$= 2\pi \int_{-1}^{1} \left(\frac{1-y}{2}\right)\sqrt{1+\left(-\frac{1}{2}\right)^2}\,dy$$
$$= 2\pi \int_{-1}^{1} \left(\frac{1-y}{2}\right)\sqrt{\frac{5}{4}}dy$$
$$= \frac{\pi\sqrt{5}}{2}\left(y - \frac{y^2}{2}\right)\Big|_{-1}^{1} = \pi\sqrt{5}.$$

The full revolved figure is a cone with added base of radius 1 (and area π). Hence, the total surface area $\pi\sqrt{5} + \pi \approx 10.1664$.

59. If $x^{2/3} + y^{2/3} = 1$, then in the first quadrant, $y = (1 - x^{2/3})^{3/2}$ and taking only the first-quadrant case (which would produce one fourth of the total length s), we have

$$y' = \frac{3}{2}(1 - x^{2/3})^{1/2}\left(-\frac{2}{3}x^{-1/3}\right)$$
$$= -x^{-1/3}(1 - x^{2/3})^{1/2},$$
$$(y')^2 = x^{-2/3}(1 - x^{2/3}) = x^{-2/3} - 1$$
$$s = 4\int_0^1 \sqrt{1 + y'^2}\,dx$$
$$= 4\int_0^1 \sqrt{x^{-2/3}}\,dx$$
$$= 4\int_0^1 x^{-1/3}\,dx = 4\left(\frac{3}{2}\right)x^{2/3}\Big|_0^1 = 6.$$

There are some technicalities in fully justifying the preceding computation, since the integrand $(x^{-1/3})$ is unbounded at $x = 0$, but the conclusion is sound.

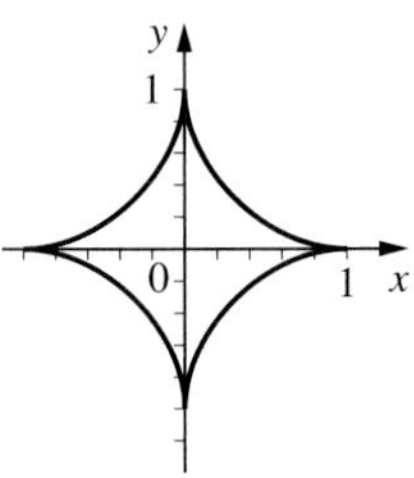

61. $x'(t) = -2 \sin t$; $y'(t) = 4 \cos t$;
t ranges from 0 to 2π.

$$s = \int_0^{2\pi} \sqrt{(-2 \sin t)^2 + (4 \cos t)^2} dt$$
$$= \int_0^{2\pi} \sqrt{4 \sin^2 t + 16 \cos^2 t} dt$$
$$\approx 19.3769$$

63. $x'(t) = 3t^2 - 4$, $y'(t) = 2t$

$$s = \int_{-2}^{2} \sqrt{(3t^2 - 4)^2 + (2t)^2} dt$$
$$\approx 15.6940$$

65. $x'(t) = -2 \sin 2t$; $y'(t) = 4 \cos 4t$;
t ranges from 0 to π.

$$s = \int_0^{\pi} \sqrt{(-2 \sin 2t)^2 + (4 \cos 4t)^2} dt$$
$$\approx \int_0^{\pi} \sqrt{4 \sin^2 2t + 16 \cos^2 4t} dt$$
$$= 9.42943$$

67. $x'(t) = \cos t - t \sin t$; $y'(t) = t \cos t + \sin t$;
t ranges from -1 to 1.

$$s = \int_{-1}^{1} \sqrt{(t \cos t + \sin t)^2 + (\cos t - t \sin t)^2} dt$$
$$= 2 \int_0^1 \sqrt{t^2 + 1} dt = \sqrt{2} + \ln(1 + \sqrt{2}) \approx 2.29559$$

69. One way to parameterize the cycloid is to mark the radius from the center to our point of concern (call it P) in red, and as the circle rolls, let t be the angle (in radians) from the red radius to the vertical. At this value of t, the length of circle which has been rolled out is $4t$ (radius times radians), and this measures the horizontal progress of the center of the circle. More precisely, the position of the center at this value of t is $(4t, 4)$. Hence, the position of our point P is

$$(x, y) = (4t + 4 \sin t, 4 + 4 \cos t).$$

Given this, we find

$$x'(t) = 4(1 + \cos t),$$
$$y'(t) = -4 \sin t,$$

and

$$x'^2 + y'^2 = 16(1 + 2 \cos t + \cos^2 t + \sin^2 t)$$
$$= 32(1 + \cos t) = 64 \cos^2(t/2).$$

One revolution of the circle occurs as t moves from 0 to 2π, hence the arc length traveled by our point P is

$$\int_0^{2\pi} 8|\cos(t/2)| dt =$$
$$16 \int_0^{\pi} \cos(t/2) dt = 32 \sin(t/2)|_0^{\pi} = 32.$$

5.4 Projectile Motion

1. $y(0) = 80$, $y'(0) = 0$

3. $y(0) = 60$, $y'(0) = 10$

5. Let $y(t)$ be the height above water at time t. Presumably, we are to assume an initial velocity of zero. Therefore from $y''(t) = -32$, we find $v(t) = y'(t) = -32t + c_1$, but $c_1 = v(0) = 0$. Continuing, we find $y(t) = -16t + c_2$, but $c_2 = y(0) = 30$. We see that y is zero when $t_2 = 30/16$, or $t = \sqrt{30}/4$. At impact, $v = y'(\sqrt{30}/4) = -32\sqrt{30}/4$. This velocity would probably be reported as a *speed* of $8\sqrt{30}$ or about 43.8 feet per second. (The minus sign indicates downward motion, but speed ignores sign.)

7. This time $v(t) = y'(t) = -32t - 32 = -32(1 + t)$. Thus $y(t) = -16t^2 - 32t + c_2$, but $c_2 = y(0) = 6$. Solving the equation $-16t^2 - 32t + 6 (= y(t)) = 0$ by the quadratic formula, we find $t = -1 + \sqrt{22}/4$

(we consider only the positive solution). The impact velocity is

$$v\left(-1+\frac{\sqrt{22}}{4}\right)=-32\left(\frac{\sqrt{22}}{4}\right)\approx -37.52,$$

which would be reported as a speed of (about) 37.52 feet per second.

9. If an object is dropped (time zero, zero initial velocity) from an initial height of y_0, then the impact moment is $t_0=\sqrt{y_0}/4$ and the impact velocity (ignoring the possible negative sign) is $v_{\text{impact}}=32t_0=8\sqrt{y_0}$. Therefore if the object is dropped from 30 ft, the impact velocity is $8\sqrt{30}\approx 43.8178$ feet per second. If dropped from 120 ft, the impact velocity is $8\sqrt{120}\approx 87.6356$ feet per second. From 3000 ft, the impact velocity is $8\sqrt{3000}\approx 438.178$ feet per second. From a height of hy_0, the impact velocity is

$$8\sqrt{hy_0}=8\sqrt{h}\sqrt{y_0}=\sqrt{h}\,(8\sqrt{y_0})\,,$$

which is to say that impact velocity increases by a factor of $\sqrt{h}$ when initial height increases by a factor of h.

11. See the solution to Exercise 9. We are told that $t_0=4$, hence $v_{\text{impact}}=(32)4=128$ feet per second, and $y_0=(v_0/8)^2=16^2=256$ feet.

13. With y the height at time t, we go back to $y''=-32$, $v(t)=y'=-32t+v(0)$, and $y=-16t^2+tv(0)+y(0)$. In this case, with launch at 64 feet per second from ground level, we have $v(0)=64$ and $y(0)=0$, hence $y=-16t2+64t=-16(t-2)^2+64$ (completing the square), and $v(t)=-32(t-2)$. This is zero at $t=2$, which is when he peaks. At time 2, $y=64$. He remains in the air until the next time $y=0$, which is at $t=4$. We see that $v(4)=-64$, which is, except for sign, the same as his launch velocity.

15. Reviewing the solution to Exercise 13, the difference is that $v(0)$ is unknown. However, we still see that $y=-16t^2+tv(0)=-t[16t-v(0)]$ (factoring, rather than completing the square). The second time that $y=0$ can be seen to occur is at time $t_2=v(0)/16$, at which time

$$v(t_2)=-32t_2+v(0)=v(0)(-2+1)=-v(0).$$

Now we see

$$v(t)=-32t+v(0)=-32t+16t_2=-16(2t-t_2).$$

The peak was therefore at time $t_2/2$, at which time the height was

$$\begin{aligned}&-(t_2/2)[16t2/2-v(0)]=-(t_2/2)[(v(0)/2)-v(0)]\\&=-(v(0)/32)[-v(0)/2]=v(0)^2/64.\end{aligned}$$

In summary, $y_{\max}=[v(0)/8]^2$ in this problem (and more generally, $y_{\max}=[v(0)/8]^2+y(0)$). If $y_{\max}=20$ inches $=5/3$ feet, then $v(0)/8=\sqrt{5/3}$, and $v(0)=8\sqrt{5/3}\approx 10.33$ feet per second. This is considerably less than Michael Jordan's initial velocity of about 17 feet per second, but the difference in velocity is not as dramatic as in height (20 inches to 54 inches).

17. Consulting the solution to Exercise 15, we see that for the flea,

$$y_{\max}=[v(0)/8]^2=(5/8)^2=25/64\text{ feet},$$

or about 4.7 inches.

19. This problem was solved in Exercise 9.

21. For the vertical component, the problem is like Exercise 17 except that the units are meters, so the starting point $y''=-g$ $(=-32$ in #17) now becomes $y''=-9.8$. In addition, $y'(0)=98\sin(\pi/3)=49\sqrt{3}$. We get

$$\begin{aligned}y(t)&=-4.9t^2+ty'(0)=-4.9t(t-[v(0)/4.9])\\&=-4.9t(t-10\sqrt{3}).\end{aligned}$$

The flight time is $10\sqrt{3}$. As to the horizontal range, we have $x'(t)$ constant and forever equal to $98\cos(\pi/3)=49$. Therefore $x(t)=49t$, and in this case the horizontal range is $49(10\sqrt{3})$ (meters).

23. For the vertical component y, the problem is again like Exercise 17. In this case, $y'(0) = 120\sin(\pi/6) = 60$. We get

$$y(t) = -16t^2 + 60t = -16t(t - 15/4).$$

The flight time is 15/4 . As to the horizontal range, we have $x'(t)$ constant and forever equal to $120\cos(\pi/6) = 60\sqrt{3}$. Therefore $x(t) = 60\sqrt{3}t$, and in this case the horizontal range is $(60\sqrt{3})(15/4) = 225\sqrt{3}$ (feet).

25. This problem modifies Example 4.5 by using a service angle of 8° (where Example 4.5 used 7°) and no other changes. Here the serve hits the net. See the solution to Exercise 27 (next) for additional quantifications.

27. In these tennis problems, the issue is purely geometric. Time is irrelevant. One can obtain valuable information by eliminating time and writing y as a function of x. For example, with service angle of θ (in degrees below the horizontal), initial speed v_0, and initial height h, one has

$$y(t) = -16t^2 - tv_0\sin\theta + h,$$
$$x(t) = tv_0\cos\theta,$$

and hence

$$y = f(x) = \frac{-16x^2}{v_0^2\cos^2\theta} - \frac{x\sin\theta}{\cos\theta} + h.$$

Now one could put $x = 60$ (the serve would be in if $f(60) < 0$), or put $x = 39$ (the serve would clear the net if $f(39) > 3$. If one were to set $f(60) = 0$ and solve for v_0, one would obtain a critical speed (call it v_1) for the given (h, θ) *above which the serve would be out.* Solving $f(39) = 3$, one would obtain a second critical speed (call it v_2) *below which the serve would hit the net.* Below we tabulate v_1 and v_2 for $h = 10$ and selected values of θ. The information in the 8° line quantifies the remarks in the solution to Exercise 25. In the 7° line, we see that it would be necessary to reduce the service speed to 149 ft/sec to get it in, and the net would not be a problem. The 7.6° line has these interesting features: the service at 176 ft/sec is out, whereas the service at 170 ft/sec is in.

h (feet)	θ (degrees)	$v1$ (ft/sec)	$v2$ (ft/sec)
10	7.0	149.0	105.7
10	7.6	171.5	117.4
10	8	193.6	127.8

29. Let $(x(t), y(t))$ be the trajectory. In this case

$$y(0) = 6,\ x(0) = 0$$
$$y'(0) = 0,\ x'(0) = 130$$
$$x''(t) \equiv 0,\ x'(t) \equiv 130$$
$$x(t) = 130t.$$

This is 60 at time $t = 6/13$. Meanwhile,

$$y''(t) = -32,\ y'(t) = -32t$$
$$y(t) = -16t^2 + 6$$
$$y\left(\frac{6}{13}\right) = -16\left(\frac{6}{13}\right)^2 + 6 = \frac{438}{169}$$
$$\approx 2.59 \text{ ft.}$$

31. Let $(x(t), y(t))$ be the trajectory. In this case 5° is converted to $\pi/36$ radians.

$$y(0) = 5,\ x(0) = 0.$$
$$y'(0) = 120\sin\frac{\pi}{36} \approx 10.46$$
$$x'(0) = 120\cos\frac{\pi}{36} \approx 119.54$$
$$x''(0) \equiv 0$$
$$x'(t) \equiv 119.54$$
$$x(t) = 119.54t.$$

This is 120 when $t = 120/119.54 = 1.00385\ldots$ Meanwhile,

$$y''(t) = -32$$
$$y'(t) = -32t + 10.46$$
$$y(t) = -16t^2 + 10.46t + 5$$
$$y(1.00385) = -16(1.00385)^2 + 10.46(1.00385) + 5$$
$$\approx -.62 \text{ ft.}$$

33. We are assuming that the height at 120 feet is the same as the release height 5. Let θ be the angle of release (above the horizontal). We have

$$y(t) = -16t^2 + 120t \sin\theta + 5,$$
$$x(t) = 120t \cos\theta.$$

Thus $x(t)$ will be 120 when $t = 1/cos\theta$, at which time $y(t)$ will be 5 only if

$$\frac{-16}{\cos^2\theta} + 120\frac{\sin\theta}{\cos\theta} = 0,$$

hence if $120 \sin\theta \cos\theta = 16$, $60 \sin 2\theta = 16$, $2\theta = \sin^{-1}(16/60) = .2699\ldots$, $\theta = .135$ (radians) or about 7.7°

To find the aim, we need the length of the vertical leg of a right triangle with opposite angle 7.7°, and adjacent leg 120 ft. Thus the player should aim

$$120\tan(7.7^\circ) \approx 120\tan(.135) \approx 16.2 \text{ ft}$$

above the first baseman's head.

35. Assuming that the ramp height h is the same as the height of the cars, this problem seems to be asking for the initial speed v_0 required to achieve a horizontal flight distance of 125 feet from a launch angle of 30° above the horizontal. We may assume $x(0) = 0$, $y(0) = h$, and we find

$$y'(0) = v_0 \sin\frac{\pi}{6} = \frac{v_0}{2}$$
$$x'(0) = v_0 \cos\frac{\pi}{6} = \frac{\sqrt{3}}{2}v_0$$
$$y''(t) \equiv -32,\ x''(t) \equiv 0$$
$$y'(t) = -32t + \frac{v_0}{2},\ x'(t) = \frac{\sqrt{3}}{2}v_0$$
$$y(t) = -16t^2 + \frac{v_0}{2}t + h,\ x(t) = \frac{\sqrt{3}}{2}v_0 t.$$

$x(t)$ will be 125 if $t = 250/\left(\sqrt{3}v_0\right)$, at which time we require that y be h. Therefore,

$$-16\left(\frac{250}{\sqrt{3}v_0}\right)^2 + \frac{v_0}{2}\left(\frac{250}{\sqrt{3}v_0}\right) = 0$$
$$v_0 = \sqrt{\frac{8000}{\sqrt{3}}} \approx 68 \text{ ft/s}.$$

37. Let $(x(t), y(t))$ be the trajectory. In this case,

$$y(0) = 256,\ x(0) = 0$$
$$y'(0) = 0,\ x'(0) = 100$$
$$y''(t) \equiv 32,\ x''(t) \equiv 0$$
$$y'(t) = -32t,\ y(t) = -16t^2 + 256$$
$$x'(t) = 100,\ x(t) = 100t.$$

y will be zero when $t = 4$, at which time x will be 400. This is the drift distance.

39. In this case with

$$\theta_0 = 0 \quad and \quad \omega = 1:$$
$$x''(t) = -25\sin(4t)$$
$$x'(0) = x(0) = 0$$
$$x'(t) = \frac{25}{4}\cos 4t - \frac{25}{4}$$
$$x(t) = \frac{25}{16}\sin 4t - \frac{25}{4}t.$$

41. In this case with

$$\theta_0 = \frac{\pi}{4} \quad and \quad \omega = 2:$$
$$x''(t) = -25\sin\left(8t + \frac{\pi}{4}\right)$$
$$x'(0) = 0 = x(0)$$
$$x'(t) = \frac{25}{8}\cos\left(8t + \frac{\pi}{4}\right) - \frac{25\sqrt{2}}{16}$$
$$x(t) = \frac{25}{64}\sin\left(8t + \frac{\pi}{4}\right) - \frac{25\sqrt{2}}{16}t - \frac{25\sqrt{2}}{128}$$

43. From Exercise 5, time of impact is $t = \frac{\sqrt{30}}{4}$ (seconds). $2\frac{1}{2}$ somersaults corresponds to 5π radians of revolution. Therefore the average angular velocity is

$$\frac{5\pi}{\sqrt{30}/4} = \frac{20\pi}{\sqrt{30}} \approx 11.47 \text{ (radians per second)}.$$

45. Let $(x(t), y(t))$ be the trajectory of the center of the basketball. We are assuming that $y(0) = 6$, $x(0) = 0$, the angle of launch θ of the shot is 52° ($\theta = \frac{13\pi}{45}$ in radians), and the initial speed is 25 feet per second. Therefore,

$$y'(0) = 25 \sin \frac{13\pi}{45} \approx 19.70$$

$$x'(0) = 25 \cos \frac{13\pi}{45} \approx 15.39$$

$$y''(t) \equiv -32,\ x''(t) \equiv 0$$

$$y'(t) = -32t + 19.70,\ x'(t) \equiv 15.39$$

$$y(t) = -16t^2 + 19.70t + 6,\ x(t) = 15.39t.$$

x will be 15 when t is about $15/15.39 = .9746\ldots$, at which time y will be about

$$-16(.9746\ldots)^2 + 19.70(.9746\ldots) + 6 \approx 10.$$

In other words, the center of the ball is at position (15, 10) and the shot is good.

More generally, with unknown θ, the number 19.70 is replaced by $25 \sin \theta$, while the number 15.39 is replaced by $25 \cos \theta$. y will be exactly 10 if

$$-16t^2 + 25t \sin \theta + 6 = 10,$$

$$t = \frac{25 \sin \theta + \sqrt{625 \sin^2 \theta - 256}}{32},$$

$$x = 25t \cos \theta.$$

As a function of θ, this last expression is too complicated to use calculus (easily) to maximize and minimize it on the θ-interval (48°, 57°), but quick spreadsheet calculations give these values:

θ (degrees)	t (seconds)	x (feet)
48.0	0.8757	14.6484
49.0	0.9021	14.7958
50.0	0.9274	14.9024
51.0	0.9516	14.9710
52.0	0.9748	15.0038
52.1	0.9771	15.0051
52.2	0.9793	15.0062
52.3	0.9816	15.0069
52.4	0.9838	15.0073
52.5	0.9861	15.0073
52.6	0.9883	15.0070
52.7	0.9905	15.0064
52.8	0.9928	15.0054
52.9	0.9950	15.0042
53.0	0.9972	15.0026
54.0	1.0187	14.9690
55.0	1.0394	14.9044
56.0	1.0594	14.8100
57.0	1.0787	14.6869

Observe that x is not a monotonic function of θ in this range. It takes its maximum when θ is between 52.4 and 52.5 degrees. The evidence is overwhelming that all the shots will be good.

47.

$$y(t) = 99.62t$$

$$x(t) = -10t^2 + 8.72t$$

$$y(t) = 90 \text{ when } t = 0.903.$$

$$x(0.903) \approx -0.29$$

The ball just barely gets into the goal.

49. Let $(x(t), y(t))$ be the trajectory of the ship. Some of our data is in feet, so we will take $g = -32$ in this problem. We have

$$y''(t) - 32$$

$$y'(t) = -32t + y'(0)$$

$$y(t) = -16t^2 + y'(0)t + y(0)$$

$$x'(t) \equiv c$$

$$x(t) = ct + x(0)$$

Solving for t, we have $\frac{1}{c}(x - x(0)) = t$. Substituting this expression for t in $y(t)$, we have

$$y - y(0) = -16\left[\frac{1}{c}(x - x(0))\right]^2 + y'(0)\left[\frac{1}{c}(x - x(0))\right].$$

Hence, the path is a parabola.

Turning to the question of the duration of weightlessness, we can assume $x(0) = 0$, and we know that $y'(t) = 0$ when $y - y(0) = 2500$. For this unknown time t_1 (the moment when y'

is zero), we have $0 = -32t_1 + y'(0)$. Therefore, $t_1 = y'(0)/32$, and

$$2500 = y(t_1) - y(0) = -16\left[\frac{y'(0)}{32}\right]^2 + y'(0)\left[\frac{y'(0)}{32}\right] = \frac{y'(0)^2}{64}.$$

Hence, $y'(0)2 = 64(2500)$, $y'(0) = 8(50) = 400$, and $t_1 = 400/32 = 25/2$. We now know that $y - y(0) = -16t^2 + 400t$ for *all* t. The second time (t_2) that $y(t) = y(0)$ (after time zero) occurs when $t = 400/16 = 25$ (seconds). This is the duration of the weightless experience. Note that $t_2 = 2t_1$. The plane must pull out of the dive soon after this time.

51. Let $y(t)$ be the height of the first ball at time t, and let v_{0y} be the initial velocity. We can assume $y(0) = 0$. As usual, we have $y'' = -32$, $y' = -32t + v_{0y}$, and $y = -16t2 + tv_{0y}$. The second return to height zero is at time $t = 16/v_{0y}$. If this is to be $5/2$, then $v_{0y} = 40$. But the maximum occurs at time $v_{0y}/32 = 5/4$, at which time the height $(y(5/4))$ is $-16(25/16) + 40(5/4) = 25$ feet. For eleven balls, the difference is that the second return to zero is to be at time $11/4$, hence $v_{0y} = 44$, and the maximum height is 30.25.

53. The student must first study the solution to Exercise 51. Here we have the additional x-component of the motion, which as in so many problems is $x(t) = tv_{0x}$. With initial *speed* of v_0 and initial angle α from the *vertical*, we have $v_{0y} = v_0 \cos\alpha$ and $v_{0x} = v_0 \sin\alpha$. The horizontal distance at elapsed time $v_{0y}/16$ (time of return to initial height) is by formula $x(v_{0y}/16) = (v_{0y}/16)v_{0x}$ which defines ω. As in #51, the maximum height occurs at time $v_{0y}/32$, and at this time the height h is

$$\begin{aligned} &-16(v_{0y}/32)^2 + v_{0y}(v_{0y}/32) = v_{0y}^2/64 \\ &= (v0y/64)(16\omega/v_{0x}) \\ &= (\omega/4)(\cos\alpha/\sin\alpha) = \omega/(4\tan\alpha). \end{aligned}$$

Thus $\omega = 4h\tan\alpha$.

55. We must use the result $\Delta\alpha \approx \dfrac{\Delta\omega}{4h}$ from Exercise 54. With $h = 25$ from the solution to Exercise 51 (ten balls) and $\omega = 1$, we get $\Delta\alpha$ about $1/100 = .01$ (radians) or about $.6°$.

57. Let $(x(t), y(t))$ be the trajectory of the arrow. We are assuming $y(0) = 0$, $x(0) = 0$, and in the metric system we have

$$\begin{aligned} &y''(t) \equiv -9.8 \\ &y'(t) = -9.8t + v_0 \sin\theta \\ &y(t) = -4.9t^2 + tv_0 \sin\theta \\ &x''(t) \equiv 0,\ x'(t) \equiv v_0\cos\theta,\ x(t) = tv_0\cos\theta \end{aligned}$$

$y'(t)$ is zero when $t = \dfrac{v_0 \sin\theta}{9.8}$. At this time,

$$x = v_0\cos\theta\left(\frac{v_0\sin\theta}{9.8}\right) = \frac{v_0^2}{9.8}\sin\theta\cos\theta,$$

$$y = \frac{v_0^2}{2(9.8)}\sin^2\theta,$$

and

$$y/x = \sin\theta/(2\cos\theta) = (\tan\theta)/2.$$

On the other hand, since $t = \dfrac{v_0\sin\theta}{9.8}$ corresponds to the arrow landing in the cauldron, we know that at this time $y = 30$ and $x = 70$. Thus $y/x = 3/7$ and

$$\theta = \tan^{-1}\left(\frac{6}{7}\right) \approx .7086 \approx 40.6°.$$

Note that a right triangle with legs 6 and 7 has hypotenuse $\sqrt{6^2+7^2} = \sqrt{85}$, so that $\sin\theta = \dfrac{6}{\sqrt{85}}$,

$$30 = y = \frac{v_0^2}{2(9.8)}\sin^2\theta = \frac{v_0^2}{2(9.8)}\frac{36}{85} = \frac{v_0^2}{4.9}\frac{9}{85},$$

hence

$$v_0^2 = \frac{(30)(85)(4.9)}{9} = \frac{4165}{3},$$

and finally

$$v_0 = \sqrt{\frac{4165}{3}} \approx 37.26 \text{ m/s}.$$

59. Let $((x(t), y(t))$ be the trajectory of the initial burst of water. If the angle of inclination of the hose is θ, we have the relations

$$\tan\theta = m, \ \sin\theta = \frac{m}{\sqrt{1+m^2}},$$
$$\cos\theta = \frac{1}{\sqrt{1+m^2}}.$$

We assume $x(0) = 0$ and $y(0) = 0$, and then find

$$y''(t) \equiv -32$$
$$y'(t) = -32t + v\sin\theta$$
$$y = y(t) = -16t^2 + tv\sin\theta = -16t^2 + \frac{tvm}{\sqrt{1+m^2}}$$
$$x'(t) \equiv v\cos\theta$$
$$x = x(t) = tv\cos\theta = \frac{tv}{\sqrt{1+m^2}}.$$

Solving the last equation in the form

$$t = \frac{x\sqrt{1+m^2}}{v}$$

and inserting this in the y-formula, we find

$$y = -16x^2\frac{(1+m^2)}{v^2} + mx.$$

61. With trajectory (x, y), and assuming $x(0) = 0$ and $y(0) = 0$, we have by now seen many times the conclusion $y = -gt^2 + tv\sin\theta$. The return to ground level occurs at time $t = 2v\sin\theta/g$, at which time the horizontal range is $x = tv\cos\theta = v^2 sin(2\theta)/g$. With $v = 60$ ft per second and $\theta = 25°$, and on earth with $g = 32$, this is about 86 feet, a short chip shot. On the moon with $g = 5$, it is about 552 ft.

63. Let $(x(t), y(t))$ be the trajectory of the paintball, and let $z(t)$ be the height of the target at time t. We do assume that $y(0) = z(0)$ (target opposite shooter at time of shot) and $y'(0) = 0$ (aiming directly at the target, hence using an initially horizontal trajectory), and as a result $y - z$ has *second derivative* zero and *initial value* zero. However, this only tells us that $y - z = [y'(0) - z'(0)]t = -z'(0)t$, *and if the target is already in motion* ($z'(0)$ not zero), the shot may miss at 20 feet or any distance. If on the other hand, the target is stationary at the moment of the shot, then the shot hits at 20 feet or any other distance.

65. It is necessary to make an assumption, such as that the acceleration is constant. (This is reasonable, since the pilot could choose to taxi at full throttle, which would generate a certain fixed force.) Then from $x''(t) = c$, we find $x'(t) = tc + k$ and $x = ct^2/2 + kt$. But $x'(0) = 0$ makes $k = 0$, and we are requiring $x'(1/120) = 240$ (measuring time in hours, distance in miles). Therefore $240 = c/120$, and $c = 2(120^2)$. Meanwhile, the distance involved is found from

$$x = ct^2/2 = [c/2]t^2 = [120^2]/(1/(120)^2 = 1 \text{ mile}.$$

5.5 Applications of Integration to Physics and Engineering

1. With x in inches,

$$F(x) = Kx$$
$$5 = K \cdot 4$$
$$K = 5/4 \text{ (lbs/inch)}$$
$$F(x) = 5x/4$$

$$W = \int_0^6 5x/4\,dx = \left.\frac{5x^2}{8}\right|_0^6 = \frac{5}{8}(36) \text{ ft-inches}.$$

The usual units for work being foot-pounds, this would be converted (by dividing by 12) to 15/8 foot-pounds.

3. With x in feet,

$$F(x) = Kx$$
$$20 = K \cdot \frac{1}{2}$$
$$K = 40 \text{ (lbs/foot)}$$
$$F(x) = 40x$$

$$W = \int_0^1 40x\,dx = 20x^2\Big|_0^1 = 20 \text{ ft-lb}.$$

5. $$W = 250 \cdot \frac{20}{12} = \frac{1250}{3} \text{ ft-lb}$$

7. $$W = 100 \cdot 3 = 300 \text{ ft-lb}$$

9. If x is between 0 and 30,000 ft, then the weight of the rocket at altitude x is $10000 - \frac{1}{15}x$. This weight is numerically the same as the lifting force, as a function of x. Therefore, the work is

$$\int_0^{30,000} \left(10,000 - \frac{x}{15}\right) dx$$
$$= \left(10,000x - \frac{x^2}{30}\right)\Bigg|_0^{30,000}$$
$$= 270,000,000 \text{ ft-lb.}$$

11. $$W = \int_0^1 800x(10x)\,dx$$
$$= \left(400x^2 - \frac{800}{3}x^3\right)\Bigg|_0^1 = \frac{400}{3} \text{ mile-lb}$$
$$= 704,000 \text{ ft-lb}$$

13. $$W = \int_0^{100} 62.4\pi(100x - x^2)(200 + x)\,dx$$
$$= 62.4\pi \int_0^{100} \left(20,000x - 100x^2 - x^3\right) dx$$
$$= 62.4\pi \left(10,000x^2 - \frac{100x^3}{3} - \frac{x^4}{4}\right)\Bigg|_0^{100}$$
$$= 8,168,140,899 \text{ ft-lb}$$

15. The difference between this exercise and Example 5.3 is that here the integral runs from 10 to 20 rather than from 0 to 20. In order to compile the number, however, it is easier to evaluate the integral from 0 to 10, getting

$$62.4\pi\left[400\frac{10^2}{2} - 40\frac{10^3}{3} + \frac{10^4}{4}\right]$$
$$\approx 1.797(10^6).$$

This number is then subtracted from the Example 5.3 solution, $2.61(10^6)$, leaving a final answer of about $8.1(10^5)$ foot-pounds.

17. We must estimate the integral

$$J = \int_0^{.0008} F(t)dt$$

on the basis of the tabulated values for F. Simpson's approximation is

$$J \approx \frac{.0008}{3(8)}\begin{bmatrix} 0 + 4(1000) + 2(2100) + \\ 4(4000) + 2(5000) + 4(5200) + \\ 2(2500) + 4(1000) + 0 \end{bmatrix}$$
$$\approx 2.133$$
$$2.13 = J = m\Delta v = .01\Delta v$$
$$\Delta v = 213.$$

The velocity after impact is $213 - 100 = 113$ ft/sec.

19. $$J \approx \frac{.6}{3(6)}\begin{bmatrix} 0 + 4(8000) + 2(16,000) + \\ 4(24,000) + 2(15,000) + \\ 4(9000) + 0 \end{bmatrix}$$
$$\approx 7533.3$$
$$7533.3 = J = m\Delta v = 200\Delta v$$
$$\Delta v = 37.7 \text{ ft/sec}$$

Since the velocity after the crash is zero, this number is the estimated original velocity.

21. With the density as given, we have

$$m = \int_0^6 \left[xe^{-x^2} + 2\right] dx = \frac{e^{-x^2}}{2} + 2x\Bigg|_0^6 = 12.5$$
$$M = \int_0^6 x\left[xe^{-x^2} + 2\right] dx = \int_0^6 x^2e^{-x^2}\,dx + 36.$$

The integral, by software, is .44311, so the center of mass is $36.44311/12.5 = 2.9155$ (meters, rounded). This is close to 3, and it would have been exactly 3 but for the exponential component in the density.

23. $\rho(x) = 4 + \dfrac{x^2}{4}$ kg/m, $-2 \le x \le 2$

$$M = \int_{-2}^2 x\left(4 + \frac{x^2}{4}\right) dx = \left(2x^2 + \frac{x^4}{16}\right)\Bigg|_{-2}^2 = 0$$
$$m = \int_{-2}^2 \left(4 + \frac{x^2}{4}\right) dx = \left(4x + \frac{x^3}{12}\right)\Bigg|_{-2}^2 = \frac{52}{3} \text{ kg}$$
$$\bar{x} = \frac{M}{m} = \frac{0}{\frac{52}{3}} = 0 \text{ m}$$

25.
$$m = \int_{-3}^{27} \left(\frac{1}{46} + \frac{x+3}{690}\right)^2 dx$$
$$= \frac{690}{3}\left(\frac{1}{46} + \frac{x+3}{690}\right)^3\Big|_{-3}^{27}$$
$$= .0614 \text{ slugs}$$
$$(.0614)(32)(16) \approx 31.5 \text{ oz}$$

This is just the baseball bat of Example 5.5 translated. It weighs the same.

27.
$$M = \int_{-3}^{27} x\left(\frac{1}{46} + \frac{x+3}{690}\right)^2 dx$$
$$= \int_{-3}^{27} x \cdot \frac{(x+18)^2}{690^2}\, dx$$
$$= \frac{1}{476{,}100}\int_{-3}^{27} (x^3 + 36x^2 + 324x)\, dx$$
$$= \frac{1}{476{,}100}\left(\frac{x^4}{4} + 12x^3 + 162x^2\right)\Big|_{-3}^{27}$$
$$\approx 1.0208$$
$$\bar{x} = \frac{M}{m} = \frac{1.0208}{.0614} \approx 16.6 \text{ in}$$

This is 3 inches less than the bat of Example 5.5, a reflection of the translation three inches to the left on the number line.

29.
$$m = \int_0^{30} .00468\left(\frac{3}{16} + \frac{x}{60}\right) dx$$
$$= .00468\left(\frac{3}{16}x + \frac{x^2}{120}\right)\Big|_0^{30}$$
$$\approx .0614 \text{ slugs}$$
$$M = \int_0^{30} .00468x\left(\frac{3}{16} + \frac{x}{60}\right) dx$$
$$= .00468\left(\frac{3x^2}{32} + \frac{x^3}{180}\right)\Big|_0^{30}$$
$$\approx 1.0969$$

weight $= m(32)(16) = 31.4$ oz

$$\bar{x} = \frac{M}{m} = \frac{1.0969}{.0614} \approx 17.8 \text{ in}$$

31. Area of the base is $\frac{1}{2}(3+1) = 2$.
Area of the body is $1 \times 4 = 4$.
Area of the tip is $\dfrac{1}{2}(1 \times 1) = \dfrac{1}{2}$.
Base:

$$m = \int_0^1 \rho(3-2x)\, dx = \rho(3x - x^2)\Big]_0^1 = 2\rho$$
$$M = \int_0^1 \rho x(3-2x)\, dx$$
$$= \rho\left(\frac{3x^2}{2} - \frac{2x^3}{3}\right)\Big|_0^1 = \frac{5}{6}\rho,$$
$$\bar{x} = \frac{M}{m} = 5/12 \approx .4167.$$

Body:

$$m = \int_1^5 \rho\, dx = \rho x|_1^5 = 4\rho$$
$$M = \int_1^5 \rho x\, dx = \rho\, \frac{x^2}{2}\Big]_1^5 = 12\rho,$$
$$\bar{x} = \frac{M}{m} = 3.$$

Tip:

$$m = \int_5^6 \rho(6-x)\, dx = \rho\left(6x - \frac{x^2}{2}\right)\Big|_5^6 = \frac{\rho}{2}.$$
$$M = \int_5^6 \rho x(6-x)\, dx$$
$$= \rho\left(3x^2 - \frac{x^3}{3}\right)\Big|_5^6 \approx 2.67\rho,$$
$$\bar{x} = \frac{M}{m} \approx 5.33.$$

33. $F'(t)$ is zero at $t = 3$, and the maximum thrust is $F(3) = 30/e$, about $11.0364\ldots$. It is implicit in the drawing that the thrust is zero after time 6 . Therefore, the impulse is

$$\int_0^6 10te^{-t/3}dt.$$

This integral requires integration by parts, yielding a value of $90 - 270e^{-2}$, about 53.55.

35. The figure is a right triangle, whose area is 12 by elementary geometry. Then, describing the equation of the hypotenuse by $y = \frac{3}{2x}$ (for use in $\bar{x}$) or by $x = \frac{2}{3y}$ (for use in $\bar{y}$), we find

$$12\bar{x} = \int_0^4 x\left(\frac{3x}{2}\right)dx = 32, \ \therefore\ \bar{x} = \frac{8}{3};$$

$$12\bar{y} = \int_0^6 y\left[4 - \frac{2}{3y}\right]dy = 24, \ \therefore\ \bar{y} = 2.$$

37. Between $y = 0$ and $y = 4 - x^2$, the area is

$$A = \int_{-2}^{2} (4 - x^2)\,dx = 32/3.$$

For this figure, $\bar{x}$ is clearly zero by considerations of symmetry. The horizontal width at height $y(0 < y < 4)$ is $2\sqrt{4-y}$, so

$$\frac{32}{3}\bar{y} = A\bar{y} = \int_0^4 y[2\sqrt{4-y}]dy.$$

The integral converts, by the substitutions

$$x = \sqrt{4-y},\ y = 4 - x^2,\ dy = -2x\,dx,$$

to

$$2\int_2^0 (4 - x^2)x(-2x\,dx)$$
$$= 4\int_0^2 (4x^2 - x^4)\,dx = 4(32)\left(\frac{2}{15}\right).$$

Therefore,

$$\bar{y} = \frac{3}{32}\left[4(32)\frac{2}{15}\right] = \frac{24}{15} = \frac{8}{5} = 1.6.$$

39. With x the depth, the horizontal width is a linear function of x, given by $x + 40$. Hence,

$$F = \int_0^{60} 62.4x(x+40)\,dx$$
$$= 62.4\left(\frac{x^3}{3} + 20x^2\right)\Big|_0^{60}$$
$$= 8{,}985{,}600 \text{ lb.}$$

41. With x the vertical deviation above the center of the window, one has horizontal width of the window given by $2\sqrt{25 - x^2}$, depth of water $40 + x$, and hydrostatic force

$$62.4\int_{-5}^{5} (x+40)2\sqrt{25-x^2}\,dx =$$
$$62.4\int_{-5}^{5} 2x\sqrt{25-x^2}\,dx +$$
$$62.4(40)\int_{-5}^{5} 2\sqrt{25-x^2}\,dx.$$

The first expression was evaluated by software, and the resulting force came to about 196,035 pounds. From another perspective, the bottom integral is recognized as the area of the circle $x^2 + y^2 = 25$. The middle integral is easily evaluated or simply recognized to be zero. Therefore, the final hydrostatic force is $(62.4)(40)(25\pi)$, numerically the same. This principle:

$$F = 62.4(\text{depth of center})(\text{area})$$

works for any submerged circular window provided the area is in square feet and the depth is in feet. The units on the 62.4 are technically pounds per cubic foot.

43. Assuming that the *center* of the circular window descends to 1000 feet, then by the previous principle, after converting the three-inch radius to 1/4 feet, we get $F = 12{,}252$ pounds. An alternate calculation in which x is the deviation downward from the top edge of the window, would be

$$F = \int_0^{.5} 62.4(999.75 + x)\cdot 2\sqrt{(.25)^2 - (.25 - x)^2}\,dx$$
$$= \int_0^{.5} 124.8(999.75 + x)\sqrt{.5x - x^2}\,dx$$
$$\approx 12,252 \text{ lb}$$

(approximated numerically).

45. With $h(t)$ the height of the vaulter at time t (time measured from the peak of his vault),

$$h(0) = 20,\ h'(0) = 0,\ h''(0) = -32$$
$$h'(t) = -32t,\ h(t) = -16t^2 + 20$$
$$h(t) = 0 \text{ for } t = \sqrt{\frac{20}{16}} = \frac{\sqrt{5}}{2}$$
$$v = h'\left(\frac{\sqrt{5}}{2}\right) = -16\sqrt{5} \approx -35.7 \text{ ft/s}$$
$$\frac{1}{2}mv^2 = \frac{1}{2}\left(\frac{200}{32}\right)\left(16^2 \cdot 5\right) = 4000 \text{ ft-lb.}$$

47.
$$\begin{aligned}&(100 \text{ tons})(20 \text{ miles/hr})\\ &= \frac{(100 \cdot 2000 \text{ lbs})(20 \cdot 5280 \text{ ft})}{3600 \text{ sec}}\\ &\approx 5{,}866{,}667 \text{ ft-lb/s}\\ &= \frac{5{,}866{,}667}{550} \text{ hp}\\ &\approx 10{,}667 \text{ hp}\end{aligned}$$

49. The bat in Exercise 29 models the bat of Example 6.5 choked up 3 in.

From Example 6.5:

$$f(x) = \left(\frac{1}{46} + \frac{x}{690}\right)^2;$$
$$\int_{-3}^{27} f(x) \cdot x^2\, dx \approx 27.22.$$

From Exercise 29:

$$f(x) = \left(\frac{1}{46} + \frac{x+3}{690}\right)^2;$$
$$\int_{-3}^{27} f(x) \cdot x^2\, dx \approx 20.54.$$

Reduction in moment:

$$\frac{27.22 - 20.54}{27.22} \approx 24.5\%.$$

51. A CAS gives

$$\int_{-a}^{a} 2\rho x^2 b\sqrt{1 - \frac{x^2}{a^2}}\, dx = \frac{1}{4}\rho\pi a^3 b.$$

53. Using the formula in Exercise 50, we find that the moments are 1323.8 for the wooden racket, 1792.9 for the midsized racket, and 2361.0 for the oversized racket. The ratios are

$$\frac{\text{mid}}{\text{wood}} \approx 1.35,\quad \frac{\text{over}}{\text{wood}} 1.78.$$

5.6 Probability

1. $f(x) = 4x^3$, $[0, 1]$

(i) $f(x) = 4x^3 \geq 0$ for $0 \leq x \leq 1$

(ii) $\int_0^1 4x^3\, dx = x^4\Big|_0^1 = 1 - 0 = 1$

3. $f(x) = x + 2x^3$, $[0, 1]$

(i) $x + 2x^3 \geq 0$ for $0 \leq x \leq 1$

(ii) $\int_0^1 (x + 2x^3)\, dx = \frac{x^2}{2} + \frac{x^4}{2}\Big|_0^1 = 1$

5. $f(x) = \frac{1}{2}\sin x$, $[0, \pi]$

(i) $\frac{1}{2}\sin x \geq 0$ for $0 \leq x \leq \pi$

(ii)
$$\begin{aligned}\int_0^{\pi} \frac{1}{2}\sin x\, dx &= -\frac{1}{2}\cos x\Big|_0^{\pi}\\ &= -\frac{1}{2}(-1) + \frac{1}{2}(1)\\ &= 1\end{aligned}$$

7. $f(x) = e^{-x/2}$, $[0, \ln 4]$

(i) $e^{-x/2} \geq 0$ for $0 \leq x \leq \ln 4$

(ii) $\int_0^{\ln 4} e^{-x/2}\, dx = -2e^{-x/2}\Big|_0^{\ln 4} = -1 + 2 = 1$

9.
$$f(x) = cx^3,\ [0, 1]$$
$$1 = \int_0^1 cx^3\, dx = \frac{cx^4}{4}\Big|_0^1 = \frac{c}{4} - 0$$
$$\therefore\ c = 4$$

11. $f(x) = ce^{-4x}$, $[0, 1]$

$$1 = \int_0^1 ce^{-4x}\,dx = -\frac{1}{4}ce^{-4x}\Big|_0^1$$
$$= -\frac{1}{4}ce^{-4} + \frac{1}{4}c$$
$$\therefore\ c = \frac{4}{1-e^{-4}}$$

13. $f(x) = 2ce^{-cx}$, $[0, 2]$

$$1 = \int_0^2 2ce^{-cx}\,dx = -2e^{-cx}\Big|_0^2$$
$$= -2e^{-2c} + 2$$
$$\therefore\ c = \frac{\ln 2}{2} \approx 0.34657\ldots.$$

15. $P(70 \le x \le 72)$

$$= \int_{70}^{72} \frac{.4}{\sqrt{2\pi}} e^{-.08(x-68)^2}\,dx \approx .157$$

17. $P(84 \le x \le 120)$

$$= \int_{84}^{120} \frac{.4}{\sqrt{2\pi}} e^{-.08(x-68)^2}\,dx$$
$$\approx 7.76 \times 10^{-11}$$

19.

$$P\left(0 \le x \le \frac{1}{4}\right) = \int_0^{1/4} 6e^{-6x}\,dx$$
$$= -e^{-6x}\Big|_0^{1/4}$$
$$= (-e^{-3/2} + 1)$$
$$\approx .77687$$

21.

$$P(1 \le x \le 2) = \int_1^2 6e^{-6x}\,dx$$
$$= -e^{-6x}\Big|_1^2$$
$$= (-e^{-12} + e^{-6})$$
$$\approx .00247$$

23.

$$P\left(\frac{1}{12} \le x \le \frac{1}{6}\right) = \int_{1/12}^{1/6} 8e^{-8x}\,dx$$
$$= -e^{-8x}\Big|_{1/12}^{1/6}$$
$$= (-e^{-4/3} + e^{-2/3})$$
$$\approx .24982$$

25.

$$P\left(0 \le x \le \frac{1}{2}\right) = \int_0^{1/2} 8e^{-8x}\,dx$$
$$= -e^{-8x}\Big|_0^{1/2}$$
$$= (-e^{-4} + 1)$$
$$\approx .9817$$

27.

$$P(0 \le x \le 1) = \int_0^1 4xe^{-2x}\,dx$$
$$= 1 - 3e^{-2} \approx .594$$

(using integration by parts and calculator, or CAS)

29. Mean: $\displaystyle\int_0^{10} x(4xe^{-2x})\,dx \approx .9999995$

31. Density $f(x) = 3x^2$, $0 \le x \le 1$.

Mean: $\displaystyle\int_0^1 x \cdot 3x^2\,dx = \frac{3}{4}x^4\Big|_0^1 = \frac{3}{4}$. For the median c:

$$1/2 = \int_0^c 3x^2\,dx = x^3\Big|_0^c = c^3$$
$$c = 1/\sqrt[3]{2} \approx .7937.$$

33. Density $f(x) = \dfrac{1}{2}\sin x$, $0 \le x \le \pi$.

Mean: $\displaystyle\int_0^{\pi} x \cdot \frac{1}{2}\sin x\,dx = \frac{\pi}{2} \approx 1.57$
(using CAS or calculator).

For the median c:

$$\frac{1}{2} = \int_0^c \frac{1}{2}\sin x\,dx$$
$$= -\frac{1}{2}\cos x\Big|_0^c$$
$$= -\frac{1}{2}\cos c + \frac{1}{2},$$
$$\therefore\ c = \frac{\pi}{2}.$$

In this case, the median and the mean are equal.

35. Density: $f(x) = \frac{1}{2}(\ln 3)e^{-kx}$, with $k = \frac{1}{3}\ln 3$, $(0 \le x \le 3)$.
Mean: $\int_0^3 x \cdot \frac{1}{2}(\ln 3)e^{-kx}\,dx \approx 1.23$ (using CAS or calculator). Regarding the median c:

$$\frac{1}{2} = \int_0^c \frac{1}{2}(\ln 3)e^{-kx}\,dx = \frac{1}{2}\ln 3\left(-\frac{1}{k}\right)e^{-kx}\Big|_0^c$$
$$= \frac{1}{2k}\ln 3\left(1 - e^{-ck}\right)$$
$$= \frac{3}{2}\left(1 - e^{-c/3\ln 3}\right).$$

So

$$\frac{1}{3} = 1 - e^{-c/3\ln 3}$$
$$\frac{2}{3} = e^{-c/3\ln 3}$$
$$-\ln\left(\frac{3}{2}\right) = \ln\left(\frac{2}{3}\right) = -\frac{c}{3}\ln 3$$
$$c = \frac{3\ln\frac{3}{2}}{\ln 3} \approx 1.1072.$$

37. Density $f(x) = \frac{4}{1-e^{-4}}e^{-4x}$, $0 \le x \le 1$.
Mean: $\int_0^1 x\frac{4}{1-e^{-4}}e^{-4x}\,dx \approx .2313\ldots$ (using CAS or calculator). Regarding the median c:

$$\frac{1}{2} = \int_0^c \frac{4}{1-e^{-4}}e^{-4x}\,dx$$
$$= \frac{-1}{1-e^{-4}}e^{-4x}\Big|_0^c$$
$$= \frac{1}{e^{-4}-1}(e^{-4c} - 1).$$

So

$$\frac{1}{2}(e^{-4} - 1) = e^{-4c} - 1$$
$$\frac{1}{2}(e^{-4} + 1) = e^{-4c}$$
$$\ln\left(\frac{1}{2}(e^{-4} + 1)\right) = -4c$$
$$-\frac{1}{4}\ln\left(\frac{e^{-4}+1}{2}\right) = c$$

$(c \approx .1687)$.

39. Density $f(x) = ce^{-4x}$, $[0, b]$, $b > 0$.

$$1 = \int_0^b ce^{-4x}\,dx = -\frac{c}{4}e^{-4x}\Big]_0^b = -\frac{c}{4}\left(e^{-4b} - 1\right)$$
$$c = \frac{4}{1-e^{-4b}}.$$

As $b \to \infty$, $c \to 4$.

41. Density $f(x) = ce^{-6x}$, $[0, b]$, $b > 0$.

$$1 = \int_0^b ce^{-6x}\,dx = \frac{-c}{6}e^{-6x}\Big|_0^b = -\frac{c}{6}\left(e^{-6b} - 1\right)$$
$$c = \frac{6}{1-e^{-6b}}.$$

As $b \to \infty$, $c \to 6$.

$$\mu = \int_0^b xce^{-6x}\,dx = \frac{ce^{-6c}}{36}(-6x - 1)\Big|_0^b$$
$$= \frac{ce^{-6b}}{36}(-6b - 1) + \frac{c}{36}.$$

As $b \to \infty$, $\mu \to \frac{1}{6}$.

43. (a)

$$P(h \le 3) = P(0) + P(1) + P(2) + P(3) = \frac{1}{256} + \frac{8}{256} + \frac{28}{256} + \frac{56}{256} = \frac{93}{256}$$

(b)

$$P(h > 4) = P(5) + P(6) + P(7) + P(8) = \frac{56}{256} + \frac{28}{256} + \frac{8}{256} + \frac{1}{256} = \frac{93}{256}$$

(c)

$$P(0) + P(8) = \frac{1}{256} + \frac{1}{256} = \frac{2}{256}$$

(d)

$$P(1) + P(3) + P(5) + P(7) = \frac{8}{256} + \frac{56}{256} + \frac{56}{256} + \frac{8}{256} = \frac{1}{2}$$

45. (a)

$$P\,(4/3) + P\,(4/2) + P\,(4/1) + P\,(4/0) = .1659 + .2073 + .2073 + .1296 = .7101$$

(b)

$$P\,(0/4) + P\,(1/4) + P\,(2/4) + P\,(3/4) = .0256 + .0615 + .0922 + .1106 = .2899$$

(c)

$$P\,(0/4) + P\,(4/0) = .0256 + .1296 = .1552$$

(d)

$$P\,(2/4) + P\,(3/4) + P\,(4/3) + P\,(4/2) = .0922 + .1106 + .1659 + .2073 = .576$$

47. $e(p) = cp^{-2}$, $1 < p < 100$. About c:

$$1 = \int_1^{100} cp^{-2}dp = -cp^{-1}\Big|_1^{100} = -c\left(\frac{1}{100} - 1\right),$$

$$\therefore\ c = \frac{100}{99}$$

The requested probability is

$$\int_{60}^{70} cp^{-2}dp = -cp^{-1}\Big|_{60}^{70} = -\frac{100}{99}\left(\frac{1}{70} - \frac{1}{60}\right) \approx .0024.$$

49.

$$f(x) = \frac{.4}{\sqrt{2\pi}}e^{-.08(x-68)^2}$$

$$f'(x) = \frac{-.064}{\sqrt{2\pi}}(x - 68)e^{-.08(x-68)^2}$$

$$f''(x) = \frac{-.064}{\sqrt{2\pi}}e^{-.08(x-68)^2}\left(1 - .16(x - 68)^2\right)$$

The second derivative is zero when

$$x - 68 = \pm 1/\sqrt{.16} = \pm 1/.4 = \pm 5/2.$$

Thus the standard deviation is $\frac{5}{2}$.

51. The prior decision that the median score will be a C gives, in isolation, *no basis for any other letter grades.* Subjectively, the scores in the collection {400, 480, 550, 620, 690} look quite different from one another, and arguably the top score deserves an A. But if the collection had been {680, 680, 685, 690, 690} the subjective evaluation might instead be that all the scores were about the same and should all get the *same grade.* This common grade would have to be C given the first rule. However, if these two collections of scores represented different groups taking the *same test,* it seems clear that the second group was a "better class," the individuals do not all deserve C's, and that justice would be better served by giving them all A's.

53.

$$f(t) = t^{-3/2}e^{0.38t-100/t}$$

$$\int_0^{40} k \cdot f(t)dt = 1 \text{ for } k = 0.000318.$$

$$\int_{20}^{30} 0.000318 \cdot f(t)dt \approx 0.0134$$

55. Because $\int_0^\infty e^{-rx}\,dx = 1/r$, any density of the form ke^{-rx} on $(0, \infty)$ must have $k = r$. Therefore in (a), $k = 2$; and in (b), $k = 4$.

57. Because $\int_0^\infty xe^{-rx}\,dx = 1/r^2$, the mean for density re^{-rx} on $(0, \infty)$ must be $r/r^2 = 1/r$.

59.
$$\int_{1/r}^\infty re^{-rx}\,dx = -e^{-rx}\Big|_{1/r}^\infty = 1/e = .3678\ldots$$

The fact that this is not $1/2$ should not surprise; the property

$$\int_c^\infty re^{-rx}\,dx = 1/2$$

is true of the *median* c, but it has never been said that the median and mean are the same. Indeed, some of the problems 31–38 show otherwise.

61. The substitutions

$$u = \sqrt{2}(x+4),$$
$$u^2 = 2(x+4)^2, \text{ and}$$
$$du = \sqrt{2}\,dx$$

convert the integral

$$\int_{-\infty}^\infty e^{-2(x+4)^2}\,dx$$

to

$$\int_{-\infty}^\infty e^{-u^2}\frac{du}{\sqrt{2}} = \frac{\sqrt{\pi}}{\sqrt{2}} = \sqrt{\frac{\pi}{2}}.$$

Therefore,

$$\sqrt{\frac{2}{\pi}}\int_{-\infty}^\infty e^{-2(x+4)^2}\,dx = 1.$$

63. This problem generalizes Exercise 61, where μ was -4 and σ was $1/2$. Here, the substitutions

$$u = \frac{x-\mu}{\sqrt{2}\sigma} \quad\text{and}\quad du = \frac{dx}{\sqrt{2}\sigma}$$

convert the integral

$$\frac{1}{\sqrt{2\pi}\sigma}\int_{-\infty}^\infty e^{-\frac{(x-\mu)^2}{2\sigma^2}}\,dx$$

to

$$\frac{1}{\sqrt{2\pi}\sigma}\int_{-\infty}^\infty e^{-u^2}(\sigma\sqrt{2}du) = \frac{1}{\sqrt{\pi}}\int_{-\infty}^\infty e^{-u^2}du = 1.$$

65. In this problem, one must keep in mind that r is the (random) variable and that t is a parameter (presumably time). Unlike the normal distribution, this r is required to be *nonnegative.* We learn from Exercise 63 with $\sigma = \sqrt{2Dt}$ that

$$\int_{-\infty}^\infty \frac{1}{2\sqrt{\pi Dt}}e^{-\frac{r^2}{4Dt}}dr = 1,$$

hence, by the symmetry of the integrand,

$$\int_0^\infty \frac{1}{2\sqrt{\pi Dt}}e^{-\frac{r^2}{4Dt}}dr = \frac{1}{2},$$

and

$$\int_0^\infty d(r)dr = \int_0^\infty \frac{1}{\sqrt{\pi Dt}}e^{-\frac{r^2}{4Dt}}dr = 1.$$

This shows that $d(r)$ is a density on $(0, \infty)$. It is not however a *normal* distribution, since all normal distributions are on $(-\infty, \infty)$. Because

$$\frac{d}{dr}\left(e^{-\frac{r^2}{4Dt}}\right) = -\frac{r}{2Dt}e^{-\frac{r^2}{4Dt}},$$

we see that

$$\int_0^\infty re^{-\frac{r^2}{4Dt}}dr = -2Dte^{-\frac{r^2}{4Dt}}\Bigg|_{r=0}^{r=\infty} = 2Dt.$$

Thus the average, or *mean*, distance from the origin of the algae at time t is

$$\int_0^\infty rd(r)dr = \int_0^\infty \frac{r}{\sqrt{\pi Dt}}e^{-\frac{r^2}{4Dt}}dr$$
$$= \frac{2Dt}{\sqrt{\pi Dt}} = 2\sqrt{\frac{Dt}{\pi}},$$

an expression which increases with time.

With $D = 1$, the three probabilities in question are (a) 1 (nothing has happened yet—all algae still concentrated at the origin), (b) .5205, (c) .3829. These must be done by some kind of integration software.

67. In the picture, although it might appear that $y > 1/2$, the conditions are that $0 \le y \le 1/2$, and the labeling in the drawing implies that the lower line is the closer. This is indeed always an allowable assumption (by turning the picture upside down if necessary).
In the right triangle whose hypotenuse is the lower half-needle, the vertical side is of length $(\sin\theta)/2$. Therefore, the needle hits the lower line if $y - (\sin\theta)/2 \le 0$, or if $y \le (\sin\theta)/2$. As to the actual probability ratio, the denominator is just $\pi/2$, while the numerator is

$$-\frac{\cos\theta}{2}\Big|_0^\pi = \frac{-\cos\pi + \cos 0}{2} = \frac{2}{2} = 1.$$

The total probability of hitting a line is thus $2/\pi \approx 63.66\%$.

Chapter 5 Review

True or False

1. False as it stands. One needs the extra condition that $g(x) \le f(x)$, and one should clarify that we are discussing only the area between $x = a$ and $x = b$.

3. true

5. False. The evaluation of arc length by exact formula is possible only for a special few functions.

7. True if we ignore air resistance

9. False. The best one can say is that the mean is not always equal to the median.

Other Problems

1. See Example 1.1.

$$y = x^2 + 2,\ y = \sin x,\ \text{ for } 0 \le x \le \pi$$

$$\int_0^\pi \left(x^2 + 2 - \sin x\right) dx$$

$$= \left(\frac{x^3}{3} + 2x + \cos x\right)\Big|_0^\pi$$

$$= \frac{\pi^3}{3} + 2\pi - 1 - (0 + 0 + 1)$$

$$= \frac{\pi^3}{3} + 2\pi - 2$$

3. See Example 1.2.

$$y = x^3,\ y = 2x^2 - x$$

$$\int_0^1 x^3 - \left(2x^2 - x\right) dx$$

$$= \left(\frac{x^4}{4} - \frac{2}{3}x^3 + \frac{x^2}{2}\right)\Big|_0^1$$

$$= \left(\frac{1}{4} - \frac{2}{3} + \frac{1}{2}\right) = \frac{1}{12}$$

5. See Example 1.3.

$$y = e^{-x},\ y = 2 - x^2$$

$$\int_{-.537}^{1.316} \left(2 - x^2 - e^x\right) dx =$$

$$\left(2x - \frac{x^3}{3} + e^{-x}\right)\Big|_{-.537}^{1.316} \approx 1.452$$

7. See Example 1.4.

$$y = x^2,\ y = 2 - x,\ y = 0$$

$$\int_0^1 x^2\,dx + \int_1^2 (2 - x)\,dx$$

$$= \frac{x^3}{3}\Big|_0^1 + \left(2x - \frac{x^2}{2}\right)\Big|_1^2$$

$$= \frac{1}{3} + (4 - 2) - \left(2 - \frac{1}{2}\right)$$

$$= \frac{5}{6}$$

9. If P is the population at time t, the equation is

$$P' = \text{birth rate} - \text{death rate}$$
$$= (10 + 2t) - (4 + t) = 6 + t.$$

Thus $P = 6t + t^2/2 + P(0)$, so at time 6, $P = 36 + 18 + 10{,}000 = 10{,}054$. Alternatively, see Example 1.1, Exercises 1.39 and 1.40, and write

$$A = \int_0^6 (10 + 2t) - (4 + t)dt$$
$$= \int_0^6 (6 + t)dt$$
$$= \left(6t + \frac{t^2}{2}\right)\Big|_0^6$$
$$= 54;$$

population $= 10{,}000 + 54 = 10{,}054$.

11. See Example 2.1.

$$V = \int_0^2 \pi(3 + x)^2\,dx$$
$$= \pi \int_0^2 (9 + 6x + x^2)\,dx$$
$$= \pi\left(9x + 3x^2 + \frac{x^3}{3}\right)\Big|_0^2$$
$$= \frac{98\pi}{3}$$

13. See Example 2.2.

$$V = 0.4\begin{pmatrix} \frac{0.4}{2} + 1.4 + 1.8 + 2.0 \\ +2.1 + 1.8 + 1.1 + \frac{0.4}{2}\end{pmatrix}$$
$$\approx 4.2$$

(This is the trapezoidal estimate.)

15. See Examples 2.6 and 2.7.

(a)
$$V = \int_{-2}^2 \pi(4)^2\,dx - \int_{-2}^2 \pi(x^2)^2\,dx$$
$$= \pi\int_{-2}^2 (16 - x^4)\,dx$$
$$= \pi\left(16x - \frac{x^5}{5}\right)\Big|_{-2}^2$$
$$= \frac{256\pi}{5}$$

(b) By the disk method, integrating in y:

$$V = \int_0^4 \pi(\sqrt{y})^2 dy = \pi\int_0^4 y dy$$
$$= \frac{\pi y^2}{2}\Big|_0^4 = 8\pi.$$

(c)
$$V = \int_0^4 \pi(2 + \sqrt{y})^2 dy -$$
$$\int_0^4 \pi(2 - \sqrt{y})^2 dy$$
$$= \pi\int_0^4 (4 + 4y^{1/2} + y)dy -$$
$$\pi\int_0^4 (4 - 4y^{1/2} + y)dy$$
$$= \pi\int_0^4 (8y^{1/2})dy$$
$$= 8\pi \cdot \frac{2}{3}y^{3/2}\Big|_0^4 = \frac{128\pi}{3}$$

(d)
$$V = \int_{-2}^2 \pi(6)^2\,dx - \int_{-2}^2 \pi(x^2 + 2)^2\,dx$$
$$= \pi\int_{-2}^2 (-x^4 - 4x^2 + 32)\,dx$$
$$= \pi\left(-\frac{x^5}{5} - \frac{4x^3}{3} + 32x\right)\Big|_{-2}^2$$
$$= \frac{1408\pi}{15}$$

17. See Examples 3.3 and 3.4.

(a)
$$V = \int_0^1 2\pi y((2 - y) - y)dy$$
$$= 2\pi\int_0^1 (2y - 2y^2)dy$$
$$= 2\pi\left(y^2 - \frac{2y^3}{3}\right)\Big|_0^1$$
$$= \frac{2\pi}{3}$$

(b)

$$\begin{aligned} V &= \int_0^1 \pi(2-y)^2 dy - \int_0^1 \pi(y)^2 dy \\ &= \pi \int_0^1 (4-4y)dy \\ &= \pi\ (4y-2y^2)\Big|_0^1 \\ &= 2\pi \end{aligned}$$

(c)

$$\begin{aligned} V &= \int_0^1 \pi((2-y)+1)^2 dy - \\ &\quad \int_0^1 \pi(y+1)^2 dy \\ &= \pi \int_0^1 (9-6y+y^2)dy - \\ &\quad \pi \int_0^1 (y^2+2y+1)dy \\ &= \pi \int_0^1 (8-8y)dy \\ &= \pi\ (8y-4y^2)\Big|_0^1 = 4\pi \end{aligned}$$

(d)

$$\begin{aligned} V &= \int_0^1 2\pi(4-y)((2-y)-y)dy \\ &= 2\pi \int_0^1 (8-10y+2y^2)dy \\ &= 2\pi \left(8y-5y^2+\frac{2y^3}{3}\right)\Big|_0^1 \\ &= \frac{22\pi}{3} \end{aligned}$$

19. See Example 3.2.

$$y = x^4,\ -1 \le x \le 1$$

$$s = \int_{-1}^1 \sqrt{1+(4x^3)^2}\, dx \approx 3.2$$

21. See Example 3.2.

$$y = e^{x/2},\ -2 \le x \le 2$$

$$\int_{-2}^2 \sqrt{1+\left(\frac{e^{x/2}}{2}\right)^2}\, dx \approx 4.767$$

23. See Example 3.5.

$$S = \int_0^1 2\pi(1-x^2)\sqrt{1+4x^2}\, dx \approx 5.48$$

25. See Example 4.1.

$$\begin{aligned} h''(t) &= -32 \\ h(0) &= 64,\ h'(0) = 0 \\ h'(t) &= -32t \\ h(t) &= -16t^2 + 64. \end{aligned}$$

This is zero when $t = 2$, at which time $h'(2) = -32(2) = -64$. The speed at impact is reported as 64 ft per second.

27. See Example 4.4.

$$\begin{aligned} & y''(t) = -32,\ x''(t) = 0, \\ & y(0) = 0,\ x(0) = 0 \\ & y'(0) = 48 \sin\left(\frac{\pi}{9}\right),\ x'(0) = 48 \cos\left(\frac{\pi}{9}\right) \\ & y'(0) \approx 16.42,\ x'(0) \approx 45.11 \\ & y'(t) = -32t + 16.42 \\ & y(t) = -16t^2 + 16.42t. \end{aligned}$$

This is zero at $t = 1.026$. Meanwhile,

$$\begin{aligned} & x'(t) \equiv 45.11 \\ & x(t) = 45.11t \\ & x(1.026) = 45.11(1.026) \approx 46.3 \text{ ft}. \end{aligned}$$

This is the horizontal range.

29. See Example 4.4.

$$\begin{aligned} & y(0) = 6,\ x(0) = 0 \\ & y'(0) = 80 \sin\left(\frac{2\pi}{45}\right) \approx 11.13 \\ & x'(0) = 80 \cos\left(\frac{2\pi}{45}\right) \approx 79.22 \end{aligned}$$

$$\begin{aligned} y''(t) &= -32,\ x''(t) = 0 \\ y'(t) &= -32t + 11.13 \\ y(t) &= -16t^2 + 11.13t + 6 \\ x'(t) &= 79.22 \\ x(t) &= 79.22t. \end{aligned}$$

This is 120 (40 yards) when t is about 1.51. At this time, the vertical height (if still in flight) would be

$$\begin{aligned} y(1.51) &= -16(1.51)^2 + 11.13(1.51) + 6 \\ &= -13.6753. \end{aligned}$$

Since this is negative, we conclude the ball is not still in flight, has hit the ground, and was not catchable.

31. See Example 4.3.

$$\begin{aligned} h''(t) &= -32 \\ h'(0) &= v_0 \\ h(0) &= 0 \\ h'(t) &= -32t + v_0. \end{aligned}$$

This is zero at $t = v_0/32$.

$$h\left(\frac{v_0}{32}\right) = -16\left(\frac{v_0^2}{32^2}\right) + \frac{v_0^2}{32} = \frac{v_0^2}{64}.$$

If this is to be 128, then clearly v_0 must be $\sqrt{(64)(128)} = 64\sqrt{2}$ (ft/sec).

Impact speed from ground to ground is the same as launch speed, which can be verified by first finding the time t of return to ground,

$$\begin{aligned} -16t^2 + v_0 t &= 0 \\ t &= v_0/16, \end{aligned}$$

and then compiling

$$h'\left(v_0/16\right) = -32(v_0/16) + v_0 = -v_0.$$

33. See Example 5.1.

$$F = kx,\ 60 = k \cdot 1,\ k = 60$$

$$W = \int_0^{2/3} 60x\,dx = 30x^2\Big|_0^{2/3}$$

$$= \frac{30 \cdot 4}{9} = \frac{40}{3} \text{ ft-lb}$$

35. With density $x^2 - 2x + 8$,

$$\begin{aligned} m &= \int_0^4 \left(x^2 - 2x + 8\right) dx \\ &= \left(\frac{x^3}{3} - x^2 + 8x\right)\Big|_0^4 = \frac{112}{3} \\ M &= \int_0^4 x\left(x^2 - 2x + 8\right) dx \\ &= \int_0^4 \left(x^3 - 2x^2 + 8x\right) dx \\ &= \left(\frac{x^4}{4} - \frac{2x^3}{3} + 4x^2\right)\Big|_0^4 \\ &= \frac{256}{3} \\ \bar{x} &= \frac{M}{m} = \frac{\frac{256}{3}}{\frac{112}{3}} = \frac{256}{112} = \frac{16}{7}. \end{aligned}$$

Center of mass is greater than 2 because the object has greater density on the right side of the interval [0, 4].

37. See Example 5.7. In this case, however, $w(x) = 140 - x$. Therefore

$$\begin{aligned} F &= \int_0^{80} 62.4x(140 - x)\,dx \\ &= 62.4 \int_0^{80} (140x - x^2)\,dx \\ &= 62.4\left(70x^2 - \frac{x^3}{3}\right)\Big|_0^{80} \\ &= 62.4(80)^2(130/3) \approx 17{,}305{,}600 \text{ lb.} \end{aligned}$$

39. See Example 5.4.

$$J \approx \frac{.0008}{3(8)} \left\{ \begin{array}{l} 0 + 4(800) + 2(1600) + \\ 4(2400) + 2(3000) + 4(3600) \\ +2(2200) + 4(1200) + 0 \end{array} \right\}$$

$$\begin{aligned} &= 1.52 \\ J &= m\Delta v \\ 1.52 &= .01\Delta v \\ \Delta v &= 152 \text{ ft/s} \\ 152 - 120 &= 32 \text{ ft/s} \end{aligned}$$

41. See Example 6.1.

$$f(x) = x + 2x^3 \text{ on } [0, 1]$$
$$f(x) \geq 0 \text{ for } 0 \leq x \leq 1$$
$$\text{and } \int_0^1 \left(x + 2x^3\right) dx = \left(\frac{x^2}{2} + \frac{x^4}{2}\right)\Big|_0^1 = 1$$

43. See Example 6.4.

$$1 = \int_1^2 \frac{c}{x^2}\, dx = \frac{-c}{x}\Big|_1^2 = \frac{-c}{2} + c = \frac{c}{2}.$$

Therefore $c = 2$.

45. See Example 6.3.

(a)
$$P(x < .5) = \int_0^{.5} 4e^{-4x}\, dx$$
$$= -e^{-4x}\Big|_0^{.5}$$
$$= 1 - e^{-2}$$
$$\approx .864$$

(b)
$$P(.5 \leq x \leq 1) = \int_{.5}^{1} 4e^{-4x}\, dx$$
$$= -e^{-4x}\Big|_{.5}^{1}$$
$$= -e^{-4} + e^{-2}$$
$$\approx .117$$

47. See Example 6.5.

(a) Regarding the mean μ:

$$\mu = \int_0^1 x\left(x + 2x^3\right) dx$$
$$= \frac{x^3}{3} + \frac{2x^5}{5}\Big|_0^1 = \frac{11}{15} \approx .7333\ldots$$

(b) Regarding the median c:

$$\frac{1}{2} = \int_0^c \left(x + 2x^3\right) dx$$
$$= \frac{x^2}{2} + \frac{x^4}{2}\Big|_0^c = \frac{c^2}{2} + \frac{c^4}{2}.$$

Therefore $c^2 + c^4 = 1$,

$$c = \sqrt{\frac{-1 + \sqrt{5}}{2}} \approx .786.$$

49. About k we require

$$1 = 2\int_0^3 ke^{-kx}\, dx = 2\, e^{-kx}\Big|_0^3 = 2(1 - e^{-3k}).$$

Therefore, $2e^{-3k} = 1$ and $k = \ln(2)/3$. The requested probability, using the fact that $e^{-3k} = 1/2$, is

$$\int_1^2 2ke^{-kx}\, dx = -2[e^{-kx}]\Big|_1^2 = 2[e^{-k} - e^{-2k}]$$
$$= 2\left[\left(\frac{1}{2}\right)^{1/3} - \left(\frac{1}{2}\right)^{2/3}\right] = 2^{2/3} - 2^{1/3} \approx .32748.$$

51. Part (a) is routine algebra. The identity in (b) is used to give the indefinite integral formula

$$\int \frac{dv}{v^2 - a^2} = \frac{1}{2a} \ln\left|\frac{v - a}{v + a}\right|$$

for any a, and in particular for our $a = \sqrt{mg/k}$. Let us now put $w(t) = (v(t) - a)/(v(t) + a)$. From (a) we find for any s, that

$$\frac{ks}{m} = \int_0^s \frac{k}{m} dt = \int_0^s \frac{v'(t)}{v^2(t) - a^2} dt =$$
$$\frac{1}{2a} \ln\left|\frac{v(t) - a}{v(t) + a}\right|\Big|_0^s = \frac{1}{2a} \ln|w(t)|\Big|_0^s =$$
$$\frac{1}{2a}(\ln|w(s)| - \ln|w(0)|) = \frac{1}{2a} \ln\left|\frac{w(s)}{w(0)}\right|.$$

We have now learned that for any s or t, we have

$$|w(s)| = |w(0)|e^{2aks/m},$$

and

$$\frac{v(t) - a}{v(t) + a} = w(t) = (\pm w(0))e^{2akt/m}$$
$$= -ce^{2akt/m}.$$

More routine algebra now gives

$$v(t) = \frac{-a(ce^{2akt/m} - 1)}{ce^{2akt/m} + 1},$$

which is the unambiguous form of (c). (It is unclear as written whether the t is under the radical or perhaps in the denominator. It is neither.) As t goes to infinity, the exponentials go to infinity, and the ratio goes to $-ac/c = -a$, so the terminal speed is $a = \sqrt{mg/k}$. (This is part (f) and the conclusion is independent of the initial velocity.) We do see that $v(0) = -a(c-1)/(c+1)$, and therefore $v(0) = 0$ if and only if $c = 1(d)$. If this is true, then we find (e)

$$\begin{aligned} v(t) &= \frac{-a(e^{2akt/m} - 1)}{e^{2akt/m} + 1} \\ &= -a\left(\frac{e^{akt/m} - e^{-akt/m}}{e^{akt/m} + e^{-akt/m}}\right) \\ &= -a \tanh\left(\frac{akt}{m}\right) = -\sqrt{\frac{mg}{k}} \tanh\left(t\sqrt{\frac{kg}{m}}\right). \end{aligned}$$

53. It is first necessary to do Exercise 52. From the general indefinite integral formula,

$$\int \tanh(x)\,dx = \int \frac{\sinh(x)\,dx}{\cosh(x)} = \ln(\cosh(x)),$$

we find from Exercise 51 the distance fallen in elapsed time s to be

$$\int_0^s |v(t)|dt = \int_0^s \sqrt{\frac{mg}{k}} \tanh\left(t\sqrt{\frac{kg}{m}}\right) dt =$$

$$\sqrt{\frac{mg}{k}}\left(\frac{1}{\sqrt{\frac{kg}{m}}}\right) \ln\left(\cosh\left(t\sqrt{\frac{kg}{m}}\right)\right)\Bigg|_0^s =$$

$$\frac{m}{k} \ln\left(\cosh\left(s\sqrt{\frac{kg}{m}}\right)\right)$$

(recalling $\ln(\cos h(0)) = \ln(1) = 0$). In the metric system with $g = 9.8$, we prepare the following table:

	weight	mg	800
	gravitation	g	9.8
	mass	m	81.63

drag factor k	terminal speed (meters/sec)	elapsed time (seconds) 2	4
		elapsed distance (meters)	
0.125	80	19.41	75.45
0.500	40	18.86	68.35

55. Presumably the flight is over when $v(t) = 0$, which occurs first at time

$$t_0 = \tan\text{h}^{-1}(v_0/7901)/.00124.$$

On the time interval $[0, t_0]$ we could take

$$u = \tanh^{-1}(v_0/7901) - .00124t,$$

a quantity decreasing (as t increases) from $u_0 = \tanh^{-1}(v_0/7901)$ to zero, and we would have $v(t) = 7901\tanh(u)$. It then follows that

$$\begin{aligned} a(t) = v'(t) &= 7901]\operatorname{sech}^2(u)(du/dt) \\ &= -7901(.00124)\operatorname{sech}^2(u). \end{aligned}$$

But $\operatorname{sech}^2(u)$ is a decreasing function of u and therefore $|a(t)|$ is maximized when $u = 0(t = t_0)$ where it has the value

$$7901(.00124) = 9.7972(\text{ m/sec}^2).$$

For contrast, at time zero,

$$\begin{aligned} |a(t)| &= 9.7972\operatorname{sech}^2(u_0) \\ &= 9.7972(1 - \tanh^2(u_0)) \\ &= 9.7972\left(1 - \left[\frac{v_0}{7901}\right]^2\right). \end{aligned}$$

Note that the hyperbolic tangent takes values in the interval $(-1, 1)$, and the *inverse* hyperbolic tangent is only defined for objects in that same interval. Therefore, we may presume $v_0 < 7901$ or the original formula is undefined.

57. For purposes of comparison, we ignore the gravitational constant. Put alternatively, we consider the integral

$$T\sqrt{g} = \int_0^1 \frac{\frac{ds}{du}}{\sqrt{-2y}} du.$$

For the line, we find the expression and value

$$\int_0^1 \frac{\sqrt{\pi^2+4}}{2\sqrt{u}} du = \sqrt{\pi^2+4} \approx 3.7242.$$

For the parabola, we find by the software Scientific Workplace, the expression and value

$$\frac{1}{2}\int_0^1 \sqrt{\frac{\pi^2+16(u-1)^2}{u(2-u)}} du \approx 3.2763.$$

For the brachistrochrome, the integrand is constantly π, which is also the value of the integral (3.1416). It appears from these examples that other things being equal, T is reduced by having a steeper initial slope. The brachistrochrome, as the picture associated to Exercise 58 shows, has a vertical tangent at the origin. This feature alone, of course, does not make it minimal for T. For example, using a quarter-ellipse, Scientific Workplace gave the value 3.651 , better than the line, but not as good as the parabola. Perhaps the ellipse is too flat near the lower end.

Chapter 6

Differential Equations

6.1 Growth and Decay Problems

1. We know $y = Ae^{kt}$ when $y' = ky$. In this case, $k = 4$ (from $y' = 4y$) and $A = y(0) = 2$ (given). So the solution is $y = 2e^{4t}$.

3. We know $y = Ae^{kt}$ when $y' = ky$. In this case, $k = -3$ (from $y' = -3y$) and $A = y(0) = 5$ (given). So the solution is $y = 5e^{-3t}$.

5. We know $y = Ae^{kt}$ when $y' = ky$. In this case, $k = 2$ (from $y' = 2y$) and by substitution $Ae^2 = Ae^k = y(1) = 2$ (given). So $A = 2e^{-2}$, and one could write the solution in the form $y = 2e^{2(t-1)}$.

7. In this problem, y' is the constant -3. Assuming an independent variable t, this gives $y = -3t + c$. If $y(0) = 3$, then $3 = -3(0) + c$. Therefore, $c = 3$ and $y = -3t + 3$.

9. We understand that this is exponential growth, and we know even without the differential equation that $y = Ae^{kt}$ for some A, k. As is typical, $A = y(0) = 100$ (given). We next solve for k using the known doubling time 4:

$$\begin{aligned} 2A &= Ae^{k \cdot 4} \\ e^{4k} &= 2 \\ 4k &= \ln 2 \\ k &= \frac{\ln 2}{4}. \end{aligned}$$

We now have

$$y = 100e^{\left(\frac{\ln 2}{4}\right)t} = 100(2^{t/4}).$$

The population will reach 6000 after t hours, if $100(2^{t/4}) = 6000$, $2^{t/4} = 60$, and $(t/4) \ln 2 = \ln 60$,

$$t = \frac{4 \ln 60}{\ln 2} \approx 23.628 \text{ hours.}$$

11. From $y = Ae^{kt}$, $(A = y(0) = 400)$, we get with $t = 1$, $800 = y(1) = 400e^k$; hence $k = \ln(2)$. This leads to $y = 400e^{(ln 2)t} = 400(e^{\ln 2})^t = 400(2^t)$. At time $t = 10$, y will be $400(2^{10}) = 400(1024) = 409{,}600$.

13. When $y = Ae^{kt}$, the doubling time t_d is *always* $(\ln 2)/k$ (from $e^{kt} = 2$, or look ahead to Exercise 17). In this case we are given $k = .44$ (per hour), so $t_d = \ln(2)/.44 = 1.575$ (hours).

15. With t measured in minutes, and $y = Ae^{kt} = 10^8 e^{kt}$ on the time interval $(0, T)$ (during which no treatment is given), the condition on T is that 10% of the population at time T (surviving after the treatment) will be the same as the initial population. In other words, $10^8 = (.1)10^8 e^{kT}$. This gives $e^{kT} = 10$ and $T = \ln(10)/k$. To get k we use the given doubling time $t_d = 20$. Since we always have $t_d = \ln(2)/k$, this leads to $k = \ln(2)/20$ and

$$T = \frac{\ln(10)}{\ln(2)/20} = \frac{20 \ln(10)}{\ln(2)} = 66.44 \text{ (minutes).}$$

17. Given $y(t) = Ae^{rt}$, the doubling time t_d obeys

$$\begin{aligned} 2A &= Ae^{rt_d} \\ 2 &= e^{rt_d} \\ rt_d &= \ln 2 \\ t_d &= \frac{\ln 2}{r} \end{aligned}$$

as desired.

19. Given $y(t) = Ae^{-1.3863t}$, the half-life t_h obeys

$$\frac{1}{2}A = Ae^{-1.3863t_h}$$
$$e^{-1.3863t_h} = \frac{1}{2}$$
$$-1.3863t_h = \ln\frac{1}{2} = -\ln 2$$
$$t_h = \frac{\ln 2}{1.3863} \approx .500 \text{ day.}$$

More generally and for future reference (also see Exercise 18), the half-life t_h is related to the decay rate r (assuming $y = Ae^{-rt}$, $r > 0$) via the law $rt_h = \ln(2)$.

21. Given the half-life of 3 hours, the decay rate r is given by $\ln(2)/3$ (See the remark attached to the solution to Exercise 19.) If the initial amount is .4, then

$$y(t) = .4e^{-\left(\frac{\ln(2)}{3}\right)t} = .4(2^{-t/3}).$$

To determine when the amount drops below .01, solve for t the equation $.4(2^{-t/3}) = .01$.
We find

$$2^{-t/3} = .01/.4 = 1/40$$
$$(-t/3)\ln(2) = \ln(1/40) = -\ln(40)$$
$$t = \frac{3\ln(40)}{\ln(2)} \approx 15.97 \text{ hours.}$$

23. From the comment after the solution to Exercise 19, the decay rate r is $\ln(2)/28$. Therefore,

$$y(t) = e^{\left(-\frac{\ln 2}{28}\right)t} = 2^{-t/28}$$

gives the percent of strontium-90 remaining after t years. After 50 years,

$$y(50) = 2^{-50/28} \approx .2900.$$

About 29% will remain.

25. With a half-life of 5730 years, the decay rate r is $\ln(2)/5730$. If 20% of the initial amount is left at a certain time t, then

$$.2 = e^{\left(-\frac{\ln 2}{5730}\right)t} = 2^{-t/5730}$$
$$(-t/5730)\ln(2) = \ln(.2) = \ln(1/5) = -\ln(5)$$
$$t = \frac{5730\ln(5)}{\ln(2)} \approx 13{,}305 \text{ (years).}$$

The fossil is about 13,305 years old.

27. We use the known solution to Newton's Law of Cooling, according to which $y(t) = Ae^{kt} + T_a$. With $T_a = 70$, we learn $200 = y(0) = A + 70$ so that $y(t) = 130e^{kt} + 70$. After one minute,

$$180 = 130e^{k\cdot 1} + 70$$
$$e^k = \frac{110}{130} = \frac{11}{13}$$
$$k = \ln\frac{11}{13}.$$

Therefore, $y(t) = 130e^{\left(\ln\frac{11}{13}\right)t} + 70$. The temperature will be 120°after t minutes where

$$120 = 130e^{\left(\ln\frac{11}{13}\right)t} + 70$$
$$e^{\left(\ln\frac{11}{13}\right)t} = \frac{50}{130} = \frac{5}{13}$$
$$\left(\ln\frac{11}{13}\right)t = \ln\frac{5}{13}$$
$$t = \frac{\ln\frac{5}{13}}{\ln\frac{11}{13}} \approx 5.720 \text{ minutes.}$$

29. Using Newton's Law of Cooling $y = Ae^{kt} + T_a$ with the ambient temperature of 70°and initial temperature of 50°,

$$50 = Ae^0 + 70$$
$$A = -20$$

so that $y(t) = -20e^{k\cdot t} + 70$. If, after two minutes, the temperature is 56°,

$$56 = -20e^{k\cdot 2} + 70$$
$$e^{2k} = \frac{14}{20} = .7$$
$$2k = \ln .7$$
$$k = \frac{1}{2}\ln .7.$$

Therefore, $y(t) = -20e^{\frac{1}{2}(\ln .7)t} + 70$.

31. Using Newton's Law of Cooling with ambient temperature 70°, initial temperature 60°, and with time t (in minutes) elapsed since 10:07, we have

$$y(t) = Ae^{kt} + 70$$
$$60 = Ae^{0k} + 70 = A + 70,$$
$$\therefore\ A = -10$$

and $y(t) = -10e^{kt} + 70$ (for the martini). Two minutes later, its temperature is 61°. Hence,

$$61 = -10e^{k\cdot 2} + 70$$
$$e^{2k} = \frac{9}{10}$$
$$2k = \ln\frac{9}{10}$$
$$k = \frac{1}{2}\ln\frac{9}{10} = \frac{1}{2}\ln .9.$$

Therefore,

$$y(t) = -10e^{\left(\frac{1}{2}\ln .9\right)t} + 70.$$

The temperature is 40°at elapsed time t only if

$$40 = -10e^{\left(\frac{1}{2}\ln .9\right)t} + 70$$
$$e^{\left(\frac{1}{2}\ln .9\right)t} = \frac{30}{10} = 3$$
$$\left(\frac{1}{2}\ln .9\right)t = \ln 3$$
$$t = \frac{2\ln 3}{\ln .9} \approx -20.854,$$

or about 21 minutes *before* 10:07. The time was 9:46. Newton's Law of Cooling *does* permit negative times.

33. With t the time elapsed since serving, with the ambient temperature 68°and if the temperature is 160°when $t = 20$, then

$$y(t) = Ae^{kt} + T_a$$
$$160 = Ae^{k\cdot 20} + 68$$
$$Ae^{20k} = 92.$$

After 22 minutes the temperature is 158°.

$$158 = Ae^{k\cdot 22} + 68$$
$$Ae^{22k} = 90$$
$$e^{2k} = \frac{Ae^{22k}}{Ae^{20k}} = \frac{90}{92}$$
$$k = \frac{1}{2}\ln\frac{90}{92}.$$

Therefore,

$$y(t) = Ae^{\frac{1}{2}\left(\ln\frac{90}{92}\right)t} + 68.$$

Using the first set of numbers,

$$Ae^{20\cdot\frac{1}{2}\ln\frac{90}{92}} = 92$$
$$A = \frac{92}{e^{10\ln\frac{90}{92}}} \approx 114.615.$$

So,

$$y(t) = 114.615e^{\frac{1}{2}\left(\ln\frac{90}{92}\right)t} + 68.$$

The serving temperature is

$$y(0) = 114.615e^{0} + 68 = 182.615,$$

or about 182.6°(more than 180o). Looking back to Exercise 32, this should not be a surprise. The friend was using linear extrapolation based on two data points. Any exponential curve such as this lies above its own (extended) chords all times *outside* the interval between the reference points.

35. Annual: $A = 100(1 + .08)^1 = \$1080.00$

Monthly: $A = 1000\left(1 + \frac{.08}{12}\right)^{12} = \$1083.00$

Daily: $A = 1000\left(1 + \frac{.08}{365}\right)^{365} = \$1083.28$

Continuous: $A = 1000e^{.08\cdot 1} = \$1083.29$

37. Person A: $A = 10{,}000e^{.12\cdot 20} = \$110{,}231.76$

Person B: $B = 20{,}000e^{.12\cdot 10} = \$66{,}402.34$

39. Let t be the number of years after 1985. Then, assuming continuous compounding at rate r,

$$\begin{aligned} 9800 &= 34e^{r\cdot 10} \\ e^{10r} &= \frac{9800}{34} \\ 10r &= \ln\left(\frac{9800}{34}\right) \\ r &= \frac{1}{10}\ln\left(\frac{9800}{34}\right) \approx .566378 \end{aligned}$$

Therefore,

$$A = 34e^{\frac{1}{10}\ln\left(\frac{9800}{34}\right)t} = 34\left(\frac{9800}{34}\right)^{t/10}.$$

In 2005, t is 20 and

$$A(20) = 34\left(\frac{9800}{34}\right)^2 = \$2,824,705.88.$$

Comment: The final number is independent of the assumption of continuous compounding. It could just as well be viewed as annual compounding at rate i, in which

$$1 + i = (9800/34)^{0.1} = 1.76187.$$

The annual interest rate 76.2% is obviously very high, and can be compared numerically with the continuous rate $r = 56.6\%$ used above. For lower rates, there is not such a notable difference.

In some circles, the continuous compounding rate is known as the *force* to distinguish it from said annual interest *rate* (called i above), which is defined by the equation $1 + i = A(t+1)/A(t)$ (t must be measured in years, but the ratio is independent of t).

In this problem, the distinction has no significance (to each rate there is a force and vice versa), but in Exercise 63 ahead, it does become an issue.

41. The problem with comparing tax rates for the income bracket [16K,20K] over a 13-year time interval is that due to inflation, the persons in this income bracket in 1988 have less purchasing power than those in the same bracket in 1975, and a lower tax rate may or may not compensate. To quantify and illustrate, assume a 5.5% annual inflation rate. This would translate into a loss of purchasing power amounting to

$$1/(1.055)^{13} = 1/(2.006) \approx 1/2,$$

which is essentially to say that in terms of comparable purchasing power, the income bracket [16K,20K] in 1988 corresponds to an income bracket of [8K,10K] in 1975. One should then go back and look at the tax rate for the latter bracket in 1975. Only if that tax rate exceeds the 1988 rate (15%) for the bracket [16K,20K] should one consider that taxes have genuinely gone down. Alternatively, one could project the income bracket [16K,20K] (in 1975) forward by a factor of two to find the comparable income bracket [32K,40K] in 1988. Only if the 1988 tax rate for this class is less than 28% would one consider taxes to have gone down. Note that we are discussing two different income equivalence classes. It is entirely possible that taxes for the poorer class went down, while taxes for the richer class went up. To give a complete answer one needs: tax rates not given in the problem; inflationary assumption; and class identification.

43. $T_1 = 30,000 \cdot .15 + (40,000 - 30,000) \cdot .28$
$= \$7300$
$T_2 = 30,000 \cdot .15 + (42,000 - 30,000) \cdot .28$
$= \$7860$
$T_1 + .05T_1 = \$7665$
The tax (T_2) on the new salary is greater than the adjusted tax ($(1.05)T_1$) on the old salary.

45. With a constant depreciation rate of 10% (as the term is used in the text), the value of the \$40,000 item after 10 years would be

$$40,000(e^{-(0.1)10}) = 40,000e^{-1} \approx 14,715,$$

and after 20 years

$$40,000(e^{-(0.1)20}) = 40,000e^{-2} \approx 5,413.$$

By the straight line method, assuming a value of zero after 20 years, the value would be 20,000 after 10 years.

47. Over an interval of elapsed time Δt, a drop from 1000 to 500 would be represented by a slope of $\frac{500}{\Delta t}$, while a drop from 10 to 5 would be represented by a slope of $\frac{5}{\Delta t}$. A drop from ln 1000 to ln 500 would result in a slope of

$$\frac{\ln 1000 \ln 500}{\Delta t} = \frac{\ln \frac{1000}{500}}{\Delta t} = \frac{\ln 2}{\Delta t},$$

while a drop from ln 10 to ln 5 would result in a slope of

$$\frac{\ln 10 - \ln 5}{\Delta t} = \frac{\ln \frac{10}{5}}{\Delta t} = \frac{\ln 2}{\Delta t}.$$

So the slopes of the logarithmic drops are the same. If a population were changing at a constant percentage rate, the graph of population versus time would appear exponential, while the graph of the logarithm of population versus time would appear linear.

49. Fitting a line to the first two data points on the plot of time vs. the natural log of the population ($y = \ln(P(x))$) produces the linear function $y = 1.468x + 0.182$, which is equivalent to fitting the original date with the exponential function

$$P(x) = e^{1.468x+0.182} \quad \text{or} \quad P(x) = 1.200e^{1.468x}.$$

Since this is only a structured approximation, we could denote it by $P_m(x)$ (the subscript "m" suggesting *model*) in order to contrast it from the original data ($P(x)$ itself). See the following solution to Exercise 51.

51. Fitting a line through the first and last data points on the time vs. ln(population) graph leads eventually to the exponential model (see the above solution of Exercise 49) $P_m(x) = 10e^{0.397x}$. We note that $P_m(1) = 14.878$, $P_m(2) = 22.136$, and $P_m(3) = 32.934$. The fit is not bad; these three numbers were supposed to be 15, 22, and 33.

53. Again fitting to the first and last data points, with x representing years elapsed since 1960, we find the exponential model

$$\frac{\text{farm population}}{\text{general population}} = .075e^{(\frac{-\ln(75/16)}{30})x}$$

$$= .075(.21053)^{x/30}.$$

This model assigns the numbers .045 to 1970 and .027 to 1980. The fit is only fair, as the actual numbers were 5.2% (.052) in 1970 and 2.5% (.025) in 1980.

55. With known conclusion $y = Ae^{-rt}$, $A = 150, t = 24$, and $r = \ln(2)/t_h$ (see the solution to Exercise 19) we find that with $t_h = 31$ we get

$$y = 150(1/2)^{(24/31)} = 87.7,$$

and with $t_h = 46$ we get

$$y = 150(1/2)^{(24/46)} = 104.5.$$

The difference is about 17 days, at 19% not a dramatically large percentage of the smaller base of 88($105/88 = 1.19$). If one had expected the two numbers to be proportional to the half-lives, one would have expected the difference to come in at 48% ($46/31 = 1.48$) and would definitely consider the 19% to be far less than anticipated.

57. See the solution above to Exercise 55. In this case, with $t_h = 4$ (hours), $A = 1$, $y = Ae^{-rt}$, and $r = \ln(2)/4$, one finds $y = (1/2)^{(t/4)}$. The curve is a typical exponential, declining from a value of 1 at $t = 0$ to $1/2^6 = 1/64 = .016$ at $t = 24$.

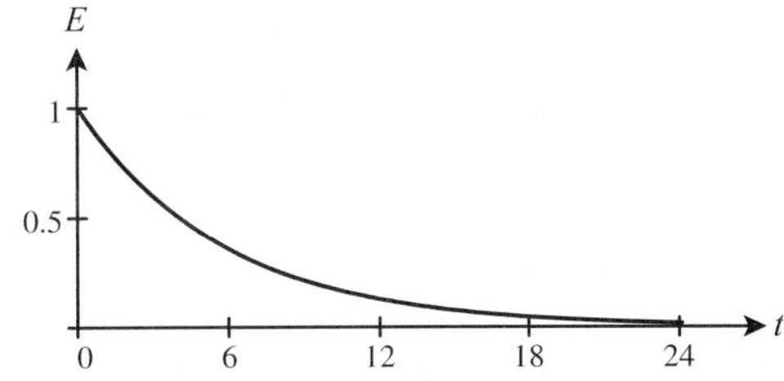

59. With r the rate of continuous compounding, the value of an initial amount X after t years is Xe^{rt}. If the goal is P, then the relation is $P = Xe^{rt}$ or $X = Pe^{-rt}$. In this case with $r = .08$, $t = 10$, and $P = 10{,}000$, we find $X = 10{,}000e^{-.8} = \$4493.29$.

61. From the two formula

$$\int_0^T e^{-rt}dt = \frac{1-e^{-rT}}{r}$$

and

$$\int_0^T te^{-rt}dt = \frac{-Te^{-rT}}{r} + \frac{1-e^{-rT}}{r^2},$$

with $r = .05$ and $T = 3$, we find for (A):60,000(20) $(1 - e^{-.15}) = 167,150(\$)$. For (B) we get the above *plus* $(3000)\{-60e^{-.15} + 400(1 - e^{-.15})\} = 12,223$ for a total of 179,373 ($). For (C), the exponentials cancel, and the answer is simply $\int_0^3 60000dt =$ 180,000($).

63. There are certain ambiguities in the question. If we assume .08 to be the annual rate of accumulation (the *annual interest rate*), and that the comparison is to be made between three years of accumulation of $1,000,000 (no spending) versus the accumulation of four annual payments of $280,000 at times 0,1,2,3 (again no spending), then the respective figures are $1,000,000(1.08)^3 = 1,259,712(\$)$ versus $280,000\{1.083 + 1.082 + 1.08 + 1\} = 1,261,711(\$)$, and one should take the annuity by a narrow margin. We get the same conclusion if we compare *present* values: The numbers come out 1,000,000 (the lump sum) versus $280,000\{1.08^{-3} + 1.08^{-2} + 1.08^{-1} + 1\} = 1,001,587(\$)$. The comparison of present values does not require the assumption of *no spending*, and it is considered theoretically preferable.

However, if we assume that the .08 is a *continuous* compounding rate (the so-called *force of interest*), the present value comparisons come out $1,000,000 (the lump sum) versus $280,000\{1 + e^{-.08} + e^{-.16} + e^{-.24}\} = 997,329(\$)$, and this time one should take the lump sum. For many numbers it would not matter whether force or rate, but .08 happens to be in a narrow range where it does matter. For lower rates (including zero), one takes the annuity because there's more money involved; $4(280,000) = 1,120,000 > 1,000,000$; for somewhat higher rates (for example, when the rate is at or above .0833 and the force is at or above .08), one takes the lump sum on the general principle "get more money sooner and it's high earnings will eventually more than make up for the smaller total." (The universal relationship between force and rate is $1 + \text{rate} = e^{\text{force}}$.)

6.2 Separable Differential Equations

1. $y' = (3x + 1)\cos y$
To separate, write $\dfrac{y'}{\cos(y)} = 1 + 3x$.

3. $y' = (3x + y)\cos y$
This is not separable.

5. $y' = x^2y + y\cos x$. To separate, write $y' = y(x^2 + \cos x)$ and $\dfrac{y'}{y} = x^2 + \cos x$.

7. $y' = x^2y - x\cos y$
This is not separable.

9. Given $y' = (x^2 + 1)y$,
separate as $\dfrac{1}{y}y' = x^2 + 1$.
Integrate as $\ln|y| = \dfrac{1}{3}x^3 + x + k$.
Exponentiate to get $|y| = e^{\frac{1}{3}x^3+x+k}$, hence $y = \pm e^{\frac{1}{3}x^3+x+k} = ce^{\frac{1}{3}x^3+x} \quad (c = \pm e^k)$.

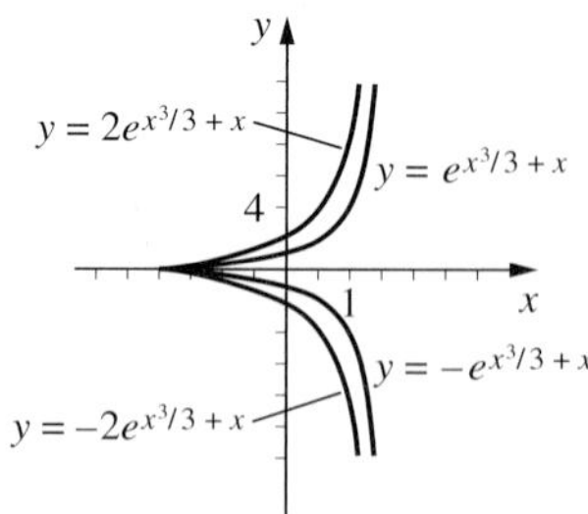

11. Given $y' = 2x^2y^2$,
separate as $\dfrac{1}{y^2}y' = 2x^2$.

Integrate as $-\frac{1}{y} = \frac{2}{3}x^3 + c$.

Solve in the form $y = -\frac{1}{\frac{2}{3}x^3 + c}$.

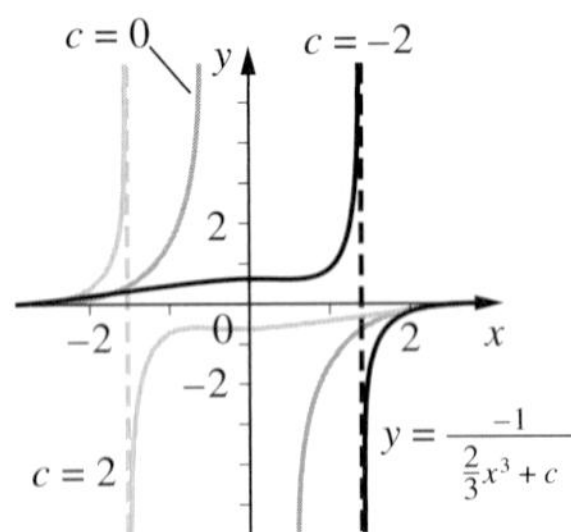

13. Given $y' = \frac{6x^2}{y(1+x^3)}$,

separate as $yy' = \frac{6x^2}{1+x^3}$.

Integrate as $\frac{1}{2}y^2 = 2\ln|1+x^3| + c$.

Solve in the form $y = \pm\sqrt{4\ln|1+x^3| + k}$ $(k = 2c)$.

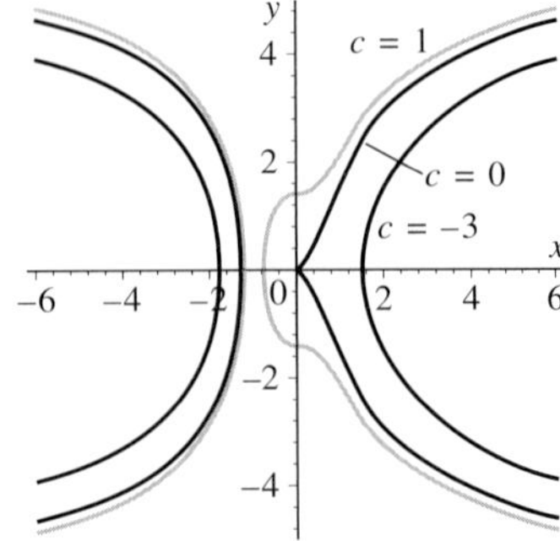

15. Given $y' = \frac{4x^2}{y^2}$,

separate as $y^2y' = 4x^2$.

Integrate as $\frac{1}{3}y^3 = \frac{4}{3}x^3 + k$, or

$y^3 = 4x^3 + 3k = 4x^3 + c \quad (c = 3k)$.

Solve as $y = (4x^3 + c)^{1/3}$.

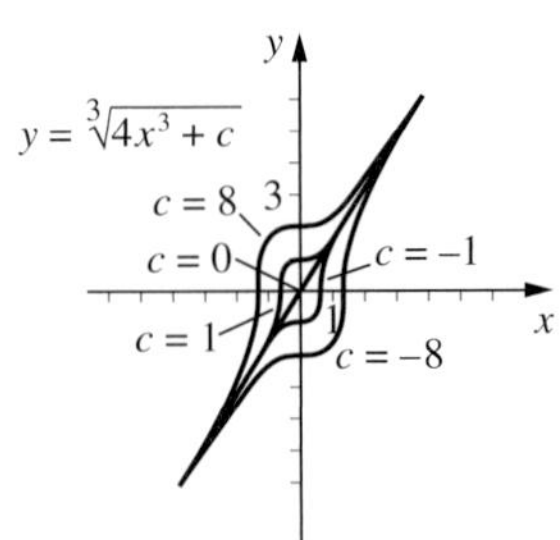

17. Given $y' = y^2 - y$,

separate as $\frac{1}{y^2 - y}y' = 1$.

By partial fractions, $\frac{1}{y^2 - y} = \frac{1}{y-1} - \frac{1}{y}$.

Integrate as $\ln|y-1| - \ln|y| = x + k$,

or $\ln\left|\frac{y-1}{y}\right| = x + k$.

Exponentiate to get $\left|\frac{y-1}{y}\right| = e^{x+k} = e^k e^x$,

hence $\frac{y-1}{y} = \pm e^k e^x = -ce^x (-c = \pm e^k)$.

Solve by writing $y - 1 = -yce^x$,

$y + yce^x = 1$, $y(1 + ce^x) = 1$, and finally

$y = \frac{1}{1 + ce^x}$.

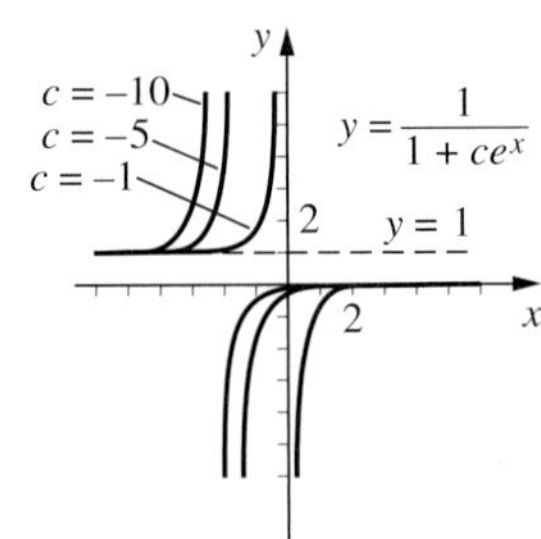

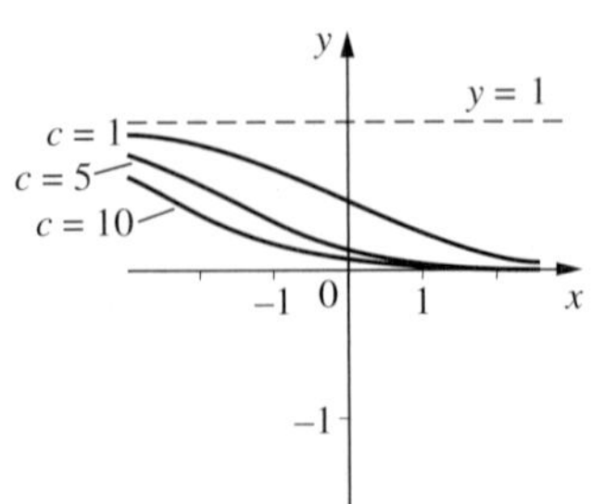

19. Given $y' = \dfrac{xy}{1+x^2}$,
separate as $\dfrac{1}{y}y' = \dfrac{x}{1+x^2}$.
Integrate as
$\ln|y| = \dfrac{1}{2}\ln(1+x^2) + k$
$= k + \ln\sqrt{1+x^2}$.
Exponentiate to get
$|y| = e^k\sqrt{1+x^2}$.
Solve in form
$y = \pm e^k\sqrt{1+x^2} = c\sqrt{1+x^2}$
($c = \pm e^k$ or zero,
any real number).

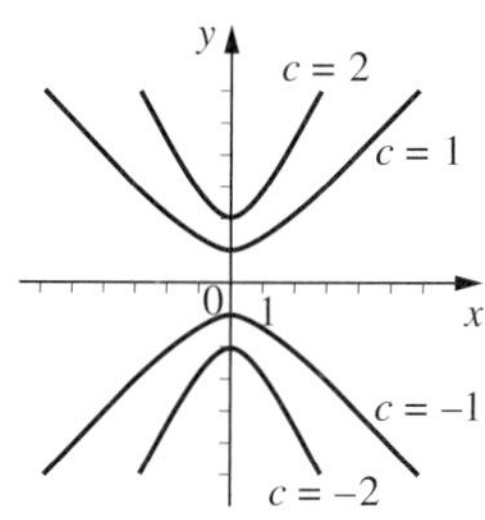

21. Given $y' = \dfrac{4\sin x}{y^3}$,
separate as $y^3y' = 4\sin x$,
integrate as $\dfrac{1}{4}y^4 = 4(-\cos x) + c$, and
solve in the form
$y = \pm\sqrt[4]{-16\cos x + c}$.

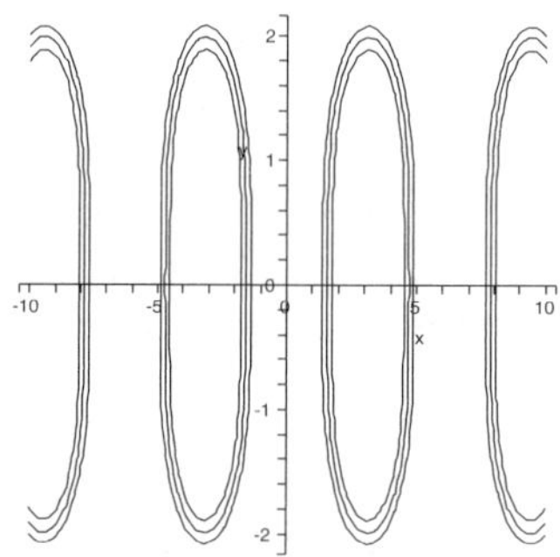

23. Given $y' = 3(x+1)^2y$,
separate as $\dfrac{1}{y}y' = 3(x+1)^2$.
Integrate to get $\ln|y| = (x+1)^3 + k$,
exponentiate to get $|y| = e^ke^{(x+1)^3}$, and
solve in the form $y = ce^{(x+1)^3}$ ($c = \pm e^k$).
Assuming $y(0) = 1$, we have
$1 = y(0) = ce^1 = ce$, so $c = \dfrac{1}{e}$. Therefore,
$y(x) = \dfrac{1}{e}e^{(x+1)^3} = e^{x^3+3x^2+3x}$.

25. Given $y' = \dfrac{4x^2}{y}$,
separate as $yy' = 4x^2$,
integrate to $\dfrac{1}{2}y^2 = \dfrac{4}{3}x^3 + k$, or
$y^2 = \dfrac{8}{3}x^3 + c$ ($c = 2k$), and
solve in the form $y = \pm\sqrt{\dfrac{8}{3}x^3 + c}$.
Assuming $y(0) = 2$, we have $2 = y(0) = \pm\sqrt{c}$, so $c = 4$ *and* we must take the positive square root, hence
$y(x) = \sqrt{\dfrac{8}{3}x^3 + 4}$.

27. Given $y' = \dfrac{4y}{x+3}$, separate as $\dfrac{1}{y}y' = \dfrac{4}{x+3}$
integrate to $\ln|y| = 4\ln|x+3| + k$, and
exponentiate to $|y| = e^{4\ln|x+3|+k}$, hence
$y = \pm e^k(x+3)^4 = c(x+3)^4$
($c = \pm e^k$ or zero, any real number).
Assuming $y(-2) = 1$, then
$1 = y(-2) = c1^4 = c$, so $c = 1$ and $y(x) = (x+3)^4$.

29. Given $y' = \dfrac{4x}{\cos y}$,
separate as $\cos y\, y' = 4x$ and
integrate to $\sin y = 2x^2 + c$.
Assuming $y(0) = 0$, then $0 = \sin(0) = \sin y(0) = 0 + c = c$.
Therefore, $c = 0$, $\sin y = 2x^2$, and $y = \arcsin(2x^2)$, $(-1/\sqrt{2} < x < 1/\sqrt{2})$.

31. With $M = 2$ and $k = 3$, equation 2.10 tells us

$$y = \frac{2Ae^{6t}}{1 + Ae^{6t}},$$

and so assuming $y(0) = 1$, we find $1 = y(0) = 2A/(1 + A)$. Therefore, $A = 1$ and

$$y = \frac{2e^{6t}}{1 + e^{6t}}.$$

33. With $M = 5$ and $k = 2$, equation 2.10 tells us

$$y = \frac{5Ae^{10t}}{1 + Ae^{10t}},$$

and so assuming $y(0) = 4$, we find $4 = y(0) = 5A/(1 + A)$. Therefore, $A = 4$ and

$$y = \frac{20e^{10t}}{1 + 4e^{10t}}.$$

35. With $M = 1$ and $k = 1$, equation 2.10 tells us

$$y = \frac{Ae^{t}}{1 + Ae^{t}},$$

and so assuming $y(0) = \frac{3}{4}$, we find $\frac{3}{4} = y(0) = A/(1 + A)$. Therefore, $A = 3$ and

$$y = \frac{3e^{t}}{1 + 3e^{t}}.$$

37. Adapting the solution

$$y = \frac{MAe^{Mkt}}{1 + Ae^{Mkt}}$$

in equation 2.10 with $r = Mk$, we find

$$y = \frac{MAe^{rt}}{1 + Ae^{rt}}.$$

In this case with $r = .71$, $M = 8 \times 10^7$, and $y(0) = 2 \times 10^7$, we find

$$2 \times 10^7 = y(0) = \frac{8 \times 10^7 A}{1 + A}.$$

Therefore,

$$\frac{A}{1 + A} = \frac{2}{8} = \frac{1}{4}, \; A = 1/3,$$

and after routine simplification we find

$$y = \frac{(8 \times 10^7)e^{.71t}}{3 + e^{.71t}}.$$

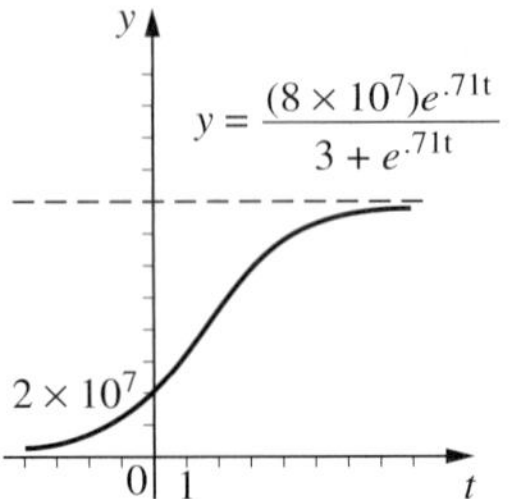

39. We should follow the solution as given up to the second equation on page 538, leaving the absolute values in place. The new version of equation 2.9 is

$$\left|\frac{y}{M - y}\right| = Ae^{Mkt}$$

with $A > 0$. Under the circumstances $y > M$, the ratio is negative, and the resolution is

$$\frac{y}{M - y} = -Ae^{Mkt}.$$

This further resolves as

$$y = -MAe^{Mkt} + yAe^{Mkt},$$

which eventually becomes

$$y = \frac{MAe^{Mkt}}{Ae^{Mkt} - 1}.$$

41. Starting from

$$y = \sqrt[3]{x^3 + \frac{21}{2}x^2 + 9x + 3c}$$

with $y(0) = 0$, we have $c = 0$. In the following graph, the focus is on the various places where $y = 0$, so we have confined the range of x to the interval $[-10, 10]$. We do see roots at $x = 0$, near $x = -1$ $(-.9416)$, and between -9 and -10 (-9.5584). At each of these roots, y exhibits a vertical tangent, of which more will be said in the solution to Exercise 43.

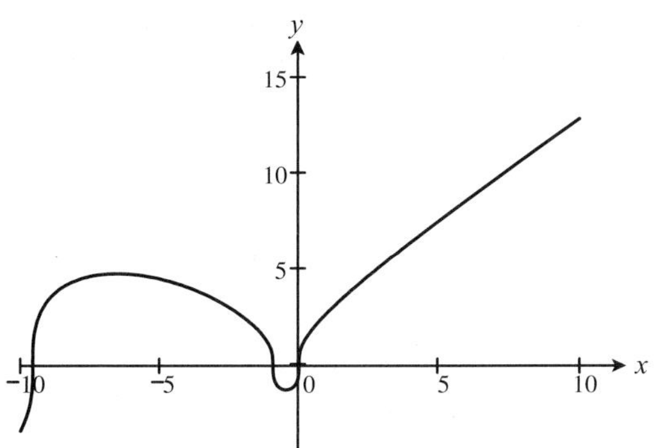

43. One would have to say more precisely, given

$$y' = \frac{x^2 + 7x + 3}{y^2},$$

that $y'(x)$ does not exist for a given x if $y(x) = 0$, unless perhaps it were true that $x^2 + 7x + 3$ was *also* zero, in which case a deeper study would be required before making a statement.

The second equation (the vanishing of the numerator in the DE) is easily solved. We will call the roots x_1 and x_2, with

$$x_1 = \frac{-7 - \sqrt{37}}{2} \approx -6.5414 \text{ and}$$

$$x_2 = \frac{-7 + \sqrt{37}}{2} \approx -.4586.$$

If one is working with the solution in the form

$$y = \sqrt[3]{x^3 + \frac{21}{2}x^2 + 9x + 3c},$$

we see that $y(x) = 0$ if

$$-3c = x^3 + \left(\frac{21}{2}\right)x^2 + 9x.$$

The cubic polynomial on the right (call it $h(x)$) has its derivative given by $h'(x) = 3x^2 + 21x + 9 = 3(x^2 + 7x + 3)$, and the roots of $h'(x)$ are our friends x_1 and x_2 above. The effect is that $h(x)$ has a relative maximum at x_1 and a relative minimum at x_2, and so the equation $-3c = h(x)$ has three solutions when $-3c$ lies between the relative minimum and the relative maximum, i.e., if $h(x_2) < -3c < h(x_1)$, or when

$$\frac{-h(x_1)}{3} < c < \frac{-h(x_2)}{3}.$$

We call the smaller c_1, and its value is exactly

$$-\left(\frac{217 + 37\sqrt{37}}{12}\right),$$

or about -36.84. We call the larger c_2, and its value is exactly

$$\left(\frac{-217 + 37\sqrt{37}}{12}\right),$$

or about .67185.

45. Continuing from the solution to Exercise 43, if c is exactly c_2, then the equation $h(x) = -3c_2$, which already has x_2 as one solution (because $-3c_2 = h(x_2)$ by definition of c_2) actually has x_2 as a *double* solution because $h'(x_2)$ is also zero (original definition of x_2). In effect, $h(x) + 3c_2$ factors into

$$(x - x_2)^2\left(x + \frac{3c_2}{(x_2)^2}\right).$$

(For those who do not remember their theory of polynomials from precalculus, all this can be verified by pure algebra.). The second solution of the equation $h(x) = -3c_2$ is evidently $-3c_2/x_2^2$, which we call x_3. Its value is about -9.6, but it is of no great importance for the way this problem is phrased. Now, in the solution $y(x)$ to the differential equation, we have

$$3y^2(x)y'(x) = \frac{d}{dx}(y^3) = \frac{d}{dx}(h(x) + 3c_2) =$$
$$h'(x) = (x - x_1)(x - x_2), \text{ while}$$
$$y^2(x) = [y^3(x)]^{2/3} = [h(x) + 3c_2]^{2/3} =$$
$$(x - x_2)^{4/3}(x - x_3)^{2/3}.$$

Now we can see that

$$y'(x) = \frac{h'(x)}{3y^2(x)} = \frac{(x - x_1)}{3(x - x_2)^{1/3}(x - x_3)^{2/3}},$$

and this will become unbounded if x approaches either x_2 or x_3. These are the two points of vertical tangency, though it would be more precise to call x_2 a *cusp*.

47. Letting A be the accumulated value at time t, the differential equation for A is $A' = .06A + 2000$, which separates as

$$\frac{A'}{.06A + 2000} = 1,$$

and integrates to

$$\frac{\ln(.06A + 2000)}{.06} = t + c,$$

or

$$.06A + 2000 = ke^{.06t}\,(k = e^{.06c}).$$

At time $t = 0$, A is the unknown initial investment P, hence $k = .06P + 2000$, and so

$$.06A + 2000 = (.06P + 2000)e^{.06t}.$$

If we want $A = 1{,}000{,}000$ at $t = 20$, we must have

$$62000 = (.06P + 2000)e^{1.2},$$

or

$$P = \frac{62000\mathrm{e}^{-1.2} - 2000}{.06} \approx 277{,}901(\$).$$

49. If we assume that the 8% is a nominal monthly interest rate, then the exact equations governing the dynamics are as follows: For any time t that represents (in years) a moment just after a monthly payment has been received, the balance due $A(t)$ obeys the initial condition $A(0) = 150{,}000$ and the law

$$A\left(t + \frac{1}{12}\right) = A(t) + \frac{.08}{12}[A(t)] - P,$$

in which the first term on the right represents the previous balance, the second term is the interest accrued on that balance in a single month, and P is the payment received at the end of the month (just before the moment $t + 1/12$). According to methods one learns in the Theory of Interest, the payment P should be set at \$1100.65 per month in order that the loan be fully repaid in 30 years. But how is the student in this course supposed to proceed?

One can move $A(t)$ to the other side of the equation and multiply by 12, coming to

$$12[A(t + 1/12) - A(t)] = .08A(t) - 12P.$$

Here, the left side is

$$\frac{A(t + 1/12) - A(t)}{1/12} \approx A'(t),$$

and so the exact relation is approximated by the DE

$$A'(t) = .08A(t) - 12P.$$

This DE is *exact* if the 8% is a *continuous* compounding rate (force) and the payment is likewise *continuous* at the rate of 12P (\$ per *year*). Since this approach is within the skill sets of this course, let us follow through to an answer and see how close we come.

We solve the differential equation by separation much in the manner of Exercise 47, coming to

$$.08A - 12P = [.08(150,000) - 12P]e^{.08t}.$$

If P is to be set so that $A = 0$ at $t = 30$, we must have

$$-12P = [12000 - 12P]e^{2.4}, \text{ or}$$

$$P = \frac{12000e^{2.4}}{12(e^{2.4} - 1)} = 1099.77$$

(\$ per *month*); not bad!!

The total paid is this number multiplied by 360, or \$395,917.20. The thirty-year interest total is the latter number less the loan amount (\$150,000), or \$245,917.20.

51. This time (see #49) we are to have $A = 0$ when $t = 15$, which leads to

$$P = \frac{1000e^{1.2}}{(e^{1.2} - 1)} = 1431.01 \text{ (\$ per month)}.$$

The total paid is \$257,581.80 and the interest alone is \$107,581.80. Cutting the time in half raises the payments by about 30% but cuts the total interest by more than half.

53. Again assuming that the 8% is a continuous compounding rate (force) and that the \$10,000 per year is paid continuously, one accumulates according to $A' = .08A + 10{,}000$ with the initial condition $A(0) = 0$. This solves in the form $.08A + 10{,}000 = 10{,}000e^{.08t}$. At time $t = 10$ we have

$$A = \frac{10{,}000(e^{.8} - 1)}{.08} = 153{,}192.62(\$).$$

This would be the amount in his fund at age 40, and without further additions (other than interest) it would accumulate in the next 25 years to

$$153{,}192.62e^{(.08)25} = 1{,}131{,}948.83(\$).$$

55. Following the conditions of Exercise 53, replacing however the 8% by an unknown force r, we come after 10 years of payment and 25 additional years of accumulation to

$$10{,}000\frac{(e^{10r} - 1)}{r}e^{25r}.$$

For contrast, if the payment rate 10,000 is replaced by 20,000, and the payment interval of 10 years is replaced by 25 years, we come to an accumulation after the 25 years of

$$20{,}000\frac{(e^{25r} - 1)}{r}.$$

This number is to be compared to the result above.

Equating the two expressions leads to

$$2(e^{25r} - 1) = e^{35r} - e^{25r} \text{ or}$$
$$3e^{25r} - 2 = e^{35r}.$$

The equation can only be solved with the help of some form of technology, but the answer of r about .105 (10.5%) can at least be checked.

57. When the given numbers are substituted for the given symbols, the differential equation becomes

$$\begin{aligned} x' &= (.4 - x)(.6 - x) - .625x^2 \\ &= \frac{3}{8}x^2 - x + \frac{6}{25} \\ &= \frac{3}{8}\left(x - \frac{12}{5}\right)\left(x - \frac{4}{15}\right). \end{aligned}$$

When separated, it takes the form

$$\frac{x'}{(x-b)(x-a)} = r$$

in which $b = 12/5$, $a = 4/15 < b$, and $r = 3/8$. By partial fractions we find

$$\frac{1}{(x-b)(x-a)} = \frac{1}{(b-a)}\left\{\frac{1}{(x-b)}\frac{1}{(x-a)}\right\},$$

and after integration we find

$$\frac{1}{(b-a)}\ln\left|\frac{x-b}{x-a}\right| = rt + c_1,$$

or in this case with $b - a = (36/15) - (4/15) = 32/15$,

$$\ln\left|\frac{x - 12/5}{x - 4/15}\right| = \frac{32}{15}\left(\frac{3}{8}t + c_1\right) = \frac{4}{5}t + c_2$$
$$\left(c_2 = \frac{32}{15}c_1\right).$$

Using the given initial condition $x = .2 = 1/5$ when $t = 0$, we find $c_2 = \ln|(11/5)/(1/15)| = \ln(33)$,

$$\ln\left|\frac{x - 12/5}{33\,(x - 4/15)}\right| = \frac{4}{5}t \text{ and}$$
$$\frac{x - 12/5}{33\,(x - 4/15)} = \pm e^{\frac{4}{5}t} = e^{\frac{4}{5}t}$$

(the choice of sign is + since the left side is 1 when $x = 1/5$). Concluding the algebra, we find

$$\frac{5x - 12}{11(15x - 4)} = e^{\frac{4}{5}t},$$

$$5x - 12 = 11(15x - 4)e^{\frac{4}{5}t},$$

$$x = \frac{12 - 44e^{\frac{4}{5}t}}{5 - 11(15)e^{\frac{4}{5}t}} = \frac{4}{5}\left\{\frac{3 - 11e^{\frac{4}{5}t}}{1 - 33e^{\frac{4}{5}t}}\right\},$$

which can also be written in the form

$$x = \frac{4}{5}\left\{\frac{11 - 3e^{-\frac{4}{5}t}}{33 - e^{-\frac{4}{5}t}}\right\}.$$

In either form, it is apparent that

$$x \to \frac{4}{15} \text{ as } t \to \infty.$$

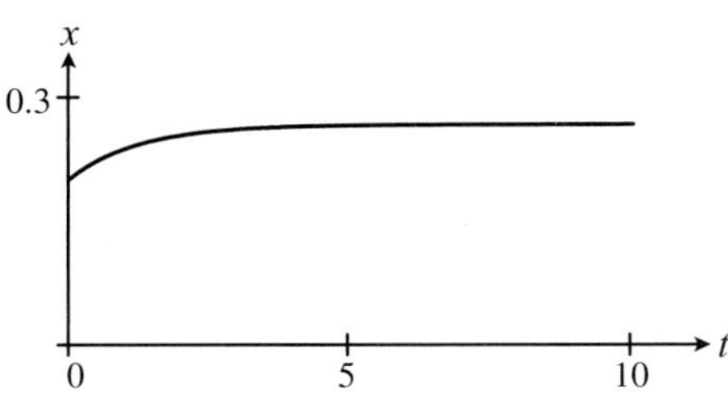

59. This problem can be done in a fashion similar to Exercise 57, after beginning

$$\begin{aligned} x' &= .6(.5 - x)(.6 - x) - .4x(0 + x) \\ &= .6(.3 - 1.1x + x^2) - .4x^2 \\ &= .2x^2 - .66x + .18 \\ &= \frac{1}{5}\left(x^2 - \frac{33}{10}x + \frac{9}{10}\right) \\ &= \frac{1}{5}(x - 3)\left(x - \frac{3}{10}\right). \end{aligned}$$

The parameters b, a, r from Exercise 57 are respectively 3, 3/10, and 1/5. We jump ahead to

$$\ln\left|\frac{x - 3}{x - 3/10}\right| = \frac{27}{10}\left(\frac{t}{5} + c_1\right) = \frac{27}{50}t + c_2.$$

In this case with $x = .2 = 1/5$ when $t = 0$ (as in Exercise 57), we find

$$c_2 = \ln\left|\frac{(1/5) - 3}{(1/5) - (3/10)}\right| = \ln\frac{14/5}{1/10} = \ln 28,$$

$$\frac{x - 3}{28\,(x - 3/10)} = \pm e^{\frac{27}{50}t} = e^{\frac{27}{50}t},$$

and the conclusion is

$$5(x - 3) = 14(10x - 3)e^{\frac{27}{50}t},$$

or

$$\begin{aligned} x &= \frac{15 - 42e^{\frac{27}{50}t}}{5 - 140e^{\frac{27}{50}t}} = \frac{42 - 15e^{-\frac{27}{50}t}}{140 - 5e^{-\frac{27}{50}t}} \\ &= \frac{3}{5}\left(\frac{14 - 5e^{-\frac{27}{50}t}}{28 - e^{-\frac{27}{50}t}}\right). \end{aligned}$$

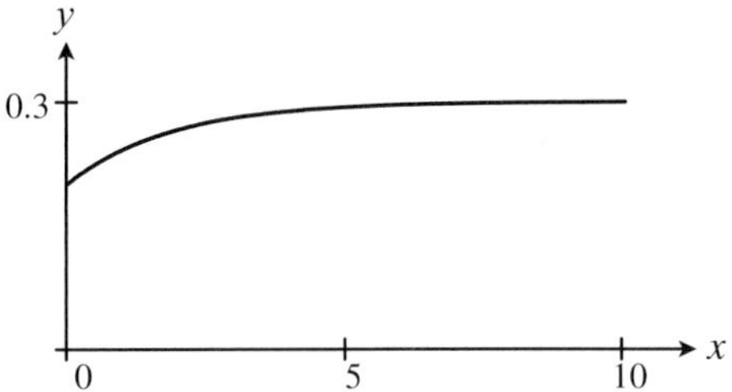

61. This time we are given outright $x' = r(a - x)(b - x)$ with $r = .4$, $a = 6$, $b = 8$. Solving as we did in Exercise 57, we find:

$$\ln\left|\frac{x - b}{x - a}\right| = (b - a)rt + c_2,$$

in this case

$$\ln\left|\frac{x - 8}{x - 6}\right| = \frac{4}{5}t + c_2,$$

and without using any initial condition,

$$\frac{x - 8}{x - 6} = ke^{\frac{4}{5}t}, \quad (k = \pm e^{c_2} \text{ or zero}),$$

$$x - 8 = (x - 6)ke^{\frac{4}{5}t},$$

$$x = \frac{8 - 6ke^{\frac{4}{5}t}}{1 - ke^{\frac{4}{5}t}} = \frac{6k - 8e^{-\frac{4}{5}t}}{k - e^{-\frac{4}{5}t}}.$$

We have been able to do all this without regard to an initial condition, but in this case the mathematics does not know all the relevant chemistry. In the original differential equation, the term $a - x$ was intended to represent the concentration of the remaining "free" substance A, but because of the two-for-one molecular exchange (one A and one B for one X) this intent is fulfilled only if the initial amount of substance X is zero, i.e., *by the nature of this problem, x obeys the initial condition $x = 0$.* Under the circumstances, then, we must have $k = 4/3$ and

$$x = \frac{24(1 - e^{-\frac{4}{5}t})}{4 - 3e^{-\frac{4}{5}t}}.$$

It is now apparent that $x \to 6$ as $t \to \infty$, and this makes perfect sense again because of the molecular exchange: The reaction continues as long as there is any "free" A and B, but in this case there is less A initially ($a = 6$), and the process essentially stops when the A is essentially gone, i.e., when $x \to a(= 6)$.

63. This problem eventually becomes another of the type found in Exercises 57, 59, and 61. We find

$$\begin{aligned} y' &= .025y(8 - y) - .2 \\ &= -.025(y^2 - 8y + 8) \\ &= -\frac{1}{40}(y - b)(y - a), \text{ in which} \\ b &= 4 + \sqrt{8}, a = 4 - \sqrt{8}. \end{aligned}$$

This leads as usual to

$$\ln\left|\frac{y-b}{y-a}\right| = -\frac{1}{40}\left(2\sqrt{8}\right)t + c_2,$$

and with $y(0) = 8$ we have

$$\ln\left|\frac{8-b}{8-a}\right| = c_2,$$

$$\ln\left|\frac{(y-b)(8-a)}{(y-a)(8-b)}\right| = \frac{-t\sqrt{8}}{20},$$

$$\frac{y-b}{y-a} = \frac{8-b}{8-a}e^{-\frac{t\sqrt{8}}{20}}.$$

We can see that as $t \to \infty$ the right side goes to zero, hence also the left side, and hence

$$y \to b = 4 + \sqrt{8} = 6.828427.$$

This represents an eventual fish population of about 682,800. In the language of Section 6.3, it is a stable equilibrium solution.

65. The equilibrium solutions are the algebraic solutions to the quadratic equation

$$\begin{aligned} .025P(8 - P) - R = 0, &\text{ or} \\ P^2 - 8P - 40R = 0. \end{aligned}$$

In the process of studying Exercise 63 ($R = .2$) we found it convenient to factor the left side (P was y at the time), and the roots were $b = 4 + \sqrt{8}$ and $a = 4 - \sqrt{8}$. In Exercise 64, the corresponding equation ($R = .6$) would be

$$0 = P^2 - 8P + 40R = P^2 - 8P + 24.$$

But *this* equation has no real roots, hence no equilibrium populations (a condition which considerably alters the more general task of solving the differential equation, not asked).

67.

$$\begin{aligned} P' &= .05P(8 - P) - .6 \\ &= -\frac{1}{20}(P^2 - 8P + 12) \\ &= -\frac{1}{20}(P - 6)(P - 2) \end{aligned}$$

Following well-established procedures, we come to

$$\ln\left|\frac{P-6}{P-2}\right| = -\frac{1}{5}t + c_2,$$

$$\frac{P-6}{P-2} = Ae^{-\frac{t}{5}}, \ (A = \pm e^{c_2} \text{ or zero}).$$

We learn from this relation that the ratio $(P - 6)/(P - 2)$ never changes sign; it is always negative if the initial condition has $P(0)$ in the interval (2,6). Clearly in this case the exponential approaches zero as $t \to \infty$ and P approaches 6. This last conclusion is true even if $P(0) > 6$.

If on the other hand $0 \le P(0) < 2$, the ratio is forever positive, and we find eventually

$$\frac{P-6}{P-2} = \frac{P(0)-6}{P(0)-2}e^{-t/5}.$$

Here the right side is a positive decreasing function of t and so must be the left side. The effect is that P itself is decreasing (not obvious) and reaches the value zero when

$$e^{-t/5} = 3\frac{P(0) - 2}{P(0) - 6} \text{ or when}$$

$$t = 5 \ln \frac{6 - P(0)}{3[2 - P(0)]} = 5 \ln \frac{6 - P(0)}{6 - 3P(0)}.$$

In the ratio inside the (second) ln, the numerator is clearly more than the denominator, which is itself positive. This is some moment of positive time, after which the population is zero and no further activity occurs.

69. The differential equation is $r'(t) = k[r(t) - S]$. This separates as $\frac{r'}{r - S} = k$, and solves as $\ln(r - S) = kt + c$.

In this case $S = 1000$, $r(0) = 14{,}000$, and $r(4) = 8000$. Putting $t = 0$, we see that the constant c is ln 13,000, and so we learn

$$\ln \frac{r - 1000}{13{,}000} = kt,$$

and putting $t = 4$,

$$\ln \frac{7}{13} = \ln \frac{7000}{13{,}000} = 4k.$$

Assembling the available information, we find

$$\ln \frac{r - 1000}{13{,}000} = kt = \frac{t}{4}(4k) = \frac{t}{4} \ln \frac{7}{13},$$

and

$$r = 1000 + 13{,}000 \left(\frac{7}{13}\right)^{t/4},$$

or equivalently $r = 1 + 13e^{-.15476t}$ (thousands).

71. If $P' = kP^{1.1}$, we separate as $P^{-1.1}P' = k$, and integrate to

$$\frac{-10}{P^{1/10}} = \frac{P^{-.1}}{(-.1)} = kt + c = kt - \frac{10}{P(0)^{1/10}}.$$

Taking reciprocals and tenth powers, we find

$$P = \left[\frac{1}{\frac{kt}{10} - \frac{1}{P(0)^{1/10}}}\right]^{10}.$$

We see that P approaches infinity as t approaches

$$\frac{10}{kP(0)^{1/10}}.$$

73. From the differential equation, with $z = y'/y$, we find $z = k(M - y)$. This is a line in the (y, z)-plane. The z-intercept is M and the slope is $-k$.

75. The differential equation

$$\frac{dv}{dt} = 32 - \frac{4v^2}{10}$$

falls into the pattern of several earlier exercises (e.g., Exercise 57) of the general type

$$\frac{dy}{dt} = k(y - b)(y - a).$$

k will be $-4/10$, and the most relevant detail here is that k is negative. The parameters b, a $(b > a)$ are the solutions to the algebraic equation

$$\frac{4}{10}v^2 = 32,$$

hence $b = \sqrt{80}$, $a = -\sqrt{80}$.

We fast-forward to the solution

$$\ln \left|\frac{v - \sqrt{80}}{v + \sqrt{80}}\right| = 2kt\sqrt{80} + c_2,$$

and given that $v(0) = 0$, we find $c2 = 0$ and

$$\frac{\sqrt{80} - v}{\sqrt{80} + v} = e^{2kt\sqrt{80}}.$$

Because $k < 0$, the right-hand side goes to zero as t goes to infinity. Therefore so must the left side, and as this happens, we must have $v \to \sqrt{80}$. This is the terminal velocity.

77. If $x' = r(10 - x)(12 - x)$ (presumably $r > 0$), we have seen many times that

$$\ln \left|\frac{x - 12}{x - 10}\right| = 2rt + c.$$

If $x(0) = 0$, then $c = \ln(12/10)$, and we come eventually to

$$\frac{10}{12}\frac{12-x}{10-x} = \pm e^{2rt} = e^{2rt},$$

in which we choose the positive sign to fit the facts when t and x are both zero. We *could* solve for t, but it works better to write

$$\frac{10-x}{12-x} = \frac{10}{12}e^{-2rt}$$

and see that $x \to 10$ as $t \to \infty$. Since the process consumes equal quantities of A and B to produce an (equal) quantity of X, it is physically reasonable that x cannot exceed the minimum of the available material, in this case 10 g of A.
The ultimate solution, by algebra, is

$$x = \frac{120(1-e^{-2rt})}{12-10e^{-2rt}}.$$

79. If $y' = ky(M-y)$, then, by the product rule,

$$y'' = k[y'(M-y) - yy'] = ky'[M-2y].$$

This will be zero when $y = M/2$. In what follows, we make exception of the two equilibrium solutions $y \equiv 0$ and $y \equiv M$. With any other solution, y is *never* 0, y is *never* M, and y' is *never* zero. Thus whatever time t_0 (if any) at which y becomes M/2 is sure to be an inflection time. Moreover, there can be *no circumstances of inflection other than* $y = M/2$. Such a time $t_0 > 0$ is bound to occur if and only if $0 < y(0) < M/2$, in which case the time t_0 is unique. The reasoning for all this takes us into topics beyond the scope of this chapter or even this book, but the solution formula does speak for itself quite nicely.

6.3 Direction Fields and Euler's Method

1.

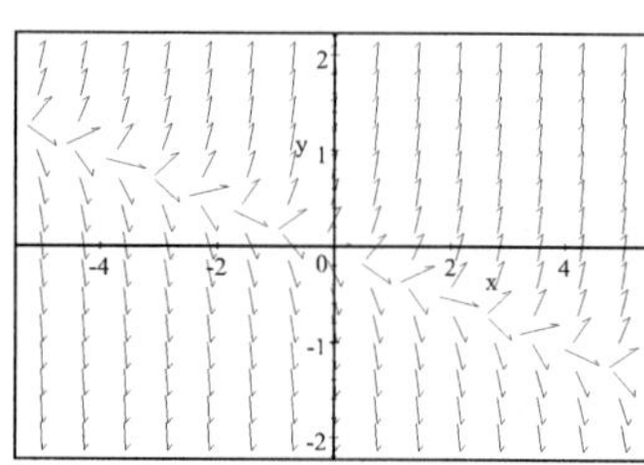

3.

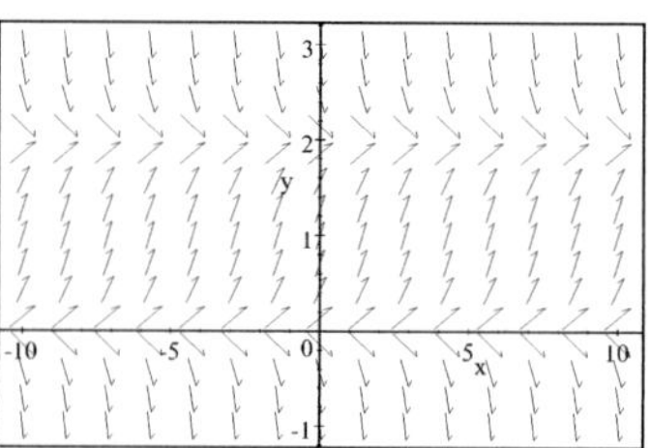

5.

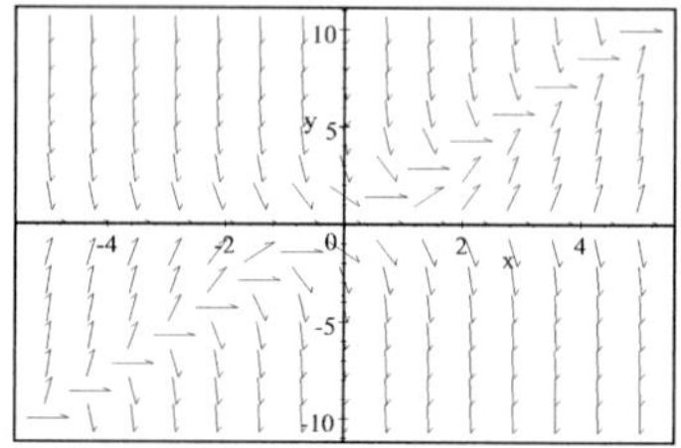

In **Exercises 7–12**, one could in principle sketch each direction field and look for the best match among the pictures. Rather than conduct this time-consuming exercise, we will examine each formula for a distinguishing feature, find the possible matches, and refine the choice if necessary.

7. One can focus on the locus of horizontal tangents ($xy = 2$ for #7 and $xy = -1/2$ for #8). Since the first is a first-and-third-quadrant locus, it is clear that C belongs to #7, and by both elimination and direct observation, B belongs to #8.

9. This formula has a vertical repeat, shared only by D. Other details fit.

The #10 formula has a horizontal repeat, shared only by F. Other details fit.

11. This formula has radial symmetry (so does #12), and (unlike #12) the slopes are never negative. This can only be A, and other details fit, e.g., the slopes getting close to zero near the origin.

The formula of #12 has mostly positive slopes, but near the origin they are negative and quite steep. This would be E.

13. We'll do the first case completely. For the other cases and other problems, we show only certain steps and the conclusions. In the spreadsheet calculation, the top entries x_o and y_o are taken from the initial conditions. y_n is code for $y(x_n)$.

The numbers in the boxes (y(1) and y(2)) are the target estimates.

DE:	$y' = f(x, y)$
Formula:	$f(x, y) = 2xy$
Step size (h):	0.1

count (n)	x_n	$y(x_n) = y_{n-1} + hf(x_{n-1}, y_{n-1})$	$f(x_n, y_n)$
0	0.0	1.0000	0.0000
1	0.1	1.0000	0.2000
2	0.2	1.0200	0.4080
3	0.3	1.0608	0.6365
4	0.4	1.1244	0.8996
5	0.5	1.2144	1.2144
6	0.6	1.3358	1.6030
7	0.7	1.4961	2.0946
8	0.8	1.7056	2.7290
9	0.9	1.9785	3.5613
10	1.0	2.3346	4.6693
11	1.1	2.8016	6.1634
12	1.2	3.4179	8.2030
13	1.3	4.2382	11.0193
14	1.4	5.3401	14.9524
15	1.5	6.8354	20.5061
16	1.6	8.8860	28.4351
17	1.7	11.7295	39.8803
18	1.8	15.7175	56.5831
19	1.9	21.3758	81.2281
			117.994
20	2.0	29.498	6

DE:	$y' = f(x, y)$
Formula:	$f(x, y) = 2xy$
Step size (h):	0.05

count (n)	x_n	$y(x_n) = y_{n-1} + hf(x_{n-1}, y_{n-1})$	$f(x_n, y_n)$
0	0.00	1.0000	0.0000
1	0.05	1.0000	0.1000
2	0.10	1.0050	0.2010
3	0.15	1.0151	0.3045
18	0.90	2.1035	3.7863
19	0.95	2.2928	4.3564
20	1.00	2.5107	5.0213
21	1.05	2.7617	5.7996
38	1.90	27.4906	3
			127.583
39	1.95	32.7138	8
			156.372
40	2.00	39.0930	0

15.

DE:	$y' = f(x, y)$
Formula:	$f(x, y) = 4y - y^2$
Step size (h):	0.1

count (n)	x_n	$y(x_n) = y_{n-1} + hf(x_{n-1}, y_{n-1})$	$f(x_n, y_n)$
0	0.0	1.0000	3.0000
1	0.1	1.3000	3.5100
9	0.9	3.7562	0.9157
10	1.0	3.8478	0.5857
11	1.1	3.9064	0.3658
19	1.9	3.9984	0.0065
20	2.0	3.9990	0.0039

DE:	$y' = f(x, y)$
Formula:	$f(x, y) = 4y - y^2$
Step size (h):	0.05

count (n)	x_n	$y(x_n) = y_{n-1} + hf(x_{n-1}, y_{n-1})$	$f(x_n, y_n)$
0	0.00	1.0000	3.0000
1	0.05	1.1500	3.2775
19	0.95	3.7766	0.8438
20	1.00	3.8188	0.6921
21	1.05	3.8534	0.5650
39	1.95	3.9972	0.0111
40	2.00	3.9978	0.0088

17.

DE: $y' = f(x, y)$
Formula: $f(x, y) = 1 - y + e^{-x}$
Step size (h): 0.1

count (n)	x_n	$y(x_n) = y_{n-1} + hf(x_{n-1}, y_{n-1})$	$f(x_n, y_n)$
0	0.0	3.0000	−1.0000
1	0.1	2.9000	−0.9952
9	0.9	2.1707	−0.7641
10	1.0	2.0943	−0.7264
11	1.1	2.0216	−0.6888
19	1.9	1.5696	−0.4200
20	2.0	1.5276	−0.3922

DE: $y' = f(x, y)$
Formula: $f(x, y) = 1 - y - e^{-x}$
Step size (h): 0.05

count (n)	x_n	$y(x_n) = y_{n-1} + hf(x_{n-1}, y_{n-1})$	$f(x_n, y_n)$
0	0.00	3.0000	−1.0000
1	0.05	2.9500	−0.9988
19	0.95	2.1365	−0.7497
20	1.00	2.0990	−0.7311
21	1.05	2.0624	−0.7125
39	1.95	1.5552	−0.4129
40	2.00	1.5345	−0.3992

19.

DE: $y' = f(x, y)$
Formula: $f(x, y) = (x + y)^{1/2}$
Step size (h): 0.1

count (n)	x_n	$y(x_n) = y_{n-1} + hf(x_{n-1}, y_{n-1})$	$f(x_n, y_n)$
0	0.0	1.0000	1.0000
1	0.1	1.1000	1.0954
9	0.9	2.2194	1.7662
10	1.0	2.3960	1.8428
11	1.1	2.5803	1.9184
19	1.9	4.3194	2.4939
20	2.0	4.5688	2.5630

DE: $y' = f(x, y)$
Formula: $f(x, y) = (x + y)^{1/2}$
Step size (h): 0.05

count (n)	x_n	$y(x_n) = y_{n-1} + hf(x_{n-1}, y_{n-1})$	$f(x_n, y_n)$
0	0.00	1.0000	1.0000
1	0.05	1.0500	1.0488
2	0.10	1.1024	1.0966
3	0.15	1.1573	1.1434
19	0.95	2.3304	1.8112
20	1.00	2.4210	1.8496
21	1.05	2.5135	1.8877
39	1.95	4.4934	2.5384
40	2.00	4.6203	2.5730

21. Referring only to Exercise 13, the DE separates to $\frac{y'}{y} = 2x$, integrates to $\ln|y| = x^2 + c$, and solves in the form $y = ke^{x^2}$, $(k = \pm e^c)$. From the IC, we find $k = 1$, and $y(1) = e = 2.7183$ (rounded), $y(2) = e^4 = 54.5982$ (rounded). These numbers can be compared to the numbers in the boxes (the y-column, for $x = 1$ and $x = 2$) in the solution to #13. The accuracy is fair only at $x = 1$, unimpressive at $x = 2$.

23.

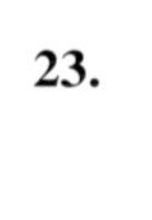

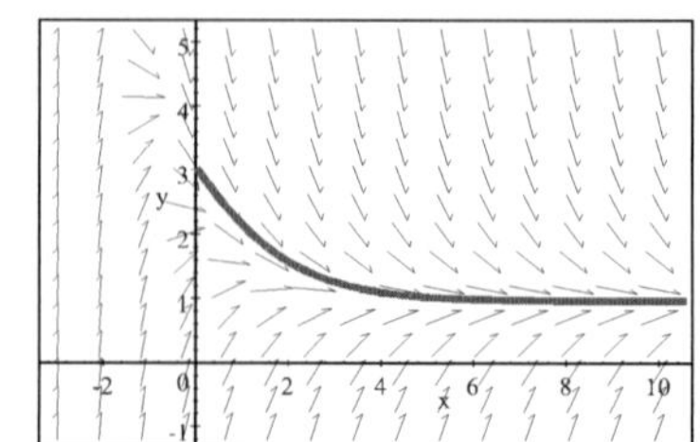

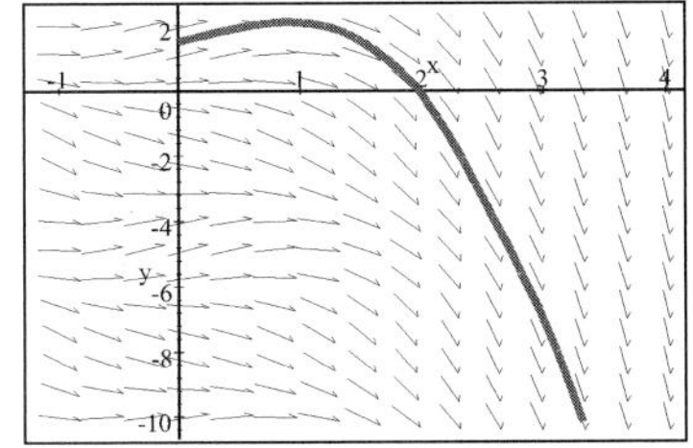

25. The equilibrium solutions are the constant solutions to the DE. If indeed $y = k$, then $0 = y' = 2k - k^2 = k(2 - k)$. Thus $k = 0$ or $k = 2$. Since y' is positive just above $y = 0$ and negative just below $y = 0$, the solution $y = 0$ is unstable. By dual reasoning, the solution $y = 2$ is stable.

27. The equilibrium solutions are the constant solutions to the DE. If indeed $y = k$, then $0 = y' = k^2 - k^4 = k^2(1 - k)(1 + k)$. Thus $k = 0$ or $k = 1$ or $k = -1$. Since y' is positive just above $y = -1$ and negative just below $y = -1$, the solution $y = -1$ is unstable. By dual reasoning, the solution $y = 1$ is stable. The solution $y = 0$ has some features of both, but on balance it must be considered unstable because solutions which start off just *above* $y = 0$ tend toward 1 as t goes to infinity.

29. The equilibrium solutions are the constant solutions to the DE. If indeed $y = k$, then $0 = y' = (1 - k)(1 + k^2)^{1/2}$. Thus $k = 1$. Since y' is negative just above $y = 1$ and positive just below $y = 1$, the solution $y = 1$ is stable.

31. The equilibrium solutions are the constant solutions to the DE. If indeed g is a certain constant k, then $0 = g' = -k + 3k^2/(1 + k^2) = -k(k^2 - 3k + 1)/(k^2 + 1)$. Thus $k = 0$ is clearly one solution, while the two roots of the quadratic in the numerator are also solutions. These are the numbers

$$a = \frac{3 - \sqrt{5}}{2} \approx .3820 \quad \text{and} \quad b = \frac{3 + \sqrt{5}}{2} \approx 2.6810.$$

Of the three, 0 and b are stable, while a is unstable. As a result of this stability feature,

$$\lim_{t\to\infty} g(t) = 0 \text{ if } 0 \le g(0) < a, \text{ while}$$

$$\lim_{t\to\infty} g(t) = b \text{ if } a < g(0).$$

As the problem evolves, g depends not only on time t, but on a certain real parameter x. We could write $g = g_x(t)$, and the dependence on x is through the initial condition:

$$g_x(0) = \frac{3}{2} + \frac{3\sin(x)}{2}.$$

With x restricted to the interval $[0, 4\pi]$ (4π being about 12.5664), the first event ($g_x(0) < a$, equivalently $\lim_{t\to\infty} g_x(t) = 0$, equivalently eventual black-stripe zone) occurs when x lies in one of the two intervals (3.9827, 5.4421) or (10.2658, 11.7253). More precisely, these are the intervals with endpoints

$$\frac{3\pi}{2} \pm \cos^{-1}\left(\frac{\sqrt{5}}{3}\right) \quad \text{and} \quad \frac{7\pi}{2} \pm \cos^{-1}\left(\frac{\sqrt{5}}{3}\right).$$

33. Some curious things happen in this problem, so we will show some parts of the Euler process and comment later.

DE:	$y' = f(x, y)$
Formula:	$f(x, y) = y^2 - 1$
Step size (h):	0.1

count (n)	x_n	$y(x_n) = y_{n-1} + hf(x_{n-1}, y_{n-1})$	$f(x_n, y_n)$
0	0.00	3.0000	8.0000
1	0.10	3.8000	13.4400
2	0.20	5.1440	25.4607
3	0.30	7.6901	58.1372
4	0.40	13.5038	181.3525
5	0.50	31.6390	1000.0295

Step size (h): 0.05

count (n)	x_n	$y(x_n) = y_{n-1} + hf(x_{n-1}, y_{n-1})$	$f(x_n, y_n)$
0	0.00	3.0000	8.0000
1	0.05	3.4000	10.5600
2	0.10	3.9280	14.4292
3	0.15	4.6495	20.6175
4	0.20	5.6803	31.2662
5	0.25	7.2436	51.4703
6	0.30	9.8172	95.3766
7	0.35	14.5860	211.7511

8	0.40	25.1735	632.7073
9	0.45	56.8089	3226.2523
10	0.50	218.1215	47576.0009

Step size (h): 0.01

count (n)	x_n	$y(x_n) = y_{n-1} + hf(x_{n-1}, y_{n-1})$	$f(x_n, y_n)$
0	0.00	3.0000	8.0000
1	0.01	3.0800	8.4864
9	0.09	3.9396	14.5203
10	0.10	4.0848	15.6855
11	0.11	4.2416	16.9915
12	0.12	4.4116	18.4619
19	0.19	6.1502	36.8248
20	0.20	6.5184	41.4900
21	0.21	6.9333	47.0711
29	0.29	13.9167	192.6740
30	0.30	15.8434	250.0139
31	0.31	18.3436	335.4861
32	0.32	21.6984	469.8214
33	0.33	26.3966	695.7823
34	0.34	33.3545	1111.5198
35	0.35	44.4697	1976.5501
36	0.36	64.2352	4125.1551
37	0.37	105.4867	11126.4453
38	0.38	216.7512	46980.0650
39	0.39	686.5518	471352.386
40	0.40	5400.0757	##########
41	0.41	297008.2385	##########
42	0.42	##########	##########
43	0.43	##########	##########
44	0.44	##########	##########
45	0.45	##########	##########
46	0.46	##########	##########
47	0.47	##########	#NUM!
48	0.48	#NUM!	#NUM!
49	0.49	#NUM!	#NUM!
50	0.50	#NUM!	#NUM!

If we compare these y-numbers with the true values for y (following chart, requested for Exercise 34) we see there is very little accuracy for any x larger than .2 and even that is only moderately accurate for the smallest step size. What else is happening in the case of the step-size .01 is that the y-numbers first become too large to be printed in the available space (####) and eventually become too large for the computer to even process (#NUM!). (These things will also happen for the larger step sizes; they just haven't happened *yet*.) Some comments in the nature of a partial explanation will be found further along.

x	y(exact)
exact formula:	$(2 + e^{2x})/(2 - e^{2x})$
0.000	3.0000
0.100	4.1374
0.200	6.8713
0.300	21.4869
0.346	1743.4049
0.347	−2345.1624
0.400	−18.7351
0.500	−6.5688

35. The DE *is* separable, and solvable by the methods of equation 6.2. The solution given by formula in Exercise 34 is not hard to derive in complete generality:

$$y = (1 + ke^{2x})/(1 - ke^{2x});$$

and it is of course easy to check. The vertical asymptote in the solution occurs when the denominator vanishes, which is to say when $e^{2x} = 1/k$, or $x = -\ln(k)/2$. In our case, with $y(0) = 3$, we have $k = 1/2$ and the vertical asymptote at $x = \ln(2)/2 = .3466$. This accounts for the change in sign in the true y-values that we observe above between the two x-values .346 and .347.

The field diagram cannot give any forewarning of the vertical asymptote. Dependent as the field equations are only on y, they can only hint at things that likewise depend on y. The location of the vertical asymptote, by its very nature an x-measurement, is instead dependent directly on the solution-parameter k and indirectly on the initial condition.

In this case, where the actual x-value does not enter the calculations, the Euler process merely generates the numbers in the recursive sequence

$$y_n = hy_{n-1}^2 + y_{n-1} - h$$

subject to an initial condition of $y_o = 3$. The numbers in such a sequence will increase to infinity, with growth rate depending on h. The simultaneous determination of x_n through the law $x_n = hn$ has nothing to do with the geometry of the solution to the differential equation. "Jumping over the asymptote" is the pseudo-event which happens when n passes from below $\frac{.3466}{h}$ to above, and has no special relation to the Euler y-numbers and no relation whatever to the solution of the DE.

37. With $y' = x^2 - 4x + 2$, the general solution is

$$y = \frac{x^3}{3} - 2x^2 + 2x + k.$$

If we are to have $y(3) = 0$, we must have $y(0) = k = 3$. This seems a reasonable outcome based on tracing the field directions back from the point (3,0). An Euler process (stepping backwards in 27 decrements of 1/9) estimates $y(0)$ to be 3.16 .

39. This DE is not solvable by any elementary techniques. We make an eyeball estimate of $y(0)$ about -7 by tracing the field directions backwards from $(5, -3)$. An Euler process with 50 decrements of .1 estimates $y(0)$ at -5.55. We cannot report any other CSI calculations for purposes of comparison.

6.4 Second-Order Equations with Constant Coefficients

In this group of solutions, we use the notations DE for "differential equation"; CP for "characteristic polynomial"; GS for "general solution" ; IC for "initial condition"; and FS for "final solution." We generally call the CP by the name $p(r)$. (If the DE is $ax'' + bx' + cx = F$, then the CP is $p(r) = ar^2 + br + c$, and the characteristic *equation* is $p(r) = 0$.)

1. DE: $y'' - 2y' - 8y = 0$
CP: $p(r) = r^2 - 2r - 8$
CP factored: $(r - 4)(r - 2)$
Real roots: $\{4, -2\}$
GS: $y = c_1e^{4t} + c_2e^{-2t}$

3. DE: $y'' - 4y' + 4y = 0$
CP: $p(r) = r^2 - 4r + 4$
CP factored: $(r - 2)^2$
Real root: 2 (double)
Two solutions: $\{e^{2t}, te^{2t}\}$
GS: $y = c_1e^{2t} + c_2te^{2t}$

5. DE: $y'' - 2y' + 5y = 0$
CP: $p(r) = r^2 - 2r + 5$
Complex roots: $\{1 \pm 2i\}$
GS: $y = c_1e^t \cos 2t + c_2e^t \sin 2t$

7. DE: $y'' - 2y' = 0$
CP: $p(r) = r^2 - 2r$
CP factored: $r(r - 2)$
Real roots: $\{0, 2\}$
GS: $y = c_1 + c_2e^{2t}$

9. DE: $y'' - 2y' - 6y = 0$
CP: $p(r) = r^2 - 2r - 6$
Real roots: $\{1 \pm \sqrt{7}\}$
GS: $y = c_1e^{(1+\sqrt{7})t} + c_2e^{(1-\sqrt{7})t}$

11. DE: $y'' - (\sqrt{5})y' + y = 0$
CP: $p(r) = r^2 - r\sqrt{5} + 1$
Real roots: $\left\{\frac{\sqrt{5} \pm 1}{2}\right\}$
GS: $y = c_1e^{(1+\sqrt{5})t/2} + c_2e^{(-1+\sqrt{5})t/2}$

13. DE: $y'' + 4y = 0$

CP: $p(r) = r^2 + 4$

Imaginary roots: $\{\pm 2i\}$

GS: $y = c_1 \cos 2t + c_2 \sin 2t$

$y' = 2c_2 \cos 2t - 2c_1 \sin 2t$

IC: $y(0) = 2,\ y'(0) = -3$

Get c's : $-3 = y'(0) = 2c_2 \therefore\ c_2 = -3/2$

$2 = y(0) = c_1$

FS: $y = 2 \cos 2t - (3/2) \sin 2t$

15. DE: $y'' - 3y' + 2y = 0$

CP: $p(r) = r^2 - 3r + 2$

CP factored: $(r - 1)(r - 2)$

Real roots: $\{1, 2\}$

GS: $y = c_1 e^t + c_2 e^{2t}$

$y' = c_1 e^t + 2c_2 e^{2t}$

IC: $y(0) = 0,\ y'(0) = 1$

Get c's: $0 = y(0) = c_1 + c_2$

$1 = y'(0) = c_1 + 2c_2$

Subtract: $c_2 = 1$

Back-substitute: $c_1 = 1 - 2c_2 = -1$

FS: $y = -e^t + e^{2t}$

17. DE: $y'' - 2y' + 5y = 0$ (same as Exercise 5) $\therefore$

GS: $y = c_1 e^t \cos 2t + c_2 e^t \sin 2t$

$= e^t (c_1 \cos 2t + c_2 \sin 2t)$

$y' = e^t (c_1 \cos 2t + c_2 \sin 2t) +$

$e^t (2c_2 \cos 2t - 2c_1 \sin 2t)$

IC: $y(0) = 2,\ y'(0) = 0$

Get c's: $2 = y(0) = c_1$

$0 = y'(0) = c_1 + 2c_2$

Back-substitute: $2c_2 = -c_1 = -2 \therefore\ c_2 = -1$

FS: $y = e^t (2 \cos 2t - \sin 2t)$

19. DE: $y'' - 2y' + y = 0$

CP: $p(r) = r^2 - 2r + 1$

CP factored: $(r - 1)^2$

Real root: 1 (double)

Two solutions: $\{e^t, te^t\}$

GS: $y = c_1 e^t + c_2 t e^t = e^t (c_1 + t c_2)$

$y' = e^t (c_1 + t c_2) + e^t c_2 = e^t (c_1 + c_2 + t c_2)$

IC: $y(0) = -1,\ y'(0) = 2$

Get c's: $-1 = y(0) = c_1$

$2 = y'(0) = c_1 + c_2$

Back-substitute: $c_2 = 2 - c_1 = 3$

FS: $y = e^t (-1 + 3t)$

21. According to trigonometry, $\sin(kt + \delta) = \sin kt \cos \delta + \cos kt \sin \delta$. Therefore, $A \sin(kt + \delta) = c_1 \cos kt + c_2 \sin kt$ if $A \sin \delta = c_1$ and A $\cos \delta = c_2$. Under these conditions,

$$c_1^2 + c_2^2 = A^2(\sin^2 kt + \cos^2 kt) = A^2,$$

and

$$\tan \delta = \frac{\sin \delta}{\cos \delta} = \frac{A \sin \delta}{A \cos \delta} = \frac{c_1}{c_2}.$$

If we have

DE: $y'' + 9y = 0$;

CP: $p(r) = r^2 + 9$;

Imaginary roots: $\{\pm 3i\}$;

GS: $y = c_1 \cos 3t + c_2 \sin 3t$,

$y' = 3c_2 \cos 3t - 3c_1 \sin 3t$,

IC: $y(0) = 3,\ y'(0) = -6$;

Get c's: $-6 = y'(0) = 3c_2 \therefore\ c_2 = -2$

$3 = y(0) = c_1$,

FS: $y = 3 \cos 2t - 2 \sin 2t$;

and if the solution is to be written in the form $y = A \sin(kt + \delta)$; then we must have

$$A = \sqrt{(-2)^2 + 3^2} = \sqrt{13}$$

(the amplitude) and

$$\delta = \tan^{-1}(-3/2) = -.983(\text{rad}) = -56.3^o$$

(the phase shift).

23. DE: $y'' + 20y = 0$
CP: $p(r) = r^2 + 20$
Imaginary roots: $\{\pm i\sqrt{20}\}$
GS: $y = c_1 \cos t\sqrt{20} + c_2 \sin t\sqrt{20}$
$y' = c_2\sqrt{20}\cos t\sqrt{20} - c_1\sqrt{20}\sin t\sqrt{20}$.
IC: $y(0) = -2$, $y'(0) = 2$
Get c's: $-2 = y(0) = c_1$
$2 = y'(0) = c_2\sqrt{20} = 2c_2\sqrt{5} \therefore\ c_2 = 1/\sqrt{5}$
FS: $y = -2\cos t\sqrt{20} + (\sin t\sqrt{20})/\sqrt{5}$.
Regarding the amplitude A, we have $A^2 = 4 + 1/5 = 21/5 \therefore\ A = \frac{\sqrt{105}}{5}$. And regarding the phase shift δ we have

$$\delta = \tan^{-1}\left(\frac{-2}{1/\sqrt{5}}\right) = -\tan^{-1}\left(2\sqrt{5}\right) =$$
$$-1.35(\text{rad}) = -77.4^o.$$

25. We are given $W = 12$ (pounds), $\Delta l = 1/2$ (feet), and with u the downward displacement from equilibrium, we have
$mu'' + ku = 0$, in which

$$m = \frac{W}{g} = \frac{12}{32} = \frac{3}{8},\ k = \frac{W}{\Delta l} = \frac{12}{(1/2)} = 24.$$

Thus, after multiplying by 8/3, the DE becomes
$u'' + 64u = 0$, with
CP: $p(r) = r^2 + 64$;
Imaginary roots: $\{\pm 8i\}$; and
GS: $u = c_1\cos 8t + c_2 \sin 8t$,
$u' = 8c_2\cos 8t - 8c_1 \sin 8t$.
The initial conditions were $u(0) = 2/3$ (downward displacement of 8 inches) and $u'(0) = 0$ (stationary release). These easily give
$c_1 = u(0) = \frac{2}{3}$,
$8c_2 = u'(0) = 0$, $\therefore\ c_2 = 0$ and
FS: $u = \left(\frac{2}{3}\right)\cos(8t)$.

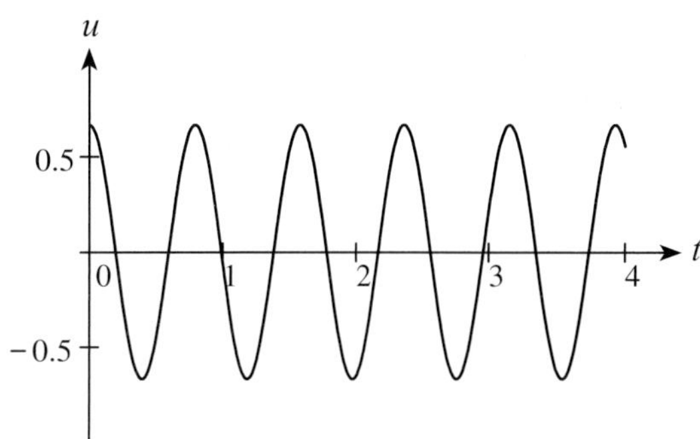

27. We are given $m = 4$ (kg), $\Delta l = .1$ (m) (10 cm), and with u the downward displacement from equilibrium, we have
$mu'' + ku = 0$, in which $m = 4$ (given), and
$k = \frac{mg}{\Delta l} = \frac{4(9.8)}{.1} = 4(98)$.
Thus, after dividing by 4, the DE becomes
$u'' + 98u = 0$, with
CP: $p(r) = r^2 + 98$;
Imaginary roots: $\{\pm 7i\sqrt{2}\}$;
GS: $u = c_1\cos(7t\sqrt{2}) + c_2\sin(7t\sqrt{2})$,
$u' = c_2 7\sqrt{2}\cos(7t\sqrt{2}) - c_1 7\sqrt{2}\sin(7t\sqrt{2})$.
The initial conditions were $u(0) = .2$ (downward displacement of 20 cm) and $u'(0) = -4$ (upward impact). These give

$$c_1 = u(0) = 2/10,$$
$$7\sqrt{2}c_2 = u'(0) = -4,\ \therefore\ c_2 = -2\sqrt{2}/7.$$

Regarding the amplitude A, we have

$$A = \sqrt{\frac{4}{100} + \frac{16}{98}}(m) \quad \text{or} \quad A \approx 45 \text{ cm}.$$

Regarding the phase shift δ, we have

$$\delta = \tan^{-1}\left(\frac{2/10}{-4/7\sqrt{2}}\right) = -\tan^{-1}\left(\frac{7\sqrt{2}}{20}\right) =$$
$$-.46(\text{rad}) = -26.3^{o}.$$

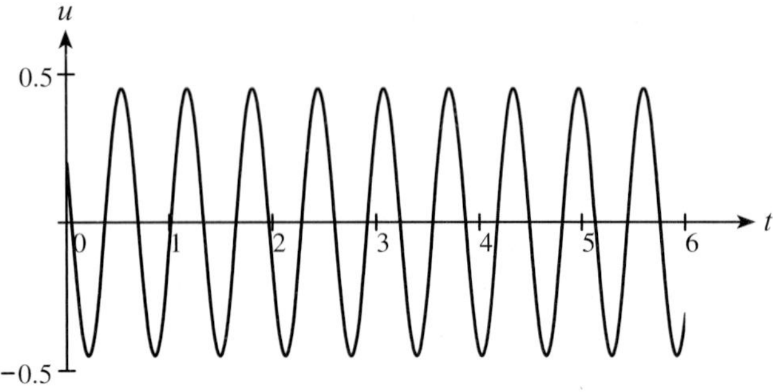

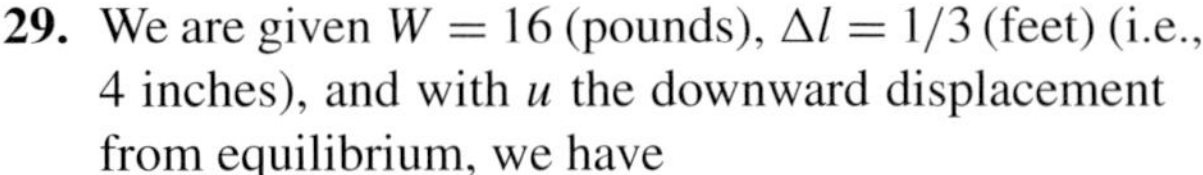

29. We are given $W = 16$ (pounds), $\Delta l = 1/3$ (feet) (i.e., 4 inches), and with u the downward displacement from equilibrium, we have

$$mu'' + cu' + ku = 0, \text{ in which}$$
$$m = \frac{W}{g} = \frac{16}{32} = \frac{1}{2}, \; k = \frac{W}{\Delta l} = \frac{16}{(1/3)} = 48.$$

With $c = 10$ (given), after multiplying by 2, the DE becomes
$u'' + 20u' + 96u = 0$, with
CP: $p(r) = r^2 + 20r + 96$;
CP factored: $(r + 12)(r + 8)$;
Real roots: $\{-8, -12\}$; and
GS: $u = c_1e^{-12t} + c_2e^{-8t}$,
$u' = 12c_1e^{-12t} - 8c_2e^{-8t}$.

The initial conditions were $u(0) = -1/2$ (upward displacement of 6 inches) and $u'(0) = 0$ (stationary release). These give

$$-1/2 = u(0) = c_1 + c_2,$$
$$0 = u'(0) = -12c_1 - 8c_20, \therefore \; c_2 = -(3/2)c_1$$
$$-(1/2) = c_1 + c_2 = c_1(1 - (3/2)) = (-1/2)c_1,$$
$$\therefore \; c_1 = 1, \; c_2 = -3/2, \text{ and}$$
$$u = e^{-12t} - (3/2)e^{-8t}.$$

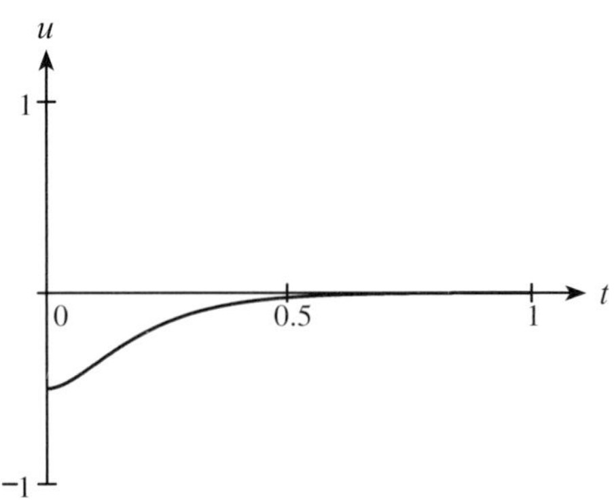

31. We are given $m = 4$ (kg), $\Delta l = 1/4$ (meters) (i.e., 25 cm), and with u the downward displacement from equilibrium, we have

$$mu'' + cu' + ku = 0, \text{ in which}$$
$$m = 4 \text{ (given)}, \; k = \frac{mg}{\Delta l} = \frac{4(9.8)}{(1/4)} = 16(9.8).$$

With $c = 2$ (given), after dividing by 4, the DE becomes
$u'' + (1/2)u' + 4(9.8)u = 0$, with
CP: $p(r) = r^2 + (1/2)r + 39.2$;
Complex roots: $\left\{-\frac{1}{4} \pm i\frac{\sqrt{15655}}{20}\right\}$;
and GS:

$$u = e^{-t/4}\left\{c_1 \cos\frac{t\sqrt{15655}}{20} + c_2 \sin\frac{t\sqrt{15655}}{20}\right\}.$$

The initial conditions were $u(0) = -1/2$ (upward displacement of half a meter) and $u'(0) = 0$ (stationary release). These give

$$-1/2 = u(0) = c_1,$$
$$0 = u'(0) = \frac{\sqrt{15655}}{20}c_2 - \frac{c_1}{4} = \frac{\sqrt{15655}}{20}c_2 + \frac{1}{8},$$
$$\therefore \; c_2 = -\frac{20}{8\sqrt{15655}} = -\frac{\sqrt{15655}}{6262} \text{ and}$$
$$u = e^{-t/4}\left\{-\frac{1}{2}\cos\frac{t\sqrt{15655}}{20} - \frac{\sqrt{15655}}{6262}\sin\frac{t\sqrt{15655}}{20}\right\}.$$

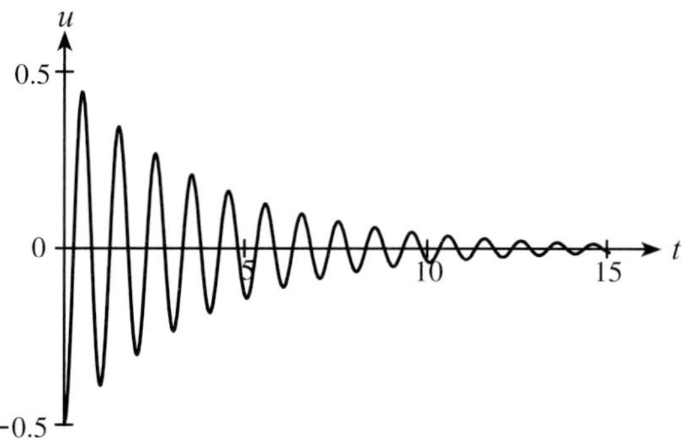

33. The model

$$LQ'' + RQ' + \frac{Q}{C} = 0$$

becomes, after substitution of the givens, the
DE: $.4Q'' + 200Q' + 10^4 Q = 0$, or
$Q'' + 500Q' + 25{,}000Q = 0$;
CP: $p(r) = r^2 + 500r + 25{,}000$;
Real roots $\{-250 \pm 50\sqrt{15}\}$; and
GS: $Q = e^{-250t}\{c_1 e^{50t\sqrt{15}} + c_2 e^{-50t\sqrt{15}}\}$.
The initial conditions were $Q(0) = 10^{-5}$ and $Q'(0) = 0$. These give
$10^{-5} = Q(0) = c_1 + c_2$,

$$\begin{aligned} 0 = Q'(0) &= 50\sqrt{15}(c_1 - c_2) - 250(c_1 + c_2) \\ &= 50\sqrt{15}(c_1 - c_2) - 250(10^{-5}), \end{aligned}$$

$$\therefore\ (c_1 - c_2) = \frac{250(10^{-5})}{50\sqrt{15}} = \frac{\sqrt{15}}{3}10^{-5}.$$

$$\therefore\ 2c_1 = \left(1 + \frac{\sqrt{15}}{3}\right)10^{-5},\ 2c_2 = \left(1 - \frac{\sqrt{15}}{3}\right)10^{-5},$$

$$c_1 = \left(\frac{3 + \sqrt{15}}{6}\right)10^{-5},\ c_2 = \left(\frac{3 - \sqrt{15}}{6}\right)10^{-5},$$

$$Q(t) = 10^{-5}e^{-250t}\left\{\frac{3 + \sqrt{15}}{6}e^{50t\sqrt{15}} + \frac{3 - \sqrt{15}}{6}e^{-50t\sqrt{15}}\right\}.$$

This is about

$$10^{-5}\{1.1455e^{-56.35t} - .1455e^{-443.65t}\},$$

and as regards the current,

$$I(t) = Q'(t) \approx 10^{-5}(.64547)\{-e^{-56.35t} + e^{-443.65t}\}.$$

35. The model $LQ'' + RQ' + \dfrac{Q}{C} = 0$ becomes, after substitution of the givens, the
DE: $.2Q'' + 10^5 Q = 0$, or
$Q'' + 500{,}000Q = 0$;
CP: $p(r) = r^2 + 500{,}000$;
Imaginary roots $\left\{\pm i500\sqrt{2}\right\}$; and
GS: $Q = c_1 \cos 500t\sqrt{2} + c_2 \sin 500t\sqrt{2}$.
The initial conditions were $Q(0) = 10^{-6}$ and $Q'(0) = 0$. These give

$$\begin{aligned} 10^{-6} &= Q(0) = c_1, \\ 0 &= Q'(0) = 500\sqrt{2}c_2, \\ &\therefore\ c_2 = 0, \\ Q(t) &= 10^{-6}\cos(500t\sqrt{2}) \\ &= 10^{-6}\sin\left(500t\sqrt{2} + \frac{\pi}{2}\right). \end{aligned}$$

The amplitude is 10^{-6} and the phase shift is 90°. As regards the current,

$$I(t) = Q'(t) = 5\sqrt{2}(10^{-4})\sin(500t\sqrt{2}).$$

37. If the DE is $ay'' + by' + cy = 0$, and if r_1 is a double root to the characteristic polynomial $p(r)$, this means that

$$ar^2 + br + c = p(r) = a(r - r_1)^2,$$

hence both

$$ar_1^2 + br_1 + c = p(r_1) = 0$$

and

$$2ar_1 + b = p'(r_1) = 0.$$

We know that $y_1 = e^{r_1 t}$ is one solution, and now we test $y_2 = te^{r_1 t}$, finding

$$\begin{aligned} y_2' &= r_1 te^{r_1 t} + e^{r_1 t}, \\ y_2'' &= r_1^2 te^{r_1 t} + 2r_1 e^{r_1 t}, \end{aligned}$$

and

$$ay_2'' + by_2' + cy_2 = (ar_1^2 + br_1 + c)te^{r_1 t} + (2ar_1 + b)e^{r_1 t}$$
$$= 0te^{r_1 t} + 0e^{r_1 t}0.$$

This shows that y_2 is also a solution.

39. The DE $u + cu' + 16u = 0$ has CP $r^2 + cr + 16$, and roots

$$\frac{\pm\sqrt{c^2 - 64}}{2}.$$

They are complex if $0 < c < 8$, and both have negative real part. There is the double root -4 if $c = 8$. The roots are real if $c > 8$, but both are negative.

In the first case, the solution shows oscillatory behavior (like the case $c = 0$) but the peaks subside to zero (*un*like the case $c = 0$). The word "underdamped" suggests that the damping is insufficient to completely eliminate oscillation. The word "overdamped" (the third case) suggests that the damping overrides the tendency to oscillate, causing the solution to decrease quickly to zero. The separating case of "critical damping" can be grouped with the overdamped class in so far as the solution subsides to zero with no tendency to oscillate.

41. Critical damping corresponds to the case in which the characteristic polynomial (in this case $mr^2 + cr + k$) has a double root. This is well known to happen when $c^2 = 4mk$, and given that $c > 0$, this makes $c = 2\sqrt{mk}$. Applying this to the case in which

$$m = \frac{W}{g} = \frac{16}{32} \quad \text{and} \quad k = \frac{W}{\Delta l} = \frac{16}{(1/4)} = 64,$$

we find critical damping when

$$c = 2\sqrt{mk} = 2\sqrt{32} = 8\sqrt{2}.$$

43. The roots

$$\frac{-b \pm \sqrt{b^2 - 4ac}}{2a}$$

(of the characteristic polynomial) are complex precisely because $b^2 - 4ac < 0$, in which case they can be written

$$\frac{-b \pm i\sqrt{4ac - b^2}}{2a},$$

and the general solution to the DE becomes

$$y = e^{-\frac{b}{2a}t}\left\{c_1 \cos\left(\frac{t\sqrt{4ac - b^2}}{2a}\right) + c_2 \sin\left(\frac{t\sqrt{4ac - b^2}}{2a}\right)\right\}.$$

The bracketed part is bounded (by $|c1| + |c2|$) and if the ratio b/a is positive, then the first part and the product goes to zero as t goes to infinity. It appears to be an implicit assumption in this problem that $a > 0$. (The assumption that $b > 0$ is explicit.)

45. The general solution is $c_1 e^{r_1 t} + c_2 t e^{r_1 t}$, and with $r_1 < 0$ it is clear that the entire expression will go to zero as t goes to infinity if $te^{r_1 t} \to 0$. This is a well-known fact that can be verified as follows: $te^{r_1 t}$ can be written in the form $\frac{t}{e^{-r_1 t}}$ in which both numerator and denominator go in infinity with t. By l'Hopital's Rule the limit is the same as the limit of

$$\frac{1}{-r_1 e^{-r_1 t}} = \frac{-e^{r_1 t}}{r_1},$$

clearly zero.

46. The roots of the characteristic polynomial are

$$\frac{-b \pm \sqrt{b^2 - 4ac}}{2a}.$$

With all three symbols positive, these roots $\{r1, r2\}$, if not complex or repeated, are both negative. In this case, the general solution $y = c_1 e^{r_1 t} + c_2 e^{r_2 t}$ clearly goes to zero as t goes to infinity. The complex and repeated cases are covered by Exercises 43, 44, or 45.

47. With the nonzero damping ($c > 0$) and the mass and spring constant intrinsically positive, Exercise 46 applies. The interpretation is that the spring eventually tends back toward its equilibrium position.

49. If the DE is $y + 2\alpha y' + \alpha^2 y$,

$$\text{the CP is } r^2 + 2\alpha r + \alpha^2 = (r + \alpha)^2.$$

This is the case of the double root $-\alpha$, and the general solution is

$$y = (c_1 + tc_2)e^{-\alpha t}.$$

From the IC $y(0) = 0$, we learn $c1 = 0$. From $y'(0) = 100$, we learn $c_2 = 100$, and so

$$y = 100te^{-\alpha t},\ y' = 100(1 - \alpha t)e^{-\alpha t}.$$

Thus

$$y^2 + y'^2 \text{ at } t = 1 \text{ is } 100^2[1 + (1 - \alpha)^2]e^{-2\alpha}.$$

This will be at most .01 if

$$[1 + (1 - \alpha)^2]e^{-2\alpha} \leq 10^{-6}.$$

The inequality cannot be solved by formula, but spreadsheet experimentation suggests that it holds if $\alpha \geq 9$. The answer in Appendix G is questionable for this one. It appears that the figure of 7.7 is required to give

$$y^2(1) + y'^2(1) < .1$$

rather than .01.

α	$(100)^2[1 + (1 - \alpha)^2]e^{-2\alpha}$
7.5	0.13230275
7.6	0.11160125
7.7	0.09409857
7.8	0.07930781
7.9	0.06681482
8.0	0.05626759
8.1	0.04736712
8.2	0.03985963
8.3	0.03352984
8.4	0.02819522
8.5	0.02370114
8.6	0.01991667
8.7	0.01673098
8.8	0.01405033
8.9	0.01179549
9.0	0.00989949
9.1	0.00830577
9.2	0.00696659
9.3	0.00584168
9.4	0.00489704
9.5	0.00410405
9.6	0.00343855
9.7	0.00288022

6.5 Nonhomogeneous Equations: Undetermined Coefficients

As we begin this section, we use the various codes established for answers in the previous Section (6.4), and in addition now use TS for "trial particular solution." In this section, we will often not rewrite the DE, but begin directly with the CP.

1. CP: $r^2 + 2r + 5$

Complex roots: $\{-1 \pm 2i\}$

GS: $u = e^{-t}[c_1 \cos(2t) + c_2 \sin(2t)] + 3e^{-2t}$

3. CP: $r^2 + 4r + 4 = (r + 2)^2$

GS: $u = c_1 e^{-2t} + c_2 te^{-2t} + t^2 - 2t + (3/2)$

5. CP: $p(r) = r^2 + 2r + 10$

Complex roots: $\{-1 \pm 3i\}$

Noting $p(-3) = 13 \neq 0$,

TS: $u_p = Ae^{-3t}$

$(u_p' = -3Ae^{-3t},\ u_p'' = 9Ae^{-3t})$.

Need: $26e^{-3t} = u_p'' + 2u_p' + 10u_p$

$$= e^{-3t}[9 - 6 + 10]A = 13Ae^{-3t}$$

$\therefore\ 13A = 26,\ A = 2.$

GS: $u = e^{-t}[c_1 \cos(3t) + c_2 \sin(3t)] + 2e^{-3t}$.

7. CP: $p(r) = r^2 + 2r + 1 = (r + 1)^2$

TS: $u_p = A\cos(t) + B\sin(t)$, with

$u_p' = B\cos(t) - A\sin(t)$ and

$u_p'' = -A\cos(t) - B\sin(t)$.

Need: $25\sin(t) = u_p + 2u'_p + u_p$

$= \cos(t)[-A + 2B + A]$

$+ \sin(t)[-B - 2A + B]$.

$\therefore\ 2B = 0,\ -2A = 25,$

$B = 0,\ A = -25/2.$

GS: $u = e^{-t}[c_1 + c_2 t] - (25/2)\cos(t)$.

9. CP: $p(r) = r^2 - 4 = (r+2)(r-2)$

TS: $u_p = At^3 + Bt^2 + Ct + D$ with

$u'_p = 3At^2 + 2Bt + C$ and

$u''_p = 6At + 2B$.

Need: $2t^3 = u''_p - 4u_p$, which is

$-4At^3 - 4Bt^2 + (6A - 4C)t + 2B - 4D;$

$\therefore\ -4A = 2(A = -1/2),$

$-4B = 0(B = 0),$

$6A - 4C = 0(4C = 6A = -3,\ C = -3/4),$

$2B - 4D = 0(4D = 2B = 0,\ D = 0).$

GS: $u = c_1 e^{2t} + c_2 e^{-2t} - \frac{1}{2}t^3 - \frac{3}{4}t$.

11. CP: $p(r) = r^2 + 2r + 10$

Complex roots: $\{-1 \pm 3i\}$

Noting $p(-1) \neq 0,\ p(3i) \neq 0,$

TS: $u_p = Ae^{-t}$

$+te^{-t}[B\cos(3t) + C\sin(3t)]$

$+D\cos(3t) + E\sin(3t)$.

13. CP: $p(r) = r^2 + 2r = r(r+2)$

Noting $p(2) \neq 0,\ p'(0) \neq 0,$

TS: $u_p = t[At^3 + Bt^2 + Ct + D] + Ee^{2t}$.

15. CP: $p(r) = r^2 + 9$

Imaginary roots: $\{\pm 3i\}$

TS: $u_p = e^t[A\cos(3t) + B\sin(3t)]$

$+t(Ct + D)\cos(3t)$

$+t(Et + F)\sin(3t)$

17. CP: $p(r) = r^2 + 4r + 4 = (r+2)^2$

TS: $u_p = e^{-2t}[At + B]\cos(t)$

$+e^{-2t}[Ct + D]\sin(t)$

$+t^2(Et^2 + Ft + G)e^{-2t}$

19. Model DE: $mu'' + cu' + ku = F$.

This case: $m = .1,\ c = .2,\ F = .1\cos(4t)$,

$\Delta l = .002$ (given),

and $k = \frac{mg}{\Delta l} = \frac{.1(9.8)}{.002} = 490$.

After multiplying by 10, we come to

DE: $u'' + 2u' + 4900u = \cos(4t)$;

CP: $r^2 + 2r + 4900$, with

complex roots $\{-1 \pm i\sqrt{4899}\}$;

TS: $A\cos(4t) + B\sin(4t)$; and

GS: $c_1 e^{-t}\cos\left(t\sqrt{4899}) + c_2 e^{-t}\sin(t\sqrt{4899}\right)$

$+A\cos(4t) + B\sin(4t)$.

After considerable struggle, one finds

$$A = \frac{1221}{4(1221^2 + 4)} = \frac{1221}{5963380},$$

$B = \dfrac{2}{5963380}$, and utilizing the IC $u(0) = 0$, we get

$$c_1 = -A = \frac{-1221}{5963380}.$$

Finally, utilizing the other IC

$u'(0) = 0$, we find

$0 = u'(0) = -c_1 + c_2\sqrt{4899} + 4B,$

$$c_2 = \frac{c_1 - 4B}{\sqrt{4899}} = \frac{-1229}{5963380\sqrt{4899}}.$$

21. Model DE: $mu'' + cu' + ku = F$.

In this case given are: $W = .4,\ c = .4,$

$F = .2e^{-t/2},\ \Delta l = 1/4$ (feet), and so

$$m = \frac{W}{g} = \frac{.4}{32},\ k = \frac{W}{\Delta l} = \frac{.4}{1/4} = 4(.4).$$

The DE then reads $\frac{.4}{32}u'' + .4u' + .4(4)u = .2e^{-t/2}$,

and after multiplication by 32 and division by .4, we come to

DE: $u'' + 32u' + 128u = 16e^{-t/2}$;

CP: $r^2 + 32r + 128$, with
complex roots $\{-16 \pm 8\sqrt{2}\}$;
TS: $u_p = Ae^{-t/2}$;
GS: $u = c_1 e^{(-16+8\sqrt{2})t} + c_2 e^{(-16-8\sqrt{2})t} + Ae^{-t/2}$;
IC: $u(0) = 0$, $u'(0) = 1$ (down impact).
In order for the trial solution to work, one finds with difficulty that $A = 64/449$. In order to fit the IC $u(0) = 0$, we must have

$$c_1 + c_2 + A = u(0) = 0,$$
$$c_1 + c_2 = -A = \frac{-64}{449}.$$

In order to fit the IC $u'(0) = 1$, we have

$$(-16 + 8\sqrt{2})c_1 + (-16 - 8\sqrt{2})c_2 - A/2$$
$$= u'(0) = 1,$$
$$-16(c_1 + c_2) + 8\sqrt{2}(c_1 - c_2)$$
$$= 1 + (A/2) = \frac{481}{449},$$

and we learn

$$c_2 - c_1 = \frac{-16(c_1 + c_2) - \frac{481}{449}}{8\sqrt{2}} = \frac{543}{8\sqrt{2}(449)}.$$

Although the work is not over, knowing both sum and difference for the c's, we find

$$c_1 = \frac{-543\sqrt{2} - 1024}{14368} \approx -.1247$$

and

$$c_2 = \frac{543\sqrt{2} - 1024}{14368} \approx -.0178.$$

23. Model DE: $LQ'' + RQ' + \dfrac{Q}{C} = E$.

This case: $.5Q'' + 2Q' + \dfrac{Q}{.05} = 3\cos(2t)$;
IC: $Q(0) = 0$, $Q'(0) = 1$.
After multiplying by 2, we come to
DE: $Q'' + 4Q' + 40Q = 6\cos(2t)$;
CP: $p(r) = r^2 + 4r + 40$;
Complex roots: $\{-2 \pm 6i\}$;
TS: $Q_p = A\cos(2t) + B\sin(2t)$, with
$Q'_p = 2B\cos(2t) - 2A\sin(2t)$ and
$Q''_p = -4A\cos(2t) - 4B\sin(2t)$.
These lead to the two conditions

$$-4A + 8B + 40A = 6,$$
$$-4B - 8A + 40B = 0.$$

These resolve with

$$A = \frac{27}{170}, \; B = \frac{6}{170} = \frac{3}{85},$$

and we come to
GS: $Q = c_1 e^{-2t}\cos(6t) + c_2 e^{-2t}\sin(6t)$
$+A\cos(2t) + B\sin(2t)$
in which A and B are already known.
The first IC gives

$$0 = Q(0) = c_1 + A, \therefore \; c_1 = -A = \frac{-27}{170},$$

while the other IC gives

$$1 = Q'(0) = -2c_1 + 6c_2 + 2B,$$
$$\therefore \; c_2 = \frac{1 + 2c_1 - 2B}{6} = \frac{26}{255}.$$

25. DE: $u'' + 2u' + 6u = 15\cos(3t)$
CP: $r^2 + 2r + 6$
Complex roots: $\{-1 \pm i\sqrt{5}\}$
Because the roots have negative real parts, the steady-state solution is the particular solution of the form
TS: $u_p = A\cos(3t) + B\sin(3t)$, with
$u'_p = 3B\cos(3t) - 3A\sin(3t)$ and
$u''_p = -9A\cos(3t) - 9B\sin(3t)$.
In order to match the left side of the DE to the right, we require

$$6A + 6B - 9A = 15, \; (2B - A = 5)$$
$$6B - 6A - 9B = 0, \; (2A + B = 0).$$

These resolve quickly as $A = -1$, $B = 2$, and we can write

$$u_p = -\cos(3t) + 2\sin(3t) = \sqrt{5}\sin(3t + \delta),$$

in which

$$\delta = \tan^{-1}(A/B) = \tan^{-1}\left(-\frac{1}{2}\right) = -.4636(\text{rad})$$

or about $-26.6°$. This is the phase shift.

27. DE: $u'' + 4u' + 8u = 15\cos(t) + 10\sin(t)$.

CP: $r^2 + 4r + 8$, which has complex roots $\{-2 \pm 2i\}$, both with negative real part. Therefore the steady-state solution is the particular solution of the form
$u_p = A\cos(t) + B\sin(t)$, with
$u'_p = B\cos(t) - A\sin(t)$ and
$u''_p = -A\cos(t) - B\sin(t)$.
For the DE to be satisfied, we must have

$$-A + 4B + 8A = 15 \text{ and}$$
$$-B - 4A + 8B = 10.$$

These resolve with $A = 1$, $B = 2$, and the steady-state solution is

$$u_p = \cos(t) + 2\sin(t) = \sqrt{5}\sin(t + \delta)$$

in which

$$\delta = \tan^{-1}(1/2) = .464 \text{ (radians)}.$$

This last is the phase shift.

29. Model DE: $mu'' + cu' + ku = F$.
This case: $W = 2$, $c = .4$, $F = 2\sin(2t)$, $\Delta l = 1/2$ (all given), and

$$\therefore\ m = \frac{W}{g} = \frac{2}{32},\ k = \frac{W}{\Delta l} = \frac{2}{1/2} = 4.$$

After multiplying by 16, we come to
DE: $u'' + 6.4u' + 64u = 32\sin(2t)$.

CP: $r^2 + 6.4r + 32$, having complex roots with negative real part. The steady-state solution is the trial solution of the form

$u_p = A\cos(2t) + B\sin(2t)$, with
$u'_p = 2B\cos(2t) - 2A\sin(2t)$, and
$u''_p = -4A\cos(2t) - 4B\sin(2t)$.

This leads to

$$-4A + 12.8B + 64A = 0 \text{ and}$$
$$-4B - 12.8A + 64B = 32.$$

After considerable struggle, one finds

$$A = \frac{-640}{5881},\ B = \frac{3000}{5881}.$$

The solution can then be written in the form $A_1\sin(2t + \delta)$, with

$$A_1 = \sqrt{A^2 + B^2} = \frac{\sqrt{3000^2 + 640^2}}{5881}$$

$\approx .5216$ (the amplitude) and
$\delta = \tan^{-1}(A/B) = -\tan^{-1}(64/300)$
$\approx -.210$ (radians, the phase shift).

31. Model DE: $LQ'' + RQ' + \dfrac{Q}{C} = E$.

This case: $Q'' + 10Q' + 2Q = .1\cos(2t)$.
CP: $p(r) = r^2 + 10r + 2$.
Real roots $\{-5 \pm \sqrt{23}\}$, both negative.
$\therefore$ the steady-state solution *is* the trial solution of the form
$Q_p = A\cos(2t) + B\sin(2t)$, with
$Q'_p = 2B\cos(2t) - 2A\sin(2t)$ and
$Q''_p = -4A\cos(2t) - 4B\sin(2t)$.
These lead to the two conditions

$$-4A + 20B + 2A = .1,$$
$$-4B - 20A + 2B = 0.$$

These resolve with

$$A = \frac{-1}{2020},\ B = \frac{10}{2020}.$$

The solution can then be written in the form $A_1\sin(2t + \delta)$, with

$$A_1 = \sqrt{A^2 + B^2} = \frac{\sqrt{101}}{2020}$$

$\approx .004975$ (the amplitude) and
$\delta = \tan^{-1}(A/B) = -\tan^{-1}(.1)$
$\approx -.0997$ (radians, the phase shift).

33. For $u'' + 3u = 4\sin(\omega t)$, the natural frequency of the system is $\sqrt{3}$. Resonance occurs if $\omega = \sqrt{3}$, while beats occur if ω is merely near $\sqrt{3}$, e.g., $\omega = 1.8$.

35. In this spring problem, we have

$$m = \frac{W}{g} = \frac{.4}{32},\ k = \frac{W}{\Delta l} = \frac{.4}{1/4} = 1.6,$$

and the DE $mu'' + ku = 2\sin(\omega t)$ becomes, after mutiplication by 80, $u'' + 128u = 160\sin(\omega t)$. The natural frequency is

$$\sqrt{\frac{k}{m}} = \sqrt{128}.$$

Resonance occurs if $\omega = \sqrt{128}$, while beats occur if ω is merely near $\sqrt{128}$, e.g., $\omega = 11$.

37. Looking first at the particular solution in the case of resonance (undamped), we find eventually $y_p = 2t\sin(3t)$, and general solution

$$y_1 = k_1\cos(3t) + k_2\sin(3t) + 2t\sin(3t).$$

In the damped case, with trial solution form

$$y_q = A\cos(3t) + B\sin(3t),$$

we find after some struggle that $A = 0$, $B = 40$. With the CP $r^2 - .1r + 9$ having complex roots

$$\left\{\frac{-.1 \pm i\sqrt{35.99}}{2}\right\} = \left\{\frac{-1 \pm i\sqrt{3599}}{20}\right\},$$

we find the general solution

$$y_2 = e^{-t/20}c_1\cos\left(\frac{t\sqrt{3599}}{20}\right) + e^{-t/20}c_2\sin\left(\frac{t\sqrt{3599}}{20}\right) + 40\sin(3t).$$

From the IC $y(0) = 1$, we learn $c_1 = 1 = k_1$, whereas from $y'(0) = 0$ we learn $k_2 = 0$ and

$$-\frac{1}{20}c_1 + \frac{\sqrt{3599}}{20}c_2 + 120 = 0,$$

$$\therefore\ c_2 = -\frac{2399}{\sqrt{3599}}.$$

The final formulae (y_1 and y_2) don't look that similar, although the numerical value

$$\frac{\sqrt{3599}}{20} \approx 2.9996$$

is very close to 3 and that helps. The graphs as seen in the answer appendix are indeed indistinguishable up to around $t = 10$. The lessons are two: (1) In the short run, a small amount of damping makes very little difference; (2) In the long run, even a little damping keeps the solution bounded.

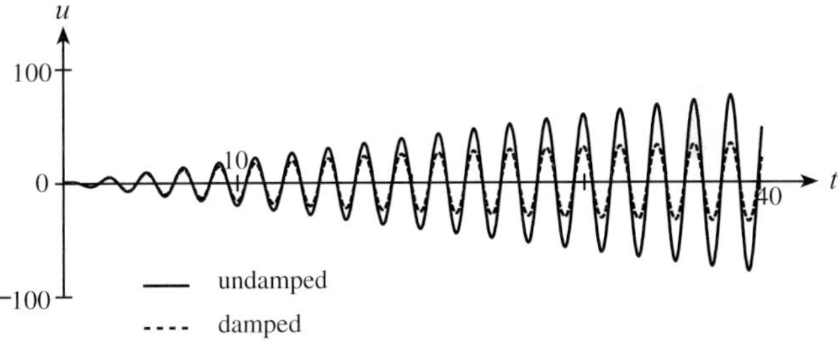

39. For the DE $x'' + 2x' + 5x = A_1\sin(\omega t)$, the CP $r^2 + 2r + 5$ has roots $\{-1 \pm 2i\}$ with negative real part, so the solution to the homogeneous problem is transient. The steady-state solution has the form $x_p = A\cos(\omega t) + B\sin(\omega t)$, with $x_p' = B\omega\cos(\omega t) - A\omega\sin(\omega t)$ and $x_p'' = -A\omega^2\cos(\omega t) - B\omega^2\sin(\omega t)$. Therefore we must have

$$-A\omega^2 + 2B\omega + 5A = 0 \text{ and}$$
$$-B\omega^2 - 2A\omega + 5B = A_1.$$

We learn

$$B = \frac{A_1(5 - \omega^2)}{(5 - \omega^2)^2 + 4\omega^2}$$

and

$$A = \frac{2B\omega}{\omega^2 - 5} = \frac{2\omega A_1}{(5 - \omega^2)^2 + 4\omega^2}.$$

If we choose to write the solution in the form $x_p = A_2 \sin(\omega t + \delta)$, we must have

$$A_2^2 = A^2 + B^2 = \frac{A_1^2[(2\omega)^2 + (5-\omega^2)^2]}{[(5-\omega^2)^2 + 4\omega^2]^2} = \frac{A_1^2}{(5-\omega^2)^2 + 4\omega^2}.$$

$$\therefore \frac{A_2}{A_1} = \frac{1}{\sqrt{(5-\omega^2)^2 + 4\omega^2}} \text{ (the } gain\text{) and}$$

$$\delta = \tan^{-1}\left(\frac{\omega^2 - 5}{2\omega}\right).$$

41. The role of the 5 and the 2 in the solution to Exercise 39 can and should be generalized: If the DE is written in the normalized form

$$x'' + \beta x' + \gamma x = A_1 \sin(\omega t)$$

and has the steady-state solution in the form $x_p = A_2 \sin(\omega t + \delta)$, then the gain g is *defined* by the ratio $\frac{A_2}{A_1}$ (both A's postive) but calculated as

$$g = \frac{1}{\sqrt{(\gamma - \omega^2)^2 + \beta^2\omega^2}}.$$

In order to maximize the gain as a function of the forcing frequency-parameter $\omega > 0$, one should minimize the radicand. Algebra is better than calculus, but calculus does work as long as one recognizes that the critical value $\omega = 0$ is a relative maximum only if $2\gamma \le \beta^2$. Assuming that $2\gamma > \beta^2$, the conclusion is that the gain is maximized by a choice of $\omega = \omega_r$, in which

$$\omega_r = \sqrt{\gamma - \frac{\beta^2}{2}},$$

and for which

$$g_{\max} = \frac{1}{\sqrt{\left(\gamma - \omega_r^2\right)^2 + \beta^2\omega_r^2}} = \frac{2}{\beta\sqrt{4\gamma - \beta^2}}.$$

We are assuming that the solutions to the homogeneous problem are transient, which amounts to $4\gamma > \beta^2$. Therefore the radicand in $g_{\max}$ is definitely positive, but this alone does not guarantee that $2\gamma > \beta^2$.

Putting these principles to work here, where $\gamma = 4$, $\beta = .4$, we find

$$\omega_r^2 = 4 - .08 = 3.92 = \frac{2(49)}{25}, \text{ and}$$

$$g_{\max} = \frac{1}{\sqrt{\omega^4 - \frac{196}{25}\omega^2 + 16}} = \frac{2}{.4\sqrt{16 - .16}} \approx 1.2563 \text{ (rounded).}$$

It appears that the answer in Appendix G may have neglected to take a certain square root. The Bode plots there are not seriously affected, but here we include new ones for Exercise 41 and Exercise 43 with focus on the resonant frequency.

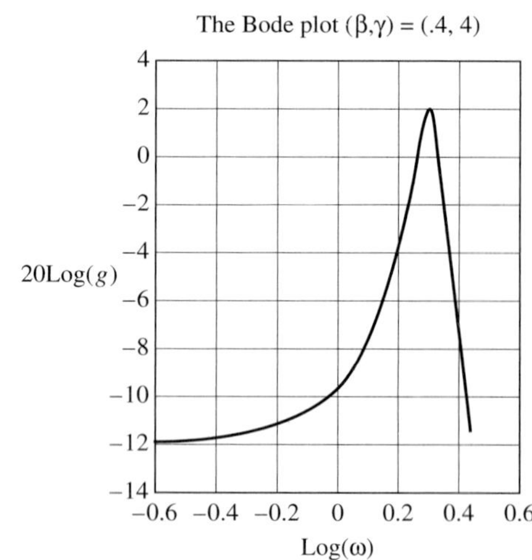

43. Putting the principles of Exercise 41 to work on this problem in which $\gamma = 4$, B $= .2$, we find $\omega_r^2 = 4 - .02 = 3.98$, and

$$g_{\max} = \frac{1}{\sqrt{\omega^4 - \frac{199}{25}\omega^2 + 16}} = \frac{2}{.2\sqrt{16 - .04}} \approx 2.503.$$

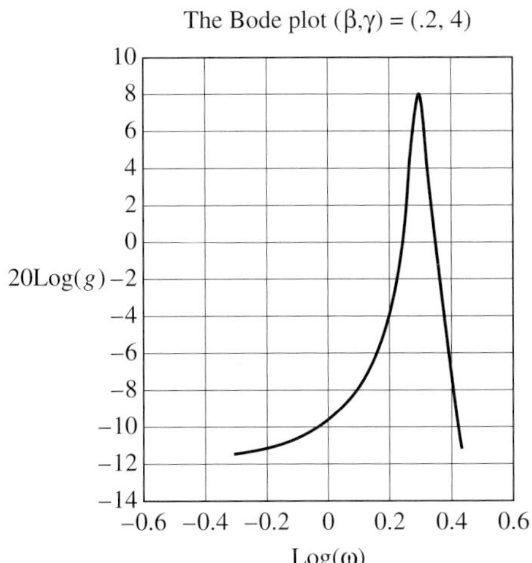

45. The computation of the gain would be the same whether the forcing function was a cosine or a sine. The presence of ω^2 as the coefficient (in the role of A_1) is irrelevant to the gain. The problem is essentially a repeat of Exercises 41 and 43 with B = 1, $\gamma = 4$. We find

$$\omega_r = \sqrt{4 - (1/2)} \approx 1.8708, \text{ and}$$

$$g_{\max} = \frac{2}{\sqrt{16-1}} \approx .5164 \text{ (rounded).}$$

From the seismological perspective, however, one might be more interested in the actual *amplitude* of the steady-state solution, which would be

$$A = \omega^2 x(\text{gain}) = \frac{\omega^2}{\sqrt{(\omega^2-\gamma)^2 + \beta^2\omega^2}}.$$

In this case, we eventually find

$$A = \frac{1}{\sqrt{1 - \frac{7}{\omega^2} + \frac{16}{\omega^4}}},$$

with maximum value

$$A_{\max} = \frac{8}{5\sqrt{39}} \approx .2562$$

when

$$\omega = 4\sqrt{\frac{2}{7}} \approx 2.138.$$

It appears that the solution in Appendix G is discussing this amplitude (squared) rather than the pure gain.

47. The form of the trial solution is $u_p = A\cos(\omega t) + B\sin(\omega t)$, with $u_p'' = -A\omega^2\cos(\omega t) - B\omega^2\sin(\omega t)$. The DE $u_p'' + 4u_p = \sin(\omega t)$ requires $A(4-\omega^2) = 0$ and $B(4-\omega^2) = 1$.

If $\omega^2 \neq 4$, then A will be zero, and the particular solution will be a pure sine. It is dangerous to allow this to become a recurrent assumption. It is only sound in an undamped problem where the external force is a pure sine and not of the same frequency as the solutions to the homogeneous DE.

49. With a lift from the discussion in the solution to Exercise 47 ($\omega = 2.1$, $B = 2/(4-\omega^2) = -200/41$), we do have a particular solution which is a pure sine, and indeed the general solution is

$$u = c_1\cos(2t) + c_2\sin(2t) - \frac{200}{41}\sin\left(\frac{21t}{10}\right).$$

From the IC $u(0) = 0$, we learn $c_1 = 0$. From $u'(0) = 0$, we learn

$$2c_2 - \frac{200(21)}{41(10)} = 0,$$

$$c_2 = \frac{210}{41}, \text{ and}$$

$$u = \frac{210}{41}\sin(2t) - \frac{200}{41}\sin\left(\frac{21t}{10}\right).$$

51. In general, we would expect

$$u = c_1\cos(2t) + c_2\sin(2t) + A\cos(\omega t) + B\sin(\omega t).$$

From the DE, we learn $A = 0$, $B = \dfrac{1}{4-\omega^2}$.
From the IC $u(0) = 0 = u'(0)$, we learn $c_1 = -A = 0$ and $2c_2 + \omega B = 0$;

$$\therefore\ c_2 = -\frac{\omega B}{2} = -\frac{\omega}{2(4-\omega^2)} \text{ and}$$

$$u = -\frac{\omega}{2(4-\omega^2)}\sin(2t) + \frac{1}{4-\omega^2}\sin(\omega t).$$

53. This problem is covered by the solution to Exericse 41, with $\beta = .1$, $\gamma = 4$, and $A_1 = 1$. Because of the last, the amplitude is the gain, and by formula this is

$$\frac{1}{\sqrt{(\gamma - \omega^2)^2 + \beta^2\omega^2}} = \frac{1}{\sqrt{(4 - \omega^2)^2 + .01\omega^2}}$$
$$= \frac{1}{\sqrt{\omega^4 - 7.99\omega^2 + 16}}.$$

55. Both the functions $\sin^{-1}$ and $\tan^{-1}$ produce, as a matter of definition, numerical values in the interval

$$\left[-\frac{\pi}{2}, \frac{\pi}{2}\right].$$

Geometrically, this is Quadrant I or IV. Thus, no angle ϕ in Quadrant III can obey $\phi = \sin^{-1}(\sin(\phi))$ or $\phi = tan^{-1}(\tan(\phi))$. The function $\cos^{-1}$, on the other hand, produces values in the interval $[0, \pi]$, and for angles ϕ in that interval, it *is* true that $\phi = \cos^{-1}(\cos(\phi))$. Thus, if indeed an angle ϕ is in Quadrant III or IV (recognized by the condition $\sin(\phi) < 0$), we *may assume* that, numerically, $-\pi \leq \theta \leq 0$. Under those circumstances, $-\theta$ lies in $[0, \pi]$. And because the cosine is an even function,

$$-\theta = \cos^{-1}(\cos(-\theta)) = \cos^{-1}(\cos(\theta)),$$

which is another way of saying that

$$\theta = -\cos^{-1}(\cos(\theta)) \quad \text{if} \quad -\pi \leq \theta \leq 0.$$

Without the proviso, this formula is correct *up to added integral multiples of* 2π *whenever* $\sin(\theta) < 0$.

57. We must be very careful how we define the gain. Presumably from its electronic interpretation, it is still the ratio of the steady-state amplitude to the forcing amplitude. The effect is that if the steady-state solution to the problem as it stands is written in the form $A_1 \sin(\omega t + \delta)$, the gain is defined by ratio A_1/A. We may normalize the math problem to

$$x'' + \frac{b}{a}x' + \frac{c}{a}x = \frac{A}{a}\sin(\omega t)$$

and apply the solution to Exercise 41 with $\beta = b/a$ and $\gamma = c/a$ to write

$$\frac{A_1}{(A/a)} = \frac{a}{\sqrt{(\gamma - \omega^2)^2 + \beta^2\omega^2}} =$$
$$\frac{1}{\sqrt{\left(\frac{c}{a} - \omega^2\right)^2 + \left(\frac{b}{a}\right)^2\omega^2}} = \frac{a}{\sqrt{(c - a\omega^2)^2 + b^2\omega^2}}.$$

There is, in this formula, a factor of a in both numerators. When that is cancelled, the conclusion follows.

Chapter 6 Review

True or False

1. False. What is constant is the rate of change as a percent of the existing amount.

3. False. Even if the calculus part (integration) can be done explicitly, the subsequent algebra may be impossible.

5. True up to the point where the number of steps becomes unmanageable.

7. False. Even if we regard $A\cos(\omega t) + B\sin(\omega t)$ as being "of the same form" as $\cos(\omega t)$ (which we do), the general form may have to be multiplied by a power of the independent variable.

Problems

1. This very standard problem in exponential growth has the solution $y = 3e^{2t}$.

3. After separation ($y'y = 2x$) and integration

$$\left(\frac{y^2}{2} = x^2 + c\right),$$

use $y(0) = 2$ to learn $c = 2$, and eventually write $y = \sqrt{2x^2 + 4}$. Only the positive square root is correct.

5. After separation

$$\left(\frac{y'}{\sqrt{y}} = \sqrt{x}\right)$$

and integration

$$\left(2\sqrt{y} = \frac{2}{3}x^{3/2} + c\right),$$

use $y(1) = 4$ to find $4 = (2/3) + c$, hence $c = 10/3$, and

$$y = \left(\frac{x^{3/2} + 5}{3}\right)^2 = \frac{x^3 + 10(\sqrt{x})^3 + 25}{9}.$$

7. With t measured in hours, we have $y = Ae^{kt}$, $A = y(0) = 10^4$. If the doubling time is 2, then $2 = e^{2k}$, $k = \ln(2)/2$, and $y = 10^4 e^{t\ \ln(2)/2} = 10^4 2^{t/2}$. To reach $y = 10^6$ at a certain unknown time t, we need

$$2^{t/2} = 100,\ t = \frac{2\ \ln(100)}{\ln(2)} \approx 13.3 \text{ (hours).}$$

9. With t measured in hours, x in milligrams, we can easily get to

$$x = 2\left(\frac{1}{2}\right)^{t/2} = \frac{2}{2^{t/2}}.$$

To get to $x = .1$ at a certain unknown time t, we need

$$2^{t/2} = \frac{2}{.1} = 20,\ t = \frac{2\ \ln(20)}{\ln(2)} \approx 8.64 \text{ (hours).}$$

11. The equation for the doubling time t_d in this case is $2 = e^{.08t_d}$, hence $t_d = \dfrac{\ln(2)}{.08} \approx 8.66$ (years).

13. For temperature T at time t and ambient temperature T_a, we have

$$\frac{T - T_a}{T(0) - T_a} = e^{kt}.$$

In this case with $T_a = 68$, $T(0) = 180$, and $T(1) = 176$, we have

$$\frac{108}{112} = \frac{176 - 68}{180 - 68} = e^k,$$

$$k = \ln\left(\frac{108}{112}\right) = \ln\left(\frac{27}{28}\right),$$

$$\frac{T - 68}{112} = e^{tk} = e^{t\ \ln(27/28)} = \left(\frac{27}{28}\right)^t,$$

$$T = 68 + 112\left(\frac{27}{28}\right)^t.$$

To reach $T = 120$ at unknown time t, we need

$$t = \frac{\ln(52/112)}{\ln(27/28)} \approx 21.1 \text{ (minutes).}$$

15. Separated: $\dfrac{y'}{y} = 2x^3$;

Integrated: $\ln|y| = \dfrac{x^4}{2} + c$;

Solved: $y = Ae^{x^4/2}$
(any $A = \pm e^c$ or zero).

17. Separated: $(y^2 + y)y' = \dfrac{4}{1 + x^2}$;

Integrated: $\dfrac{y^3}{3} + \dfrac{y^2}{2} = (4\tan^{-1} x) + c$;

Solved: Impossible, illustrating # 3 in the T/F review problems.

19. The equilibrium solutions are $y = 0$ and $y = 2$. The second is stable. This can be seen by looking at the field diagram or by separating, integrating, and solving the DE in the form

$$\left(-\frac{1}{2}\right)\ln\left|\frac{2 - y}{y}\right| = 3t + k,$$

$$y = \frac{2}{1 + Ae^{-6t}}(A = \pm e^{-2k}),$$

valid if $y(0)$ is not 0. In this case, one can see that $\lim_{t\to\infty} y(t) = 2$.

21. The only equilibrium solution is $y = 0$. This is stable, easily tipped off by the field diagram. With considerable effort, we reached the general solution

$$y = \frac{2Ae^t}{A^2e^{2t} - 1}$$

for some constant A with $|A| > 1$. From this it can be seen that $\lim_{t\to\infty} y(t) = 0$.

23.

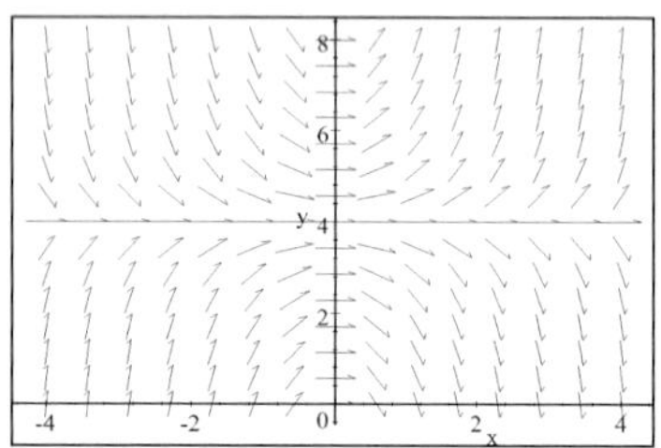

25.

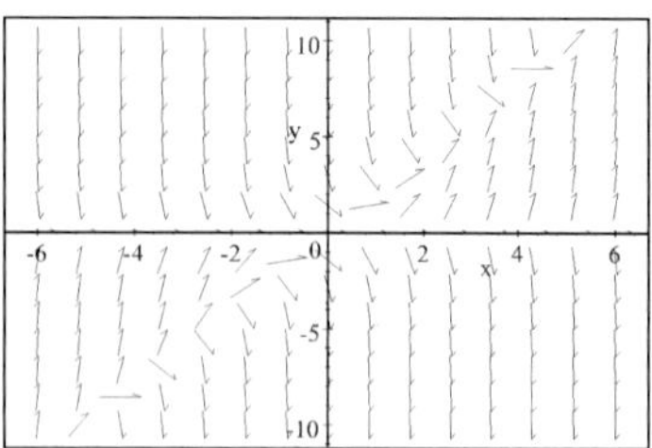

27. We are borrowing the methods from Exercise 57 in Section 6.2. The differential equation is

$$\begin{aligned} x' &= (.3 - x)(.4 - x) - .25x^2 \\ &= .12 - .7x + .75x^2 \\ &= \frac{3}{4}(x^2 - \frac{14}{15}x + \frac{4}{25}) \\ &= \frac{3}{4}(x - r)(x - s), \end{aligned}$$

in which

$$r = \frac{7 + \sqrt{13}}{15},\ s = \frac{7\sqrt{13}}{15},\ r - s = \frac{2\sqrt{13}}{15}.$$

When separated it takes the form

$$\frac{x'}{(x - r)(x - s)} = k$$

in which $k = 3/4$. By partial fractions we find

$$\frac{1}{(x - r)(x - s)} = \frac{1}{(r - s)}\left\{\frac{1}{(x - r)}\frac{1}{(x - s)}\right\},$$

and after integration we find

$$\frac{1}{(r - s)} \ln\left|\frac{x - r}{x - s}\right| = kt + c_1,$$

or in this case

$$\ln\left|\frac{x - r}{x - s}\right| = \frac{2\sqrt{13}}{15}\left(\frac{3}{4}t + c_1\right) = wt + c_2$$

$$(w = \frac{\sqrt{13}}{10} \approx .36056,\ c_2 = \frac{2\sqrt{13}}{15}c_1).$$

Using the initial condition $x(0) = c$, we find $c_2 = \ln|(c - r)/(c - s)|$,

$$\ln\left|\frac{(c - s)(x - r)}{(c - r)(x - s)}\right| = wt \text{ and}$$

$$\frac{x - r}{x - s} = \pm\frac{c - r}{c - s}e^{wt} = \frac{c - r}{c - s}e^{wt},$$

$$x = \frac{s(r - c)e^{wt} + r(c - s)}{(r - c)e^{wt} + (c - s)} = \frac{r\left(\frac{c - s}{r - c}\right)e^{-wt} + s}{\left(\frac{c - s}{r - c}\right)e^{-wt} + 1}.$$

The choice of sign is + since the left side of the middle equation is $(c - r)/(c - s)$ when $t = 0$ and $x = c$. The last expression is one of many possible ways to normalize. It is apparent that $x \to s \approx .22630$ as $t \to \infty$.

(a) Numerically, when $c = .1$, this comes to

$$x = \frac{.22630 - .14710e^{-.36056t}}{1 - .20806e^{-.36056t}}.$$

(b) The student can make the substitution $c = .4$ if he/she wishes, and the mathematics works out fine, but the chemistry makes no sense. In the original formulation from Exercise 57 from Section 6.2, we had the model

$$\begin{aligned} x' &= k_1(a + c - x)(b + c - x) \\ &\quad - k_{-1}x(d - c + x) \end{aligned}$$

in which a, b, c, and d are nonnegative initial concentrations for various substances ($c = x(0)$). Apparently, then, in this problem $a + c = .3$ and $b + c = .4$ (or vice versa). If $c = .4$, one of the other two (a or b) will be negative.

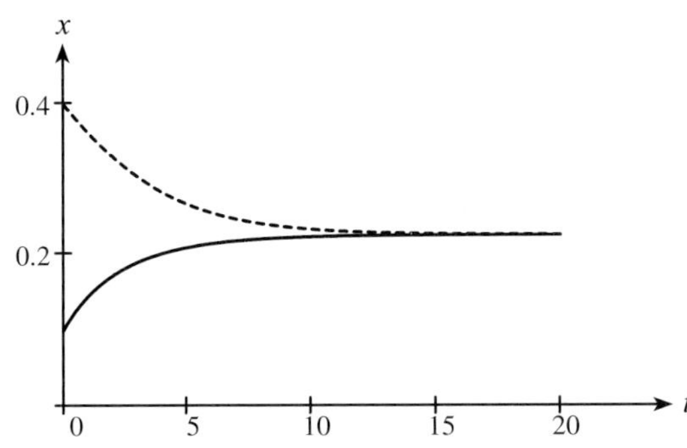

29. The DE $\dfrac{x'}{(x-a)^2} = r$ integrates immediately to

$$\frac{-1}{x-a} = rt + c.$$

In problems of this type, recall that an IC of $x(0) = 0$ is presumed. Therefore,

$$c = \frac{1}{a},\ x - a = \frac{-1}{c+rt} = \frac{-a}{1+art},$$
$$x = a\left(1 - \frac{1}{1+art}\right) = \frac{a^2rt}{1+art}.$$

One can see that all values of x lie between 0 and a, and that $\lim_{t\to\infty} x(t) = a$. All the initial amounts of the A and B substances (both a in this case) will eventually be converted to the X substance, which ultimately will have the same concentration as the original concentrations of the other two substances.

31. With A the amount in the account at time t, the DE is $A'(t) = .10A + 20{,}000$ with an IC of $A(0) = 100{,}000$. The DE separates and integrates easily, yielding

$$10 \ln |.10A + 20{,}000| = t + c$$
$$c = 10 \ln(30{,}000),$$
$$.10A + 20{,}000 = 30{,}000e^{t/10}.$$

If the fortune is to reach 1,000,000 at unknown time t, we must have

$$120{,}000 = 30{,}000e^{t/10}$$
$$\frac{t}{10} = \ln\frac{12}{3} = \ln(4),$$
$$t = 10 \ln(4) \approx 13.86 \text{ years}.$$

33. DE: $y'' + y' - 12y = 0$

CP: $p(r) = r^2 + r - 12$

CP factored: $(r + 4)(r - 3)$

Real roots: $\{-4, 3\}$

GS: $y = c_1e^{-4t} + c_2e^{3t}$

35. DE: $y'' + y' + 3y = 0$

CP: $p(r) = r^2 + r + 3$

Complex roots: $\left\{\dfrac{-1 \pm i\sqrt{11}}{2}\right\}$

GS: $y = c_1e^{-t/2}\cos\left(\dfrac{t\sqrt{11}}{2}\right) +$

$c_2e^{-t/2}\sin\left(\dfrac{t\sqrt{11}}{2}\right)$

37. DE: $y'' - y' - 6y = e^{3t} + t^2 + 1$

CP: $p(r) = r^2 - r - 6$

CP factored: $(r + 2)(r - 3)$

Real roots: $\{-2, 3\}$

TS: $y_p = Ate^{3t} + Bt^2 + Ct + D$, with

$y_p' = A(3t + 1)e^{3t} + 2Bt + C$ and

$y_p'' = A(9t + 6)e^{3t} + 2B.$

For this to work, we require

$5A = 1\ (\therefore\ A = 1/5)$

$-6B = 1\ (\therefore\ B = -1/6)$

$-2B - 6C = 0\ (\therefore\ C = -B/3 = 1/18)$

$2B - C - 6D = 1$

$\left(\therefore\ D = \dfrac{2B - C - 1}{6} = \dfrac{-50}{216}\right)$ and

GS: $y = c_1e^{3t} + c_2e^{-2t} + y_p =$

$$\left(c_1 + \frac{t}{5}\right)e^{3t} + c_2e^{-2t} + \frac{-36t^2 + 12t - 50}{216}.$$

39. DE: $y'' + 2y' - 8y = 0$

CP: $p(r) = r^2 + 2r - 8$

CP factored: $(r + 4)(r - 2)$

Real roots: $\{2, -4\}$

GS: $y = c_1e^{2t} + c_2e^{-4t}$, with

$y' = 2c_1e^{2t} - 4c_2e^{-4t}$

From the IC $y(0) = 5$,
we learn $5 = c_1 + c_2$,
while from the IC $y'(0) = -2$,
we learn $-2 = 2c_1 - 4c_2$.
From these we find $c_2 = 2$, $c_1 = 3$, and the FS $y = 3e^{2t} + 2e^{-4t}$.

41. DE: $y'' + 4y = 3\cos(t)$

CP: $p(r) = r^2 + 4$

Imaginary roots: $\{\pm 2i\}$

TS: $y_p = A\cos(t) + B\sin(t)$, with

$y'_p = B\cos(t) - A\sin(t)$ and

$y''_p = A\cos(t) - B\sin(t)$

For this to work, we require
$3A = 3$, $3B = 0$, $\therefore$ $A = 1$, $B = 0$ and the

GS $y = c_1\cos(2t) + c_2\sin(2t) + \cos(t)$,

with $y' = -2c_1\sin(2t) + 2c_2\cos(2t) - \sin(t)$.
From $y(0) = 1$ we learn $1 = c_1 + 1$ ($\therefore$ $c_1 = 0$).
From $y'(0) = 2$ we learn $2 = 2c_2$ ($\therefore$ $c_2 = 1$).
And the FS is $y = \sin(2t) + \cos(t)$.

43. We are given $W = 4$ (pounds), $\Delta l = 1/3$ (feet) (i.e., 4 inches), and with u the downward displacement from equilibrium, we have $mu'' + ku = 0$, in which

$$m = \frac{W}{g} = \frac{4}{32} = \frac{1}{8},\ k = \frac{W}{\Delta l} = \frac{4}{(1/3)} = 12.$$

Thus, after multiplying by 8, the DE becomes $u'' + 96u = 0$.

CP: $p(r) = r^2 + 96$;

Imaginary roots: $\left\{\pm i\sqrt{96} = \pm 4i\sqrt{6}\right\}$;

GS: $u = c_1\cos\left(4t\sqrt{6}\right) + c_2\sin\left(4t\sqrt{6}\right)$.

From $u(0) = 2/12 = 1/6$ (feet) (i.e., 2 inches), and $u'(0) = 0$ (stationary release), we learn $c_1 = 1/6$, $c2 = 0$, and come to the *FS*

$$u = \frac{\cos(4t\sqrt{6})}{6}.$$

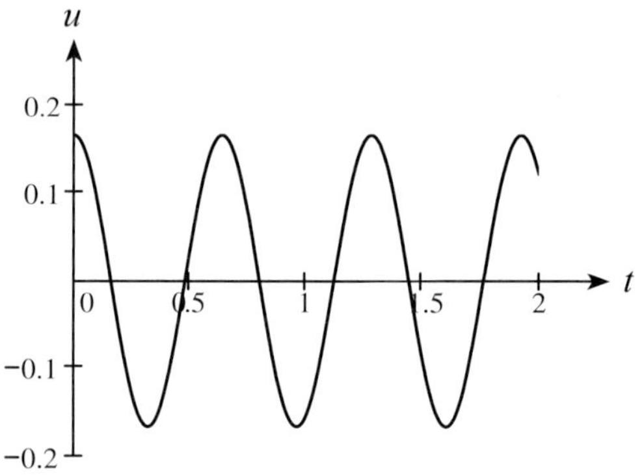

u
0.2
0.1
0
−0.1
−0.2
0.5
1
1.5
2
t

45. The model $LQ'' + RQ' + \frac{Q}{C} = 0$ becomes, after substitution of the givens, the

DE $.2Q'' + 160Q' + 100Q = 0$, or

$Q'' + 800Q' + 500Q = 0$.

CP: $p(r) = r^2 + 800r + 500$;

Real roots $\left\{-400 \pm \sqrt{159500} = -400 \pm 10\sqrt{1595}\right\}$;

GS: $Q = e^{-400t}\left\{c_1e^{10t\sqrt{1595}} + c_2e^{-10t\sqrt{1595}}\right\}$.

From $Q(0) = 10^{-4}$, we learn $c_1 + c_2 = .0001$.
From $Q'(0) = 0$, we learn

$$c_1(-400 + 10\sqrt{1595}) + c_2(-400 - 10\sqrt{1595}) = 0,$$

$$10\sqrt{1595}(c_1 - c_2) = 400(c_1 + c_2) = .04,$$

$$c_1 - c_2 = \frac{.004}{\sqrt{1595}} = \frac{4\sqrt{1595}}{1{,}595{,}000}.$$

The struggle is far from over, but knowing both sum and difference of the c's, we eventually come to

$$c_1 = \frac{1595 + 40\sqrt{1595}}{31{,}900{,}000} = \frac{319 + 8\sqrt{1595}}{6{,}380{,}000} \approx .000100,$$

$$c_2 = \frac{1595 - 40\sqrt{1595}}{31{,}900{,}000} = \frac{319 - 8\sqrt{1595}}{6{,}380{,}000} \approx 0 \text{ (six places)}.$$

As to the current, this would be

$$Q'(t) = c_1(-400 + 10\sqrt{1595})e^{(-400+10\sqrt{1595})t} - c_2(400 + 10\sqrt{1595})e^{-(400+10\sqrt{1595})t}.$$

With the c's as given, this is

$$I(t) = -\frac{\sqrt{1595}}{638{,}000}(e^{(-400+10\sqrt{1595})t} - e^{-(400+10\sqrt{1595})t}).$$

47. DE: $u'' + 2u' + 5u = 2e^{-t}\sin(2t) + 4t^3 - 2\cos(2t)$

CP: $p(r) = r^2 + 2r + 5$

Complex roots: $\{-1 \pm 2i\}$

The form of the trial solution would be

$$te^{-t}[A\cos(2t) + B\sin(2t)] + C_o + C_1 t + C_2 t^2 + C_3 t^3 + D\cos(2t) + E\sin(2t).$$

49. We are given $W = 4$ (pounds), $\Delta l = 1/3$ (feet) (i.e., 4 inches), and with u the downward displacement from equilibrium, we have

$$mu'' + cu' + ku = F,$$

in which

$$m = \frac{W}{g} = \frac{4}{32} = \frac{1}{8}, \; k = \frac{W}{\Delta l} = \frac{4}{(1/3)} = 12,$$

$F = 2\sin(2t)$. With $c = .4$ (given), after multiplying by 8, the DE becomes

$u'' + 3.2u' + 96u = 16\sin(2t)$, with

CP $p(r) = r^2 + 3.2r + 96$.

Complex roots are $\{-1.6 \pm i\sqrt{93.44}\}$ with negative real part. Therefore, the steady-state solution is the particular solution of the trial form

$u_p = A\cos(2t) + B\sin(2t)4$, with

$u'_p = 2B\cos(2t) - 2A\sin(2t)$ and

$u''_p = -4A\cos(2t) - 4B\sin(2t)$.

If it is to work, we must have

$$-4A + 6.4B + 96A = 0(92A + 6.4B = 0)$$
$$-4B - 6.4A + 96B = 16(-6.4A + 92B = 16).$$

These resolve to give

$$A = \frac{-160}{13289}, \; B = \frac{2300}{13289}$$
$$u_p = \frac{-160\cos(2t) + 2300\sin(2t)}{13289}.$$

Returning to the matter of setting up the initial value problem, the general solution will be of the form

$$u = e^{-1.6t}\left[c_1\cos(t\sqrt{93.44}) + c_2\sin(t\sqrt{93.44})\right] + u_p \text{ (in which } u_p \text{ is given above).}$$

The equations that must be solved to get the c's are from the IC

$$\frac{1}{6} = u(0) = c_1 + u_p(0) = c_1 + A = c_1 - \frac{160}{13289}.$$

We will get $c_1(\approx .1787)$ from this alone.
But we also have

$$2 = u'(0) = -1.6c_1 + c_2\sqrt{93.44} + 2B,$$

and with B and c_1 known, we can finally get $c_2(\approx .2007)$.

Chapter 7

Infinite Series

7.1 Sequences of Real Numbers

1. $1, \frac{3}{4}, \frac{5}{9}, \frac{7}{16}, \frac{9}{25}, \frac{11}{36}$

3. $4, 2, \frac{2}{3}, \frac{1}{6}, \frac{1}{30}, \frac{1}{180}$

5. $\lim_{n\to\infty} \frac{1}{n^3} = 0$ (As n gets large, n^3 gets large, so $1/n^3$ goes to 0.)

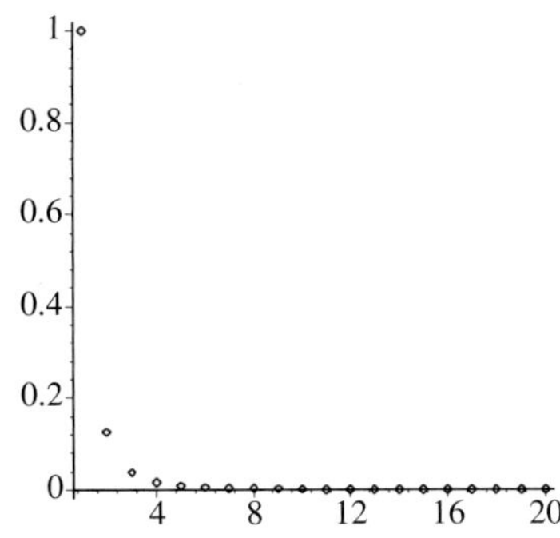

7. $\lim_{n\to\infty} \frac{n}{n+1} = \lim_{n\to\infty} \frac{1}{1+\frac{1}{n}} = 1$

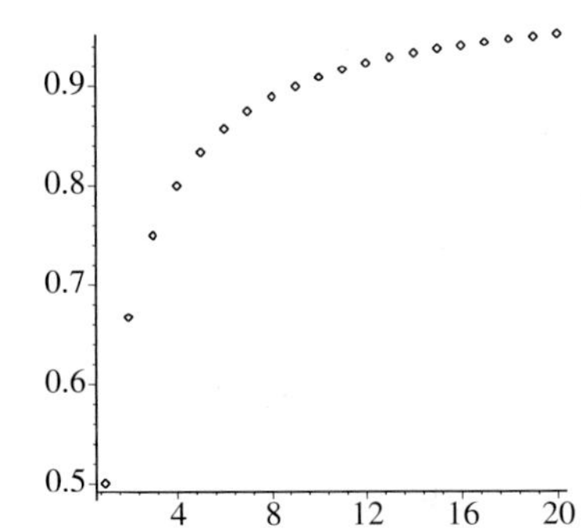

9. $\lim_{n\to\infty} \frac{2}{\sqrt{n}} = 0$ (As n gets large, $\sqrt{n}$ gets large, so $\frac{2}{\sqrt{n}}$ goes to 0.)

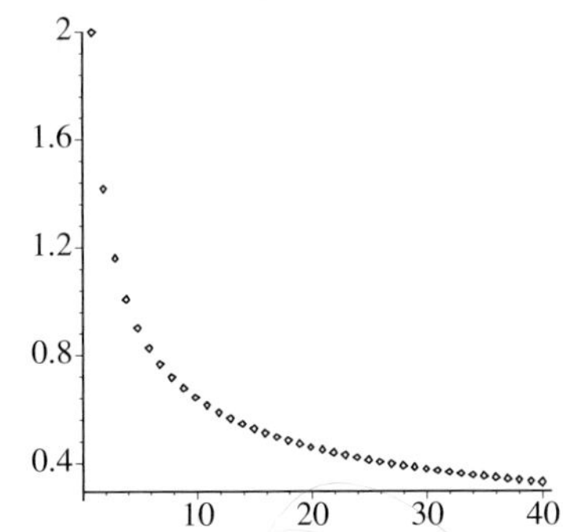

11. $\lim_{n\to\infty} \frac{3n^2+1}{2n^2-1} = \lim_{n\to\infty} \frac{3+\frac{1}{n^2}}{2-\frac{1}{n^2}} = \frac{3}{2}$; converges to $\frac{3}{2}$

13. $\lim_{n\to\infty} \frac{n^2+1}{n+1} = \lim_{n\to\infty} \frac{n+\frac{1}{n}}{1+\frac{1}{n}} = \infty$; diverges

15. $\lim_{n\to\infty} \frac{n+2}{3n-1} = \lim_{n\to\infty} \frac{1+\frac{2}{n}}{3-\frac{1}{n}} = \frac{1}{3}$; converges to $\frac{1}{3}$

17. $\lim_{n\to\infty} (-1)^n \frac{n+2}{3n-1} = \lim_{n\to\infty} (-1)^n \frac{1+\frac{2}{n}}{3-\frac{1}{n}} = \pm\frac{1}{3}$, so $\lim_{n\to\infty} (-1)^n \frac{n+2}{3n-1}$ does not exist; diverges.

19.
$$\lim_{n\to\infty} \left|(-1)^n \frac{n+2}{n^2+4}\right| = \lim_{n\to\infty} \frac{n+2}{n^2+4} = \lim_{n\to\infty} \frac{\frac{1}{n}+\frac{2}{n^2}}{1+\frac{4}{n^2}} = 0,$$

so by Corollary 1.1, $\lim_{n\to\infty} \frac{(-1)^2(n+2)}{n^2+4} = 0$; converges to 0.

21. $\{\cos \pi n\}_{n=1}^{\infty} = \{-1, 1, -1, 1, \ldots\}$; diverges

23. $\lim_{x\to\infty} \frac{x}{e^x} = \lim_{x\to\infty} \frac{1}{e^x}$ by l'Hopital's Rule and

$\lim_{x\to\infty} \frac{1}{e^x} = 0$, so by Theorem 1.2

$\lim_{n\to\infty} \frac{n}{e^n} = 0$; converges to 0.

25. $\lim_{n\to\infty} \frac{e^n+2}{e^{2n}-1} = \lim_{n\to\infty} \frac{\frac{1}{e^n}+\frac{2}{e^{2n}}}{1-\frac{1}{e^{2n}}} = 0$; converges to 0

27. For $n \geq 1$, $e^n + 1 < 2e^n$,

so $\frac{3^n}{e^n+1} > \frac{1}{2} \cdot \frac{3^n}{e^n} = \frac{1}{2} \cdot \left(\frac{3}{e}\right)^n$.

Since $\frac{3}{e} > 1$, $\lim_{n\to\infty} \frac{1}{2}\left(\frac{3}{e}\right)^n = i\infty$;

so $\lim_{n\to\infty} \frac{3^n}{e^n+1} = \infty$; diverges.

29. $-1 \leq \cos n \leq 1 \Rightarrow \frac{-1}{n!} \leq \frac{\cos n}{n!} \leq \frac{1}{n!}$ for all n,

and $\lim_{n\to\infty} \frac{-1}{n!} = \lim_{n\to\infty} \frac{1}{n!} = 0$, so by the Squeeze Theorem, $\lim_{n\to\infty} \frac{\cos n}{n!} = 0$; converges to 0.

31. $-1 \leq \cos n \leq 1 \Rightarrow \frac{-1}{n^2} \leq \frac{\cos n}{n^2} \leq \frac{1}{n^2}$ for all n, and

$\lim_{n\to\infty} \frac{-1}{n^2} = \lim_{n\to\infty} \frac{1}{n^2} = 0$, so by the Squeeze Theorem

$\lim_{n\to\infty} \frac{\cos n}{n^2} = 0$; converges to 0.

33. $0 \leq |a_n| = \frac{1}{ne^n} \leq \frac{1}{n}$ and $\lim_{n\to\infty} \frac{1}{n} = 0$, so by the Squeeze Theorem and Corollary 1.1,

$\lim_{n\to\infty} \frac{(-1)^n}{ne^n} = 0$; converges to 0.

35. $\frac{a_{n+1}}{a_n} = \left(\frac{n+4}{n+3}\right) \cdot \left(\frac{n+2}{n+3}\right) = \frac{n^2+6n+8}{n^2+6n+9} < 1$

for all n, so $a_{n+1} < a_n$ for all n, so $\{a_n\}_{n=1}^{\infty}$ is decreasing.

37. $\frac{a_{n+1}}{a_n} = \left(\frac{e^{n+1}}{n+1}\right) \cdot \left(\frac{n}{e^n}\right) = \frac{e \cdot n}{n+1} > 1$

for all n, so $a_{n+1} > a_n$ for all n, so $\{a_n\}_{n=1}^{\infty}$ is increasing.

39. $\frac{a_{n+1}}{a_n} = \left(\frac{2^{n+1}}{(n+2)!}\right) \cdot \left(\frac{(n+1)!}{2^n}\right) = \frac{2}{n+2} < 1$

for all n, so $a_{n+1} < a_n$ for all n, so $\{a_n\}_{n=1}^{\infty}$ is decreasing.

41. $\left(\frac{a_{n+1}}{a_n}\right) = \left(\frac{10^{n+1}}{(n+1)!}\right) \cdot \left(\frac{n!}{10^n}\right) = \frac{10}{n+1} < 1$

for $n > 9$; so $a_{n+1} < a_n$ for $n > 9$, so as $n \to \infty$, $\{a_n\}_{n=1}^{\infty}$ is decreasing.

43. $|a_n| = \left|\frac{3n^2-2}{n^2+1}\right| = \frac{3n^2-2}{n^2+1} < \frac{3n^2}{n^2+1} < \frac{3n^2}{n^2} = 3$,

so $\{a_n\}_{n=1}^{\infty}$ is bounded by 3; $|a_n| < 3$.

45. $|a_n| = \left|\frac{\sin(n^2)}{n+1}\right| \leq \frac{1}{n+1} \leq \frac{1}{2}$ for $n \geq 1$, so

$\{a_n\}_{n=1}^{\infty}$ is bounded by $\frac{1}{2}$; $|a_n| \leq \frac{1}{2}$.

47. $a_n = \left(1+\frac{1}{n}\right)^n$; $a_{1000} \approx 2.716924$; $e \approx 2.718282$

$b_n = \left(1-\frac{1}{n}\right)^n$; $b_{1000} \approx 0.367695$; $e^{-1} \approx 0.367879$

49. A plausible argument goes like this:

$$\lim_{n\to\infty}\left(1+\frac{r}{n}\right)^n = \lim_{m\to\infty}\left(1+\frac{r}{r\cdot m}\right)^{r\cdot m} = \lim_{m\to\infty}\left(\left(1+\frac{1}{m}\right)^m\right)^r.$$

Note that $m = n/r$ will generally not be an integer, so we need to proceed as follows: Let $y = x/r$. Then

$$\lim_{x\to\infty}\left(1+\frac{r}{x}\right)^x = \lim_{y\to\infty}\left(1+\frac{1}{y}\right)^{ry}$$
$$= \left[\lim_{y\to\infty}\left(1+\frac{1}{y}\right)^y\right]^r$$
$$= e^r.$$

Now apply Theorem 1.2 to show that

$$\lim_{n\to\infty}\left(1+\frac{r}{n}\right)^n = e^r.$$

51. The first 10 terms are: 1.414214, 1.681793, 1.834008, 1.915207, 1.957144, 1.978456, 1.989199, 1.994592, 1.997294, 1.998647. The limit appears to be 2.

53. If $|p|$ is > 1, then the absolute value of the denominator gets large and so the sequence converges to 0. If $|p| < 1$, then the denominator goes to 0 and so the sequence diverges. If $p = 1$, then $a_n = 1$ and the sequence converges to 1. If $p = -1$, then $a_n = (-1)n$ and so diverges. Hence, the sequence converges for p in the range $(-\infty, -1) \cup [1, \infty)$.

55. If side $s = 12''$ and the diameter $D = \frac{12}{n}$, then the number of disks that fit along one side is

$$\frac{12}{\left(\frac{12}{n}\right)} = n.$$

Thus, the total number of disks is

$$\frac{12}{\left(\frac{12}{n}\right)} \cdot \frac{12}{\left(\frac{12}{n}\right)} = n \cdot n = n^2.$$

$a_n =$ wasted area in box with n^2 disks, so
$a_n =$ area of box$-$area of disks;
$a_n = 12 \cdot 12 - n^2\left(\frac{6}{n}\right)^2 \pi = 144 - 36\pi$
$a_n \approx 30.9.$

57.
$$a_0 = 3.049q$$
$$a_1 = 3.049 + .005(3.049)^{2.01} = 3.096$$
$$a_2 = 3.096 + .005(3.096)^{2.01} = 3.144$$
$$a_3 = 3.144 + .005(3.144)^{2.01} = 3.194$$
$$a_{10} = 3.594 < 3.721 = \text{population in 1970}$$
$$a_{20} = 4.376 < 4.473 = \text{population in 1980}$$
$$a_{30} = 5.589 > 5.333 = \text{population 1990}$$

Estimated population in 2035 $= a_{75} = 4.131 \times 10^{53}$.

59. In the 3rd month, only the adult rabbits have newborns, so $a_3 = 2 + 1 = 3$. In the 4th month, only the 2 pairs of adult rabbits from a_2 can have newborns, so $a_4 = 3 + 2 = 5$. In general, $a_n = a_{n-1} + a_{n-2}$.

61. Using the distance formula, we see that $(c_2 - c_1)^2 + (r_2 - r_1)^2$ is the square of the distance between the centers of the circles C_1 and C_2. Since these circles are tangent, the distance between the centers is the sum $r_1 + r_2$ of the radii. Hence,

$$(c_2 - c_1)^2 + (r_2 - r_1)^2 = (r_1 + r_2)^2.$$

Thus,

$$(c_2 - c_1)^2 = (r_1 + r_2)^2 - (r_2 - r_1)^2 = 4r_1r_2,$$

so $|c_2 - c_1| = 2\sqrt{r_1r_2}$.
By the same reasoning,

$$|c_3 - c_1| = 2\sqrt{r_1r_3} \quad \text{and} \quad |c_2 - c_3| = 2\sqrt{r_2r_3}.$$

If the circles are arranged with C_1 on the left, C_2 on the right, and C_3 in the middle, then in fact

$$c_2 - c_1 = 2\sqrt{r_1r_2},$$
$$c_2 - c_3 = 2\sqrt{r_2r_3}, \text{ and}$$
$$c_3 - c_1 = 2\sqrt{r_1r_3}.$$

Adding the last two equations, we get

$$2\sqrt{r_1r_3} + 2\sqrt{r_2r_3} = c_3 - c_1 + c_2 - c_3$$
$$= c_2 - c_1 = 2\sqrt{r_1r_2}.$$

So

$$\sqrt{r_1r_3} + \sqrt{r_2r_3} = \sqrt{r_1r_2}.$$

Squaring this last equation, we get

$$r_1r_3 + r_2r_3 + 2r_3\sqrt{r_1r_2} = r_1r_2,$$

or

$$r_3(r_1 + r_2 + 2\sqrt{r_1r_2}) = r_1r_2,$$
$$r_3(\sqrt{r_1} + \sqrt{r_2})^2 = r_1r_2,$$
$$r_3 = \frac{r_1r_2}{(r_1 + r_2)^2},$$
$$\sqrt{r_3} = \frac{\sqrt{r_1r_2}}{(r_1 + r_2)}.$$

63. Let the center of the circle be $(0, c)$ and suppose the circle is tangent to the parabola at the point (p, q), with $q = p^2$. Then the slope of the tangent line to the parabola at that point will be $2p$. For the moment let's assume, as in the picture, that the point of tangency is not the origin. Now the line L from the center of the circle to the point of tangency will be perpendicular to the tangent line, and so will have slope $-1/(2p)$. Since slope = rise/run, we have $-1/(2p) = (q - c)/p$, so $c - q = \frac{1}{2}$. The radius r of the circle is the distance from $(0, c)$ to (p, q), which is

$$\sqrt{p^2 + (c - q)^2} = \sqrt{p^2 + (1/4)}.$$

Note that this implies that $r > \frac{1}{2}$. Then

$$r^2 = p^2 + \frac{1}{4} = q + \frac{1}{4} = \left(c - \frac{1}{2}\right) + \frac{1}{4} = c - \frac{1}{4}.$$

So this process correctly constructs a circle of radius r, tangent to the parabola, whose center $(0, c)$ is given by $c = (1/4) + r^2$. (If $r \le 1/2$, the circle turns out to be tangent to the parabola at the origin.)

65.

$$\lim_{n\to\infty} \frac{\ln n}{n} = \lim_{n\to\infty} \frac{\frac{1}{n}}{1} = \lim_{n\to\infty} \frac{1}{n} = 0$$

by l'Hopital's Rule, so

$$\lim_{n\to\infty} \sqrt[n]{n} = \lim_{n\to\infty} e^{(1/n)\ln n} = e^0 = 1,$$

so the sequence converges to 1.

67. Assuming the limit exists, and letting $L = \lim_{n\to\infty} a_{n+1} = \lim_{n\to\infty} a_n$, we have

$$L = \frac{1}{2}\left(L + \frac{c}{L}\right) \Rightarrow 2L^2 = L^2 + c \Rightarrow L^2 = c$$
$$\Rightarrow L = \sqrt{c}.$$

7.2 Infinite Series

1. $\sum_{k=0}^{\infty} 3\left(\frac{1}{5}\right)^k$ is a geometric series with $a = 3$ and $|r| = \frac{1}{5} < 1$, so it converges to $\dfrac{3}{1 - \frac{1}{5}} = \dfrac{15}{4}$.

3. $\sum_{k=0}^{\infty} \frac{1}{2}\left(-\frac{1}{3}\right)^k$ is a geometric series with $a = \frac{1}{2}$ and $|r| = \frac{1}{3} < 1$, so it converges to $\dfrac{\frac{1}{2}}{1 - \left(-\frac{1}{3}\right)} = \dfrac{3}{8}$.

5. $\sum_{k=0}^{\infty} \frac{1}{2}(3)^k$ is a geometric series with $|r| = 3 > 1$, so it diverges.

7. Using partial fractions,

$$S_n = \sum_{k=1}^{n} \frac{4}{k(k+2)} = \sum_{k=1}^{n} \left(\frac{2}{k} - \frac{2}{k+2}\right)$$
$$= \left(2 - \frac{2}{3}\right) + \left(1 - \frac{2}{4}\right) + \left(\frac{2}{3} - \frac{2}{5}\right)$$
$$+ \cdots + \left(\frac{2}{n-1} - \frac{2}{n+1}\right) + \left(\frac{2}{n} - \frac{2}{n+2}\right)$$
$$= 2 + 1 - \frac{2}{n+1} - \frac{2}{n+2}$$
$$= 3 - \frac{4n+6}{n^2 + 3n + 2}$$

and

$$\lim_{n\to\infty} S_n = \lim_{n\to\infty} \left(3 - \frac{4n+6}{n^2 + 3n + 2}\right) = 3.$$

Thus, the series converges to 3.

9. $\lim\limits_{k\to\infty} \dfrac{3k}{k+4} = 3 \neq 0$, so by the kth-Term Test for Divergence, the series diverges.

11. $\sum\limits_{k=1}^{\infty} \dfrac{2}{k} = 2\sum\limits_{k=1}^{\infty} \dfrac{1}{k}$ and from Example 2.7, $\sum\limits_{k=1}^{\infty} \dfrac{1}{k}$ diverges, so $2\sum\limits_{k=1}^{\infty} \dfrac{1}{k}$ diverges.

13. Using partial fractions,

$$S_n = \sum_{k=1}^{n} \frac{2k+1}{k^2(k+1)^2}$$

$$= \sum_{k=1}^{n} \left[\frac{1}{k^2} - \frac{1}{(k+1)^2}\right]$$

$$= \left(1 - \frac{1}{4}\right) + \left(\frac{1}{4} - \frac{1}{9}\right) + \left(\frac{1}{9} - \frac{1}{16}\right) + \cdots$$

$$+ \left(\frac{1}{(n-1)^2} - \frac{1}{n^2}\right) + \left(\frac{1}{n^2} - \frac{1}{(n+1)^2}\right)$$

$$= 1 - \frac{1}{(n+1)^2} \text{ and}$$

$$\lim_{n\to\infty} S_n = \lim_{n\to\infty} \left[1 - \frac{1}{(n+1)^2}\right]$$

$$= 1.$$

Thus the series converges to 1.

15. $\sum\limits_{k=0}^{\infty} \dfrac{1}{3^k}$ is a geometric series with $a = 1$, and $|r| = \frac{1}{3} < 1$, so it converges to $\dfrac{1}{1-\frac{1}{3}} = \dfrac{3}{2}$.

17.
$$\sum_{k=0}^{\infty} \left(\frac{1}{2^k} - \frac{1}{k+1}\right) = \sum_{k=0}^{\infty} \frac{1}{2^k} - \sum_{k=0}^{\infty} \frac{1}{k+1}.$$

The first series is a convergent geometric series, but the second series is the divergent harmonic series, so the original series diverges.

19.
$$\sum_{k=0}^{\infty} \left(\frac{2}{3^k} + \frac{1}{2^k}\right) = \sum_{k=0}^{\infty} \frac{2}{3^k} + \sum_{k=0}^{\infty} \frac{1}{2^k}.$$

The first series is a geometric series with $a = 2$ and $|r| = \dfrac{1}{3} < 1$, so it converges to $\dfrac{2}{1-\frac{1}{3}} = 3$. The second series is a geometric series with $a = 1$ and $|r| = \dfrac{1}{2} < 1$, so it converges to $\dfrac{1}{1-\frac{1}{2}} = 2$. Thus

$$\sum_{k=0}^{\infty} \left(\frac{2}{3^k} + \frac{1}{2^k}\right)$$

converges to $3 + 2 = 5$.

21. $\sum\limits_{k=0}^{\infty} 3\left(-\dfrac{1}{2}\right)^k$ is a geometric series with $a = 3$ and $|r| = \dfrac{1}{2} < 1$, so it converges to $\dfrac{3}{1-\left(-\frac{1}{2}\right)} = 2$.

23. $\lim\limits_{k\to\infty} |a_k| = \lim\limits_{k\to\infty} \dfrac{3k}{k+1} = 3 \neq 0$, so by the kth-Term Test for Divergence, the series diverges.

25.

n	$S_n = \sum_{k=1}^{n} \frac{1}{k^2}$
1	1
2	1.25
3	1.3611
4	1.4236
5	1.4636
6	1.4914
7	1.5118
8	1.5274
9	1.5398
10	1.5498
11	1.5580
12	1.5650
13	1.5709
14	1.5760
15	1.5804
16	1.5843
17	1.5878
18	1.5909
19	1.5937
20	1.5962

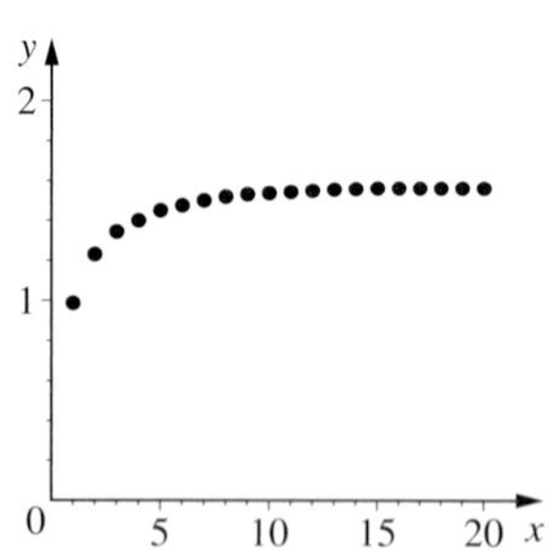

Appears to converge.

27.

n	$S_n = \sum_{k=1}^{n} \frac{3}{k!}$
1	3
2	4.5
3	5.0
4	5.125
5	5.150
6	5.15417
7	5.15476
8	5.15484
9	5.154844
10	5.154845

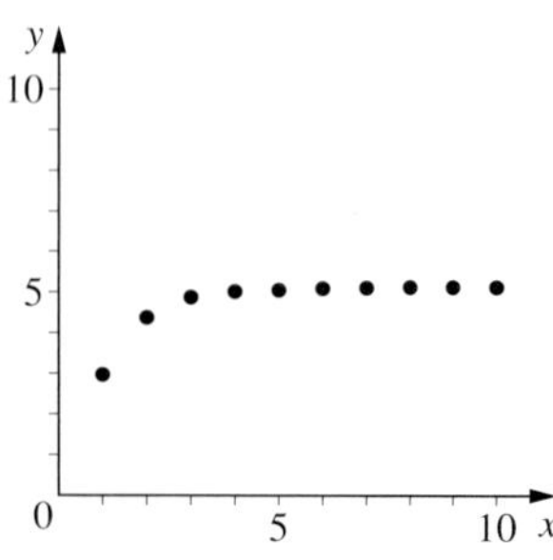

Appears to converge.

29.

n	$S_n = \sum_{k=1}^{n} \frac{4^k}{k^2}$
1	4
2	8
3	15.1111
4	31.1111
5	72.0711
6	185.8489
7	520.2162
8	1544.2162
9	4780.5619
10	15,266.3219

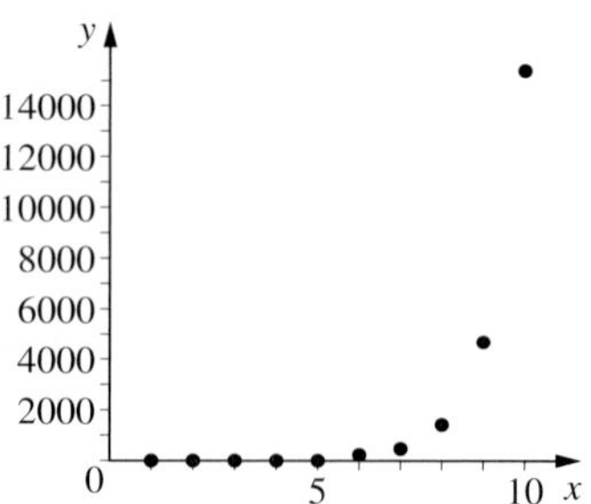

Appears to diverge.

31. Assume $\sum_{k=1}^{\infty} a_k$ converges to L. Then for any m,

$$L = \sum_{k=1}^{\infty} a_k = \sum_{k=1}^{m-1} a_k + \sum_{k=m}^{\infty} a_k = S_{m-1} + \sum_{k=m}^{\infty} a_k.$$

So $\sum_{k=m}^{\infty} a_k = L - S_{m-1}$, and thus converges.

33. Assume $\sum_{k=1}^{\infty} a_k$ converges to A and $\sum_{k=1}^{\infty} b_k$ converges to B. Then the sequences of partial sums converge, and letting

$$S_n = \sum_{k=1}^{n} a_k \quad \text{and} \quad T_n = \sum_{k=1}^{n} b_k,$$

we have

$$\lim_{n\to\infty} S_n = A \quad \text{and} \quad \lim_{n\to\infty} T_n = B.$$

Let $Q_n = \sum_{k=1}^{n} (a_k + b_k)$, the sequence of partial sums for $\sum_{k=1}^{\infty} (a_k + b_k)$. Since S, T, and Q are all finite sums, $Q_n = S_n + T_n$.

Then by Theorem 1.1(i),

$$\begin{aligned} A + B &= \lim_{n\to\infty} S_n + \lim_{n\to\infty} T_n \\ &= \lim_{n\to\infty} (S_n + T_n) \\ &= \lim_{n\to\infty} Q_n \\ &= \sum_{k=1}^{\infty} (a_k + b_k). \end{aligned}$$

The proofs for $\sum_{k=1}^{\infty} (a_k - b_k)$ and $\sum_{k=1}^{\infty} ca_k$ are similar.

35. If $\sum_{k=1}^{\infty} a_k$ converges, then $\lim_{k\to\infty} a_k = 0$. But then the sequence $\frac{1}{a_k}$ diverges, so by the kth-Term Test for Divergence, $\sum_{k=1}^{\infty} \frac{1}{a_k}$ diverges.

37. Let $S_n = \sum_{k=1}^{\infty} \frac{1}{k}$. Then $S_1 = \frac{1}{1} = 1$ and $S_2 = 1 + \frac{1}{2} = \frac{3}{2}$. Since $S_8 > \frac{5}{2}$, we have

$$\begin{aligned} S_{16} &= S_8 + \frac{1}{9} + \frac{1}{10} + \cdots + \frac{1}{16} > S_8 + 8\left(\frac{1}{16}\right) \\ &= S_8 + \frac{1}{2} > \frac{5}{2} + \frac{1}{2} = 3. \text{ So } S_{16} > 3. \\ S_{32} &= S_{16} + \frac{1}{17} + \frac{1}{18} + \cdots + \frac{1}{32} > S_{16} + 16\left(\frac{1}{32}\right) \\ &= S_{16} + \frac{1}{2} > 3 + \frac{1}{2} = \frac{7}{2}. \text{ So } S_{32} > \frac{7}{2}. \end{aligned}$$

If $n = 64$, then $S_{64} > 4$. If $n = 256$, then $S_{256} > 5$. If $n = 4^{m-1}$, then $S_n > m$.

39. $.9 + .09 + .009 + \cdots = \sum_{k=0}^{\infty} .9(.1)^k$, which is a geometric series with $a = .9$ and $|r| = .1 < 1$, so it converges to $\frac{.9}{1 - .1} = 1$.

41. The amount of overhang is $\sum_{k=0}^{n-1} \frac{L}{2(n-k)}$. So if n = 8, then

$$\sum_{k=0}^{7} \frac{L}{2(8-k)} = 1.3589L.$$

When $n = 4$,

$$L \sum_{k=0}^{n-1} \frac{1}{2(n-k)} = L \sum_{k=0}^{3} \frac{1}{2(4-k)} = 1.0417L > L.$$

$$\begin{aligned} \lim_{n\to\infty} \sum_{k=0}^{n-1} \frac{L}{2(n-k)} &= \lim_{n\to\infty} \sum_{k=1}^{n} \frac{L}{2k} \\ &= \frac{L}{2} \lim_{n\to\infty} \sum_{k=1}^{n} \frac{1}{k} \\ &= \infty. \end{aligned}$$

43. If $0 < 2r < 1$, then

$$1 + 2r + (2r)^2 + \cdots = \sum_{k=0}^{\infty} (2r)^k,$$

which is a geometric series with $a = 1$ and $|2r| = 2r < 1$, so it converges to $\frac{1}{1-2r}$.

If $r = \frac{1}{1000}$, then

$$\begin{aligned} 1 + .002 + .000004 + \cdots &= \frac{1}{1 - 2\left(\frac{1}{1000}\right)} \\ &= \frac{500}{499} \\ &= 1.002004008\ldots \end{aligned}$$

45.

$$\begin{aligned} &p^2 + 2p(1-p)p^2 + [2p(1-p)]^2 p^2 + \cdots \\ &= \sum_{k=0}^{\infty} p^2 [2p(1-p)]^k \end{aligned}$$

is a geometric series with $a = p^2$ and $|r| = 2p(1-p) < 1$ because $2p(1-p)$ is a probability and therefore must be between 0 and 1. So the series converges to

$$\frac{p^2}{1-[2p(1-p)]} = \frac{p^2}{1-2p+2p^2}.$$

If $p = .6$, $\dfrac{.6^2}{1-2(.6)+2(.6)^2} = .692 > .6$.

If $p > \dfrac{1}{2}$, $\dfrac{p^2}{1-2p(1-p)} > p$.

47.
$$\begin{aligned} 30 + 15 + \frac{15}{2} + \cdots &= \sum_{k=0}^{\infty} 30\left(\frac{1}{2}\right)^k \\ &= \frac{30}{1-\frac{1}{2}} \\ &= 60 \text{ miles} \end{aligned}$$

The bikes meet after 1 hour. In that time, a fly flying 60 mph will have traveled 60 miles.

49. $\displaystyle\sum_{k=0}^{\infty} 100{,}000\left(\frac{3}{4}\right)^k = \frac{100{,}000}{1-\frac{3}{4}} = 400{,}000;$
\$400,000

51. $d + de^{-r} + de^{-2r} + \cdots = \displaystyle\sum_{k=0}^{\infty} d(e^{-r})^k$, which is a geometric series with $a = d$ and $\left|e^{-r}\right| = e^{-r} < 1$ if $r > 0$. So

$$\sum_{k=0}^{\infty} d(e^{-r})^k = \frac{d}{1-e^{-r}}.$$

If $r = .1$, $\displaystyle\sum_{k=0}^{\infty} d(e^{-.1})^k = \frac{d}{1-e^{-.1}} = 2$,
so $d = 2(1 - .905) \approx .19$.

53. $\displaystyle\sum_{k=1}^{\infty} \frac{1}{k}$ and $\displaystyle\sum_{k=1}^{\infty} \frac{-1}{k}$

7.3 The Integral Test and Comparison Tests

1. $\displaystyle\sum_{k=1}^{\infty} \frac{4}{\sqrt[3]{k}}$ is a divergent p-series $\left(p = \frac{1}{3} < 1\right)$; diverges.

3. $\displaystyle\sum_{k=1}^{\infty} \frac{1}{k^{11/10}}$ is a convergent p-series $\left(p = \frac{11}{10} > 1\right)$; converges.

5. Using the Limit Comparison Test, let

$$a_k = \frac{k+1}{k^2+2k+3} \quad \text{and} \quad b_k = \frac{1}{k},$$

so

$$\begin{aligned} \lim_{k\to\infty} \frac{a_k}{b_k} &= \lim_{k\to\infty} \left(\frac{k+1}{k^2+2k+3}\right)\left(\frac{k}{1}\right) \\ &= \lim_{k\to\infty} \frac{k^2+k}{k^2+2k+3} \\ &= 1 > 0, \end{aligned}$$

and since $\displaystyle\sum_{k=1}^{\infty} \frac{1}{k}$ is the divergent harmonic series, $\displaystyle\sum_{k=0}^{\infty} \frac{k+1}{k^2+2k+3}$ diverges.

7. Using the Limit Comparison Test, let

$$a_k = \frac{4}{2+4k} \quad \text{and} \quad b_k = \frac{1}{k},$$

so

$$\begin{aligned} \lim_{k\to\infty} \frac{a_k}{b_k} &= \lim_{k\to\infty} \left(\frac{4}{2+4k}\right)\left(\frac{k}{1}\right) \\ &= \lim_{k\to\infty} \frac{4k}{2+4k} \\ &= 1 > 0, \end{aligned}$$

and since $\displaystyle\sum_{k=1}^{\infty} \frac{1}{k}$ is the divergent harmonic series, $\displaystyle\sum_{k=1}^{\infty} \frac{4}{2+4k}$ diverges.

9. Let $f(x) = \dfrac{2}{x \ln x}$. Then f is continuous and positive on $[2, \infty]$ and

$$f'(x) = \frac{-2(1+\ln x)}{x^2(\ln x)^2} < 0 \quad \text{for} \quad x \in [2, \infty),$$

so f is decreasing. So we can use the Integral Test,

$$\begin{aligned}\int_2^{\infty} \frac{2}{x \ln x}\,dx &= \lim_{R\to\infty} \int_2^{R} \frac{2}{x \ln x}\,dx \\ &= 2 \lim_{R\to\infty} [\ln(\ln x)]_2^R \\ &= 2 \lim_{R\to\infty} [\ln(\ln R) - \ln(\ln 2)] \\ &= \infty, \text{ so } \sum_{k=2}^{\infty} \frac{2}{k \ln k} \text{ diverges.}\end{aligned}$$

11. Using the Limit Comparison Test, let

$$a_k = \frac{2k}{k^3+1} \quad \text{and} \quad b_k = \frac{1}{k^2}.$$

Then

$$\begin{aligned}\lim_{k\to\infty} \frac{a_k}{b_k} &= \lim_{k\to\infty} \left(\frac{2k}{k^3+1}\right) \cdot \left(\frac{k^2}{1}\right) \\ &= \lim_{k\to\infty} \frac{2k^3}{k^3+1} \\ &= 2 > 0\end{aligned}$$

and $\sum_{k=1}^{\infty} \frac{1}{k^2}$ is a convergent p-series ($p = 2 > 1$), so $\sum_{k=1}^{\infty} \frac{2k}{k^3+1}$ converges.

13. Let $f(x) = \frac{e^{1/x}}{x^2}$. Then f is continuous and positive on $[1, \infty)$ and

$$f'(x) = \frac{-e^{1/x}(1+2x)}{x^4} < 0 \quad \text{for all} \quad x \in [1, \infty),$$

so f is decreasing. Therefore, we can use the Integral Test.

$$\begin{aligned}\int_1^{\infty} \frac{e^{1/x}}{x^2}\,dx &= \lim_{R\to\infty} \int_1^{R} \frac{e^{1/x}}{x^2}\,dx \\ &= \lim_{R\to\infty} \left[-e^{1/x}\right]_1^R \\ &= \lim_{R\to\infty} \left[e - e^{1/R}\right] \\ &= e - 1.\end{aligned}$$

So the series $\sum_{k=1}^{\infty} \frac{e^{1/k}}{k^2}$ converges.

15. Let $f(x) = \frac{e^{-\sqrt{x}}}{\sqrt{x}}$. Then f is continuous and positive on $[1, \infty)$ and

$$f'(x) = \frac{-(\sqrt{x}-1)}{2x^{3/2}e^{\sqrt{x}}} < 0 \quad \text{for} \quad x \in [1, \infty).$$

So f is decreasing. Therefore, we can use the Integral Test.

$$\begin{aligned}\int_1^{\infty} \frac{e^{-\sqrt{x}}}{\sqrt{x}}\,dx &= \lim_{R\to\infty} \int_1^{R} \frac{e^{-\sqrt{x}}}{\sqrt{x}}\,dx \\ &= \lim_{R\to\infty} \left[-2e^{-\sqrt{x}}\right]_1^R \\ &= \lim_{R\to\infty} \left[\frac{2}{e} - \frac{2}{e^{\sqrt{R}}}\right] = \frac{2}{e}.\end{aligned}$$

So $\sum_{k=1}^{\infty} \frac{e^{\sqrt{k}}}{\sqrt{k}}$ converges.

17. Using the Limit Comparison Test, let

$$a_k = \frac{3k}{k^{3/2}+2} \quad \text{and} \quad b_k = \frac{1}{k^{1/2}}.$$

Then

$$\begin{aligned}\lim_{k\to\infty} \frac{a_k}{b_k} &= \lim_{k\to\infty} \left(\frac{3k}{k^{3/2}+2}\right) \left(\frac{k^{1/2}}{1}\right) \\ &= \lim_{k\to\infty} \frac{3k^{3/2}}{k^{3/2}+2} \\ &= 3 > 0\end{aligned}$$

and $\sum_{k=1}^{\infty} \frac{1}{k^{1/2}}$ is a divergent p-series $\left(p = \frac{1}{2} < 1\right)$, so $\sum_{k=1}^{\infty} \frac{3k}{k^{3/2}+2}$ diverges.

19. Using the Limit Comparison Test, let

$$a_k = \frac{2}{\sqrt{k^2+4}} \quad \text{and} \quad b_k = \frac{1}{k}.$$

Then

$$\begin{aligned}\lim_{k\to\infty} \frac{a_k}{b_k} &= \lim_{k\to\infty} \left(\frac{2}{\sqrt{k^2+4}}\right) \left(\frac{k}{1}\right) \\ &= \lim_{k\to\infty} \frac{2k}{\sqrt{k^2+4}} \\ &= 2 > 0\end{aligned}$$

and $\sum_{k=1}^{\infty} \frac{1}{k}$ is the divergent harmonic series,

so $\sum_{k=0}^{\infty} \frac{2}{\sqrt{k^2+4}}$ diverges.

21. Using the Limit Comparison Test, let

$$a_k = \frac{k^2+1}{\sqrt{k^5+1}} \quad \text{and} \quad b_k = \frac{1}{k^{1/2}}.$$

Then

$$\begin{aligned}\lim_{k\to\infty} \frac{a_k}{b_k} &= \lim_{k\to\infty} \left(\frac{k^2+1}{\sqrt{k^5+1}}\right)\left(\frac{k^{1/2}}{1}\right) \\ &= \lim_{k\to\infty} \frac{k^{5/2}+k^{1/2}}{\sqrt{k^5+1}} \\ &= 1 > 0\end{aligned}$$

and $\sum_{k=1}^{\infty} \frac{1}{k^{1/2}}$ is a divergent p-series $\left(p = \frac{1}{2} < 1\right)$,

so $\sum_{k=0}^{\infty} \frac{k^2+1}{\sqrt{k^5+1}}$ diverges.

23. Let $f(x) = \frac{\tan^{-1} x}{1+x^2}$, which is continuous and positive on $[1, \infty)$, and

$$f'(x) = \frac{1-2x\tan^{-1} x}{(1+x^2)^2} < 0 \quad \text{for} \quad x \in [1, \infty),$$

so f is decreasing. So we can use the Integral Test.

$$\begin{aligned}\int_1^{\infty} \frac{\tan^{-1} x}{1+x^2}\, dx &= \lim_{R\to\infty} \int_1^R \frac{\tan^{-1} x}{1+x^2}\, dx \\ &= \lim_{R\to\infty} \left[\frac{1}{2}(\tan^{-1} x)^2\right]_1^R \\ &= \lim_{R\to\infty} \left[\frac{1}{2}(\tan^{-1} R)^2 - \frac{1}{2}(\tan^{-1} 1)^2\right] \\ &= \frac{1}{2}\left(\frac{\pi}{2}\right)^2 - \frac{1}{2}\left(\frac{\pi}{4}\right)^2 \\ &= \frac{3\pi^2}{32},\end{aligned}$$

so $\sum_{k=1}^{\infty} \frac{\tan^{-1} k}{1+k^2}$ converges.

25. Since $\left|\cos^2 k\right| \le 1$, $\frac{1}{\cos^2 k} > 1$, so by the kth-Term Test for Divergence, $\sum_{k=1}^{\infty} \frac{1}{\cos^2 k}$ diverges.

27. $-1 \le \sin k \le 1 \Rightarrow 0 < \frac{\sin k + 2}{k^2} \le \frac{3}{k^2}$, and $\sum_{k=1}^{\infty} \frac{1}{k^2}$ is a convergent p-series ($p = 2 > 1$), so by the Comparison Test, $\sum_{k=1}^{\infty} \frac{\sin k + 2}{k^2}$ converges.

29. Let $f(x) = \frac{\ln x}{x}$, which is continuous and positive on $[2, \infty)$, and $f'(x) = \frac{1-\ln x}{x^2} < 0$ for $x \in [2, \infty)$, so f is decreasing. Therefore, we can use the Integral Test.

$$\begin{aligned}\int_2^{\infty} \frac{\ln x}{x}\, dx &= \lim_{R\to\infty} \int_2^R \frac{\ln x}{x}\, dx \\ &= \lim_{R\to\infty} \left[\frac{(\ln x)^2}{2}\right]_2^R \\ &= \lim_{R\to\infty} \left[\frac{(\ln R)^2}{2} - \frac{(\ln 2)^2}{2}\right] \\ &= \infty,\end{aligned}$$

so $\sum_{k=2}^{\infty} \frac{\ln k}{k}$ diverges.

31. Using the Limit Comparison Test, let

$$a_k = \frac{k^4+2k-1}{k^5+3k^2+1} \quad \text{and} \quad b_k = \frac{1}{k}.$$

Then

$$\begin{aligned}\lim_{k\to\infty} \frac{a_k}{b_k} &= \lim_{k\to\infty} \left(\frac{k^4+2k-1}{k^5+3k^2+1}\right)\left(\frac{k}{1}\right) \\ &= \lim_{k\to\infty} \frac{k^5+2k^2-k}{k^5+3k^2+1} \\ &= 1 > 0.\end{aligned}$$

Since $\sum_{k=1}^{\infty} \frac{1}{k}$ is the divergent harmonic series, $\sum_{k=1}^{\infty} \frac{k^4+2k-1}{k^5+3k^2+1}$ diverges.

33. $\lim_{k\to\infty} \frac{k+1}{k+2} = 1 \neq 0$, so by the kth-Term Test for Divergence, $\sum_{k=1}^{\infty} \frac{k+1}{k+2}$ diverges.

35. Using the Limit Comparison Test, let

$$a_k = \frac{k+1}{k^3+2} \quad \text{and} \quad b_k = \frac{1}{k^2}.$$

Then

$$\begin{aligned}\lim_{k\to\infty} \frac{a_k}{b_k} &= \lim_{k\to\infty} \left(\frac{k+1}{k^3+2}\right)\left(\frac{k^2}{1}\right) \\ &= \lim_{k\to\infty} \frac{k^3+k^2}{k^3+2} \\ &= 1 > 0\end{aligned}$$

and $\sum_{k=1}^{\infty} \frac{1}{k^2}$ is a convergent p-series ($p = 2 > 1$), so $\sum_{k=1}^{\infty} \frac{k+1}{k^3+2}$ converges.

37. We are only concerned with what happens to the terms as $k \to \infty$, so the first few terms don't matter.

39. Since $\lim_{k\to\infty} \frac{a_k}{b_k} = 0$, there exists N, so that for all $k > N$, $\left|\frac{a_k}{b_k} - 0\right| < 1$. Since a_k and b_k are positive, $\frac{a_k}{b_k} < 1$, so $a_k < b_k$. Thus, since $\sum_{k=1}^{\infty} b_k$ converges, $\sum_{k=1}^{\infty} a_k$ converges by the Comparison Test.

41. If $p \leq 1$, then the series diverges because $\frac{1}{k(\ln k)^p} \geq \frac{1}{k \ln k}$ (at least for $k > 2$) and $\sum_{k=2}^{\infty} \frac{1}{k \ln k}$ diverges (from Exercise 9). If $p > 1$, then let $f(x) = \frac{1}{x(\ln x)^p}$. f is continuous and positive on $[2, \infty)$, and

$$f'(x) = \frac{-(\ln(x))^{-p-1}(p + \ln x)}{x^2} < 0,$$

so f is decreasing. Thus, we can use the Integral Test.

$$\begin{aligned}&\int_2^{\infty} \frac{1}{x(\ln x)^p}\, dx \\ &= \lim_{R\to\infty} \int_2^R \frac{1}{x(\ln x)^p}\, dx \\ &= \lim_{R\to\infty} \left[\frac{1}{(1-p)(\ln x)^{p-1}}\right]_2^R \\ &= \lim_{R\to\infty} \left[\frac{1}{(1-p)(\ln R)^{p-1}} - \frac{1}{(1-p)(\ln 2)^{p-1}}\right] \\ &= -\frac{1}{(1-p)(\ln 2)^{p-1}},\end{aligned}$$

so $\sum_{k=2}^{\infty} \frac{1}{k(\ln k)^p}$ converges when $p > 1$.

43. If $p \leq 1$, then $k^p \leq k$ for all $k \geq 1$, so $\frac{1}{k^p} \geq \frac{1}{k}$ for all $k \geq 1$, and $\frac{\ln k}{k^p} \geq \frac{\ln k}{k} \geq \frac{1}{k}$ for all $k > 2$, and $\sum_{k=2}^{\infty} \frac{1}{k}$ diverges.
So by the Comparison Test, $\sum_{k=2}^{\infty} \frac{\ln k}{k^p}$ diverges. If $p > 1$, then let $f(x) = \frac{\ln x}{x^p}$. Then f is continuous and positive on $[2, \infty)$ and

$$f'(x) = \frac{k^{p-1}(1 - p \ln k)}{k^2 p} < 0$$

for $k > 2$, so f is decreasing. Thus, we can use the Integral Test.

$$\begin{aligned}\int_2^{\infty} \frac{\ln x}{x^p}\, dx &= \lim_{R\to\infty} \int_2^R \frac{\ln x}{x^p}\, dx \\ &= \lim_{R\to\infty} \left[\frac{\ln x}{(1-p)x^{p-1}} - \frac{1}{(1-p)^2 x^{p-1}}\right]_2^R \\ &= \lim_{R\to\infty} \left[\frac{\ln R}{(1-p)R^{p-1}} - \frac{1}{(1-p)^2 R^{p-1}}\right] \\ &\quad - \frac{\ln 2}{(1-p)2^{p-1}} + \frac{1}{(1-p)^2 2^{p-1}} \\ &= \frac{1}{(1-p)^2 2^{p-1}} - \frac{\ln 2}{(1-p)2^{p-1}}\end{aligned}$$

because

$$\lim_{R\to\infty} \frac{\ln R}{(1-p)R^{p-1}} = \lim_{R\to\infty} \frac{\frac{1}{R}}{-(1-p)^2 R^{p-2}}$$
$$= \lim_{R\to\infty} \frac{-1}{(1-p)^2 R^{p-1}} = 0$$

by l'Hopital's Rule and because $p > 1$, and

$$\lim_{R\to\infty} \frac{1}{(1-p)^2 R^{p-1}} = 0$$

because $p > 1$. Thus by the Integral Test, $\sum_{k=R}^{\infty} \frac{\ln k}{k^p}$ converges when $p > 1$.

45. $\frac{1}{3 \cdot 100^3} = 3.33 \times 10^{-7}$

47. $\frac{6}{7 \cdot 50^7} = 1.097 \times 10^{-12}$

49. $< \frac{1}{2}e^{-40^2}$. To estimate $\left(\frac{1}{2}e^{-40^2}\right)$, take the logarithm $\log\left(\frac{1}{2}e^{-40^2}\right) \approx -695.2$, so the error is less than 10^{-695}.

51. We need $\frac{3}{3n^2} < 10^{-6}$, so $n > 100$ will do.

53. We need $\frac{1}{2}e^{-n^2} < 10^{-6}$. Taking the natural logarithm, we need $n > \sqrt{\ln 500000}$ or $n \geq 4$.

55. **(a)** can't tell
(b) converge
(c) converges
(d) can't tell

57.

$$1 + \frac{1}{3} + \frac{1}{5} + \frac{1}{7} + \cdots = \sum_{k=0}^{\infty} \frac{1}{2k+1}.$$

Using the Limit Comparison Test, let

$$a_k = \frac{1}{2k+1} \quad \text{and} \quad b_k = \frac{1}{k}.$$

Then

$$\lim_{k\to\infty} \frac{a_k}{b_k} = \lim_{k\to\infty} \left(\frac{1}{2k+1}\right)\left(\frac{k}{1}\right)$$
$$= \lim_{k\to\infty} \frac{k}{2k+1}$$
$$= \frac{1}{2} > 0,$$

and since $\sum_{k=1}^{\infty} \frac{1}{k}$ is the divergent harmonic series, then $\sum_{k=1}^{\infty} \frac{1}{2k+1}$ diverges.

59. If $x > 1$, $\zeta(x) = \sum_{k=1}^{\infty} \frac{1}{k^x}$ is a convergent p-series. If $x \leq 1$, $\zeta(x)$ is a divergent p-series.

61.

$$\sum_{k=1}^{25} \frac{1}{k^4} = 1 + \frac{1}{2^4} + \frac{1}{3^4} + \frac{1}{4^4} + \frac{1}{5^4}$$
$$+ \cdots + \frac{1}{25^4} \approx 1.0823$$
$$x = 4$$

63.

$$\sum_{k=1}^{7} \frac{1}{k^8} = 1 + \frac{1}{2^8} + \frac{1}{3^8} + \frac{1}{4^8}$$
$$+ \frac{1}{5^8} + \frac{1}{6^8} + \frac{1}{7^8} \approx 1.0041$$
$$x = 8$$

65. $\int_1^{\infty} \frac{x}{2^x}\,dx =$

$$\lim_{R\to\infty} \int_1^R \frac{x}{2^x}\,dx = \lim_{R\to\infty} \frac{1}{2^x}\left(-\frac{1}{\ln^2 2} - \frac{x}{\ln 2}\right)\Bigg|_2^R$$

This integral converges since $\lim_{R\to\infty} \frac{R}{2^R} = 0$ by l'Hopital's rule.

$$\sum_{k=1}^{20} k\left(\frac{1}{2}\right)^k \approx 2.0.$$

67. $$\sum_{k=1}^{\infty} \frac{9k}{10^k} = 9\sum_{k=1}^{\infty} k\frac{1}{10^k} =$$

$$9\left(\frac{1}{10} + \frac{1}{10^2} + \frac{1}{10^2} + \frac{1}{10^3} + \frac{1}{10^3} + \frac{1}{10^3} + \frac{1}{10^4} + \frac{1}{10^4} + \frac{1}{10^4} + \frac{1}{10^4} + \frac{1}{10^5} + \frac{1}{10^5} + \frac{1}{10^5} + \frac{1}{10^5} + \frac{1}{10^5}\cdots\right)$$

$$= 9\left(\frac{1}{10} + \frac{1}{10^2} + \frac{1}{10^3} + \frac{1}{10^4} + \frac{1}{10^5} + \cdots + \frac{1}{10^2} + \frac{1}{10^3} + \frac{1}{10^4} + \frac{1}{10^5} + \cdots + \frac{1}{10^3} + \frac{1}{10^4} + \frac{1}{10^5}\cdots\right)$$

$$= 9\left(\sum_{k=1}^{\infty} \frac{1}{10^k} + \sum_{k=2}^{\infty} \frac{1}{10^k} + \sum_{k=3}^{\infty} \frac{1}{10^k} + \cdots\right)$$

Each of these sums is a geometric series with $r = 1/10$, so we get

$$9\left(\frac{1/10}{1-\frac{1}{10}} + \frac{1/10^2}{1-\frac{1}{10}} + \frac{1/10^3}{1-\frac{1}{10}} + \cdots\right)$$
$$= \frac{9}{9/10}\sum_{k=1}^{\infty}\frac{1}{10^k} = \left(\frac{9}{9/10}\right)\left(\frac{1/10}{1-\frac{1}{10}}\right)$$
$$= \left(\frac{9}{9/10}\right)\left(\frac{1/10}{9/10}\right) = \frac{10}{9}.$$

69. $1 + \frac{10}{9} + \frac{10}{8} + \cdots + \frac{10}{1} \approx 29.29$

71. For n cards in the set, you will need an average of

$$\frac{n}{n-1} + \frac{n}{n-2} + \cdots \frac{n}{1}$$

attempts. The ratio is $\sum_{k=1}^{n-1} \frac{1}{k}$.

7.4 Alternating Series

1.
$$\lim_{k\to\infty} a_k = \lim_{k\to\infty} \frac{3}{k} = 0, \text{ and}$$
$$0 < a_{k+1} = \frac{3}{k+1} \le \frac{3}{k} = a_k$$

for all $k \ge 1$, so by the Alternating Series Test,

$$\sum_{k=1}^{\infty} (-1)^{k+1}\frac{3}{k}$$

converges.

3.
$$\lim_{k\to\infty} a_k = \lim_{k\to\infty} \frac{2}{k^2} = 0 \text{ and}$$
$$\frac{a_{k+1}}{a_k} = \frac{2}{(k+1)^2}\cdot\frac{k^2}{2} = \frac{2k^2}{2(k^2+2k+1)} < 1,$$

so $a_{k+1} \le a_k$ for all $k \ge 1$. Thus by the Alternating Series Test,

$$\sum_{k=1}^{\infty} (-1)^k \frac{2}{k^2}$$

converges.

5.
$$\lim_{k\to\infty} a_k = \lim_{k\to\infty} \frac{k^2}{k+1} = \infty,$$

so by the kth-Term Test for Divergence,

$$\sum_{k=1}^{\infty} (-1)^{k+1}\frac{k^2}{k+1}$$

diverges.

7.
$$\lim_{k\to\infty} a_k = \lim_{k\to\infty} \frac{k}{k^2+2} = 0 \text{ and}$$
$$\frac{a_{k+1}}{a_k} = \frac{k+1}{(k+1)^2+2}\cdot\frac{k^2+2}{k}$$
$$= \frac{k^3+k^2+2k+2}{k^3+2k^2+3k} \le 1$$

for all $k \ge 1$, so $a_{k+1} < a_k$ for all $k \ge 1$. Thus by the Alternating Series Test,

$$\sum_{k=1}^{\infty} (-1)^k \frac{k}{k^2+2}$$

converges.

9. $$\lim_{k\to\infty} a_k = \lim_{k\to\infty} \frac{k}{2^k} = \lim_{k\to\infty} \frac{1}{2^k \ln 2} = 0$$

(by l'Hopital's Rule) and

$$\frac{a_{k+1}}{a_k} = \frac{k+1}{2^{k+1}} \cdot \frac{2^k}{k} = \frac{k+1}{2k} 1 \text{ for all } k \geq 1,$$

so $a_{k+1} \leq a_k$ for all $k \geq 1$. Thus by the Alternating Series Test,

$$\sum_{k=1}^{\infty} (-1)^{k+1} \frac{k}{2^k}$$

converges.

11. $$\begin{aligned} \lim_{k\to\infty} a_k &= \lim_{k\to\infty} \frac{4^k}{k^2} \\ &= \lim_{k\to\infty} \frac{4^k \ln 4}{2k} \\ &= \lim_{k\to\infty} \frac{4^k (\ln 4)^2}{2} \\ &= \infty \end{aligned}$$

(by l'Hopital's Rule), so by the kth-Term Test for Divergence,

$$\sum_{k=1}^{\infty} (-1)^k \frac{4^k}{k^2}$$

diverges.

13. $$\lim_{k\to\infty} a_k = \lim_{k\to\infty} \frac{2k}{k+1} = 2,$$

so by the kth-Term Test for Divergence,

$$\sum_{k=1}^{\infty} \frac{2k}{k+1}$$

diverges.

15. $$\lim_{k\to\infty} a_k = \lim_{k\to\infty} \frac{3}{\sqrt{k+1}} = 0 \text{ and}$$

$$\frac{a_{k+1}}{a_k} = \frac{3}{\sqrt{k+2}} \cdot \frac{\sqrt{k+1}}{3} = \frac{\sqrt{k+1}}{\sqrt{k+2}} < 1 \text{ for all } k \geq 1,$$

so $a_{k+1} < a_k$ for all $k \geq 1$. Thus by the Alternating Series Test,

$$\sum_{k=1}^{\infty} (-1)^k \frac{3}{\sqrt{k+1}}$$

converges.

17. $$\lim_{k\to\infty} a_k = \lim_{k\to\infty} \frac{2}{k!} = 0 \text{ and}$$

$$\frac{a_{k+1}}{a_k} = \frac{2}{(k+1)!} \cdot \frac{k!}{2} = \frac{1}{k+1} < 1 \text{ for all } k \geq 1,$$

so $a_{k+1} < a_k$ for all $k \geq 1$. Thus by the Alternating Series Test,

$$\sum_{k=1}^{\infty} (-1)^{k+1} \frac{2}{k!}$$

converges.

19. $$a_k = \frac{k!}{2^k} = \frac{1 \cdot 2 \cdot 3 \ldots k}{2 \cdot 2 \cdot 2 \ldots 2} \geq \frac{1}{2} \cdot \frac{k}{2} = \frac{k}{4} \text{ and}$$

$$\lim_{k\to\infty} = k = \infty,$$

so by the kth-Term Test for Divergence,

$$\sum_{k=1}^{\infty} (-1)^k \frac{k!}{2^k}$$

diverges.

21. $$\lim_{k\to\infty} a_k = \lim_{k\to\infty} 2e^{-k} = 0 \text{ and}$$

$$\frac{a_{k+1}}{a_k} = \frac{2e^{-(k+1)}}{2e^{-k}} = \frac{2e^{-1}}{2} = \frac{1}{e} < 1 \text{ for all } k \geq 0,$$

so $a_{k+1} < a_k$ for all $k \geq 0$. Thus by the Alternating Series Test,

$$\sum_{k=0}^{\infty} (-1)^{k+1} 2e^{-k}$$

converges.

23. $\lim_{k\to\infty} a_k = \lim_{k\to\infty} \ln k = \infty$, so by the kth-Term Test for Divergence,

$$\sum_{k=2}^{\infty} (-1)^k \ln k$$

diverges.

25. $$\lim_{k\to\infty} a_k = \lim_{k\to\infty} \frac{1}{2^k} = 0 \text{ and}$$

$$\frac{a_{k+1}}{a_k} = \frac{1}{2^{k+1}} \cdot \frac{2^k}{1} = \frac{1}{2} < 1 \text{ for all } k \geq 0,$$

so $a_{k+1} < a_k$ for all $k \geq 0$. Thus by the Alternating Series Test,

$$\sum_{k=0}^{\infty} (-1)^{k+1} \frac{1}{2^k}$$

converges.

27. $|S - S_k| \leq a_{k+1} = \dfrac{4}{(k+1)^3} \leq .01,$

so $400 \leq (k+1)^3$, then $\sqrt[3]{400} \leq k+1$,

so $k \geq \sqrt[3]{400} - 1 \approx 6.37$. Thus $k \geq 7$.

So $S \approx S_7 = 3.61$.

29. $|S - S_k| \leq a_{k+1} = \dfrac{k+1}{2^{k+1}} \leq .01$

If $k = 9$, $a_{10} = \dfrac{10}{2^{10}} \approx .00977 < .01$.

So $S \approx S_9 = -.23$.

31. $|S - S_k| \leq a_{k+1} = \dfrac{3}{(k+1)!} \leq .01$

If $k = 5$, $a_6 = \dfrac{3}{6!} \approx .0042 < .01$.

So $S \approx S_5 = 1.10$.

33. $|S - S_k| \leq a_{k+1} = \dfrac{4}{(k+1)^4} \leq .01,$

so $400 \leq (k+1)^4$, then $\sqrt[4]{400} \leq k+1$,

so $k \geq \sqrt[4]{400} - 1 \approx 3.47$, so $k \geq 4$.

Thus $S \approx S_4 = 3.78$.

35. $|S - S_k| \leq a_{k+1} = \dfrac{2}{k+1} \leq .0001,$

so $k + 1 \geq 20{,}000$, then $k \geq 19{,}999$.

Thus $k = 20{,}000$.

37. $|S - S_k| \leq a_{k+1} = \dfrac{2^{k+1}}{(k+1)!} \leq .0001$

If $k = 10$, $a_{11} = \dfrac{2^{11}}{11!} \approx .00005 < .0001$.

So $k = 10$, which is 11 terms (because k starts at 0).

39. $|S - S_k| \leq a_{k+1} = \dfrac{(k+1)!}{(k+1)^{k+1}} \leq .0001$

If $k = 11$, $a_{12} = \dfrac{12!}{12^{12}} \approx .000054 < .0001$.

So $k = 11$, which is 11 terms.

41. If the derivative of a function $f(k) = a_k$ is negative, it means that the function is decreasing, i.e., each successive term is smaller than the one before. If

$$f(k) = a_k = \frac{k}{k^2+2},$$

then

$$f'(k) = \frac{-k^2+2}{(k^2+2)^2} < 0 \text{ for all } k \geq 2,$$

so a_k is decreasing.

43. The sum of the odd terms is

$$1 + \frac{1}{3} + \frac{1}{5} + \frac{1}{7} + \cdots,$$

which diverges to ∞ (see Exercise 57 in Section 7.3). The sum of the even terms is

$$-\sum_{k=1}^{\infty} \frac{1}{(2k)^2},$$

which is a convergent p-series. So the series diverges to ∞.

45. $$\frac{3}{4} - \frac{3}{4}\left(\frac{3}{4}\right) + \frac{3}{4}\left(\frac{3}{4}\right)^2 - \cdots = \sum_{k=0}^{\infty} \left(\frac{3}{4}\right)\left(-\frac{3}{4}\right)^k,$$

which is a geometric series with $a = \dfrac{3}{4}$ and $|r| = \dfrac{3}{4} < 1$. So

$$\sum_{k=0}^{\infty} \left(\frac{3}{4}\right)\left(-\frac{3}{4}\right)^k = \frac{\frac{3}{4}}{1+\frac{3}{4}} = \frac{3}{7}.$$

The person ends up $\dfrac{3}{7}$ of the distance from home.

47. The sum of the first million terms is 0.69314668; $\ln 2 \approx 0.69314718$.

7.5 Absolute Convergence and the Ratio Test

1. By the Ratio Test,

$$\lim_{k\to\infty}\left|\frac{a_{k+1}}{a_k}\right| = \lim_{k\to\infty}\frac{3}{(k+1)!}\cdot\frac{k!}{3} = \lim_{k\to\infty}\frac{1}{k+1} = 0 < 1,$$

so

$$\sum_{k=0}^{\infty}(-1)^k\frac{3}{k!}$$

converges absolutely.

3. By the Ratio Test,

$$\lim_{k\to\infty}\left|\frac{a_{k+1}}{a_k}\right| = \lim_{k\to\infty}\frac{2^{k+1}}{2^k} = \lim_{k\to\infty}2 = 2 > 1,$$

so

$$\sum_{k=0}^{\infty}(-1)^k 2^k$$

diverges. (Or use the kth-Term Test.)

5. By the Alternating Series Test,

$$\lim_{k\to\infty}a_k = \lim_{k\to\infty}\frac{k}{k^2+1} = 0 \text{ and}$$

$$\frac{a_{k+1}}{a_k} = \frac{k+1}{(k+1)^2+1}\cdot\frac{k^2+1}{k} = \frac{k^3+k^2+k+1}{k^3+2k^2+2k} < 1$$

for all $k \geq 1$, so $a_{k+1} < a_k$ for all $k \geq 1$, so the series converges.

But by the Limit Comparison Test, letting

$$a_k = \frac{k}{k^2+1} \quad\text{and}\quad b_k = \frac{1}{k},$$

$$\lim_{k\to\infty}\frac{a_k}{b_k} = \lim_{k\to\infty}\frac{k^2}{k^2+1} = 1 > 0,$$

and $\sum_{k=1}^{\infty}\frac{1}{k}$ is the divergent harmonic series. Therefore,

$$\sum_{k=1}^{\infty}|a_k| = \sum_{k=1}^{\infty}\frac{k}{k^2+1}$$

diverges. Thus,

$$\sum_{k=1}^{\infty}(-1)^{k+1}\frac{k}{k^2+1}$$

converges conditionally.

7. By the Ratio Test,

$$\lim_{k\to\infty}\left|\frac{a_{k+1}}{a_k}\right| = \lim_{k\to\infty}\frac{3^{k+1}}{(k+1)!}\cdot\frac{k!}{3^k} = \lim_{k\to\infty}\frac{3}{k+1} = 0 < 1,$$

so

$$\sum_{k=0}^{\infty}(-1)^k\frac{3^k}{k!}$$

converges absolutely.

9. $\lim_{k\to\infty}a_k = \lim_{k\to\infty}\frac{k}{2k+1} = \frac{1}{2}$, so by the kth-Term Test for Divergence,

$$\sum_{k=1}^{\infty}(-1)^{k+1}\frac{k}{2k+1}$$

diverges.

11. Using the Ratio Test,

$$\lim_{k\to\infty}\left|\frac{a_{k+1}}{a_k}\right| = \lim_{k\to\infty}\frac{(k+1)2^{k+1}}{3^{k+1}}\cdot\frac{3^k}{k2^k} = \lim_{k\to\infty}\frac{2(k+1)}{3k} = \frac{2}{3} < 1,$$

so

$$\sum_{k=1}^{\infty}(-1)^k\frac{k2^k}{3^k}$$

converges absolutely.

13. Using the Root Test,

$$\lim_{k\to\infty} \sqrt[k]{|a_k|} = \lim_{k\to\infty} \sqrt[k]{\left(\frac{4k}{5k+1}\right)^k} = \lim_{k\to\infty} \frac{4k}{5k+1} = \frac{4}{5} < 1,$$

so

$$\sum_{k=1}^{\infty} \left(\frac{4k}{5k+1}\right)^k$$

converges absolutely.

15. $\sum_{k=1}^{\infty} \frac{1}{k}$ is the divergent harmonic series; so

$$\sum_{k=1}^{\infty} \frac{-2}{k}$$

diverges.

17. Using the Alternating Series Test,

$$\lim_{k\to\infty} a_k = \lim_{k\to\infty} \frac{\sqrt{k}}{k+1} = 0 \text{ and}$$

$$\frac{a_{k+1}}{a_k} = \frac{\sqrt{k+1}}{k+2} \cdot \frac{k+1}{\sqrt{k}} = \frac{(k+1)^{3/2}}{k^{3/2} + 2k^{1/2}} < 1 \text{ for all } k \ge 1,$$

so $a_{k+1} < a_k$ for all $k \ge 1$ so the series converges. But by the Limit Comparison Test, letting

$$a_k = \frac{\sqrt{k}}{k+1} \quad \text{and} \quad b_k = \frac{1}{k^{1/2}},$$

$$\lim_{k\to\infty} \frac{a_k}{b_k} = \lim_{k\to\infty} \frac{\sqrt{k}}{k+1} \cdot \frac{k^{1/2}}{1} = \lim_{k\to\infty} \frac{k}{k+1} = 1 > 0,$$

and $\sum_{k=1}^{\infty} \frac{1}{k^{1/2}}$ is a divergent p-series $\left(p = \frac{1}{2} < 1\right)$. Therefore,

$$\sum_{k=1}^{\infty} |a_k| = \sum_{k=1}^{\infty} \frac{\sqrt{k}}{k+1}$$

diverges. So

$$\sum_{k=1}^{\infty} (-1)^{k+1} \frac{\sqrt{k}}{k+1}$$

converges conditionally.

19. Using the Ratio Test,

$$\lim_{k\to\infty} \left|\frac{a_{k+1}}{a_k}\right| = \lim_{k\to\infty} \frac{(k+1)^2}{e^{k+1}} \cdot \frac{e^k}{k^2} = \lim_{k\to\infty} \frac{(k+1)^2}{ek^2} = \frac{1}{e} < 1,$$

so $\sum_{k=1}^{\infty} \frac{k^2}{e^k}$ converges absolutely.

21. Using the Root Test,

$$\lim_{k\to\infty} \sqrt[k]{|a_k|} = \lim_{k\to\infty} \sqrt[k]{\left(\frac{e^3}{k^3}\right)^k} = \lim_{k\to\infty} \frac{e^3}{k^3} = 0 < 1,$$

so

$$\sum_{k=2}^{\infty} \frac{e^{3k}}{k^{3k}}$$

converges absolutely.

23. Since $|\sin k| \le 1$ for all k,

$$\left|\frac{\sin k}{k^2}\right| = \frac{|\sin k|}{k^2} \le \frac{1}{k^2},$$

and $\sum_{k=1}^{\infty} \frac{1}{k^2}$ is a convergent p-series ($p = 2 > 1$), so by the Comparison Test,

$$\sum_{k=1}^{\infty} \left|\frac{\sin k}{k^2}\right|$$

converges, and by Theorem 5.1,

$$\sum_{k=1}^{\infty} \frac{\sin k}{k^2}$$

converges absolutely.

25. Since $\cos k\pi = (-1)^k$ for all k,

$$\left|\frac{\cos k\pi}{k}\right| = \left|\frac{(-1)^k}{k}\right| = \frac{1}{k},$$

and $\sum_{k=1}^{\infty} \frac{1}{k}$ is the divergent harmonic series, so

$$\sum_{k=1}^{\infty} \left|\frac{\cos k\pi}{k}\right|$$

diverges by the Comparison Test. But, using the Alternating Series Test,

$$\lim_{k\to\infty} a_k = \lim_{k\to\infty} \frac{1}{k} = 0 \text{ and}$$

$$\frac{a_{k+1}}{a_k} = \frac{1}{k+1} \cdot \frac{k}{1} = \frac{k}{k+1} < 1 \text{ for all } k \geq 1,$$

so $a_{k+1} < a_k$ for all $k \geq 1$, so

$$\sum_{k=1}^{\infty} \frac{(-1)^k}{k}$$

converges. Thus

$$\sum_{k=1}^{\infty} \frac{(-1)^k}{k} = \sum_{k=1}^{\infty} \frac{\cos k\pi}{k}$$

converges conditionally.

27. Using the Alternating Series Test,

$$\lim_{k\to\infty} a_k = \lim_{k\to\infty} \frac{1}{\ln k} = 0 \text{ and}$$

$$\frac{a_{k+1}}{a_k} = \frac{1}{\ln(k+1)} \cdot \frac{\ln k}{1} = \frac{\ln k}{\ln(k+1)} < 1 \text{ for all } k \geq 2,$$

so $a_{k+1} < a_k$ for all $k \geq 2$. So

$$\sum_{k=2}^{\infty} \frac{(-1)^k}{\ln k}$$

converges. But by the Comparison Test, because $\ln k < k$ for all $k \geq 2$, $\frac{1}{\ln k} > \frac{1}{k}$ for all $k \geq 2$ and $\sum_{k=2}^{\infty} \frac{1}{k}$ is the divergent harmonic series. Therefore,

$$\sum_{k=2}^{\infty} |a_k| = \sum_{k=2}^{\infty} \frac{1}{\ln k}$$

diverges. Thus

$$\sum_{k=2}^{\infty} \frac{(-1)^k}{\ln k}$$

converges conditionally.

29. $\sum_{k=1}^{\infty} |a_k| = \sum_{k=1}^{\infty} \frac{1}{k^{3/2}}$, which is a convergent p-series $\left(p = \frac{3}{2} > 1\right)$, so by Theorem 5.1,

$$\sum_{k=1}^{\infty} \frac{(-1)^k}{k\sqrt{k}}$$

converges absolutely.

31. Consider $\sum_{k=1}^{\infty} \frac{1}{k^k}$. Using the Root Test,

$$\lim_{k\to\infty} \sqrt[k]{|a_k|} = \lim_{k\to\infty} \sqrt[k]{\left(\frac{1}{k}\right)^k} = \lim_{k\to\infty} \frac{1}{k} = 0 < 1,$$

so $\sum_{k=1}^{\infty} \frac{1}{k^k}$ converges absolutely, thus

$$3\sum_{k=1}^{\infty} \frac{1}{k^k} = \sum_{k=1}^{\infty} \frac{3}{k^k}$$

converges absolutely.

33. Using the Ratio Test,

$$\begin{aligned} \lim_{k\to\infty} \left|\frac{a_{k+1}}{a_k}\right| &= \lim_{k\to\infty} \frac{(k+1)!}{4^{k+1}} \cdot \frac{4^k}{k!} \\ &= \lim_{k\to\infty} \frac{k+1}{4} \\ &= \infty > 1, \end{aligned}$$

so

$$\sum_{k=1}^{\infty} (-1)^{k+1} \frac{k!}{4^k}$$

diverges.

35. Using the Ratio Test,

$$\lim_{k\to\infty}\left|\frac{a_{k+1}}{a_k}\right| = \lim_{k\to\infty}\frac{(k+1)^{10}}{(2k+2)!}\cdot\frac{2k!}{k^{10}}$$
$$= \lim_{k\to\infty}\frac{(k+1)^{10}}{k^{10}(2k+1)(2k+2)}$$
$$= \lim_{k\to\infty}\frac{(k+1)^{10}}{4k^{12}+6k^{11}+2k^{10}}$$
$$= 0 < 1,$$

so

$$\sum_{k=1}^{\infty}(-1)^{k+1}\frac{k^{10}}{(2k)!}$$

converges absolutely.

37. Using the Ratio Test,

$$\lim_{k\to\infty}\left|\frac{a_{k+1}}{a_k}\right| = \lim_{k\to\infty}\left|\frac{(-2)^{k+1}(k+2)}{5^{k+1}}\cdot\frac{5^k}{(2)^k(k+1)}\right|$$
$$= \lim_{k\to\infty}\frac{2(k+2)}{5(k+1)}$$
$$= \frac{2}{5} < 1,$$

so

$$\sum_{k=0}^{\infty}\frac{(-2)^k(k+1)}{5^k}$$

converges absolutely.

39.
$$\frac{\sqrt{8}}{9801}\sum_{k=0}^{0}\frac{(4k)!(1103+26390k)}{(k!)^4 396^{4k}}$$
$$= \frac{\sqrt{8}}{9801}(1103) \approx .318309878 \approx \frac{1}{\pi}$$

so $\pi \approx 3.14159273$

$$\frac{\sqrt{8}}{9801}\sum_{k=0}^{1}\frac{(4k)!(1103+26390k)}{(k!)^4 396^{4k}}$$
$$= \frac{\sqrt{8}}{9801}(1103) + \frac{\sqrt{8}}{9801}\cdot\frac{4!(27,493)}{396^4}$$
$$\approx .318309886183791 \approx \frac{1}{\pi}$$

For comparison, the value of $1/\pi$ to 15 places is 0.318309886183791, so two terms of the series give this value correct to 15 places.

41. Using the Ratio Test,

$$\lim_{k\to\infty}\left|\frac{a_{k+1}}{a^k}\right| = \lim_{k\to\infty}\frac{(k+1)!}{(k+1)^{k+1}}\cdot\frac{k^k}{k!}$$
$$\lim_{k\to\infty}\frac{k^k}{(k+1)^k} = \lim_{k\to\infty}\left(\frac{k}{1+k}\right)^k = \frac{1}{e} < 1,$$

so

$$\sum_{k=1}^{\infty}\frac{k!}{k^k}$$

converges absolutely.

7.6 Power Series

1.
$$f(x) = \frac{2}{1-x} = 2\left(\frac{1}{1-x}\right) = 2\sum_{k=0}^{\infty}x^k = \sum_{k=0}^{\infty}2x^k,$$

and this is a geometric series that converges when $|x| < 1$, so $-1 < x < 1$, thus the interval of convergence is $(-1, 1)$ and $r = 1$.

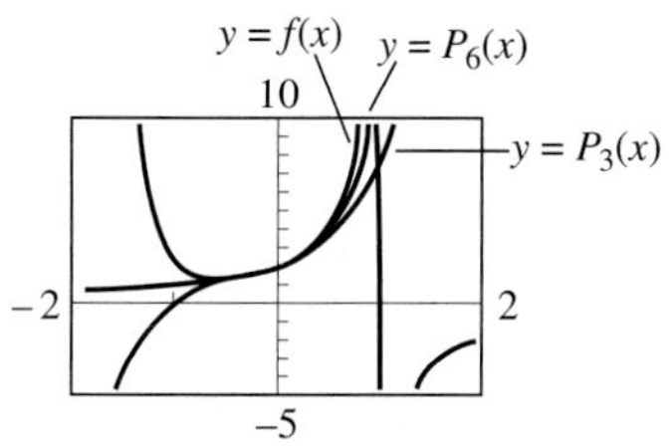

3.
$$f(x) = \frac{3}{1+x^2}$$
$$= 3\left(\frac{1}{1-(-x^2)}\right)$$
$$= 3\sum_{k=0}^{\infty}(-x^2)^k$$
$$= \sum_{k=0}^{\infty}(-1)^k 3x^{2k},$$

and this is a geometric series that converges when $|-x^2| < 1$, so $x^2 < 1$, then $|x| < 1$ or $-1 < x < 1$, so the interval of convergence is $(-1, 1)$ and $r = 1$.

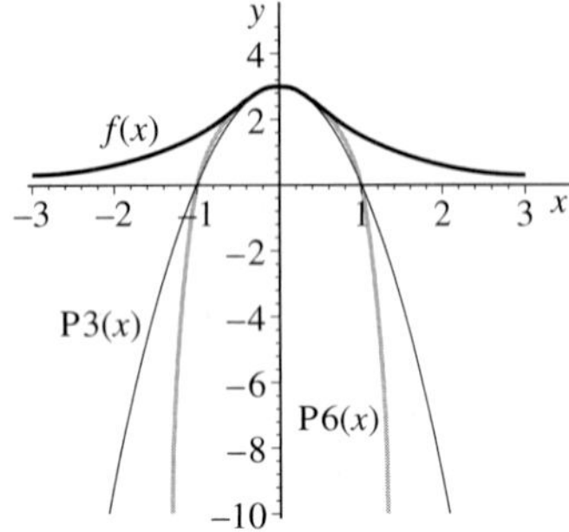

5.

$$\begin{aligned} f(x) &= \frac{2x}{1-x^3} \\ &= 2x\left(\frac{1}{1-x^3}\right) \\ &= 2x\sum_{k=0}^{\infty}(x^3)^k \\ &= \sum_{k=0}^{\infty} 2x^{3k+1}, \end{aligned}$$

and this is geometric series that converges when $x^3 < 1$, so $|x| < 1$, or $-1 < x < 1$, so the interval of convergence is $(-1, 1)$ and $r = 1$.

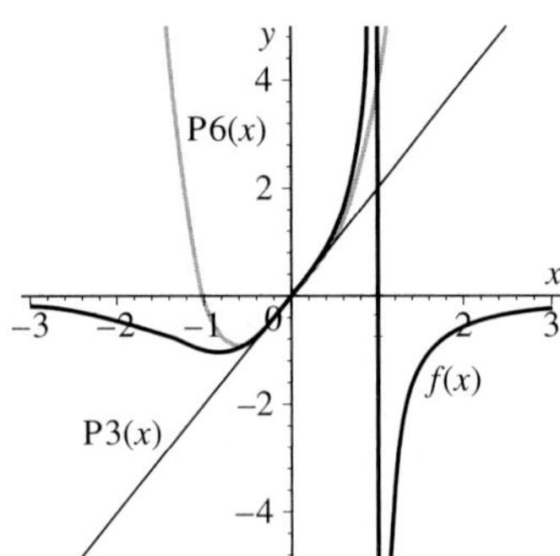

7.

$$\begin{aligned} f(x) &= \frac{4}{1+4x} \\ &= 4\left(\frac{1}{1-(4-x)}\right) \\ &= 4\sum_{k=0}^{\infty}(-4x)^k \\ &= \sum_{k=0}^{\infty}(-1)^k 4^{k+1} x^k, \end{aligned}$$

which is a geometric series that converges when $|-4x| < 1$ so $-1 < -4x < 1$ or $1 > 4x > -1$, so the interval of convergence is $\left(-\frac{1}{4}, \frac{1}{4}\right)$ and $r = \frac{1}{4}$.

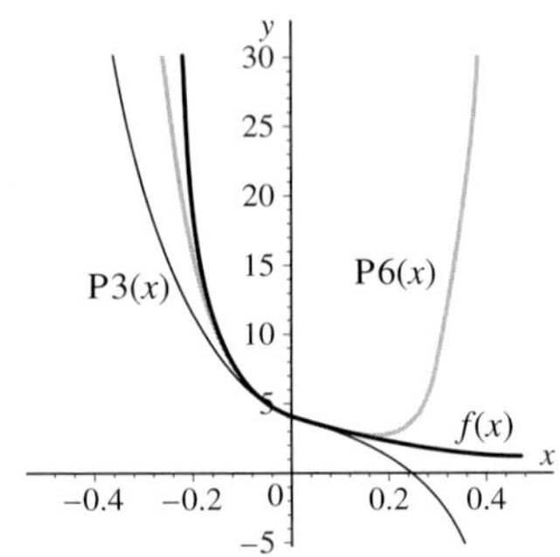

9.

$$\begin{aligned} f(x) &= \frac{2}{4+x} \\ &= \frac{\frac{1}{2}}{1+\frac{x}{4}} \\ &= \frac{1}{2}\left(\frac{1}{1-(-\frac{x}{4})}\right) \\ &= \frac{1}{2}\sum_{k=0}^{\infty}\left(-\frac{x}{4}\right)^k \\ &= \sum_{k=0}^{\infty}\frac{(-1)^k x^k}{2^{2k+1}}, \end{aligned}$$

which is a geometric series that converges when

$$\left|-\frac{x}{4}\right| < 1, \text{ so } -1 < \frac{-x}{4} < 1 \text{ or, } 1 > \frac{x}{4} > 1,$$

thus the interval of convergence is $(-4, 4)$ and $r = 4$.

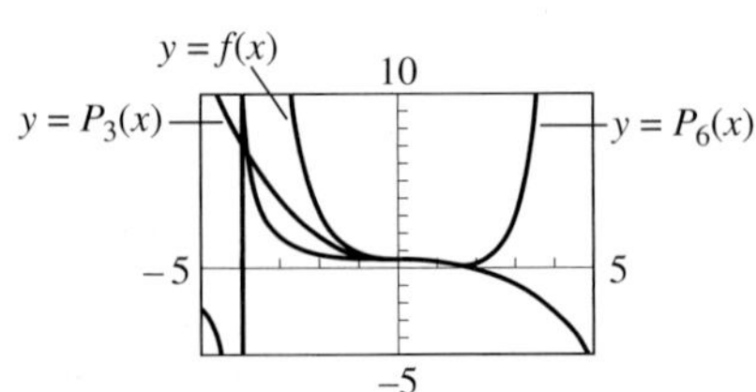

11. This is a geometric series

$$\sum_{k=0}^{\infty}(x+2)^k = \frac{1}{1-(x+2)} = \frac{-1}{1+x} = f(x),$$

which converges for $|x+2| < 1$, so $-1 < x+2 < 1$ or $-3 < x < -1$, thus the interval of convergence is $(-3, -1)$ and $r = 1$.

13. This is a geometric series

$$\sum_{k=0}^{\infty}(2x-1)^k = \frac{1}{1-(2x-1)} = \frac{1}{2-2x} = f(x),$$

which converges for $|2x-1| < 1$, so $-1 < 2x-1 < 1$ or $0 < x < 1$, thus the interval of convergence is $(0, 1)$ and $r = \frac{1}{2}$.

15. This is a geometric series

$$\begin{aligned}\sum_{k=0}^{\infty}(-1)^k\left(\frac{x}{2}\right)^k &= \sum_{k=0}^{\infty}\left(\frac{-x}{2}\right)^k \\ &= \frac{1}{1+\frac{x}{2}} \\ &= \frac{2}{2+x} \\ &= f(x),\end{aligned}$$

which converges for $\left|\frac{-x}{2}\right| < 1$, so $1 < \frac{-x}{2} < 1$ or $2 > x > -2$, thus the interval of convergence is $(-2, 2)$ and $r = 2$.

17. Using the Ratio Test,

$$\begin{aligned}\lim_{k\to\infty}\left|\frac{a_{k+1}}{a_k}\right| &= \lim_{k\to\infty}\left|\frac{2^{k+1}k!(x-2)^{k+1}}{2^k(k+1)!(x-2)^k}\right| \\ &= 2|x-2|\lim_{k\to\infty}\frac{1}{k+1} \\ &= 0 \text{ and } 0 < 1 \text{ for all } x,\end{aligned}$$

so the series converges absolutely for $x \in (-\infty, \infty)$. The interval of convergence is $(-\infty, \infty)$ and $r = \infty$.

19. Using the Ratio Test,

$$\begin{aligned}\lim_{k\to\infty}\left|\frac{a_{k+1}}{a_k}\right| &= \lim_{k\to\infty}\left|\frac{(k+1)4^k x^{k+1}}{k4^{k+1}x^k}\right| \\ &= \frac{|x|}{4}\lim_{k\to\infty}\frac{k+1}{k} \\ &= \frac{|x|}{4} \text{ and } \frac{|x|}{4} < 1 \text{ when } |x| < 4\end{aligned}$$

or $-4 < x < 4$. So the series converges absolutely for $x \in (-4, 4)$. When $x = 4$,

$$\sum_{k=0}^{\infty}\frac{k}{4^k}4^k = \sum_{k=0}^{\infty}k \quad \text{and} \quad \lim_{k\to\infty}k = \infty,$$

so the series diverges by the kth-Term Test for Divergence. When $x = -4$,

$$\sum_{k=0}^{\infty}\frac{k}{4^k}(-4)^k = \sum_{k=0}^{\infty}(-1)^k k,$$

which diverges by the kth-Term Test for Divergence. Thus, the interval of convergence is $(-4, 4)$ and $r = 4$.

21. Using the Ratio Test,

$$\begin{aligned}\lim_{k\to\infty}\left|\frac{a_{k+1}}{a_k}\right| &= \lim_{k\to\infty}\left|\frac{(-1)^{k+1}k3^k(x-1)^{k+1}}{(-1)^k(k+1)3^{k+1}(x-1)^k}\right| \\ &= \frac{|x-1|}{3}\lim_{k\to\infty}\frac{k}{k+1} \\ &= \frac{|x-1|}{3} \text{ and } \frac{|x-1|}{3} < 1\end{aligned}$$

when $-3 < x-1 < 3$ or $-2 < x < 4$, so the series converges absolutely for $x \in (-2, 4)$. When $x = -2$,

$$\sum_{k=0}^{\infty}\frac{(-1)^k}{k3^k}(-3)^k = \sum_{k=0}^{\infty}\frac{1}{k},$$

which is the divergent harmonic series. When $x = 4$,

$$\sum_{k=0}^{\infty}\frac{(-1)^k}{k3^k}3^k = \sum_{k=0}^{\infty}\frac{(-1)^k}{k},$$

which converges by the Alternating Series Test (Example 4.2). So the interval of convergence is $(-2, 4]$ and $r = 3$.

23. Using the Ratio Test,

$$\lim_{k\to\infty}\left|\frac{a_{k+1}}{a_k}\right| = \lim_{k\to\infty}\left|\frac{(k+1)!(x+1)^{k+1}}{k!(x+1)^k}\right|$$
$$= \lim_{k\to\infty}[(k+1)\,|x+1|]$$
$$= \begin{cases} 0 & \text{if } x=-1 \\ \infty & \text{if } x\neq -1, \end{cases}$$

so this series converges absolutely for $x=-1$ and $r=0$.

25. Using the Ratio Test,

$$\lim_{k\to\infty}\left|\frac{a_{k+1}}{a_k}\right| = \lim_{k\to\infty}\left|\frac{k(x-1)^{k+1}}{(k+1)(x-1)^k}\right|$$
$$= |x-1|\lim_{k\to\infty}\frac{k}{k+1}$$
$$= |x-1| \text{ and } |x-1|<1$$

when $-1<x-1<1$ or $0<x<2$, so the series converges absolutely for $x\in(0,2)$.

When $x=0$, $\sum_{k=0}^{\infty}\frac{(-1)^k}{k}$, which converges by the Alternating Series Test (Example 4.2). If $x=2$, $\sum_{k=1}^{\infty}\frac{1}{k}$, which is the divergent harmonic series. So the interval of convergence is $[0,2)$ and $r=1$.

27. Using the Ratio Test,

$$\lim_{k\to\infty}\left|\frac{a_{k+1}}{a_k}\right| = \lim_{k\to\infty}\left|\frac{(k+1)^2(x-3)^{k+1}}{k^2(x-3)^k}\right|$$
$$= |x-3|\lim_{k\to\infty}\frac{(k+1)^2}{k^2}$$
$$= |x-3|<1$$

when $-1<x-3<1$ or $2<x<4$, so the series converges absolutely for $x\in(2,4)$. If $x=2$, $\sum_{k=0}^{\infty}k^2(-1)^k$ and $\lim_{k\to\infty}k^2=\infty$, so the series diverges by the kth-Term Test for Divergence. If $x=4$, $\sum_{k=0}^{\infty}k^2$ and $\lim_{k\to\infty}k^2=\infty$, so the series diverges by the kth-Term Test for Divergence. Thus, the interval of convergence is $(2,4)$ and $r=1$.

29. Using the Ratio Test,

$$\lim_{k\to\infty}\left|\frac{a_{k+1}}{a_k}\right| = \lim_{k\to\infty}\left|\frac{(k+1)!(2k)!x^{k+1}}{k!(2k+2)!x^k}\right|$$
$$= |x|\lim_{k\to\infty}\frac{k+1}{(2k+2)(2k+1)}$$
$$= 0 \text{ and } 0<1$$

for all x, so the series converges for all $x\in(-\infty,\infty)$. Thus, the interval of convergence is $(-\infty,\infty)$ and $r=\infty$.

31. Using the Ratio Test,

$$\lim_{k\to\infty}\left|\frac{a_{k+1}}{a_k}\right| = \lim_{k\to\infty}\left|\frac{2^{k+1}k^2(x+2)^{k+1}}{2^k(k+1)^2(x+2)^k}\right|$$
$$= 2\,|x+2|\lim_{k\to\infty}\frac{k^2}{(k+1)^2}$$
$$= 2\,|x+2| \text{ and } 2\,|x+2|<1$$

when

$$-\frac{1}{2}<x+2<\frac{1}{2} \quad\text{or}\quad -\frac{5}{2}<x<-\frac{3}{2},$$

so the series converges absolutely for

$$x\in\left(-\frac{5}{2},-\frac{3}{2}\right).$$

If $x=-\frac{5}{2}$, then

$$\sum_{k=0}^{\infty}\frac{2^k}{k^2}\left(-\frac{1}{2}\right)^k=\sum_{k=0}^{\infty}\frac{(-1)^k}{k^2}$$

and

$$\lim_{k\to\infty}\frac{1}{k^2}=0$$

and

$$\frac{a_{k+1}}{a_k}=\frac{k^2}{(k+1)^2}<1 \text{ for all } k\geq 0,$$

so $a_{k+1}<a_k$ for all $k\geq 0$, so the series converges by the Alternating Series Test. If $x=-\frac{3}{2}$, then

$$\sum_{k=0}^{\infty}\frac{2^k}{k^2}\left(\frac{1}{2}\right)^k=\sum_{k=0}^{\infty}\frac{1}{k^2},$$

which is a convergent p-series ($p = 2 > 1$). So the interval of convergence is $\left[-\frac{5}{2}, -\frac{3}{2}\right]$ and $r = \frac{1}{2}$.

33. Using the Ratio Test,

$$\lim_{k\to\infty}\left|\frac{a_{k+1}}{a_k}\right| = \lim_{k\to\infty}\left|\frac{4^{k+1}k^{1/2}x^{k+1}}{4^k(k+1)^{1/2}x^k}\right|$$
$$= 4|x| \lim_{k\to\infty}\frac{k^{1/2}}{(k+1)^{1/2}}$$
$$= 4|x| \text{ and } 4|x| \le 1$$

when $-\frac{1}{4} < x < \frac{1}{4}$, so the series converges absolutely for $x \in \left(-\frac{1}{4}, \frac{1}{4}\right)$. If $x = -\frac{1}{4}$, then

$$\sum_{k=0}^{\infty}\frac{4^k}{\sqrt{k}}\left(-\frac{1}{4}\right)^k = \sum_{k=0}^{\infty}\frac{(-1)^k}{\sqrt{k}}$$

and

$$\lim_{k\to\infty}\frac{1}{\sqrt{k}} = 0$$

and

$$\frac{a_{k+1}}{a_k} = \frac{\sqrt{k}}{\sqrt{k+1}} < 1 \text{for all } k \ge 0,$$

so $a_{k+1} < a_k$ for all $k \ge 0$, therefore the series converges by the Alternating Series test. If

$$x = \frac{1}{4}, \sum_{k=0}^{\infty}\frac{4^k}{\sqrt{k}}\left(\frac{1}{4}\right)^k = \sum_{k=0}^{\infty}\frac{1}{\sqrt{k}},$$

which is a divergent p-series $\left(p = \frac{1}{2} < 1\right)$. Thus the interval of convergence is $\left[-\frac{1}{4}, \frac{1}{4}\right)$ and $r = \frac{1}{4}$.

35. From Exercise 3, $\frac{3}{1+x^2} = \sum_{k=0}^{\infty}(-1)^k 3x^{2k}$ with $r = 1$.
Integrating both sides gives

$$\int \frac{3}{1+x^2}\,dx = \sum_{k=0}^{\infty} 3(-1)^k \int x^{2k}\,dx,$$

$$3\tan^{-1}x = \sum_{k=0}^{\infty}\frac{3(-1)^k x^{2k+1}}{2k+1} + c.$$

Taking $x = 0$,

$$3\tan^{-1}0 = \sum_{k=0}^{\infty}\frac{3(-1)^k(0)^{2k+1}}{2k+1} + c = c \text{ so that}$$

$c = 3\tan^{-1}(0) = 0$. Thus,

$$3\tan^{-1}x = \sum_{k=0}^{\infty}\frac{3(-1)^k x^{2k+1}}{2k+1} \text{ with } r = 1.$$

37. From Exercise 4, $\frac{2}{1-x^2} = \sum_{k=0}^{\infty} 2x^{2k}$ with $r = 1$.
Taking the derivative of both sides gives

$$\frac{d}{dx}\left[\frac{2}{1-x^2}\right] = \sum_{k=0}^{\infty} 2\frac{d}{dx}\left[x^{2k}\right],$$

$$\frac{4x}{(1-x^2)^2} = \sum_{k=0}^{\infty} 2\cdot 2kx^{2k-1},$$

$$\frac{2x}{(1-x^2)^2} = \sum_{k=0}^{\infty} 2kx^{2k-1} \text{ with } r = 1.$$

39. From Exercise 6, $\frac{3x}{1+x^2} = \sum_{k=0}^{\infty}(-1)^k 3x^{2k+1}$ with $r = 1$.
Integrating both sides gives

$$\int \frac{3x}{1+x^2}\,dx = \sum_{k=0}^{\infty}(-1)^k 3\int x^{2k+1}\,dx,$$

$$\frac{3}{2}\ln(1+x^2) = \sum_{k=0}^{\infty}\frac{(-1)^k 3x^{2k+2}}{2k+2} + c,$$

$$\ln(1+x^2)\sum_{k=0}^{\infty}\frac{(-1)^k x^{2k+2}}{k+1} + c.$$

Taking $x = 0$,

$$\ln(1) = \sum_{k=0}^{\infty}\frac{(-1)^k(0)^{2k+2}}{k+1} + c = c \text{ so that}$$

$c = \ln(1) = 0$. Thus,

$$\ln(1+x^2) = \sum_{k=0}^{\infty}\frac{(-1)^k x^{2k+2}}{k+1} \text{ with } r = 1.$$

41. From Exercise 7, $\frac{4}{1+4x} = \sum_{k=0}^{\infty}(-1)^k 4^{k+1}x^k$ with $r = \frac{1}{4}$.

Taking the derivative of both sides gives

$$\frac{d}{dx}\left[\frac{4}{1+4x}\right]=\sum_{k=0}^{\infty}(-1)^k 4^{k+1}\frac{d}{dx}\left[x^k\right],$$

$$\frac{-16}{(1+4x)^2}=\sum_{k=0}^{\infty}(-1)^k 4^{k+1} k x^{k-1},$$

$$\frac{1}{(1+4x)^2}=\sum_{k=0}^{\infty}(-1)^{k+1} k(4x)^{k-1} \text{ with } r=\frac{1}{4}.$$

43. Since $|\cos(k^3x)|\le 1$ for all x,

$$\left|\frac{\cos(k^3x)}{k^2}\right|\le\frac{1}{k^2} \text{ for all } x,$$

and $\sum_{k=0}^{\infty}\frac{1}{k^2}$ is a convergent p-series, so by the Comparison Test, $\sum_{k=0}^{\infty}\frac{\cos(k^3x)}{k^2}$ converges absolutely for all x. So the interval of convergence is $(-\infty,\infty)$ and $r=\infty$. The series of derivatives is

$$\sum_{k=0}^{\infty}\frac{d}{dx}\left[\frac{\cos(k^3x)}{k^2}\right]=\sum_{k=0}^{\infty}(-k)\sin(k^3x),$$

and $\lim_{k\to\infty}(-k)\sin(k^3x)$ diverges if $x\ne 0$, while

$$\sum_{k=0}^{\infty}(-k)\sin(k^3(0))=\sum_{k=0}^{\infty}0=0,$$

thus the series converges absolutely if $x=0$.

45. Using the Ratio Test,

$$\lim_{k\to\infty}\left|\frac{a_{k+1}}{a_k}\right|=\lim_{k\to\infty}\left|\frac{e^{(k+1)x}}{e^{kx}}\right|=e^x\lim_{k\to\infty}1=e^x$$

and $e^x<1$ when $x<0$, so the series converges absolutely for all $x\in(-\infty,0)$. When $x=0$,

$$\sum_{x=0}^{\infty}e^0=\sum_{x=0}^{\infty}1,$$

which diverges by the kth-Term Test for Divergence because $\lim_{k\to\infty}1=1$. So the interval of convergence is $(-\infty,0)$. The series of derivatives is

$$\sum_{x=0}^{\infty}\frac{d}{dx}\left[e^{kx}\right]=\sum_{x=0}^{\infty}ke^{kx}.$$

Using the Ratio Test,

$$\begin{aligned}\lim_{k\to\infty}\left|\frac{a_{k+1}}{a_k}\right|&=\lim_{k\to\infty}\left|\frac{(k+1)e^{(k+1)x}}{ke^{kx}}\right|\\&=e^x\lim_{k\to\infty}\frac{k+1}{k}\\&=e^x \text{ and } e^x<1 \text{ when } x<0,\end{aligned}$$

so the series converges absolutely for all $x\in(-\infty,0)$. When $x=0$,

$$\sum_{k=0}^{\infty}ke^0=\sum_{k=0}^{\infty}k,$$

which diverges by the kth-Term Test for Divergence because $\lim_{k\to\infty}k=\infty$. So the interval of convergence is $(-\infty,0)$.

47. Using the Ratio Test,

$$\begin{aligned}\lim_{k\to\infty}\left|\frac{a_{k+1}}{a_k}\right|&=\lim_{k\to\infty}\left|\frac{(x-a)^{k+1}b^k}{b^{k+1}(x-a)^k}\right|\\&=\frac{|x-a|}{b}\lim_{k\to\infty}1\\&=\frac{|x-a|}{b} \text{ and } \frac{|x-a|}{b}<1\end{aligned}$$

when $-b<x-a<b$ or $a-b<x<a+b$. So the series converges absolutely for $x\in(a-b,a+b)$. If $x=a-b$,

$$\sum_{k=0}^{\infty}\frac{(a-b-a)^k}{b^k}=\sum_{k=0}^{\infty}\frac{(-1)^kb^k}{b^k}=\sum_{k=0}^{\infty}(-1)^k,$$

which diverges by the kth-Term Test for Divergence because $\lim_{k\to\infty}1=1$. If $x=a+b$,

$$\sum_{k=0}^{\infty}\frac{(a+b-a)^k}{b^k}=\sum_{k=0}^{\infty}\frac{b^k}{b^k}=\sum_{k=0}^{\infty}1$$

which diverges by the kth-Term Test for Divergence because $\lim_{k\to\infty}1=1$. So the interval of convergence is $(a-b,a+b)$ and $r=b$.

49. If the radius of convergence of $\sum_{k=0}^{\infty} a_k x^k$ is r, then $-r < x < r$. For any constant c, $-r - c < x - c < r - c$. Thus, the radius of convergences of $\sum_{k=0}^{\infty} a_k(x-c)^k$ is

$$\frac{(r-c)-(-r-c)}{2} = r.$$

51. $\dfrac{1}{1-x} = \sum_{k=0}^{\infty} x^k$

$$\frac{d}{dx}\left[\frac{1}{1-x}\right] = \frac{1}{(1-x)^2} = \sum_{k=0}^{\infty} \frac{d}{dx}[x^k] = \sum_{k=0}^{\infty} kx^{k-1}$$

$$\frac{x+1}{(1-x)^2} = \frac{x}{(1-x)^2} + \frac{1}{(1-x)^2}$$

$$= \sum_{k=0}^{\infty} kx^k + \sum_{k=0}^{\infty} kx^{k-1} = \sum_{k=0}^{\infty} (2k+1)x^k$$

$$= 1 + 3x + 5x^2 + 7x^3 + \cdots$$

Since $\sum_{k=0}^{\infty} x^k$ converges for

$$|x| < 1,\ \frac{d}{dx}\left[\sum_{k=0}^{\infty} x^k\right] \quad \text{and} \quad x\frac{x}{dx}\left[\sum_{k=0}^{\infty} k^x\right]$$

converge for $|x| < 1$. Hence, $\dfrac{x+1}{(1-x)^2}$ converges for $|x| < 1$, so $r = 1$. If $x = \dfrac{1}{1000}$, then

$$\frac{1{,}001{,}000}{998{,}000} = 1 + \frac{3}{1000} + \frac{5}{(1000)^2} + \frac{7}{(1000)^3} + \cdots$$

$1.003005007\ldots$

53. If $x = 1, \ldots + 1 + 1 + 1 + 1 + 1 + 1 + \cdots \neq 0$.

$\dfrac{1}{1-\frac{1}{x}} = \sum_{k=0}^{\infty}\left(\dfrac{1}{x}\right)^k$ is a geometric series that converges for $\dfrac{1}{|x|} < 1$ or $|x| > 1$.

$\dfrac{x}{1-x} = \sum_{k=0}^{\infty} x^{k+1}$ is a geometric series that converges for $|x| < 1$. Euler's mistake was that there are no x's for which both series converge.

55.

$$E(x) = kq\left(\frac{1}{(x-1)^2} - \frac{1}{(x+1)^2}\right)$$

$$= kq\frac{4x}{(x^2-1)^2}$$

Now the integral of $kq\dfrac{4x}{(x^2-1)^2}$ is $2kq\dfrac{1}{1-x^2}$. We will find the power series for $2kq\dfrac{1}{1-x^2}$ and then differentiate the result. We know $\dfrac{1}{1-x^2} = \sum_{k=0}^{\infty} x^j$. If we replace x by x^2, we get

$$2kq\frac{1}{1-x^2} = 2kq\sum_{j=0}^{\infty} x^{2j}.$$

Finally if we differentiate this last power series, we get

$$E(x) = 2kq\frac{d}{dx}\sum_{j=0}^{\infty} x^{2j} = 2kq\sum_{j=0}^{\infty} 2jx^{2j-1}$$

$$= 4kq\sum_{j=0}^{\infty} (j+1)x^{2j+1}.$$

7.7 Taylor Series

1.

$$f(x) = \cos x,\ f'(x) = -\sin x$$

$$f''(x) = -\cos x,\ f'''(x) = \sin x$$

$$f(0) = 1,\ f'(0) = 0$$

$$f''(0) = -1,\ f'''(0) = 0$$

$$\sum_{k=0}^{\infty} \frac{f^{(k)}(0)}{k!}(x-0)^k = 1 - \frac{1}{2}x^2 +$$

$$\frac{1}{4!}x^4 - \frac{1}{6!}x^6 + \cdots = \sum_{k=0}^{\infty} \frac{(-1)^k}{(2k)!}x^{2k}$$

And, using the Ratio Test,

$$\lim_{k\to\infty}\left|\frac{a_{k+1}}{a_k}\right| = \lim_{k\to\infty}\left|\frac{(-1)^{k+1}(2k)!x^{2k+2}}{(-1)^k(2k+2)!x^{2k}}\right|$$

$$= \left|x^2\right| \lim_{k\to\infty} \frac{1}{(2k+1)(2k+2)} = 0.$$

So the interval of convergence is $(-\infty, \infty)$.

3. $f(x) = \frac{3}{x-2}$, $f'(x) = -\frac{3}{(x-2)^2}$

$f''(x) = 2\frac{3}{(x-2)^3}$, $f'''(x) = -(2)(3)\frac{(3)}{(x-2)^4}$

$f(0) = -\frac{3}{2}$, $f'(0) = -\frac{3}{2^2}$

$f''(0) = -2\frac{3}{2^3}$, $f'''(0) = -(2)(3)\frac{3}{2^4}$

$$\sum_{k=0}^{\infty} f^{(k)}(0)\frac{x^k}{k!} = \sum_{k=0}^{\infty} \frac{(-3)(k!)}{2^{k+1}}\frac{x^k}{k!} = \sum_{k=0}^{\infty} \frac{(-3)}{2^{k+1}}x^k$$

And, using the Ratio Test,

$$\lim_{k\to\infty}\left|\frac{a_{k+1}}{a_k}\right| = \lim_{k\to\infty}\left|\frac{(-3)2^{k+1}x^{k+1}}{(-3)2^{k+2}x^k}\right| = \left|\frac{x}{2}\right| \text{ and}$$

$$\left|\frac{x}{2}\right| < 1 \text{ when } -2 < x < 2.$$

So the series converges absolutely for all $x \in (-2, 2)$. When $x = 2$,

$$\sum_{k=0}^{\infty} \frac{(-3)}{2^{k+1}}x^k = \sum_{k=0}^{\infty} \frac{(-3)}{2^{k+1}}2^k = \sum_{k=0}^{\infty} \frac{(-3)}{2},$$

which diverges by the kth-Term Test. The same argument shows that series diverges for $x = -2$, so the interval of convergence is $(-2, 2)$.

5. $f(x) = \ln(1+x)$, $f'(x) = \frac{1}{1+x}$,

$f''(x) = \frac{-1}{(1+x)^2}$, $f'''(x) = \frac{2}{(1+x)^3}$,

$f(0) = 0$, $f'(0) = 1$

$f''(0) = -1$, $f'''(0) = 2$

$$\sum_{k=0}^{\infty} \frac{f^{(k)}(0)}{k!}x^k = x - \frac{1}{2}x^2 + \frac{2!}{3!}x^3 - \frac{3!}{4!}x^4 + \cdots$$

$$= \sum_{k=0}^{\infty} \frac{(-1)^k x^{k+1}}{k+1}$$

And, using the Ratio Test,

$$\lim_{k\to\infty}\left|\frac{a_{k+1}}{a_k}\right| = \lim_{k\to\infty}\left|\frac{(-1)^{k+1}(k+1)x^{k+2}}{(-1)^k(k+2)x^{k+1}}\right|$$

and $|x| < 1$ when $-1 < x < 1$. So the series converges absolutely for all $x \in (-1, 1)$. When $x = 1$,

$$\sum_{k=0}^{\infty} \frac{(-1)^k(1)^{k+1}}{k+1} = \sum_{k=0}^{\infty} \frac{(-1)^k}{k+1}$$

and

$$\lim_{k=0} \frac{1}{k+1} = 0 \quad \text{and} \quad \frac{a_{k+1}}{a_k} = \frac{k}{k+1} < 1$$

for all $k \geq 0$, so $a_{k+1} > a_k$, for all $k \geq 0$, so the series converges by the Alternating Series Test. When $x = -1$,

$$\sum_{k=0}^{\infty} \frac{(-1)^k(-1)^{k+1}}{k+1} = \sum_{k=0}^{\infty} \frac{-1}{k+1},$$

which is the negative of the harmonic series and so diverges. Hence, the interval of convergence is $(-1, 1]$.

7. $f(x) = (1+x)^2$, $f'(x) = -2(1+x)^{-3}$

$f''(x) = 6(1+x)^{-4}$, $f'''(x) = -24(1+x)^3 + \cdots$

$f(0) = 1$, $f'(0) = -2$

$f''(0) = 3!$, $f'''(0) = -4!$

$$\sum_{k=0}^{\infty} \frac{f^{(k)}(0)}{k!}x^k = 1 - 2x + \frac{3!}{2!}x^2 - \frac{4!}{3!}x^3 + \cdots$$

$$= \sum_{k=0}^{\infty} (-1)^k(k+1)x^k$$

And, using the Ratio Test,

$$\lim_{k\to\infty}\left|\frac{a_{k+1}}{a_k}\right| = \lim_{k\to\infty}\left|\frac{(k+2)x^{k+1}}{(k+1)x^k}\right|,$$

$$|x| \lim_{k\to\infty} \frac{k+2}{k+1} = |x|,$$

and $|x| < 1$ when $-1 < x < 1$. So the series converges absolutely for $x \in (-1, 1)$. When $x = -1$,

$$\sum_{k=0}^{\infty} (k+1) \quad \text{and} \quad \lim_{k\to\infty} (k+1) = \infty,$$

so the series diverges by the kth-Term Test for Divergence. When $x = 1$,

$$\sum_{k=0}^{\infty}(-1)(k+1)^k \quad \text{and} \quad \lim_{k\to\infty} k+1 = \infty,$$

so the series diverges by the kth-Term Test for Divergence. So the interval of convergence is $(-1, 1)$.

9. $f(x) = e^{x-1}, \quad f'(x) = e^{x-1}, \quad f''(x) = e^{x-1}$

$f(1) = 1, \quad f'(1) = 1, \quad f''(1) = 1$

$$\sum_{k=0}^{\infty}\frac{f^{(k)}(1)}{k!}(x-1)^k = 1 + (x-1) + \frac{1}{2!}(x-1)^2 + \cdots$$

$$= \sum_{k=0}^{\infty}\frac{(x-1)^k}{k!}.$$

And, using the Ratio Test,

$$\lim_{k\to\infty}\left|\frac{a_{k+1}}{a_k}\right| = \lim_{k\to\infty}\left|\frac{k!(x-1)^{k+1}}{(k+1)!(x-1)^k}\right|,$$

$$= |x-1|\lim_{k\to\infty}\frac{1}{k+1} = 0,$$

and $0 < 1$ for all x, so the series converges absolutely for $x \in (-\infty, \infty)$. Thus, the interval of convergence is $(-\infty, \infty)$.

11. $f(x) = \ln x, \ f'(x) = \dfrac{1}{x}$

$f''(x) = -\dfrac{1}{x^2}, \ f'''(x) = \dfrac{2}{x^3}$

$f(e) = 1, \ f'(e) = \dfrac{1}{e}$

$f''(e) = -\dfrac{1}{e^2}, \ f'''(e) = \dfrac{2!}{e^3}$

$$\sum_{k=0}^{\infty}\frac{f^{(k)}(e)}{k!}(x-e)^k = 1 + \frac{1}{e}(x-e) - \frac{1}{e^2 2}(x-e)^2$$

$$+ \frac{2!}{e^3 3!}(x-e)^3 + \cdots = 1 + \sum_{k=1}^{\infty}(-1)^{k+1}\frac{e^{-k}}{k}(x-e)^k$$

Using the Ratio Test,

$$\lim_{k\to\infty}\left|\frac{a_{k+1}}{a_k}\right| = \lim_{k\to\infty}\left|\frac{e^{-k-1}k(x-e)^{k+1}}{e^{-k}(k+1)(x-e)^k}\right|$$

$$= \frac{|x-e|}{x}\lim_{k\to\infty}\frac{k}{k+1} = \frac{|x-e|}{x}$$

and $\dfrac{|x-e|}{e} < 1$ when $0 < x < 2e$. When $x = 0$,

$$\sum_{k=1}^{\infty}\frac{(-1)^k e^{-k}(-e)^k}{k} = \sum_{k=1}^{\infty}\left(-\frac{1}{k}\right),$$

which is the divergent harmonic series. When $x = 2e$,

$$\sum_{k=1}^{\infty}\frac{(-1)^{k+1}e^{-k}e^k}{k} = \sum_{k=1}^{\infty}\frac{-(1)^{k+1}}{k}$$

converges (Example 4.2). So the interval of convergence is $(0, 2e]$.

13. $f(x) = \dfrac{1}{x}, \ f'(x) = \dfrac{-1}{x^2},$

$f''(x) = \dfrac{2}{x^3}, \ f'''(x) = \dfrac{-6}{x^4}$

$f(1) = 1, \ f'(1) = -1$

$f''(1) = 2!, \ f'''(1) = -3!$

$$\sum_{k=0}^{\infty}\frac{f^{(k)}(1)}{k!}(x-1)^k = 1 - (x-1)$$

$$+ (x-1)^2 - (x-1)^3 + \cdots = \sum_{k=0}^{\infty}(-1)^k(x-1)^k$$

Using the Ratio Test,

$$\lim_{k\to\infty}\left|\frac{a_{k+1}}{a_k}\right| = \lim_{k\to\infty}\left|\frac{(-1)^{k+1}(x-1)^{k+1}}{(-1)^k(x-1)^k}\right|$$

$$= |x-1|\lim_{k\to\infty} 1 = |x-1|$$

and $|x-1| < 1$ when $0 < x < 2$. So the series converges absolutely for $x \in (0, 2)$. When $x = 0$,

$$\sum_{k=0}^{\infty}(-1)^k(-1)^k = \sum_{k=0}^{\infty} 1,$$

which diverges by the kth-Term Test for Divergence because $\lim_{k\to\infty} 1 = 1$. When $x = 2$,

$$\sum_{k=0}^{\infty}(-1)^k(1)^k = \sum_{k=0}^{\infty}(-1)^k,$$

which diverges by the kth-Term Test for Divergence. So the interval of convergence is $(0, 2)$.

15. $f(x) = \cos x$ about $c = 0$

$n = 5$;

$$\sum_{k=0}^{2}\frac{(-1)^k}{(2k)!}x^{2k} = 1 - \frac{x^2}{2!} + \frac{x^4}{4!}$$

$n = 9$;

$$\sum_{k=0}^{4}\frac{(-1)^k}{(2k)!}x^{2k} = 1 - \frac{x^2}{2} + \frac{x^4}{4!} - \frac{x^6}{6!} + \frac{x^8}{8!}$$

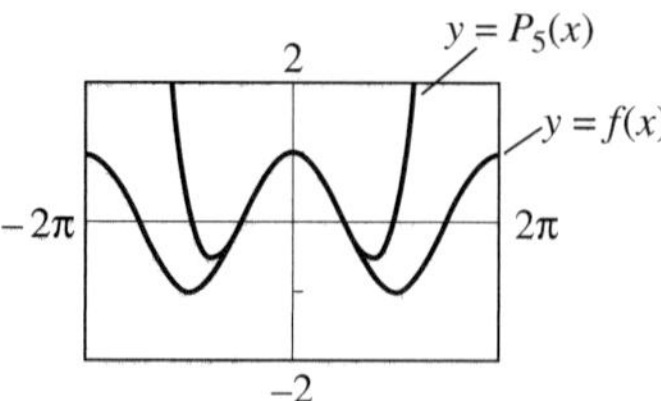

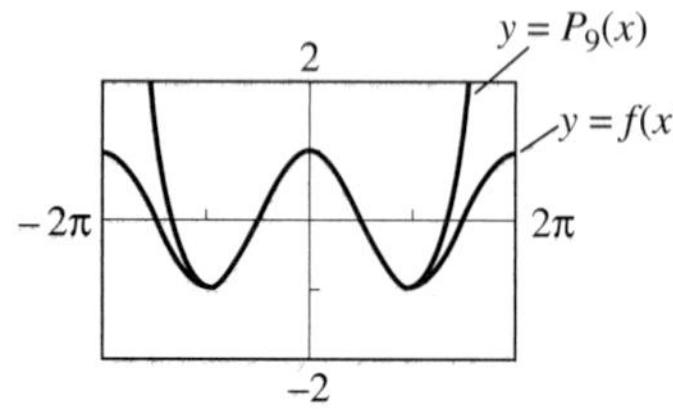

17. $f(x) = \sqrt{x}$ about $c = 1$

$$f(x) = \sqrt{x},\ f'(x) = \frac{1}{2}x^{-1/2},$$

$$f''(x) = \frac{-1}{4}x^{-3/2},\ f'''(x) = \frac{3}{8}x^{-5/2}$$

$$f(1) = 1,\ f'(1) = \frac{1}{2},$$

$$f''(1) = -\frac{1}{4},\ f'''(1) = \frac{3}{8}$$

$n = 3$;

$$\sum_{k=0}^{3}\frac{f^{(k)}(1)}{k!}(x-1)^k = 1 + \frac{1}{2}(x-1) - \frac{1}{8}(x-1)^2 + \frac{1}{16}(x-1)^3$$

$n = 6$;

$$\sum_{k=0}^{6}\frac{f^{(k)}(1)}{k!}(x-1)^k = 1 + \frac{1}{2}(x-1) - \frac{1}{8}(x-1)^2 + \frac{1}{16}(x-1)^3 - \frac{5}{2^7}(x-1)^4 + \frac{7}{2^8}(x-1)^5 - \frac{21}{2^{10}}(x-1)^6$$

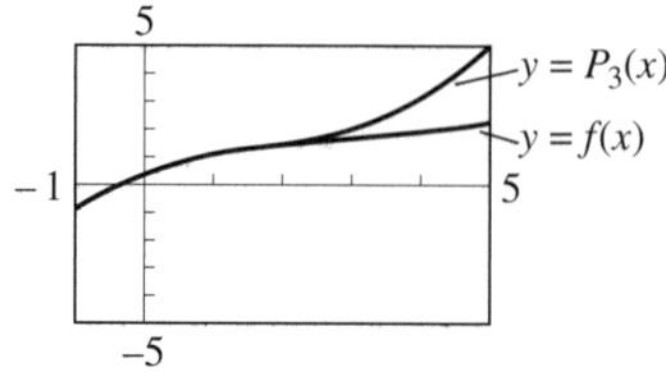

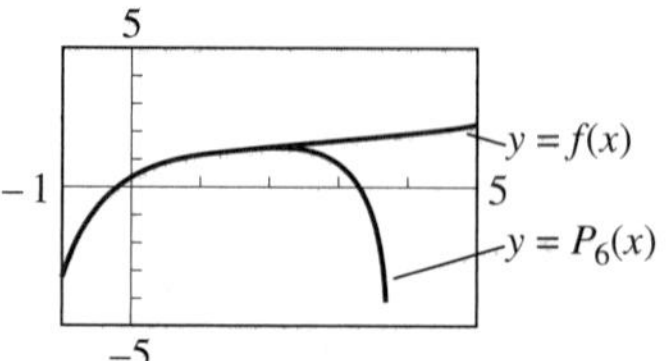

19. $f(x) = e^x$ about $c = 2$

$$f(x) = e^x \quad f'(x) = e^x \quad f''(x) = e^x$$

$$f(2) = e^2 \quad f'(2) = e^2 \quad f''(2) = e^2$$

$n = 3$;

$$\sum_{k=0}^{3}\frac{e^2}{k!}(x-2)^k = e^2\left[1 + (x-2) + \frac{(x-2)^2}{2} + \frac{(x-2)^3}{3!}\right]$$

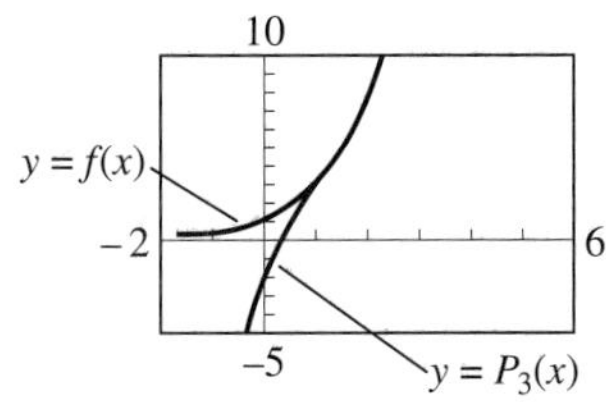

$n = 6$;

$$\sum_{k=0}^{6} \frac{e^2}{k!}(x-2)^k = e^2\left[1 + (x-2) + \frac{(x-2)^2}{2} + \frac{(x-2)^3}{3!} + \frac{(x-2)^4}{4!} + \frac{(x-2)^5}{5!} + \frac{(x-2)^6}{6!}\right]$$

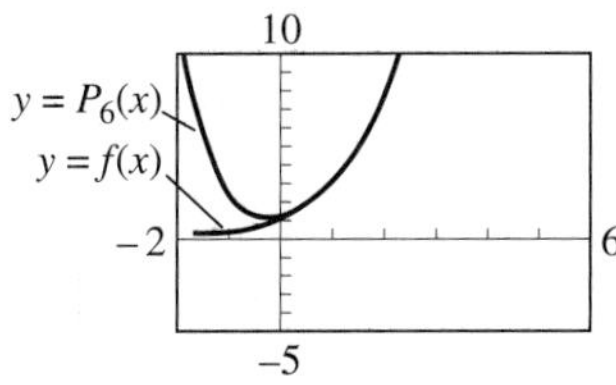

21. $f(x) = x^{-1/2}$ about $c = 4$

$f(x) = x^{-1/2}$, $f'(x) = -\frac{1}{2}x^{-3/2}$,

$f''(x) = \frac{3}{4}x^{-5/2}$, $f'''(x) = -\frac{15}{8}x^{-7/2}$,

$f(4) = \frac{1}{2}$, $f'(4) = -\frac{1}{16}$,

$f''(4) = \frac{3}{2^7}$, $f'''(4) = \frac{-15}{2^{10}}$,

$n = 2$;

$$\sum_{k=0}^{2} \frac{f^{(k)}(4)}{k!}(x-4)^k = \frac{1}{2} - \frac{1}{2^4}(x-4) + \frac{3}{2^8}(x-4)^2$$

$n = 4$;

$$\sum_{k=0}^{4} \frac{f^{(k)}(4)}{k!}(x-4)^k = \frac{1}{2} - \frac{1}{2^4}(x-4) \frac{3}{2^8}(x-4)^2 - \frac{5}{2^{11}}(x-4)^3 + \frac{35}{2^{16}}(x-4)^4$$

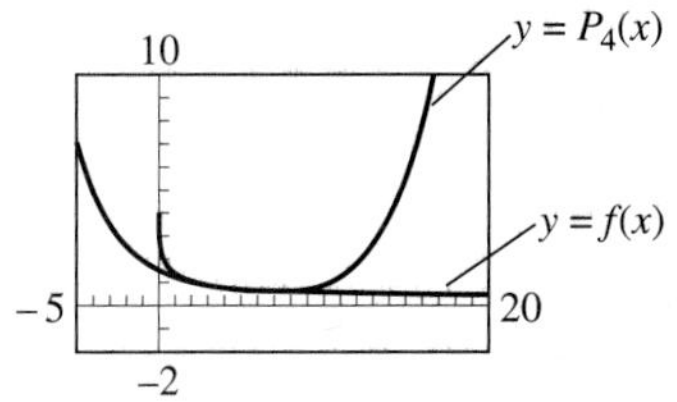

23. $R_n(x) = \dfrac{f^{(n+1)}(z)}{(n+1)!}x^{n+1}$. We have

$$f^{(n+1)}(z) = \begin{cases} \pm\cos z \text{ if } n \text{ is even} \\ \pm\sin z \text{ if } n \text{ is odd} \end{cases} \text{ so}$$

$\left|f^{(n+1)}(z)\right| \le 1$ for all n. So

$$\left|R_n(x)\right| = \left|\frac{f^{(n+1)}(z)}{(n+1)!}x^{n+1}\right| \le \frac{|x|^{n+1}}{(n+1)!}, \text{ so}$$

consider $\sum_{n=0}^{\infty} \frac{|x|^{n+1}}{(n+1)!}$. Using the Ratio Test, we have

$$\lim_{k\to\infty}\left|\frac{a_{n+1}}{a_n}\right| = \lim_{n\to\infty}\frac{|x|^{n+2}(n+1)!}{|x|^{n+1}(n+2)!} = |x| \lim_{n\to\infty}\frac{1}{n+2} = 0,$$

and $0 < 1$ for all x. So $\sum_{n=0}^{\infty} \frac{|x|^{n+1}}{(n+1)!}$ converges absolutely for all x. Thus, by the kth-Term Test for Divergence, $\lim_{n\to\infty} \frac{|x|^{n+1}}{(n+1)!} = 0$, and since $\left|R_n(x)\right| \le \frac{|x|^{n+1}}{(n+1)!}$, $\lim_{n\to\infty} R_n(x) = 0$ for all x.

25. $R_n(x) = \dfrac{f^{(n+1)}(z)}{(n+1)!}x^{n+1}$. We have

$$f^{(n+1)}(z) = \begin{cases} \dfrac{n!}{z^{n+1}} & \text{if } n \text{ is even} \\ -\dfrac{n!}{z^{n+1}} & \text{if } n \text{ is odd} \end{cases} \quad \text{so}$$

$\left|f^{(n+1)}(z)\right| = \dfrac{n!}{|z|^{n+1}}$ for all n. So

$$\left|R_n(x)\right| = \frac{n!}{(n+1)!}\left|\frac{x-1}{z}\right|^{n+1} = \frac{1}{n+1}\left|\frac{x-1}{z}\right|^{n+1},$$

and $1 < z < 2$, so $\dfrac{1}{z} < \dfrac{1}{1} = 1$. So

$$\left|R_n(x)\right| < \frac{|x-1|^{n+1}}{n+1}.$$

For $1 < x < 2$, the numerator $|x-1|^{n+1}$ is bounded by 1 and the denominator tends to ∞ with n, so $\lim_{n\to\infty} R_n(x) = 0$.

27. (a) Expand $f(x) = \ln x$ into a Taylor series about $c = 1$. Recall

$$\ln(x) = \sum_{k=0}^{\infty} \frac{(-1)^{k+1}}{k}(x-1)^k$$

(from Example 7.6). So

$$P_4(x) = \sum_{k=0}^{4} \frac{f^{(k)}(1)}{k!}(x-1)^k$$
$$= (x-1) - \frac{1}{2}(x-1)^2 + \frac{1}{3}(x-1)^3 - \frac{1}{4}(x-1)^4.$$

Letting $x = 1.05$ gives $\ln(1.05) \approx P_4(1.05) = .04879$.

(b)
$$\begin{aligned}|\text{Error}| &= |\ln(1.05) - P_4(1.05)| \\ &= |R_n(1.05)| \\ &= \left|\frac{f^{(4+1)}(x)}{(4+1)!}(1.05-1)^{4+1}\right| \\ &= \frac{4!|z|^{-5}(0.05)^5}{5!},\end{aligned}$$

so because $1 < z < 1.05$, then $\dfrac{1}{z} < \dfrac{1}{1} = 1$. Thus we have

$$|\text{Error}| = \frac{(.05)^5}{5z^5} < \frac{(.05)^5}{5(1)^5} = \frac{(.05)^5}{5}.$$

(c) From part (b), we have for $1 < z < 1.05$,

$$\begin{aligned}\left|R_n(1.05)\right| &= \left|\frac{f^{n+1}(z)}{(n+1)!}(1.05-1)^{n+1}\right| \\ &= \frac{n!(.05)^{n+1}}{(n+1)!z^{n+1}} \\ &= \frac{(.05)^{n+1}}{(n+1)z^{n+1}} \\ &< \frac{(.05)^{n+1}}{n+1}\end{aligned}$$

since $1 < z < 1.05$ gives $\dfrac{1}{z} < \dfrac{1}{1} = 1$. So

$$\left|R_n(1.05)\right| < \frac{(.05)^{n+1}}{n+1} < 10^{-10} \text{ if } n = 7.$$

29. (a) Expand $f(x) = \sqrt{x}$ into a Taylor series about $c = 1$. So, as shown in Exercise 17,

$$\sqrt{x} = 1 + \sum_{k=1}^{\infty} \frac{(-1)^{k-1}1\cdot 3\cdot 5\ldots(2n-3)}{2^k k!}(x-1)^k.$$

Then

$$\begin{aligned}&P_4(x) \\ &= \sum_{k=0}^{4} \frac{f^{(k)}(1)}{k!}(x-1)^k \\ &= 1 + \frac{(x-1)}{2} - \frac{(x-1)}{8} + \frac{(x-1)^3}{16} - \frac{5(x-1)^4}{128}.\end{aligned}$$

Letting $x = 1.1$ gives $\sqrt{1.1} \approx P_4(1.1) = 1.0488$.

(b)
$$\begin{aligned}|\text{Error}| &= \left|\sqrt{1.1} - P_4(1.1)\right| \\ &= \left|R_4(1.1)\right| \\ &= \left|\frac{f^{4+1}(z)}{(4+1)!}(1.1-1)^{4+1}\right| \\ &= \frac{3\cdot 5\cdot 7(.1)^5}{2^5\cdot 5!z^{9/2}}\end{aligned}$$

Because $1 < z < 1.1$, $\frac{1}{z} < \frac{1}{1} = 1$, we have

$$|\text{Error}| = \frac{21(.1)^5}{2^5 4! z^{9/2}} < \frac{21(.1)^5}{2^5 4!} = \frac{7(.1)^5}{256}.$$

(c) From part (b), we have for $1 < z < 1.1$,

$$\begin{aligned} |R_n(1.1)| &= \left| \frac{f^{(n+1)}(z)}{(n+1)!}(1.1-1)^{n+1} \right| \\ &= \frac{1 \cdot 3 \cdot 5 \ldots (2n-1)(.1)^{n+1}}{2^{n+1}(n+1)! z^{\left(\frac{2n+1}{2}\right)}} \\ &< \frac{1 \cdot 3 \cdot 5 \ldots (2n-1)(.1)^{n+1}}{2^{n+1}(n+1)!} \end{aligned}$$

because $1 < z < 1.1$ gives $\frac{1}{z} < \frac{1}{1} = 1$. So

$$\begin{aligned} |R_n(1.1)| &< \frac{1 \cdot 3 \cdot 5 \ldots (2n-1)(.1)^{n+1}}{2^{n+1}(n+1)!} \\ &< 10^{-10} \text{ if } n = 8. \end{aligned}$$

31. (a) Expand $f(x) = e^x$ into Taylor series about $c = 0$. So

$$e^x = \sum_{k=0}^{\infty} \frac{x^k}{k!}$$

(from Example 7.1). Then

$$\begin{aligned} P_4(x) &= \sum_{k=0}^{4} \frac{f^{(k)}(0)}{k!} x^k \\ &= 1 + x + \frac{x^2}{2!} + \frac{x^3}{4!} + \frac{x^4}{4!}. \end{aligned}$$

Letting $x = .1$ gives $e^{.1} \approx P_4(.1) = 1.10517$.

(b)

$$\begin{aligned} |\text{Error}| &= \left| e^{.1} - p_4(.1) \right| \\ &= |R_4(.1)| \\ &= \left| \frac{f^{(4+1)}(z)}{(4+1)!}(.1)^{4+1} \right| \\ &= \frac{e^z(.1)^5}{5!} \end{aligned}$$

Because $0 < z < .1$ gives $e^z < e^{.1} < 1.5$, we have

$$|\text{Error}| = \frac{e^z(.1)^5}{5!} = \frac{(1.5)(.1)^5}{5!}.$$

(c) From part (b), we have for $0 < z < .1$,

$$\begin{aligned} |R_n(.1)| &= \left| \frac{f^{(n+1)}(z)}{(n+1)!}(.1)^{n+1} \right| \\ &= \frac{e^z(.1)^{n+1}}{(n+1)!} \\ &< \frac{(1.5)(.1)^{n+1}}{(n+1)!} \end{aligned}$$

because $0 < z < .1$ gives $e^z < e^{.1} < 1.5$. So

$$|R_n(.1)| < \frac{(1.5)(.1)^{n+1}}{(n+1)!} < 10^{-10} \text{ if } n = 6.$$

33. Since

$$\sum_{k=0}^{\infty} \frac{1}{k!} x^k = e^x,$$

by replacing x with 2, we have

$$\sum_{k=0}^{\infty} \frac{(2)^k}{k!} = e^2;$$

e^x with $x = 2$.

35. Since

$$\sum_{k=0}^{\infty} \frac{(-1)^k x^{2k+1}}{(2k+1)!} = \sin x,$$

by replacing x with π, we have

$$\sum_{k=0}^{\infty} \frac{(-1)^k \pi^{2k+1}}{(2k+1)!} = \sin \pi = 0;$$

$\sin x$ with $x = \pi$.

37. Since

$$\sum_{k=0}^{\infty} \frac{(-1)^k}{2k+1} x^{2k+1} = \tan^{-1} x,$$

by replacing x with 1, we have

$$\sum_{k=0}^{\infty} \frac{(-1)^k}{2k+1} = \tan^{-1}(1) = \frac{\pi}{4};$$

$\tan^{-1}(x)$ with $x = 1$.

39. Since

$$e^x = \sum_{k=0}^{\infty} \frac{x^k}{k!}$$

with $r = \infty$, by replacing x with $-2x$, we have

$$e^{-2x} = \sum_{k=0}^{\infty} \frac{(-1)^k (2x)^k}{k!}$$

with $r = \infty$.

41. Since

$$\sin x = \sum_{k=0}^{\infty} \frac{x^k}{k!}$$

with $r = \infty$, by replacing x with $-x^2$ and multiplying by x, we have

$$xe^{-x^2} = \sum_{k=0}^{\infty} \frac{(-1)^k x^{2k+1}}{k!}$$

with $r = \infty$.

43. Since

$$\sin x = \sum_{k=0}^{\infty} \frac{(-1)^k}{(2k+1)!} x^{2k+1}$$

with $r = \infty$, by replacing x with x^2, we have

$$\sin x^2 = \sum_{k=0}^{\infty} \frac{(-1)^k x^{4k+2}}{(2k+1)!}$$

with $r = \infty$.

45. Since

$$\cos x = \sum_{k=0}^{\infty} \frac{(-1)^k x^{2k}}{(2k)!}$$

with $r = \infty$, by replacing x with $3x$, we have

$$\cos 3x = \sum_{k=0}^{\infty} \frac{(-1)^k 9^k x^{2k}}{(2k)!}$$

with $r = \infty$.

47. If $f(x) = \begin{cases} e^{-1/x^2} & \text{if } x \neq 0 \\ 0 & \text{if } x = 0 \end{cases}$, then

$$f'(x) = \begin{cases} \dfrac{2}{x^3} e^{-1/x^2} & \text{if } x \neq 0 \\ 0 & \text{if } x = 0 \end{cases}, \text{ so } f'(0) = 0,$$

$$\text{and } f''(x) = \begin{cases} \dfrac{4}{x^6} e^{-1/x^2} - \dfrac{6}{x^4} e^{-1/x^2} & \text{if } x \neq 0 \\ 0 & \text{if } x = 0 \end{cases},$$

so $f''(0) = 0$.

49. Let's rewrite $R(c)$ as

$$R(c) = \frac{hc}{.55h + .45c}$$

to avoid division by 0. Then

$$R'(c) = \frac{.55h^2}{(.55h + 45c)^2}$$

$$R''(c) = \frac{(-2)(.45)(.55)h^2}{(.55h + .45x)^3}$$

$$R'''(c) = \frac{6(.45)^2(.55)h^2}{(.55h + .45c)^4}$$

$$R(0) = 0$$

$$R'(0) = \frac{.55}{(.55)^2} = \frac{20}{11}$$

$$R''(0) = \frac{(-2)(.45)}{(.55)^2 h} = \frac{360}{121h}$$

$$R'''(0) = \frac{6(.45)^2}{(.55)^3 h^2} = \frac{9720}{1331h^2}$$

$$\text{Hence, } R(c) \approx \frac{20}{11}c - \frac{360}{121h^2}\frac{c^2}{2} + \frac{9720}{1331h^2}\frac{c^3}{6}$$

$$= \frac{20}{11}c - \frac{180}{121h}c^2 + \frac{1620}{1331h^2}c^3.$$

If the city rating improves, so does R.

51. The 5th term of the series is $\frac{1}{9!}$, which is less than 10^{-5}. So 4 terms will give the desired accuracy. In Simpson's Rule, the error is bounded by $\frac{1}{180n^4}$. Solving $\frac{1}{180n^4} \leq 10^{-5}$, we get $n > 4.9$, which means n must be 6, since n is an even integer.

53. Let's find the Maclauren series for $g(z) = \sqrt{c+z} - \sqrt{c-z}$, and then replace c by a^2 and z by x^2. In fact, let's write $g(z) = h(z) - k(z)$, where $h(z) = \sqrt{c+z}$ and $k(z) = \sqrt{c-z}$. It is easy to find the derivatives of $h(z)$:

$$h'(z) = \frac{1}{2}(c+z)^{-1/2}$$
$$h''(z) = \left(-\frac{1}{2}\right)\left(\frac{1}{2}\right)(c+z)^{-3/2}$$
$$h'''(z) = \left(\frac{3}{2}\right)\left(\frac{1}{2}\right)\left(\frac{1}{2}\right)(c+z)^{-5/2}$$
$$h^{(4)}(z) = \left(-\frac{5}{2}\right)\left(\frac{3}{2}\right)\left(\frac{1}{2}\right)\left(\frac{1}{2}\right)(c+z)^{-7/2}$$

and so on. Then

$$h(0) = \frac{1}{2}c$$
$$h'(0) = \frac{1}{2}(c)^{-1/2}$$
$$h''(0) = \left(-\frac{1}{2}\right)\left(\frac{1}{2}\right)(c)^{-3/2}$$
$$h'''(0) = \left(\frac{3}{2}\right)\left(\frac{1}{2}\right)\left(\frac{1}{2}\right)(c)^{-5/2}$$
$$h^{(4)}(0) = \left(-\frac{5}{2}\right)\left(\frac{3}{2}\right)\left(\frac{1}{2}\right)\left(\frac{1}{2}\right)(c)^{-7/2}$$

and so on.

Now what about the derivatives of $k(z)$? They will be similar to the derivatives computed above, except that the expression $(c+z)$ is replaced by $(c-z)$, and all the odd terms have an additional minus sign, which comes from differentiating $(c-z)$ an odd number of times. So when we evaluate these derivatives at $z = 0$, the even ones will be the same as above and the odd ones will contain a negative sign. So if we calculate $g^{(n)}(0) = h^{(n)}(0) - k^{(n)}(0)$, when n is even we get 0, and when n is odd we get twice $h^{(n)}(0)$. That is,

$$g(0) = 0$$
$$g'(0) = (c)^{-1/2}$$
$$g''(0) = 0$$
$$g'''(0) = \left(\frac{3}{2}\right)\left(\frac{1}{2}\right)(c)^{-5/2}$$
$$g^{(4)}(0) = 0$$
$$g^{(5)}(0) = \left(\frac{7}{2}\right)\left(\frac{5}{2}\right)\left(\frac{3}{2}\right)\left(\frac{1}{2}\right)(c)^{-9/2}.$$

In general, for $k > 0$,

$$\begin{aligned} g^{2k+1}(0) &= \left(\frac{4k-1}{2}\right)\left(\frac{4k-3}{2}\right) \\ &\quad \cdots \left(\frac{5}{2}\right)\left(\frac{3}{2}\right)\left(\frac{1}{2}\right)(c)^{-(4k+1)/2} \\ &= \frac{1}{2^{2k}}\frac{(4k-1)!}{2^{2k-1}(2k-1)!}c^{-(4k+1)/2} \\ &= \frac{1}{2^{4k-1}}\frac{(4k-1)!}{(2k-1)!}c^{-(4k+1)/2}. \end{aligned}$$

So the McLauren series for $g(z)$ is

$$zc^{-1/2} + \sum_{k=1}^{\infty} \frac{1}{2^{4k-1}}\frac{(4k-1)!}{(2k-1)!}c^{-(4k+1)/2}\frac{z^{2k+1}}{(2k+1)!}.$$

Finally, we get the McLauren series for $f(x)$ by replacing c by a^2 and z by x^2:

$$\frac{x^2}{a} + \sum_{k=1}^{\infty} \frac{1}{2^{4k-1}}\frac{(4k-1)!}{(2k-1)!}a^{-(4k+1)}\frac{x^{4k+1}}{(2k+1)!}.$$

55. $$f(t) = \sum_{k=0}^{3} \frac{f^{(0)}(0)}{k!}t^k = 10 + 10t + t^2 - \frac{t^3}{3!}$$

$$f(2) = 10 + 10(2) + (2)^2 - \frac{(2)^3}{6} = \frac{98}{3} \text{ miles}$$

57. $f(x) = e^x$ and $f'(x) = e^x$, so $f(c) = e^c = f'(c)$, so $e^x = \sum_{k=0}^{\infty} \frac{f^{(k)}(c)}{k!}(x-c)^k = \frac{e^c}{k!}(x-c)^k$

59. To generate the series, differentiate $(1+x)^r$ repeatedly. The series terminates if r is a positive integer, since all derivatives after the rth derivative will be 0. Otherwise the series is infinite.

61.

$$1+\sum_{k=1}^{\infty}\frac{\left(\frac{1}{2}\right)\left(-\frac{1}{2}\right)\cdots\left(\frac{1}{2}-k+1\right)}{k!}x^k$$

63.

$$\begin{aligned}\cosh x &= 1+\frac{1}{2}x^2+\frac{1}{24}x^4+\frac{1}{720}x^6+\cdots\\ &=\sum_{k=1}^{\infty}\frac{x^{2k}}{(2k)!}\\ \sinh x &= x+\frac{1}{6}x^3+\frac{1}{120}x^5+\frac{1}{5040}x^7+\cdots\\ &=\sum_{k=1}^{\infty}\frac{x^{2k+1}}{(2k+1)!}\end{aligned}$$

These match the cosine and sine series except that here all signs are positive.

7.8 Applications of Taylor Series

1. Using the first three terms of the sine series around the center $\pi/2$, we get $\sin 1.61 \approx 0.999231634426433$.

3. Using the first five terms of the cosine series around the center 0, we get $\cos 0.34 \approx 0.94275466553403$.

5. Using the first nine terms of the exponential series around the center 0, we get $e^{0.2} \approx 0.818730753079365$.

7. Since

$$\begin{aligned}\cos x^2 &= \sum_{k=0}^{\infty}\frac{(-1)^k x^{4k}}{(2k)!}\\ &= 1-\frac{x^4}{2!}+\frac{x^8}{4!}-\frac{x^{12}}{6!}+\frac{x^{16}}{8!}\ldots \text{ then}\end{aligned}$$

$$\begin{aligned}\lim_{k\to\infty}\frac{\cos x^2-1}{x^4} &= \lim_{k\to\infty}\frac{\left(-\frac{x^4}{2!}+\frac{x^8}{4!}-\frac{x^{12}}{6!}\ldots\right)-1}{x^4}\\ &= \lim_{k\to\infty}\left(-\frac{1}{2}+\frac{x^4}{4!}-\frac{x^8}{6!}\right)\\ &= -\frac{1}{2}.\end{aligned}$$

9. Since

$$\begin{aligned}\ln x &= \sum_{k=1}^{\infty}\frac{(-1)^k(x-1)^k}{k}\\ &= (x-1)-\frac{1}{2}(x-1)^2+\frac{1}{3}(x-1)^3-\ldots, \text{ then}\end{aligned}$$

$$\begin{aligned}&\lim_{k\to\infty}\frac{\ln x-(x-1)}{(x-1)^2}\\ &= \lim_{k\to\infty}\frac{\begin{bmatrix}(x-1)-\frac{1}{2}(x-1)^2\\ +\frac{1}{3}(x-1)^3-\ldots,\end{bmatrix}-(x-1)}{(x-1)^2}\\ &= \lim_{k\to\infty}\left(-\frac{1}{2}+\frac{1}{3}(x-1)-\frac{1}{4}(x-1)^2+\cdots\right)\\ &= -\frac{1}{2}.\end{aligned}$$

11. Since

$$e^x=\sum_{k=0}^{\infty}\frac{x^k}{k!}=1+x+\frac{1}{2}x^2+\frac{1}{3!}x^3+\cdots,$$

then

$$\lim_{x\to 0}\frac{e^x-1}{x}=\lim_{x\to 0}\frac{\left(1+x+\frac{x^2}{2}+\frac{x^3}{3!}+\cdots\right)-1}{x}$$

$$\lim_{x\to 0}\left(1+\frac{x}{2}+\frac{x^2}{3}+\cdots\right)=1.$$

13. Since

$$\sin x=\sum_{k=0}^{\infty}\frac{(-1)^k x^{2k+1}}{(2k+1)!},$$

then when $n = 3$, $\sin x \approx x - \frac{x^3}{3!} + \frac{x^5}{5!}$ so

$$\int_{-1}^{1} \frac{\sin x}{x}\,dx \approx \int_{-1}^{1} \left(1 - \frac{x^2}{6} + \frac{x^4}{120}\right) dx$$
$$= \left[x - \frac{x^3}{18} + \frac{x^5}{600}\right]_{-1}^{1}$$
$$= \frac{1703}{900} \approx 1.8922.$$

15. Since

$$e^{-x^2} = \sum_{k=0}^{\infty} \frac{(-1)^k x^{2k}}{k!},$$

then when $n = 5$, $e^{-x^2} \approx 1 - x^2 + \frac{x^4}{2} - \frac{x^6}{6} + \frac{x^8}{24}$, so

$$\int_{-1}^{1} e^{-x^2}\,dx \approx \int_{-1}^{1} \left(1 - x^2 + \frac{x^4}{2} - \frac{x^6}{6} + \frac{x^8}{24}\right) dx$$
$$= \left[x - \frac{x^3}{3} + \frac{x^5}{10} - \frac{x^7}{42} + \frac{x^9}{(24)(9)}\right]_{-1}^{1}$$
$$= \frac{5651}{3780} \approx 1.495.$$

17. Since

$$\ln x = \sum_{k=1}^{\infty} \frac{(-1)^{k+1}}{k}(x-1)^k,$$

then when $n = 5$,

$$\ln x \approx (x-1) - \frac{1}{2}(x-1)^2$$
$$+ \frac{1}{3}(x-1)^3 - \frac{1}{4}(x-1)^4 + \frac{1}{5}(x-1)^5,$$

so

$$\int_1^2 \ln x\,dx$$
$$\approx \int_1^2 (x-1) - \frac{1}{2}(x-1)^2$$
$$+ \frac{1}{3}(x-1)^3 - \frac{1}{4}(x-1)^4 + \frac{1}{5}(x-1)^5\,dx$$
$$= \left(\frac{(x-1)^2}{2} - \frac{(x-1)^3}{6} + \right.$$
$$\left. \frac{(x-1)^4}{12} - \frac{(x-1)^5}{20} + \frac{(x-1)^6}{30}\right)\Bigg|_1^2$$
$$= \frac{2}{5}.$$

19.

$$\left|\frac{a_{n+1}}{a_n}\right| \le \frac{x^2}{2^2(n+1)(n+2)}$$

Since this ratio tends to 0, the radius of convergence for $J_1(x)$ is infinite.

21. We need the first neglected term to be less than 0.04. The kth term is bounded by

$$\frac{10^{2k+1}}{2^{2k+2}k\,(k+2)!},$$

which is equal to 0.0357 for $k = 12$. Thus, we will need the terms up through $k = 11$, that is, the first 12 terms of the series.

23. Let $f(x) = \frac{1}{\sqrt{1-x}}$. Then

$$f(x) \approx f(0) + f'(0)x = 1 + \frac{1}{2}x.$$

Now let $x = \frac{v^2}{c^2}$. Then

$$\frac{1}{\sqrt{1 - \frac{v^2}{c^2}}} \approx 1 + \frac{v^2}{2c^2}.$$

Thus,

$$m(v) = m_0/\sqrt{1 - v^2/c^2} \approx m_0\left(1 + \frac{v^2}{2c^2}\right).$$

To increase mass by 10%, we want $1 + \frac{v^2}{2c^2} = 1.1$. Solving for v, we have $\frac{1}{2c^2}v^2 = .1$, so $v^2 = .2c^2$, thus $v = \sqrt{.2}c \approx 83{,}000$ miles per second.

25. $w(x) = \dfrac{mgR^2}{(R+x)^2}$ with $w(0) = mg$ and

$w'(x) = \dfrac{-2mgR^2}{(R+x)^3}$ with $w'(0) = \dfrac{-2mg}{R}$,

so $w(x) \approx \sum_{k=0}^{1} \dfrac{f^{(k)}(0)}{k!}x^k = mg\left(1 - \dfrac{2x}{R}\right)$.

To reduce weight by 10%, we want $1 - \dfrac{2x}{R} = 9$.

Solving for x, we have $\dfrac{-2x}{R} = -.1$, so $2x = .1R$,

thus $x = \dfrac{R}{20}$.

27. No, since x is much smaller than R for high-altitude locations on Earth. You have to go out to an altitude of more than 100 miles before you weigh significantly less.

29. The first neglected term is negative, so this estimate is too large.

31. Use the first two terms of the series for tanh: $\tanh x \approx x - \frac{1}{3}x^3$. Substitute $\sqrt{\dfrac{g}{40m}}t$ for x, multiply by $\sqrt{40mg}$, and simplify. The result is $gt - \dfrac{g^2}{120m}t^3$.

33. Using the series for $(1+x)^{3/2}$ around the center $x = 0$, we get

$$S(d) \approx \frac{8\pi c^2}{3}\left(\frac{3}{32}\cdot\frac{d^2}{c^2} + \frac{3}{2048}\cdot\frac{d^4}{c^4} - \frac{1}{65536}\cdot\frac{d^6}{c^6}\right).$$

If we ignore the d^4 and d^6 terms, this simplifies to $\dfrac{\pi d^2}{4}$.

35. $f(\lambda) = \lambda^5(e^{hc/\lambda kT} - 1)$

$$= (1 + (hc/\lambda kT) + \frac{1}{2}\left(\frac{hc}{\lambda kT}\right)^2 + \cdots - 1).$$

Accurate for $\lambda > \dfrac{hc}{kT}$.

7.9 Fourier Series

1. $f(x) = x$

$$a_0 = \frac{1}{\pi}\int_{-\pi}^{\pi} f(x)\,dx$$
$$= \frac{1}{\pi}\int_{-\pi}^{\pi} x\,dx = \frac{1}{\pi}\left[\frac{x^2}{2}\right]_{-\pi}^{\pi}$$
$$= 0$$

$$a_k = \frac{1}{\pi}\int_{-\pi}^{\pi} x\cos(kx)\,dx$$
$$= \frac{1}{\pi}\left[\frac{x}{k}\sin(kx) + \frac{1}{k^2}\cos(kx)\right]_{-\pi}^{\pi}$$
$$= \frac{1}{\pi}\left[\begin{array}{l}\frac{\pi}{k}\sin(k\pi) + \frac{1}{k^2}\cos(k\pi)\\ -\left(-\frac{\pi}{k}\sin(-k\pi)\right) + \frac{1}{k^2}\cos(-k\pi)\end{array}\right]$$
$$= \frac{1}{\pi}\left(\frac{1}{k^2}\cos k\pi - \frac{1}{k^2}\cos(-k\pi)\right)$$
$$= 0$$

$$b_k = \frac{1}{\pi}\int_{-\pi}^{\pi} x\sin(kx)\,dx$$
$$= \frac{1}{\pi}\left[\frac{-x}{k}\cos kx + \frac{1}{k^2}\sin kx\right]_{-\pi}^{\pi}$$
$$= \frac{1}{\pi}\left[\begin{array}{l}-\frac{\pi}{\cos}(k\pi) + \frac{1}{k^2}\sin(k\pi)\\ -\left(\frac{\pi}{\cos}(-k\pi) + \frac{1}{k^2}\sin(-k\pi)\right)\end{array}\right]$$
$$= \frac{1}{\pi}\left[\frac{-2\pi}{k}\cos(k\pi)\right]$$
$$= \frac{-2}{k}(-1)^k$$

So

$$f(x) = \sum_{k=1}^{\infty}(-1)^{k+1}\frac{2}{k}\sin(kx).$$

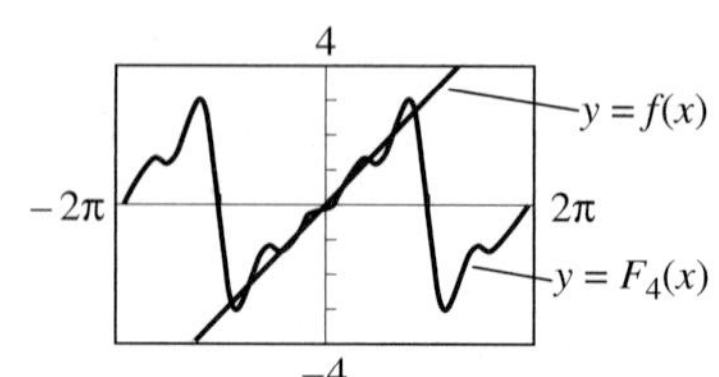

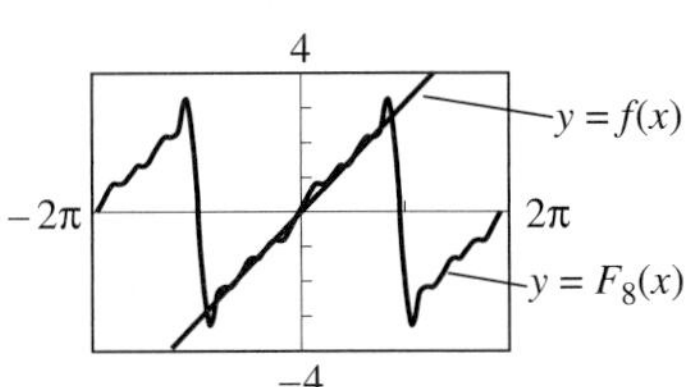

3. $f(x) = 2|x| = \begin{cases} -2x & \text{if } -\pi < x < 0 \\ 2x & \text{if } 0 < x < \pi \end{cases}$

$$\begin{aligned}
a_0 &= \frac{1}{\pi}\int_{-\pi}^{\pi} f(x)\,dx \\
&= \frac{1}{\pi}\int_{-\pi}^{0} (-2x)\,dx + \frac{1}{\pi}\int_{0}^{\pi} 2x\,dx \\
&= \frac{1}{\pi}\left[-x^2\right]_{-\pi}^{0} + \frac{1}{\pi}\left[x^2\right]_{0}^{\pi} \\
&= \frac{1}{\pi}\left[0+\pi^2\right] + \left[\pi^2 - 0\right] \\
&= 2\pi
\end{aligned}$$

$$\begin{aligned}
a_k &= \frac{1}{\pi}\int_{-\pi}^{\pi} f(x)\cos(kx)\,dx \\
&= \frac{1}{\pi}\int_{-\pi}^{0} -2x\cos(kx)\,dx + \frac{1}{x}\int_{0}^{\pi} 3x\cos(kx)\,dx \\
&= \frac{1}{\pi}\left[-\frac{2x}{k}\sin(kx) - \frac{2}{k^2}\cos(kx)\right]_{-\pi}^{0} \\
&\quad + \frac{1}{\pi}\left[\frac{2x}{k}\sin(kx) + \frac{2}{k^2}\cos(kx)\right]_{0}^{\pi} \\
&= \frac{1}{\pi}\left[\begin{array}{l} 0 - \frac{2}{k^2}\cos(0) \\ -\left(\frac{2\pi}{k}\sin(-k\pi) - \frac{2}{k^2}\cos(-k\pi)\right) \end{array}\right] \\
&\quad + \frac{1}{\pi}\left[\begin{array}{l} \frac{2\pi}{k}\sin(k\pi) + \frac{2}{k^2}\cos(k\pi) \\ -\left(0 - \frac{2}{k^2}\cos(0)\right) \end{array}\right] \\
&= \frac{-2}{k^2\pi} + \frac{2}{k^2\pi}(-1)^k + \frac{2}{k^2\pi}(-1)^k - \frac{2}{k^2\pi} \\
&= \frac{4}{k^2\pi}(-1)^k - \frac{4}{k^2\pi} \\
&= \frac{4}{k^2\pi}[(-1)^k - 1 \\
&= \begin{cases} 0 & \text{if } k \text{ is even} \\ \dfrac{-8}{k^2\pi} & \text{if } k \text{ is odd} \end{cases}
\end{aligned}$$

So $a_{2k} = 0$ and $a_{(2k-1)} = \dfrac{-8}{(2k-1)^2\pi}$.

$$\begin{aligned}
b_k &= \frac{1}{\pi}\int_{-\pi}^{\pi} f(x)\sin(kx)\,dx \\
&= \frac{1}{\pi}\int_{-\pi}^{0} -2x(kx)\,dx + \frac{1}{\pi}\int_{0}^{\pi} 2x\sin(kx)\,dx \\
&= \frac{1}{\pi}\left[\frac{2x}{k}\cos(kx) - \frac{2}{k^2}\sin(kx)\right]_{-\pi}^{0} \\
&\quad + \frac{1}{\pi}\left[\frac{-2x}{k}\cos(kx) + \frac{2}{k^2}\sin(kx)\right]_{0}^{\pi} \\
&= \frac{1}{\pi}\left[\begin{array}{l} 0 - \frac{2}{k^2}\sin(0) \\ -\left(\frac{-2\pi}{k}\cos(-k\pi) - \frac{2}{k^2}\sin(-k\pi)\right) \end{array}\right] \\
&\quad + \frac{1}{\pi}\left[\begin{array}{l} \frac{-2\pi}{k}\cos(k\pi) + \frac{2}{k^2}\sin(k\pi) \\ -\left(0 + \frac{2}{k^2}\sin(0)\right) \end{array}\right] \\
&= \frac{2}{k}\cos(k\pi) - \frac{2}{k}\cos k\pi \\
&= 0
\end{aligned}$$

So

$$\begin{aligned}
f(x) &= \frac{2\pi}{2} + \sum_{k=1}^{\infty} \frac{-8}{(2k-1)^2\pi}\cos(2k-1)x \\
&= \pi - \sum_{k=1}^{\infty} \frac{8}{(2k-1)^2\pi}\cos(2k-1).
\end{aligned}$$

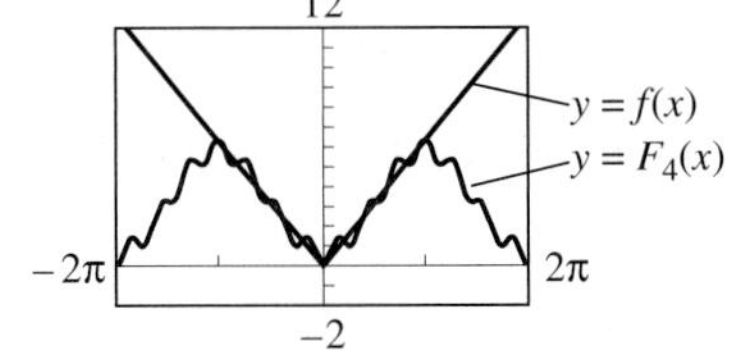

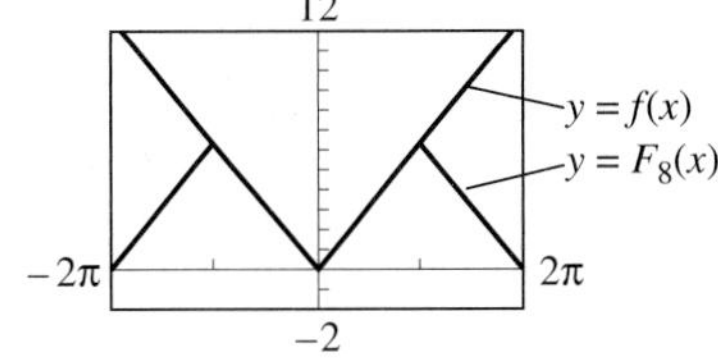

5. $f(x) = \begin{cases} 1 & \text{if } -\pi < x < 0 \\ -1 & \text{if } 0 < x < \pi \end{cases}$

$$\begin{aligned} a_0 &= \frac{1}{\pi}\int_{-\pi}^{0} 1\,dx + \frac{1}{\pi}\int_{0}^{\pi}(-1)\,dx \\ &= \frac{1}{\pi}[x]_{-\pi}^{0} + \frac{1}{\pi}[-x]_0^{\pi} \\ &= \frac{1}{\pi}(0-(-\pi)) + \frac{1}{\pi}(-\pi-0) = 0 \end{aligned}$$

$$\begin{aligned} a_k &= \frac{1}{\pi}\int_{-\pi}^{0}\cos(kx)\,dx + \frac{1}{\pi}\int_{0}^{\pi}[-\cos(kx)]\,dx \\ &= \frac{1}{\pi}\left[\frac{1}{k}\sin kx\right]_{-\pi}^{0} - \frac{1}{\pi}\left[\frac{1}{k}\sin kx\right]_{0}^{\pi} \\ &= \frac{1}{\pi}\left[\frac{1}{k}\sin 0 - \frac{1}{k}\sin(-k\pi)\right] \\ &\quad - \frac{1}{\pi}\left[\frac{1}{k}\sin k\pi - \frac{1}{k}\sin 0\right] \\ &= 0 \end{aligned}$$

$$\begin{aligned} b_k &= \frac{1}{\pi}\int_{\pi}^{0}[\sin(kx)]\,dx + \frac{1}{\pi}\int_{0}^{\pi}[-\cos(kx)]\,dx \\ &= \frac{1}{\pi}\left[-\frac{1}{k}\cos kx\right]_{-\pi}^{0} - \frac{1}{\pi}\left[-\frac{1}{k}\cos kx\right]_{0}^{\pi} \\ &= \frac{1}{\pi}\left[-\frac{1}{k}\cos 0 + \frac{1}{k}\cos(-k\pi)\right] \\ &\quad - \frac{1}{\pi}\left[-\frac{1}{k}\cos k\pi + \frac{1}{k}\cos(0)\right] \\ &= \frac{-2}{k\pi} + \frac{2}{k\pi}(-1)^k \\ &= \frac{2}{k\pi}\left[(-1)^k - 1\right] \\ &= \begin{cases} 0 & \text{if } k \text{ is even} \\ \dfrac{-4}{k\pi} & \text{if } k \text{ is odd} \end{cases} \end{aligned}$$

So $b_{2k} = 0$ and $b_{2k-1} = \dfrac{-4}{(2k-1)\pi}$. Thus,

$$f(x) = \sum_{k=1}^{\infty} \frac{-4}{(2k-1)\pi}\sin(2k-1)x.$$

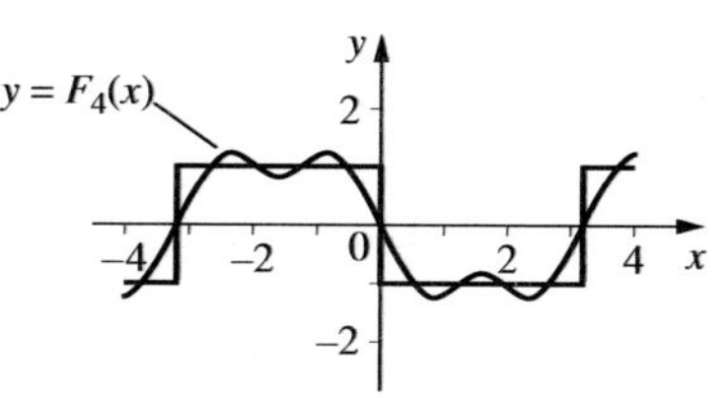

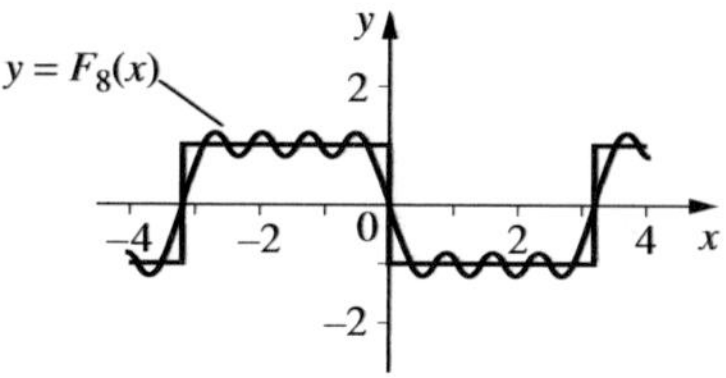

7. $f(x) = 3\sin 2x$ is already periodic on $[-2\pi, 2\pi]$.

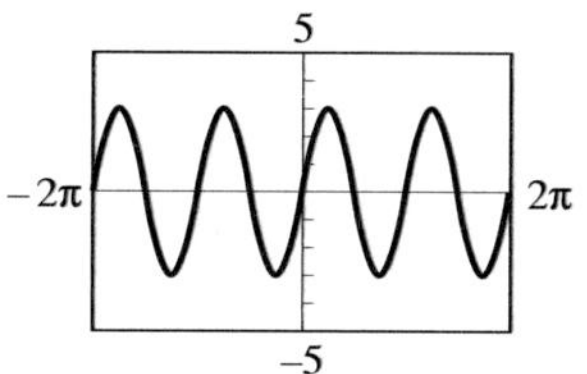

9. $f(x) = -x$ on $[-1, 1]$; so $l = 1$.

$$\begin{aligned} a_0 &= \int_{-1}^{1}(-x)\,dx = \left[\frac{-x^2}{2}\right]_{-1}^{1} \\ &= \frac{-1}{2} - \frac{-1}{2} \\ &= 0 \end{aligned}$$

$$\begin{aligned} a_k &= \int_{-1}^{1}(-x)\cos(k\pi x)\,dx \\ &= \left[\frac{-x}{k\pi}\sin k\pi x - \frac{1}{(k\pi)^2}\cos k\pi x\right]_{-1}^{1} \\ &= \frac{-1}{k\pi}\sin k\pi - \frac{1}{(k\pi)^2}\cos k\pi \\ &\quad - \left[\frac{1}{k\pi}\sin(-k\pi) - \frac{1}{(k\pi)^2}\cos(-k\pi)\right] \\ &= 0 \end{aligned}$$

$$b_k = \int_{-1}^{1} (-x)\sin(k\pi x)\,dx$$
$$= \left[\frac{x}{k\pi}\cos k\pi x - \frac{1}{(k\pi)^2}\sin k\pi x\right]_{-1}^{1}$$
$$= \frac{1}{k\pi}\cos k\pi - \frac{1}{(k\pi)^2}\sin k\pi - \left[\frac{-1}{k\pi}\cos(-k\pi) - \frac{1}{(k\pi)^2}\sin(-k\pi)\right]$$
$$= \frac{2}{k\pi}\cos k\pi$$
$$= \frac{2}{k\pi}(-1)^k$$

So $f(x) = \sum_{k=1}^{\infty}(-1)^k \frac{2}{k\pi}\sin k\pi x.$

11. $f(x) = x^2$ on $[-1, 1]$ so $l = 1$

$$a_0 = \frac{1}{1}\int_{-1}^{1} x^2\,dx = \left.\frac{x^3}{3}\right]_{-1}^{1}$$
$$= \frac{1}{3} - \left(-\frac{1}{3}\right) = \frac{2}{3} \text{ and } \frac{a_0}{2} = \frac{1}{3}$$
$$a_k = \frac{1}{1}\int_{-1}^{1} x^2\cos k\pi x\,dx$$
$$= \left[\frac{x^2}{k\pi}\sin k\pi x + \frac{2x}{(k\pi)^2}\cos k\pi x - \frac{2}{(k\pi)^3}\sin k\pi x\right]_{-1}^{1}$$
$$= \left[\frac{2}{(k\pi)^2}\cos k\pi - \left(-\frac{2}{(k\pi)^2}\cos(-k\pi)\right)\right]_{-1}^{1}$$
since $\sin(\pm k\pi x) = 0$
$$= \frac{4}{(k\pi)^2}\cos k\pi$$
$$= \frac{4}{(k\pi)^2}(-1)^k$$

$$b_k = \frac{1}{1}\int_{-1}^{1} x^2\sin(k\pi x)\,dx$$
$$= \left[\frac{x^2}{k\pi}\cos k\pi x + \frac{2x}{(k\pi)^2}\sin k\pi x + \frac{2}{(k\pi)^2}\cos k\pi x\right]_{-1}^{1}$$
$$= \left[\frac{-1}{k\pi}\cos k\pi + \frac{2}{(k\pi)^3}\cos k\pi - \left(-\frac{1}{k\pi}\cos(-k\pi) + \frac{2}{(k\pi)^3}\cos(-k\pi)\right)\right]$$
$$= 0$$

So

$$f(x) = \frac{1}{3} + \sum_{k=1}^{\infty}\frac{(-1)^k 4}{(k\pi)^2}\cos k\pi x.$$

13. $f(x) = \begin{cases} 0 & \text{if } -1 < x < 0 \\ x & \text{if } 0 < x < 1 \end{cases}$ so $l = 1$

$$a_0 = \int_{-1}^{1} f(x)\,dx = \int_{-1}^{0} 0\,dx + \int_{0}^{1} x\,dx = \left[\frac{x^2}{2}\right]_0^1 = \frac{1}{2}$$

and $\frac{a_0}{2} = \frac{1}{4}$

$$a_k = \int_{-1}^{0} 0\,dx + \int_{0}^{1} x\cos(k\pi x)\,dx$$
$$= \left[\frac{x}{k\pi}\sin(k\pi x) + \frac{1}{(k\pi)^2}\cos(k\pi x)\right]_0^1$$
$$= \frac{1}{k\pi}\sin k\pi + \frac{1}{(k\pi)^2}\cos k\pi - 0 - \frac{1}{(k\pi)^2}\cos(0)$$
$$= \frac{1}{(k\pi)^2}[(-1)^k - 1]$$
$$= \begin{cases} 0 & \text{if } k \text{ is even} \\ \dfrac{-2}{(k\pi)^2} & \text{if } k \text{ is odd} \end{cases}$$

So $a_{2k} = 0$ and $a_{2k-1} = \frac{-2}{2(k-1)^2\pi^2}.$

$$b_k = \int_{-1}^{0} 0\,dx + \int_{0}^{1} x\cos(k\pi x)\,dx$$
$$= \left[\frac{-x}{k\pi}\cos(k\pi x) + \frac{1}{(k\pi)^2}\sin(k\pi x)\right]_0^1$$
$$= -\frac{1}{k\pi}\cos k\pi + \frac{1}{k\pi}\sin k\pi - 0 + \frac{1}{k\pi}\sin(0)$$
$$= \frac{-1}{k\pi}(-1)^k$$

So

$$f(x) = \frac{1}{4} + \sum_{k=1}^{\infty}\left[\begin{array}{l}\frac{-2}{(2k-1)^2\pi^2}\cos(2k-1)\pi x \\ +\frac{(-1)^{k+1}}{k\pi}\sin k\pi x\end{array}\right].$$

15. $f(x) = x$ on $[-2, 2]$

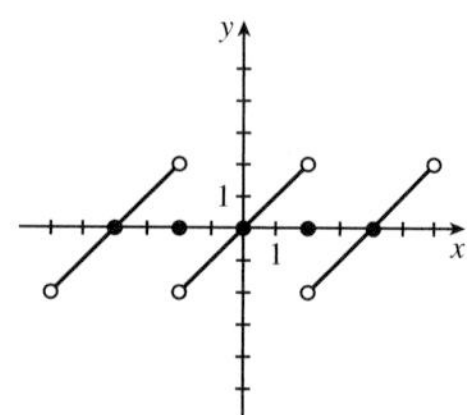

17. $f(x) = \begin{cases} -x & \text{if } -1 < x < 0 \\ 0 & \text{if } 0 < x < 1 \end{cases}$

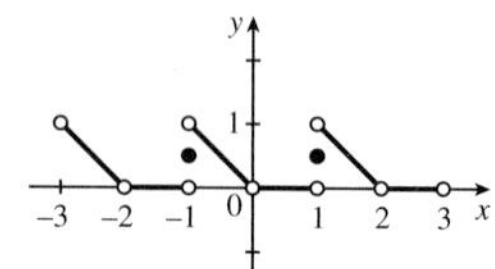

19. $f(x) = \begin{cases} -1 & \text{if } -2 < x < -1 \\ 0 & \text{if } -1 < x < 1 \\ 1 & \text{if } 1 < x < 2 \end{cases}$

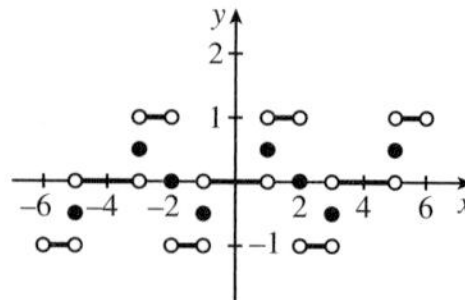

21. Since the function $f(x) = x$ and its derivative are continuous on $(-\pi, \pi)$, the Fourier series will converge to that function on that interval. At the endpoints, the Fourier series will converge to the average of $f(\pi)$ and $f(-\pi)$, which is 0.

23. Since the function and its derivative are continuous except for one jump discontinuity on the interval $(-\pi, \pi)$, the Fourier series will converge to -1 on $(0, \pi)$ and will converge to 1 on $(-\pi, 0)$. At the points, π, 0, and $-\pi$, the Fourier series will converge to the average of 1 and -1, which is 0.

25. $$f(x) = x^2 = \frac{1}{3} + \sum_{k=1}^{\infty}(-1)^k\frac{4}{k^2\pi^2}\cos k\pi x$$
$$f(1) = 1 = \frac{1}{3} + \sum_{k=1}^{\infty}(-1)^k\frac{4}{k^2\pi^2}(-1)^k$$
$$\frac{2}{3} = \sum_{k=1}^{\infty}\frac{4}{k^2\pi^2} \text{ so } \frac{2}{3} = \frac{4}{\pi^2}\sum_{k=1}^{\infty}\frac{1}{k^2} \text{ so } \frac{\pi^2}{6} = \sum_{k=1}^{\infty}\frac{1}{k^2}$$

27. $$f(x) = |x| = \begin{cases} -x & \text{if } -\pi \le x < 0 \\ x & \text{if } 0 \le x \le \pi \end{cases} \text{ gives us}$$
$$f(x) = \frac{\pi}{2} - \sum_{k=1}^{\infty}\frac{4}{(2k-1)^2\pi}\cos(2k-1)x$$
$$f(0) = 0 = \frac{\pi}{2} - \sum_{k=1}^{\infty}\frac{4}{(2k-1)^2\pi}\cos 0$$
$$\frac{\pi}{2} = \sum_{k=1}^{\infty}\frac{4}{(2k-1)^2\pi} = \frac{4}{\pi}\sum_{k=1}^{\infty}\frac{1}{(2k-1)^2}$$
$$\frac{\pi^2}{8} = \sum_{k=1}^{\infty}\frac{1}{(2k-1)^2}$$

29. If $f(x) = \cos x$, then

$$f(-x) = \cos(-x) = cosx = f(x),$$

so $\cos x$ is even. If $f(x) = \sin x$, then

$$f(-x) = \sin(-x) = -\sin x = -f(x),$$

so $\sin x$ is odd. If $f(x) = \cos x + \sin x$, then

$$f(-x) = \cos(-x) + \sin(-x)$$
$$= \cos x - \sin x \neq f(x),$$

and $\cos x - \sin x \neq -f(x)$, so $\cos x + \sin x$ is neither even nor odd.

31. If $f(x)$ is odd and $g(x) = f(x)\cos x$, then

$$g(-x) = f(-x)\cos(-x) = -f(x)\cos x = -g(x)$$

is odd. Also, if $h(x) = f(x)\sin x$, then

$$\begin{aligned} h(-x) &= f(-x)\sin(-x) \\ &= -f(x)(-\sin x) \\ &= f(x)\sin x \\ &= h(x), \end{aligned}$$

so $h(x)$ is even.

33. If f is even and g is odd, then

$$f(-x)g(-x) = f(x)[-g(x)] = -f(x)g(x),$$

so fg is odd.

35. We have the Fourier series expansion of f with period $2l$ is

$$f(x) = \frac{a_0}{2} + \sum_{k=1}^{\infty}\left[a_k\cos\left(\frac{k\pi x}{l}\right) + b_k\sin\left(\frac{n\pi x}{l}\right)\right].$$

Multiply both sides of this equation by $\cos\left(\frac{n\pi x}{l}\right)$ and integrate with respect to x on the interval $[-l, l]$ to get

$$\int_{-l}^{l} f(x)\cos\left(\frac{n\pi x}{l}\right)dx = \int_{-\ell}^{\ell}\frac{a_0}{2}\cos\left(\frac{n\pi x}{l}\right)dx$$

$$+\sum_{k=1}^{\infty}\left[\begin{array}{l} a_k\int_{-l}^{l}\cos\left(\frac{k\pi x}{l}\right)\cos\left(\frac{n\pi x}{l}\right)dx \\ +b_k\int_{-l}^{l}\sin\left(\frac{k\pi x}{l}\right)\cos\left(\frac{n\pi x}{l}\right)dx \end{array}\right].$$

Well,

$$\int_{-l}^{1}\cos\left(\frac{n\pi x}{l}\right)dx = 0 \text{ for all } n.$$

And,

$$\int_{-l}^{l}\sin\left(\frac{k\pi x}{l}\right)\cos\left(\frac{n\pi x}{l}\right)dx = \begin{cases} 0 & \text{if } n \neq k \\ l & \text{if } n = k \end{cases}.$$

So when $n = k$, we have

$$\int_{-l}^{l} f(x)\cos\left(\frac{k\pi x}{l}\right)dx = a_k l,$$

so

$$a_k = \frac{1}{l}\int_{-l}^{l} f(x)\cos\left(\frac{k\pi x}{l}\right)dx.$$

Now multiply both sides of the original equation by $\sin\left(\frac{n\pi x}{l}\right)$ and integrate on $[-l, l]$, so we have

$$\int_{-l}^{l} f(x)\sin\left(\frac{n\pi x}{l}\right)dx = \int_{-l}^{l}\frac{a_0}{2}\sin\left(\frac{n\pi x}{l}\right)dx$$

$$+\sum_{k=1}^{\infty}\left[\begin{array}{l} a_k\int_{-l}^{l}\cos\left(\frac{k\pi x}{l}\right)\sin\left(\frac{n\pi x}{l}\right)dx \\ b_k\int_{-l}^{l}\sin\left(\frac{k\pi x}{l}\right)\sin\left(\frac{n\pi x}{l}\right)dx \end{array}\right].$$

Well,

$$\int_{-l}^{l}\sin\left(\frac{n\pi x}{l}\right)dx = 0.$$

And

$$\int_{-l}^{l}\cos\left(\frac{k\pi x}{l}\right)\sin\left(\frac{n\pi x}{l}\right)dx = 0.$$

Also,

$$\int_{-l}^{l}\sin\left(\frac{k\pi x}{l}\right)\sin\left(\frac{k\pi x}{l}\right)dx = \begin{cases} 0 & \text{if } n \neq k \\ l & \text{if } n = k \end{cases}.$$

So when $n = k$, we have

$$\int_{-l}^{l} f(x)\sin\left(\frac{k\pi x}{l}\right)dx = b_k l.$$

So

$$b_k = \frac{1}{l}\int_{-l}^{l} f(x)\sin\left(\frac{k\pi x}{l}\right)dx.$$

37.

$$\begin{aligned} f(x) &= x^3 \\ f(-x) &= (-x)^3 = -x^3 \end{aligned}$$

So f is odd and the Fourier Series will contain only sine.

39.

$$f(x) = e^x,\ f(-x) = e^{-x} \neq f(x),$$
$$\text{and } e^{-x} \neq -f(x).$$

So f is neither even nor odd, so the Fourier Series will contain both.

41. $$f(x) = \begin{cases} 0 & \text{if } -1 < x < 0 \\ x & \text{if } 0 < x < 1, \end{cases}$$

$f(-x) = -x = -f(x)$ if $x > 0$ or $x < 0$,
and $f(-0) = 0 = f(x)$.

So f is neither even nor odd so the Fourier Series will contain both.

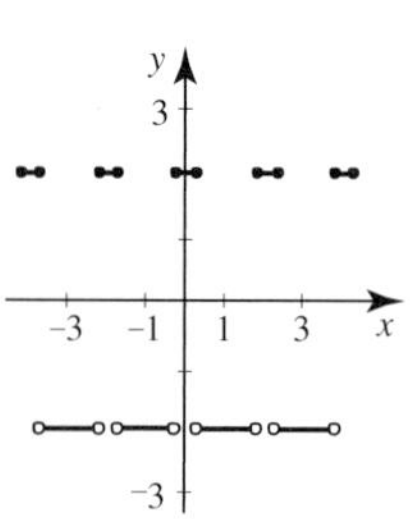

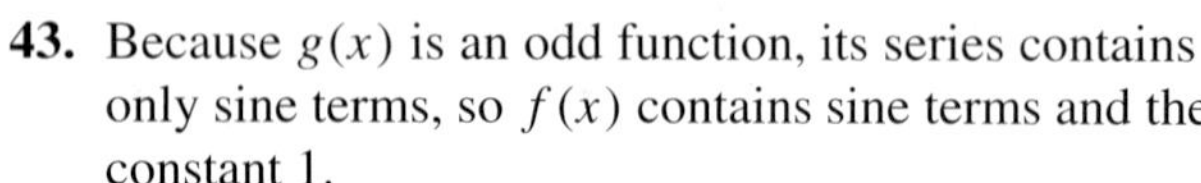

43. Because $g(x)$ is an odd function, its series contains only sine terms, so $f(x)$ contains sine terms and the constant 1.

45. $$f(x) = -\text{on}\ [-4, 4]$$

$$f(x) = \begin{cases} -2 & \frac{1}{3} < |x| < 1 \\ 2 & |x| \le \frac{1}{3} \end{cases}$$

$$f(x) = \begin{cases} -2 & \frac{1}{4} < |x| < 1 \\ 2 & |x| \le \frac{1}{4} \end{cases}$$

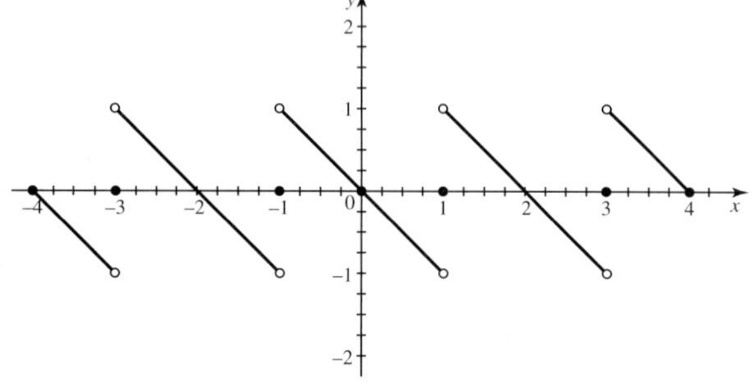

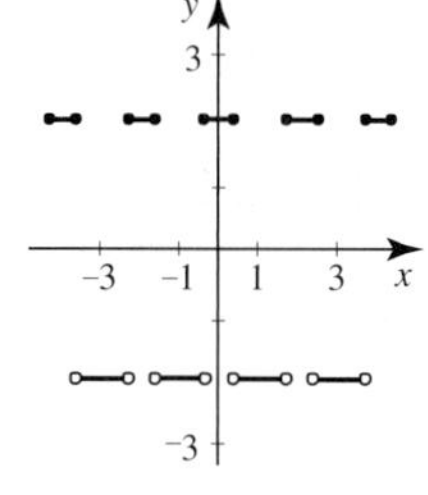

47. The cutoff frequency, n, corresponds to the partial sum $F_n(x)$ because in both, everything after the nth term is zero.

$$f(x) = -x \text{ on } [-1, 1], \text{ so}$$

$$f(x) = \sum_{k=1}^{\infty} \frac{(-1)^k 2}{k\pi} \sin k\pi x, \text{ thus}$$

$$F_2(x) = \frac{2}{\pi}\left[-\sin \pi x + \frac{1}{2}\sin 2\pi x\right].$$

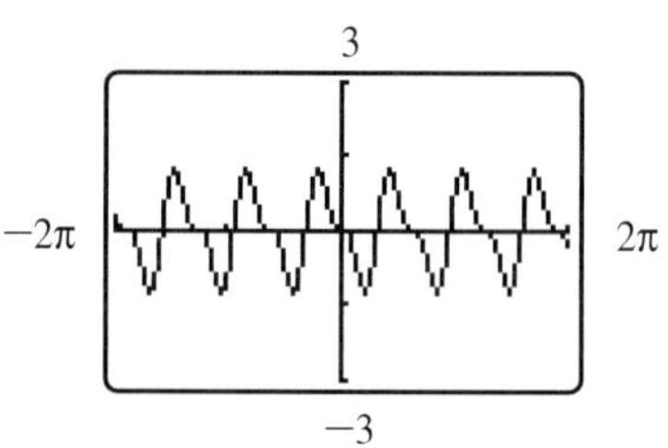

$F_4(x) =$

$$\frac{2}{\pi}\left[-\sin \pi x + \frac{1}{2}\sin 2\pi - \frac{1}{3}\sin 3\pi x + \frac{1}{4}\sin 4\pi x\right].$$

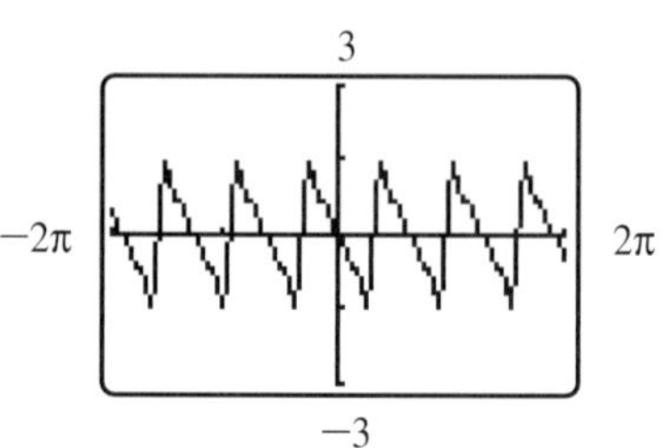

$$f(x) = \begin{cases} -2 & -1 < x \le 0 \\ 2 & 0 < x \le 1 \end{cases}, \text{ so}$$

$$f(x) = \sum_{k=1}^{\infty} \frac{8}{(2k-1)\pi} \sin[(2k-1)\pi x],$$

thus

$$F_2(x) = \frac{8}{\pi}\left[\sin \pi x + \frac{1}{3}\sin 3\pi x\right].$$

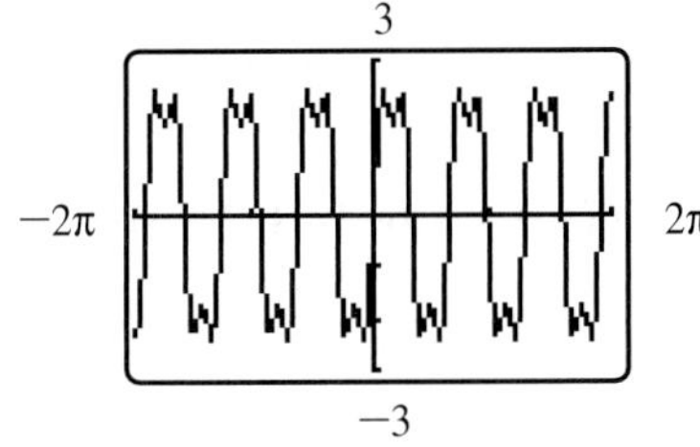

$F_4(x) =$

$$\frac{8}{\pi}\left[\sin \pi x + \frac{1}{3}\sin 3\pi x + \frac{1}{5}\sin 5\pi x + \frac{1}{7}\sin 7\pi x\right].$$

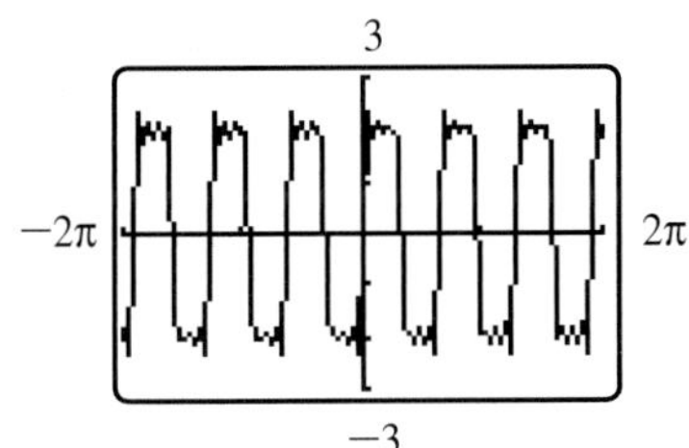

Because the wave is smoother.

49. The halo effect could be caused by an overshoot at the edges.

51. If $A = -1$ and $B = 4\pi$, then $-\sin 4\pi t = f(t)$, so

$$f(0) = 0,\ f\left(\frac{1}{3}\right) = \frac{\sqrt{3}}{2},$$

$$f\left(\frac{2}{3}\right) = \frac{\sqrt{3}}{2},\ f(1) = 0.$$

If $A = 1$ and $B = 2\pi$, then $\sin 2\pi t = f(t)$, so

$$f(0) = 0,\ f\left(\frac{1}{3}\right) = \frac{\sqrt{3}}{2},$$

$$f\left(\frac{2}{3}\right) = \frac{\sqrt{3}}{2},\ f(1) = 0.$$

7.10 Using Series to Solve Differential Equations

1. The answer in Appendix G is incorrect.

$$0 = y'' + 2xy' + 4y$$

$$= \sum_{n=2}^{\infty} n(n-1)a_n x^{n-2} + 2\sum_{n=1}^{\infty} n a_n x^{n-1} + 4\sum_{n=0}^{\infty} a_n x^n$$

$$= \sum_{n=0}^{\infty} (n+2)(n+1)a_{n+2}x^n$$

$$+ 2\sum_{n=0}^{\infty} n a_n x^n + 4\sum_{n=0}^{\infty} a_n x^n$$

$$= \sum_{n=0}^{\infty} \left[(n+2)(n+1)a_{n+2} + 1(n+2)a_n\right] x^n.$$

So $(n+2)(n+1)a_{n+2} + 2(n+2)a_n = 0$, or

$$a_{n+2} = -\frac{2(n+2)}{(n+2)(n+1)}a_n = -\frac{2}{n+1}a_n.$$

So

$$a_{2n} = \frac{(-1)^n 2^{2n-1}(n-1)!}{(2n-1)!}a_0$$

and $a_{2n+1} = \dfrac{(-1)^n}{n!}a_1$.

$$y = \sum_{n=0}^{\infty} a_n x^n = \sum_{n=0}^{\infty} a_{2n}x^{2n} + \sum_{n=0}^{\infty} a_{2n+1}x^{2n+1}$$

$$= a_0 \sum_{n=0}^{\infty} \frac{(-1)^n 2^{2n-1}(n-1)!}{(2n-1)!}x^{2n}$$

$$+ a_1 \sum_{n=0}^{\infty} \frac{(-1)^n}{n!}x^{2n+1}.$$

3. $0 = y'' - xy' - y$

$$= \sum_{n=2}^{\infty} n(n-1)a_n x^{n-2} - x\sum_{n=1}^{\infty} na_n x^{n-1} - \sum_{n=0}^{\infty} a_n x^n$$

$$= \sum_{n=0}^{\infty} (n+2)(n+1)a_{n-2}x^n - x\sum_{n=0}^{\infty} na_n x^n - \sum_{n=0}^{\infty} a_n x^n$$

$$= \sum_{n=0}^{\infty} \left[(n+2)(n+1)a_{n+2} - (n+1)a_n\right] x^n$$

So $(n+2)(n+1)a_{n+2} - (n+1)a_n$, or

$$a_{n+2} = \frac{n+2}{(n+2)(n+1)}a_n = \frac{1}{n+2}a_n,$$

$$a_{2n} = \frac{1}{2^n n!}a_0, \quad a_{2n+1} = \frac{2^n n!}{(2n+1)!}a_1.$$

$$y = \sum_{n=0}^{\infty} a_n x^n = \sum_{n=0}^{\infty} a_{2n}x^{2n} + \sum_{n=0}^{\infty} a_{2n+1}x^{2n+1}$$

$$= a_0 \sum_{n=0}^{\infty} \frac{1}{2^n n!} + a_1 \sum_{n=0}^{\infty} \frac{2nn!}{(2n+1)!}x^{2n+1}.$$

5. $0 = y'' - xy' =$

$$\sum_{n=2}^{\infty} n(n-1)a_n x^{n-2} - x\sum_{n=1}^{\infty} na_n x^{n-1}$$

$$= \sum_{n=0}^{\infty} (n+2)(n+1)a_{n+2}x^n - \sum_{n=0}^{\infty} na_n x^n$$

$$= \sum_{n=0}^{\infty} \left[(n+2)(n+1)a_{n+2} - na_n\right] x^n$$

So $(n+2)(n+1)a_{n+2} - na_n = 0$, or

$$a_{n+2} = \frac{n}{(n+2)(n+1)}a_n,$$

$$a_{2n} = 0 \text{ for } n > 0, \quad a_{2n+2} = \frac{a_1}{(2n+1)n\,2^n}.$$

$$y = \sum_{n=0}^{\infty} a_n x^n = \sum_{n=0}^{\infty} a_{2n}x^{2n} + \sum_{n=0}^{\infty} a_{2n+1}x^{2n+1}$$

$$= a_0 + a_1 \sum_{n=0}^{\infty} \frac{x^{2n+1}}{(2n+2)n!2^n}.$$

7. $0 = y'' - x^2 y'$

$$= \sum_{n=2}^{\infty} a_n n(n-1)x^{n-2} - x^2\sum_{n=1}^{\infty} na_n x^{n-1}$$

$$= \sum_{n=2}^{\infty} a_n n(n-1)x^{n-2} - \sum_{n=1}^{\infty} na_n x^{n+1}$$

$$= 2a_2 + \sum_{n=3}^{\infty} a_n n(n-1)x^{n-2} - \sum_{n=1}^{\infty} na_n x^{n+1}$$

$$= 2a_2 + \sum_{n=0}^{\infty} a_{n+3}n(n+3)(n+2)x^{n+1} - \sum_{n=0}^{\infty} na_n x^{n+1}$$

$$= 2a_2 + \sum_{n=0}^{\infty} \left[a_{n+3}(n+3)(n+2) - na_n\right] x^{n+1}.$$

Hence, $a_2 = 0$, $a_{n+3} = \dfrac{na_n}{(n+3)(n+2)}$.
Thus, $a_2 = a_5 = a_8 = a_{11} = \ldots = 0$
and $a_3 = a_6 = a_9 = a_{12} = \ldots = 0$,
while $a_{3n+1} = \dfrac{a_1}{(3n+1)(n!3^n}$.

Therefore, $y = a_0 + a_1 \sum_{n=0}^{\infty} \dfrac{x^{3n+1}}{(3n+1)n!3^n}$.

9. Rewrite the equation as $y'' - (x-1)y' - y = 0$. Then

$$0 \sum_{n=2}^{\infty} a_n n(n-1)(x-1)^{n-2}$$

$$-(x-1)\sum_{n=1}^{\infty} a_n n(x-1)^{n-1} - \sum_{n=0}^{\infty} a_n n(x-1)^n$$

$$= \sum_{n=0}^{\infty} a_{n+2}(n+2)(n+1)(x-1)^n - \sum_{n=0}^{\infty} a_n n(x-1)^n$$

$$- \sum_{n=0}^{\infty} a_n n(x-1)^n.$$

So $a_{n+2} = (n+2)(n+1) - a_n n - a_n = 0$.
That is, $a_{n+2} = \dfrac{a_n}{n+2}$. So

$$a_{2n} = \frac{a_0}{2^n n!}, \quad a^{2n+1} = \frac{2^n n! a_1}{(2n+1)!}.$$

Thus,

$$y = a_0 \sum_{n=0}^{\infty} \frac{(x-1)^{2n}}{2^n n!} + a_1 \sum_{n=0}^{\infty} \frac{2^n n!(x-1)^{2n+1}}{(2n+1)!}.$$

11. $0 = y'' - (x-1)y - y$

$$= \sum_{n=2}^{\infty} a_n n(n-1)(x-1)^{n-2}$$

$$- (x-1) \sum_{n=0}^{\infty} a_n (x-1)^n$$

$$- \sum_{n=0}^{\infty} a_n (x-1)^n$$

$$= \sum_{n=0}^{\infty} a_{n+2}(n+2)(n+1)(x-1)^n$$

$$- \sum_{n=0}^{\infty} a_n (x-1)^{n+1}$$

$$- \sum_{n=0}^{\infty} a_n (x-1)^n$$

$$= a_2(2)(1) - a_0$$

$$+ \sum_{n=1}^{\infty} a_{n+2}(n+2)(n+1)(x-1)^n$$

$$- \sum_{n=1}^{\infty} a_{n-1}(x-1)^n - \sum_{n=1}^{\infty} a_{n-1}(x-1)^n$$

$$= 2a_2 - a_0$$

$$+ \sum_{n=1}^{\infty} \left[a_{n+2}(n+2)(n+1) - a_{n-1} - a_n\right](x-1)^n$$

So a_0 and a_1 are arbitrary, $a_2 = \frac{a_0}{2}$ and

$$a_{n+2}(n+1)(n+1) - a_{n-1} - a_n n = 0$$

for $n \geq 1$. That is

$$a_{n+2} = \frac{a_{n-1} + a_n}{(n+2)(n+1)}.$$

13. The book should say "See Example 10.2," not "See Exercise 1." According to Example 10.2, the general solution is

$$y = a_0 \sum_{n=0}^{\infty} \frac{(-1)^n x^{2n}}{n!} + a_1 \sum_{n=0}^{\infty} \frac{(-1)^n 2^{2n} n! x^{2n+1}}{(2n+1)!}.$$

Then $y(0) = a_0 = 5$ $y'(0) = a_1 = -7$, so the solution to the initial value problem is

$$y = 5 \sum_{n=0}^{\infty} \frac{(-1)^n x^{2n}}{n!} - 7 \sum_{n=0}^{\infty} \frac{(-1)^n 2^{2n} n! x^{2n+1}}{(2n+1)!}.$$

15. The general solution is

$$y = a_0 \sum_{n=0}^{\infty} \frac{(x-1)^{2n}}{2^n n!} + a_1 \sum_{n=0}^{\infty} \frac{2^n n!(x-1)^{2n+1}}{(2n+1)!}.$$

Then $y(1) = a_0 = -3$ and $y'(1) = a_1 = 12$, so the solution to the initial value problem is

$$y = -3 \sum_{n=0}^{\infty} \frac{(x-1)^{2n}}{2^n n!} + 12 \sum_{n=0}^{\infty} \frac{2^n n!(x-1)^{2n+1}}{(2n+1)!}.$$

17. Using the Ratio Test, we can look separately at

$$\sum_{n=0}^{\infty} \frac{1}{2^n n!} x^{2n} \quad \text{and} \quad \sum_{n=0}^{\infty} \frac{2^n n!}{(2n+1)!} x^{2n+1}.$$

Let

$$a_n = \frac{x^{2n}}{2^n n!}.$$

Then

$$\lim_{n\to\infty} \left|\frac{a_{n+1}}{a_n}\right|$$

$$= \lim_{n\to\infty} \left|\frac{x^{2n+2}}{2^{n+1}(n+1)!} \frac{2^n n!}{x^{2n}}\right|$$

$$= \lim_{n\to\infty} \left|\frac{x^2}{2(n+1)}\right| = 0$$

for all x, so this series converges for all x. Similarly, let

$$a_n = \frac{2^n n! x^{2n+1}}{(2n+2)!}.$$

Then

$$\lim_{n\to\infty}\left|\frac{a_{n+1}}{a_n}\right|$$
$$=\lim_{n\to\infty}\left|\frac{(2n+1)!}{2^n n!\,x^{(2n+1)}}\;\frac{2^{n+1}(n+1)!\,x^{2n+3}}{(2n+3)!}\right|$$
$$=\lim_{n\to\infty}\left|\frac{x^2}{2n+3}\right|=0$$

for all x, so this series also converges for all x. Hence, the sum of these two series converges for all x, so the radius of convergence of the power series solutions is $r=\infty$.

19. Using the Ratio Test, we can look separately at

$$\sum_{n=0}^{\infty}\frac{(x-1)^{2n}}{2^n n!} \quad\text{and}\quad \sum_{n=0}^{\infty}\frac{2^n n!(x-1)^{2n+1}}{(2n+1)!}.$$

These are the same series we looked at in Exercise 17, except that x is replaced by $x-1$. So the radius of convergence is again ∞.

21. $0=x^2y''+xy'+x^2y$

$$=x^2\sum_{n=2}^{\infty}a_n n(n-1)x^{n-2}+x\sum_{n=1}^{\infty}a_n n x^{n-1}$$
$$+x^2\sum_{n=0}^{\infty}a_n x^n$$
$$=\sum_{n=2}^{\infty}a_n n(n-1)x^n+\sum_{n=1}^{\infty}a_n n x^n$$
$$+\sum_{n=0}^{\infty}a_n x^{n+2}$$
$$=\sum_{n=0}^{\infty}a_{n+2}(n+2)(n+1)x^{n+2}$$
$$+\sum_{n=-1}^{\infty}a_{n+2}(n+2)x^{n+2}+\sum_{n=0}^{\infty}a_n x^{n+2}$$
$$=a_1x+$$
$$\sum_{n=0}^{\infty}\left[a_{n+2}(n+2)(n+1)+a_{n+2}(n+2)+a_n\right]x^{n+2}$$

Hence, $a_1=0$ and

$$a_{n+2}(n+2)(n+1)+a_{n+2}(n+2)+a_n=0,$$

which means

$$a_{n+2}=\frac{-a_n}{(n+2)^2}.$$

Thus, $0=a_1=a_3=a_5=\ldots$ and

$$a_{2n}=\frac{(-1)^n a_0}{2^{2n}(n!)^2}.$$

So one solution to the differential equation is

$$\sum_{n=0}^{\infty}a_{2n}x^{2n}=a_0\sum_{n=0}^{\infty}\frac{(-1)^n x^{2n}}{2^{2n}(n!)^2}.$$

23. Let y_1 and y_2 be the two solutions described in Example 10.3. Then y_1 is of the form $\sum_{n=0}^{\infty} b_{3n}x^{3n}$, where the coefficients b_n satisfy the recurrence relation

$$b_{n+3}=\frac{1}{(n+3)(n+2)}b_n.$$

We apply the Ratio Test to the series $\sum_{n=0}^{\infty} b_{3n}x^{3n}$.

$$\lim_{n\to\infty}\left|\frac{b_{3n+3}x^{3n+3}}{b_{3n}x^{3n}}\right|=\lim_{n\to\infty}\left|\frac{x^3}{(3n+3)(3n+2)}\right|=0$$

for all x, so this series converges for all x. The same argument applies to the series for y_2, so the radius of convergence of the series solution is $r=\infty$.

25. We are given

$$y''+2xy'-xy=0,$$
$$y(0)=2,\ y'(0)=-5.$$

Substituting $x=0$, we get

$$y''(0)+2\cdot 0\cdot y'(0)-0\cdot y(0)=0,$$

so $y''(0)=0$. Now differentiate the equation

$$y''=-2xy'+xy \text{ to get}$$
$$y'''=-2y'-2xy''+y+xy'$$
$$=-2xy''+(x-2)y'+y.$$

Then

$$y'''(0) = -2(0)y''(0) + (0-2)y'(0) + y(0)$$
$$= 0 - 2(-5) + 2 = 12.$$

Differentiating

$$y''' = -2xy'' + (x-2)y' + y,$$

we get

$$y^{(4)} = -2y'' - 2xy''' + y' + (x-2)y'' + y'$$
$$= -2xy''' + (x-4)y'' + 2y'.$$

Then

$$y^{(4)}(0) = -2(0)y'''(0) + (0-4)y''(0) + 2y'(0)$$
$$= 0 - 4(0) + 2(-5) = -10.$$

Differentiating

$$y^{(4)} = -2xy''' + (x-4)y'' + 2y',$$

we get

$$y^{(5)} = -2y''' - 2xy^{(4)} + y'' + (x-4)y''' + 2y''$$
$$= -2xy^{(4)} + (x-6)y''' + 3y''.$$
$$y^{(5)}(0) = -2(0)y^{(4)}(0) + (0-6)y'''(0),$$

so $+ 3y''(0)$

$$= 0 - 6(12) + 3(0) = -72.$$

Thus,

$$P_5(x) = y(0) + y'(0)x + y''(0)\frac{x^2}{2}$$
$$+ y'''(0)\frac{x^3}{3!} + y^{(4)}(0)\frac{x^4}{4!} + y^{(5)}(0)\frac{x^5}{5!}$$
$$= 2 + (-5)x + (0)\frac{x^2}{2} + (12)\frac{x^3}{3!}$$
$$+ (-10)\frac{x^4}{4!} + (-72)\frac{x^5}{5!}$$
$$= 2 - 5x + 2x^3 - \frac{5x^4}{12} - \frac{3x^5}{5}.$$

27. We are given

$$y'' = -e^x y' + (\sin x)y.$$

Differentiating, we get

$$y''' = -e^x y' - e^x y'' + (\cos x)y + (\sin x)y'$$
$$= -e^x y'' + (\sin x - e^x)y' + (\cos x)y.$$

Differentiating again,

$$y^{(4)} = -e^x y'' - e^x y''' + (\cos x - e^x)y'$$
$$+ (\sin x - e^x)y'' + (-\sin x)y + (\cos x)y'$$
$$= -e^x y''' + (\sin x - 2e^x)y' - (\sin x)y.$$

Differentiating once more,

$$y^{(5)} = -e^x y''' - e^x y^{(4)} + (\cos x - 2e^x)y''$$
$$+ (\sin x - 2e^x)y''' + (-2\sin x - e^x)y'$$
$$+ (2\cos x - e^x)y'' - (\cos x)y - (\sin x)y'$$
$$= -e^x y^{(4)} + (\sin x - 3e^x)y'''$$
$$+ (3\cos x - 3e^x)y'' + (-3\sin x - e^x)y'$$
$$- (\cos x)y.$$

Now starting from $y(0) = -2$, $y'(0) = 1$, we get

$$y''(0) = -e^0 y'(0) + (\sin 0)y(0),$$
$$= -1 \cdot 1 + 0 = -1,$$

so $y''(0) = -1 \cdot 1 + 0 = -1$. Then

$$y'''(0) = -e^0 y''(0) + (\sin 0 - e^0)y'(0)$$
$$+ (\cos 0)y(0)$$
$$= (-1)(-1) + (-1)(1) + (1)(-2) = -2,$$
$$y^{(4)}(0) = -e^0 y'''(0) + (\sin 0 - 2e^0)y''(0)$$
$$+ (2\cos 0 - e^0)y'(0) - (\sin 0)y(0)$$
$$= (-1)(-2) + (-2)(-1) + (1)(1) - 0$$
$$= 5,$$
$$y^{(5)}(0) = -e^0 y^{(4)}(0) + (\sin 0 - 3e^0)y'''0(0) +$$
$$(3\cos 0 - 3e^0)y''(0) + (-3\sin 0 - e^0)y'(0)$$
$$- (\cos 0)y(0)$$
$$= (-1)(5) + (-3)(-2) + (0)(-1) +$$
$$(-1)(1) - (1)(-2)$$
$$= 2.$$

Hence,

$$P_5(x) = y(0) + y'(0)x + y''(0)\frac{x^2}{2} + y'''(0)\frac{x^3}{3!}$$
$$+ y^{(4)}(0)\frac{x^4}{4!} + y^{(5)}(0)\frac{x^5}{5!}$$
$$= -2 + (1)x + (-1)\frac{x^2}{2} + (-2)\frac{x^3}{3!}$$
$$+ (5)\frac{x^4}{4!} + (2)\frac{x^5}{5!}$$
$$= -2 + x - \frac{x^2}{2} - \frac{x^3}{3} + \frac{5x^4}{24} + \frac{x^5}{60}.$$

29. We are given

$$y'' = -xy' - (\sin x)y.$$

Differentiating,

$$y''' = -y' - xy'' - (\cos x)y - (\sin x)y'$$
$$= -xy'' + (-1 - \sin x)y' - (\cos x)y.$$

Differentiating again,

$$y^{(4)} = -y'' - xy''' + (-\cos x)y'$$
$$+ (-1 - \sin x)y'' + (\sin x)y - (\cos x)y'$$
$$= xy''' + (-2 - \sin x)y'' + (-2\cos x)y'$$
$$+ (\sin x)y.$$

Differentiating once more,

$$y^{(5)} = -y''' - xy^{(4)}$$
$$+ (-\cos x)y'' + (-2 - \sin x)y'''$$
$$+ (2\sin x)y' + (-2\cos x)y''$$
$$+ (\cos x)y + (\sin x)y'$$
$$= -xy^{(4)} + (-3 - \sin x)y'''$$
$$+ (-3\cos x)y'' + (3\sin x)y'$$
$$+ (\cos x)y.$$

Now using $y(\pi) = 0$, $y'(\pi) = 4$, we get

$$y''(\pi) = -\pi y'(\pi) - (\sin \pi)y(\pi)$$
$$= (-\pi)(4) + 0 = -4\pi.$$

Then

$$y'''(\pi) = -\pi y''(\pi) + (-1 - \sin\pi)y'\pi$$
$$- (\cos\pi)y(\pi)$$
$$= (-\pi)(-4\pi) + (-1)(4) - (-1)(0)$$
$$= 4\pi^2 - 4.$$

And then

$$y^{(4)}(\pi) = (-\pi)y'''(\pi) + (-2 - \sin\pi)y''(\pi)$$
$$+ (-2\cos\pi)y'(\pi) + (\sin\pi)y(\pi)$$
$$= (-\pi)(4\pi^2 - 4) + (-2)(-4\pi)$$
$$+ (2)(4) + 0$$
$$= -4\pi^4 + 12\pi + 8.$$

Finally,

$$y^{(5)} = (-\pi)y^{(4)}(\pi) + (-3 - \sin\pi)y'''(\pi)$$
$$+ (-3\cos\pi)y''(\pi) + (3\sin\pi)y'(\pi)$$
$$+ (\cos\pi)y(\pi)$$
$$= (-\pi)(-4\pi^3 + 12\pi + 8)$$
$$+ (-3)(4\pi^2 - 4) + (3)(-4\pi) + 0 + 0$$
$$= 4\pi^4 - 24\pi^2 - 20\pi + 12.$$

Thus,

$$P_5(x)$$
$$= 0 + 4(x - \pi) + (-4\pi)\frac{(x-\pi)^2}{2}$$
$$+ (4\pi^2 - 4)\frac{(x-3)^3}{3!} +$$
$$(-4\pi^3 + 12\pi + 8)\frac{(x-\pi)^4}{4!}$$
$$+ (4\pi^4 - 24\pi^2 - 20\pi + 12)\frac{(x-\pi)^5}{5!}$$
$$= 4(x-\pi) - 2\pi(x-\pi)^2$$
$$+ (2\pi^2 - 2)\frac{(x-\pi)^3}{3}$$
$$+ (-\pi^3 + 3\pi + 2)\frac{(x-\pi)^4}{6}$$
$$+ (\pi^4 - 6\pi^2 - 5\pi + 3)\frac{(x-\pi)^5}{30}.$$

Chapter 7 Review

1. See Example 1.2.

$$\lim_{n\to\infty} \frac{4}{3+n} = \lim_{n\to\infty} \frac{\frac{4}{n}}{\frac{3}{n}+1} = 0$$

converges to 0.

3. See Example 1.7.

$$\lim_{n\to\infty} |a_n| = \lim_{n\to\infty} \frac{n}{n^2+4} = \lim_{n\to\infty} \frac{\frac{1}{n}}{1+\frac{4}{n^2}} = 0,$$

so original sequence converges to 0.

5. See Example 1.8.

$$0 < \frac{4^n}{n!} = \frac{4\cdot 4\cdot 4\ldots 4}{1\cdot 2\cdot 3\ldots n} <$$
$$4\cdot\frac{4}{3}\cdot\frac{4}{4}\cdot\frac{4}{n} = \frac{128}{3n}$$

and $\lim_{n\to\infty} \frac{128}{3n} = 0$, so by the Squeeze Theorem,

$$\left\{\frac{4^n}{n!}\right\}_{n=1}^{\infty}$$

converges to 0.

7. See Example 1.4.
$\{\cos \pi n\}_{n=1}^{\infty} = \{-1, 1, -1, \ldots\}$ diverges.

For Exercises 9 through 17, see Sections 7.2 and 7.3.

9. diverges

11. can't tell

13. diverges

15. converges

17. converges

19. See Example 2.4.
$\sum_{k=0}^{\infty} 4\left(\frac{1}{2}\right)^k$ is a geometric series with $a = 4$ and $|r| = \frac{1}{2} < 1$, so the series converges to

$$\frac{a}{1-r} = \frac{4}{1-\frac{1}{2}} = 8.$$

21. See Example 2.4.
$\sum_{k=0}^{\infty} \left(\frac{1}{2}\right)^k$ is a geometric series with $a = 1$ and $|r| = \frac{1}{4} < 1$, so the series converges to

$$\frac{a}{1-r} = \frac{1}{1-\frac{1}{4}} = \frac{4}{3};$$

converges to $\frac{4}{3}$.

23. See Example 4.6.

$$|S - S_k| \le a_{k+1} = \frac{k+1}{(k+1)^4+1} \le .01$$

when $k = 4$, so $S \approx S_4 \approx -0.40$.

25. See Example 2.6.

$$\lim_{k\to\infty} \frac{2k}{k+3} = \lim_{k\to\infty} \frac{2}{1+\frac{3}{k}} = 2 \ne 0,$$

so by the kth-Term Test for Divergence,

$$\sum_{k=0}^{\infty} \frac{2k}{k+3}$$

diverges.

27. See Example 4.3.

$$\lim_{k\to\infty} |a_k| = \lim_{k\to\infty} \frac{4}{(k+1)^{1/2}} = 0$$

and

$$\frac{a_{k+1}}{a_k} = \frac{4}{(k+2)^{1/2}} \cdot \frac{(k+1)^{1/2}}{4} = \frac{(k+1)^{1/2}}{(k+2)^{1/2}} < 1$$

for all $k \geq 0$, so $a_{k+1} < a_k$ for all $k \geq 0$, so by the Alternating Series Test,

$$\sum_{k=0}^{\infty} \frac{(-1)^x 4}{\sqrt{k+1}}$$

converges.

29. See Example 3.2.
$\sum_{k=1}^{\infty} \frac{3}{k^{7/8}}$ is a divergent p-series $\left(p = \frac{7}{8} < 1\right)$, therefore $\sum_{k=1}^{\infty} \frac{3}{k^{7/8}}$ diverges.

31. See Example 3.9.
Using the Limit Comparison Test, let

$$a_k = \frac{\sqrt{k}}{k^3+1} \quad \text{and} \quad b_k = \frac{1}{k^{5/2}}.$$

Then

$$\lim_{k\to\infty} \frac{a_k}{b_k} = \lim_{k\to\infty} \frac{\sqrt{k}}{k^3+1} \cdot \frac{k^{5/2}}{1}$$
$$= \lim_{k\to\infty} \frac{k^3}{k^3+1} = 1 > 0,$$

so because

$$\sum_{k=1}^{\infty} \frac{1}{k^{5/2}}$$

is a convergent p-series $\left(p = \frac{5}{2} > 1\right)$,

$$\sum_{k=1}^{\infty} \frac{k}{\sqrt{k^3+1}}$$

converges.

33. See Example 4.3.
Using the Alternating Series Test,

$$\lim_{k\to\infty} \frac{4^k}{k!} = 0$$

(as in Exercise 5) and

$$\frac{a_{k+1}}{a_k} = \frac{4k+1}{(k+1)!} \cdot \frac{k!}{4} = \frac{4}{k+1} \leq 1$$

for $k \geq 3$, so $a_{k+1} \leq a_k$ for $k \geq 3$, thus

$$\sum_{k=0}^{\infty} (-1)^k \frac{4^k}{k!}$$

converges.

35. See Example 4.2.
Using the Alternating Series Test,
$\lim_{k\to\infty} |a_k| = \lim_{k\to\infty} \ln\left(1 + \frac{1}{k}\right) = \ln 1 = 0$ and
$a_{k+1} = \ln\left(\frac{k+2}{k+1}\right) < \ln\left(\frac{k+1}{k}\right) = a_k$ for all
$k \geq 1$, so $\sum_{k=1}^{\infty} (-1)^k \ln\left(1 + \frac{1}{k}\right)$ converges.

37. See Example 3.9.
Using the Limit Comparison Test, let

$$a_k = \frac{2}{(k+3)^2} \quad \text{and} \quad b_k = \frac{1}{k^2},$$

so

$$\lim_{k\to\infty} \frac{a_k}{b_k} = \lim_{k\to\infty} \frac{2k^2}{(k+3)^2}$$
$$= \lim_{k\to\infty} \frac{2k^2}{k^2+6k+9}$$
$$= 2 > 0,$$

so because

$$\sum_{k=1}^{\infty} \frac{1}{k^2}$$

is a convergent p-series $(p = 2 > 1)$,

$$\sum_{k=1}^{\infty} \frac{2}{(k+3)^2}$$

converges.

39. See Example 2.6.

$$\frac{k!}{3^k} = \frac{1 \cdot 2 \cdot 3 \ldots k}{3 \cdot 3 \cdot 3 \ldots 3} > \frac{1}{3} \cdot \frac{2}{3} \cdot \frac{k}{3} = \frac{2k}{27},$$

and $\lim_{k\to\infty} \frac{2k}{27} = \infty$, $\frac{k!}{3^k} > \frac{2k}{27}$, $\lim_{k\to\infty} \frac{k!}{3^k} = \infty$, and by the kth-Term Test for Divergence,

$$\sum_{k=1}^{\infty} \frac{k!}{3^k}$$

diverges.

41. See Example 3.5.
Using the Comparison Test, $e^{1/k} \le e$ for all $k \ge 1$, so

$$\frac{e^{1/k}}{k^2} \le \frac{e}{k^2},$$

for all $k \ge 1$ and

$$\sum_{k=1}^{\infty} \frac{e}{k^2} = e \sum_{k=1}^{\infty} \frac{1}{k^2},$$

which is a convergent p-series ($p = 2 > 1$). Thus

$$\sum_{k=0}^{\infty} \frac{e^{1/k}}{k^2}$$

converges.

43. See Example 5.4.
Using the Ratio Test,

$$\begin{aligned} \lim_{k\to\infty} \left| \frac{a_{k+1}}{a_k} \right| &= \lim_{k\to\infty} \left| \frac{4^{k+1} k!^2}{4^k (k+1)!^2} \right| \\ &= \lim_{k\to\infty} \frac{4}{(k+1)^2} = 0 < 1, \end{aligned}$$

so the series converges absolutely. Thus,

$$\sum_{k=0}^{\infty} \frac{4^k}{(k!)^2}$$

converges.

45. See Example 5.2.
Using the Alternating Series Test,

$$\lim_{k\to\infty} |a_k| = \lim_{k\to\infty} \frac{k}{k^2+1} = 0 \text{ and}$$

$$\frac{a_{k+1}}{a_k} = \frac{(k+1)(k^2+1)}{k\left[(k+1)^2+1\right]} = \frac{k^3+k^2+k+1}{k^3+2k^2+2k} < 1$$

for all $k \ge 1$, so $a_{k+1} < a_k$ for all $k \ge 1$, so the series converges. But letting

$$a_k = \frac{k}{k^2+1} \quad \text{and} \quad b_k = \frac{1}{k},$$

$$\sum_{k=1}^{\infty} |a_k| = \sum_{k=1}^{\infty} \frac{k}{k^2+1}$$

diverges using the Limit Comparison Test, so

$$\lim_{k\to\infty} \frac{a_k}{b_k} = \lim_{k\to\infty} \frac{k^2}{k^2+1} = 1 > 0.$$

Since $\sum_{k=1}^{\infty} \frac{1}{k}$ is the divergent harmonic series,

$$\sum_{k=0}^{\infty} \frac{k}{k^2+1}$$

diverges. Thus,

$$\sum_{k=1}^{\infty} \frac{(-1)^k k}{k^2+1}$$

converges conditionally.

47. See Example 5.3.

$$\left| \frac{\sin k}{k^{3/2}} \right| = \frac{|\sin k|}{k^{3/2}} \le \frac{1}{k^{3/2}}$$

because $|\sin k| \le 1$ for all k. And

$$\sum_{k=1}^{\infty} \frac{1}{k^{3/2}}$$

is a convergent p-series $\left(p = \frac{3}{2} > 1\right)$. So by the Comparison Test,

$$\sum_{k=1}^{\infty} \left| \frac{\sin k}{k^{3/2}} \right|$$

converges, so

$$\sum_{k=1}^{\infty} \frac{\sin k}{k^{3/2}}$$

converges absolutely.

49. See Examples 3.2 and 3.8.
Using the Limit Comparison Test, we have

$$a_k = \frac{2}{(3+k)^p} \quad \text{and} \quad b_k = \frac{1}{k^p}, \text{ so}$$

$$\lim_{k\to\infty} \frac{a_k}{b_k} = \lim_{k\to\infty} \frac{2k^p}{(3+k)^p} = 2 > 0,$$

so the series

$$\sum_{k=1}^{\infty} \frac{2}{(3+k)^p}$$

converges if and only if the p-series

$$\sum_{k=1}^{\infty} \frac{1}{k^p}$$

converges, which happens when $p > 1$.

51. See Example 4.6.

$$|S - S| \le a_{n+1} = \frac{3}{(k+1)^2} \le 10^{-6},$$

so $3{,}000{,}000 \le (k+1)^2$, thus $1732.05 \le k+1$, so $1731.05 \le k$. Hence, $k = 1732$ terms.

53. See Example 6.6.

$$\begin{aligned} f(x) &= \frac{1}{4+x} \\ &= \frac{1}{4}\left(\frac{1}{1-(-\frac{x}{4})}\right) \\ &= \frac{1}{4}\sum_{k=0}^{\infty}\left(-\frac{x}{4}\right)^k \\ &= \sum_{k=0}^{\infty} \frac{(-1)^k x^k}{4^{k+1}}, \end{aligned}$$

which is a geometric series that converges when

$$\left|-\frac{x}{4}\right| < 1, \text{ or } -4 < x < 4.$$

Thus, $r = 4$.

55. See Example 6.6.

$$\begin{aligned} f(x) = \frac{3}{3+x^2} &= \frac{1}{1-\left(-\frac{x^2}{3}\right)} \\ &= \sum_{k=0}^{\infty}\left(-\frac{x^2}{3}\right)^k \\ &= \sum_{k=0}^{\infty} \frac{(-1)^k x^k}{4^{k+1}}, \end{aligned}$$

which is a geometric series that converges for $\left|-\frac{x^2}{3}\right| < 1$, so $x^2 < 3$ or $|x| < \sqrt{3}$, so $-\sqrt{3} < x < \sqrt{3}$. Thus, $r = \sqrt{3}$.

57. See Example 6.6.
From Exercise 53,

$$\frac{1}{4+x} = \sum_{k=0}^{\infty} \frac{(-1)^k x^k}{4^{k+1}}$$

with $r = 4$. By integrating both sides, we get

$$\int \frac{1}{4+x}\,dx = \sum_{k=0}^{\infty} \frac{(-1)^k}{4^{k+1}} \int x^k\,dx, \text{ so}$$

$$\ln(4+x) = \sum_{k=0}^{\infty} \frac{(-1)^k x^{k+1}}{(k+1)4^{k+1}} + \ln 4$$

with $r = 4$.

59. See Example 6.3.
Using the Ratio Test, we have

$$\lim_{k\to\infty}\left|\frac{a_{k+1}}{a_k}\right| = \lim_{k\to\infty}\left|\frac{2x^{k+1}}{2x^k}\right| = |x| \lim_{k\to\infty} 1 = |x|$$

and $|x| < 1$ when $-1 < x < 1$. So the series converges absolutely for $x \in (-1, 1)$. When $x = -1$,

$$\sum_{k=0}^{\infty}(-1)^k 2(-1)^k = \sum_{k=1}^{\infty} 2$$

and $\lim_{k\to\infty} 2 = 2$, so the series diverges by the kth-Term Test for Divergence. When $x = 1$, $\sum_{k=0}^{\infty}(-1)^k 2$ and $\lim_{k\to\infty} 2 = 2$, so the series diverges by the

k^{th}-Term Test for Divergence. So the interval of convergence is $(-1, 1)$.

61. See Example 6.3.
Using the Ratio Test, we have

$$\lim_{k\to\infty}\left|\frac{a_{k+1}}{a_k}\right| = \lim_{k\to\infty}\left|\frac{2kx^{k+1}}{2(k+1)x^k}\right| \text{ and } |x| < 1$$

$$= |x| \lim_{k\to\infty}\frac{k}{k+1} = |x|$$

when $-1 < x < 1$. So the series converges absolutely for $x \in (01, 1)$. When $x = -1$,

$$\sum_{k=0}^{\infty}(-1)^k\frac{2}{k}(-1)^k = \sum_{k=1}^{\infty}\frac{2}{k},$$

which is the divergent harmonic series. When $x = 1$,

$$\sum_{k=0}^{\infty}(-1)^k\frac{2}{k} \quad \text{and} \quad \lim_{k\to\infty}\frac{2}{k} = 0$$

and

$$\frac{a_{k+1}}{a_k} = \frac{k}{k+1} < 1 \text{ for all } k \geq 0,$$

so $a_{k+1} < a_k$ for all $k \geq 0$, so this series converges by the Alternating Series Test. Thus, the interval of convergence is $(-1, 1]$.

63. See Example 6.2.
Using the Ratio Test, we have

$$\lim_{k\to\infty}\left|\frac{a_{k+1}}{a_k}\right| = \lim_{k\to\infty}\left|\frac{4k!(x-2)^{k+1}}{(k+1)!4(x-2)^k}\right|$$

$$= |x-2| \lim_{k\to\infty}\frac{1}{k+1}$$

$$= 0,$$

and $0 < 1$ for all x, so the series converges absolutely for all $x \in (-\infty, \infty)$. Thus, the interval of convergence is $(-\infty, \infty)$.

65. See Example 6.3.
Using the Ratio Test, we have

$$\lim_{k\to\infty}\left|\frac{a_{k+1}}{a_k}\right| = \lim_{k\to\infty}\left|\frac{3^{k+1}(x-2)^{k+1}}{3^k(x-2)^k}\right|$$

$$= 3|x-2| \lim_{k\to\infty} 1$$

$$= 3|x-2|$$

and $3|x-2| < 1$ when $-\frac{1}{3} < x - 2 < \frac{1}{3}$ or $\frac{5}{3} < x < \frac{7}{3}$, so the series converges for all $x \in \left(\frac{5}{3}, \frac{7}{3}\right)$.

When $x = \frac{5}{3}$,

$$\sum_{k=0}^{\infty} 3^k\left(-\frac{1}{3}\right)^k = \sum_{k=0}^{\infty}(-1)^k,$$

which diverges by the kth-Term Test for Divergence because $\lim_{k\to\infty} 1 = 1$. When $x = \frac{7}{3}$,

$$\sum_{k=1}^{\infty} 3^k\left(\frac{1}{3}\right)^k = \sum_{k=1}^{\infty} 1,$$

which diverges by the kth-Term Test for Divergence because $\lim_{k\to\infty} 1 = 1$. So the interval of convergence is $\left(\frac{5}{3}, \frac{7}{3}\right)$.

67. See Example 7.1.

$f(x) = \sin x$ about $c = 0$.

$f(x) = \sin x$, $f'(x) = \cos x$, $f''(x) = -\sin x$,

$f'''(x) = -\cos x$,

$f(0) = 0$, $f'(0) = 1$, $f''(0) = 0$, $f'''(0) = -1$, so

$$\sum_{k=0}^{\infty}\frac{f^{(k)}(0)}{k!}x^k = x - \frac{1}{3!}x^3 + \frac{1}{5!}x^5 - \ldots$$

$$= \sum_{k=0}^{\infty}\frac{(-1)^k x^{2k+2}}{(2k+1)!}$$

69. See Example 7.6.

$f(x) = \ln x$ about $c = 1$.

$$f(x) = \ln x,\ f'(x) = \frac{1}{x},\ f''(x) = -\frac{1}{2},$$

$$f'''(x) = \frac{2}{x^3},\ f^{(4)}(x) = -\frac{6}{x^4}$$

$$f(1) = 0,\ f'(1) = 1,\ f''(1) = -1,$$
$$f'''(1) = 2!,\ f^{(4)}(1) = -3!$$

$$P_4(x) = \sum_{k=0}^{4} \frac{f^{(k)}(1)}{k!}(x-1)^k$$
$$= (x-1) - \frac{1}{2}(x-1)^2 + \frac{1}{3}(x-1)^3 - \frac{1}{4}(x-1)^4$$

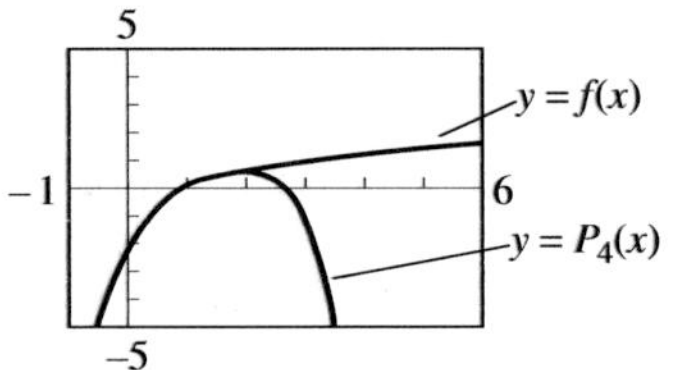

71. See Examples 7.6 and 7.7.
From Exercise 69,

$$P_4(x) = (x-1) - \frac{1}{2}(x-1)^2 + \frac{1}{3}(x-1)^3 - \frac{1}{4}(x-1)^4.$$

Letting $x = 1.2$ gives $\ln(1.2) \approx P_4(1.2) = .1823$.
Also,

$$|R_n(1.2)| = \left|\frac{f^{(n+1)}(z)}{(n+1)!}(1.2-1)^{n+1}\right|$$
$$= \frac{n!(.2)^{n+1}}{(n+1)!|z|^{n+1}} = \frac{(.2)^{n+1}}{(n+1)z^{n+1}} < \frac{(.2)^{n+1}}{n+1}$$

because $1 < z < 1.2$, so $\frac{1}{z} < \frac{1}{1} = 1$. So

$$|R_n(1.2)| < \frac{(.2)^{n+1}}{n+1} < 10^{-8}$$

if $n = 10$.

73. See Example 7.8.
Since

$$e^x = \sum_{k=0}^{\infty} \frac{x^k}{k!}$$

with $r = \infty$, by replacing x with $-3x^2$, we have

$$e^{-3x^2} = \sum_{k=0}^{\infty} \frac{(-3x^2)^k}{k!} = \sum_{k=0}^{\infty} \frac{(-1)^k 3^k x^{2k}}{k!}$$

with radius of convergence ($r = \infty$).

75. See Examples 7.8 and 8.4.
Since

$$\tan^{-1} x = \sum_{k=0}^{\infty} \frac{(-1)^k}{2k+1} x^{2k+1},$$

$$\tan^{-1} x \approx x - \frac{1}{3}x^3 + \frac{1}{5}x^5 - \frac{1}{7}x^7 + \frac{1}{9}x^9, \text{ so}$$

$$\int_0^1 \tan^{-1} x\, dx$$
$$= \int_0^1 \left(x - \frac{1}{3}x^3 + \frac{1}{5}x^5 - \frac{1}{7}x^7 + \frac{1}{9}x^9\right) dx$$
$$= \left[\frac{x^2}{2} - \frac{x^4}{12} + \frac{x^6}{30} - \frac{x^8}{58} + \frac{x^{10}}{90}\right]$$
$$= \frac{1117}{2520} \approx 0.4433.$$

77. See Example 9.3.

$$f(x) = x,\ -2 \le x \le 2,\ \text{so } l = 2$$

$$a_0 = \frac{1}{2}\int_{-2}^{2} x\, dx = \frac{1}{2}\left[\frac{x^2}{x}\right]_{-2}^{2} = \frac{1}{2}\left[\frac{4}{2} - \frac{4}{2}\right] = 0$$

$$a_k = \frac{1}{2}\int_{-2}^{2} x\cos\left(\frac{k\pi x}{2}\right)dx$$
$$= \frac{1}{2}\left[\frac{2x}{k\pi}\sin\frac{k\pi x}{2} + \frac{4}{(k\pi)^2}\cos\frac{k\pi x}{2}\right]_{-2}^{2}$$
$$= \frac{1}{2}\left(\begin{array}{l} -\frac{4}{k\pi}\sin k\pi + \frac{4}{(k\pi)^2}\cos k\pi \\ -\left[\frac{4}{k\pi}\sin(-k\pi) + \frac{4}{(k\pi)^2}\cos(-k\pi)\right]\end{array}\right)$$
$$= 0$$

$$b_k = \frac{1}{2}\int_{-2}^{2} x\sin\left(\frac{k\pi x}{2}\right)dx$$
$$= \frac{1}{2}\left[-\frac{2x}{k\pi}\cos\frac{k\pi x}{2} + \frac{4}{(k\pi)^2}\sin\frac{k\pi x}{2}\right]_{-2}^{2}$$
$$= \frac{1}{2}\left(\begin{array}{l} -\frac{4}{k\pi}\cos k\pi + \frac{4}{(k\pi)^2}\sin k\pi \\ -\left[\frac{4}{k\pi}\cos(-k\pi) + \frac{4}{(k\pi)^2}\sin(-k\pi)\right]\end{array}\right)$$
$$= \frac{1}{2}\left(-\frac{8}{k\pi}\cos k\pi\right)$$
$$= -\frac{4}{k\pi}(-1)^k$$

So

$$f(x) = \sum_{k=1}^{\infty}\frac{(-1)^{k+1}4}{k\pi}\sin\left(\frac{k\pi x}{2}\right).$$

79. See Theorem 9.1.
$f(x) = x^2,\ -1 \le x \le 1$

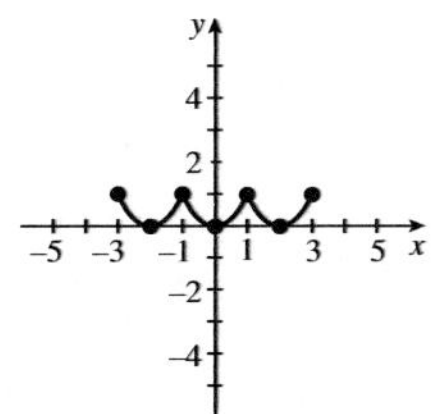

81. See Theorem 9.1.

$$f(x)\begin{cases} -1 & \text{if } -1 < x \le 0 \\ 1 & \text{if } 0 < x \le 1\end{cases}$$

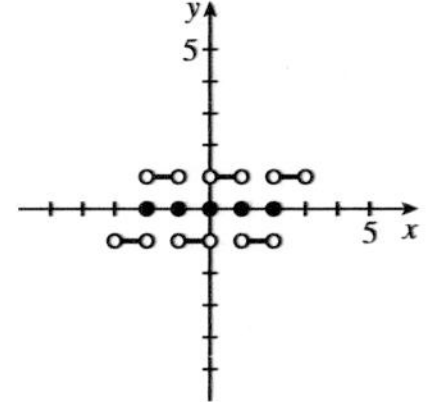

83. See Example 2.4.

$$\frac{1}{2} + \frac{1}{8} + \frac{1}{32} = \cdots = \sum_{k=0}^{\infty}\frac{1}{2}\left(\frac{1}{4}\right)^4,$$

which is a geometric series with $a = \frac{1}{2}$ and $|r| = \frac{1}{4} < 1$, so it converges to

$$\left(\frac{\frac{1}{2}}{1-\frac{1}{4}}\right) = \left(\frac{\frac{1}{2}}{\frac{3}{4}}\right) = \frac{2}{3}.$$

85. See Example 10.2.

$$0 = y'' - 2xy' - 4y$$
$$= \sum_{n=2}^{\infty} n(n-1)a_n x^{n-2}$$
$$- 2x\sum_{n=1}^{\infty} na_n x^{n-1} - 4\sum_{n=0}^{\infty} a_n x^n$$
$$= \sum_{n=0}^{\infty}(n+2)(n+1)a_{n+2}x^n$$
$$- 2x\sum_{n=0}^{\infty} na_n x^n - 4\sum_{n=0}^{\infty} a_n x^n$$
$$= \sum_{n=0}^{\infty}\left[(n+2)(n+1)a_{n+2} - 2(n+2)a_n\right]x^n$$

So $(n+2)(n+1)a_{n+2} - 2(n+2)a_n = 0$, or

$$a_{n+2} = \frac{2(n+2)}{(n+2)(n+1)}a_n = \frac{2}{n+1}a_n$$

$$a_{2n} = \frac{2^{2n}n!}{(2n)!}a_0 \quad a_{2n+1} = \frac{1}{n!}a_1$$

$$y = \sum_{n=0}^{\infty} a_n x^n = \sum_{n=0}^{\infty} a_{2n}x^{2n} + \sum_{n=0}^{\infty} a_{2n+2}x^{2n+1}$$

$$= a_0 \sum_{n=0}^{\infty} \frac{2^{2n}n!}{(2n)!}x^{2n} + a_1 \sum_{n=0}^{\infty} \frac{1}{n!}x^{2n+1}.$$

87. See Example 10.2.

$$0 = y'' - 2xy' - 4y$$
$$= y'' - 2(x-1)y' - 2y' - 4y$$
$$= \sum_{n=2}^{\infty} n(n-1)a_n(x-1)^{n-2}$$
$$- 2(x-1)\sum_{n=1}^{\infty} na_n(x-1)^{n-1}$$
$$- 2\sum_{n=1}^{\infty} na_n(x-1)^{n-1}$$
$$- 4\sum_{n=0}^{\infty} a_n(x-1)^n$$
$$= \sum_{n=0}^{\infty}(n+2)(n+1)a_{n+2}(x-1)^n - 2\sum_{n=0}^{\infty} na_n(x-1)^n$$
$$- 2\sum_{n=0}^{\infty}(n+1)a_{n+1}(x-1)^n - 4\sum_{n=0}^{\infty} a_n(x-1)^n$$
$$= \sum_{n=0}^{\infty} \big[(n+2)(n+1)a_{n+2}$$
$$-2na_n - 2(n+1)a_{n+1} - 4a_n\big](x-1)^n$$

So $(n+2)(n+1)a_{n+2} - 2na_n - 2(n+1)a_{n+1}$
$-4a_n = 0$,
or $(n+2)(n+1)a_{n+2}$
$= (2n+4)a_n + 2(n+1)a_{n+1}$

$$a_{n+2} = \frac{2a_n}{n+1} + \frac{2a_{n+1}}{n+2}$$

89. In Exercise 85, set $a_0 = 4$ and $a_1 = 2$ to get

$$4\sum_{n=0}^{\infty} \frac{2^{2n}n!}{(2n)!}x^{2n} + 2\sum_{n=0}^{\infty} \frac{1}{n!}x^{2n+1}.$$

Notes

Notes

Notes

Notes

Notes

Notes